ARIADNE'S THREAD

Ariadne's Thread

Ariadne, the daughter of King Minos of Crete, possessed a magic ball of thread that had the power to lead one into and back out of the otherwise impenetrable Labyrinth wherein the Minotaur was kept. This ball – or clew – she gave to the young Athenian, Theseus, so that he might safely slay the monster and escape from the maze.

The Search for New Modes of Thinking

In 1946, in a letter to prominent Americans, Albert Einstein wrote the following words:

"Our world faces a crisis as yet unperceived by those possessing the power to make great decisions for good or evil. The unleashed power of the atom has changed everything save our modes of thinking, and thus we drift towards unparalleled catastrophe . . . a new type of thinking is essential if mankind is to survive."

Quoted by O. Nathan and H. Nordin (eds), *Einstein on Peace* (New York: Simon & Schuster, 1960) p. 376

Ariadne's Thread

The Search for New Modes of Thinking

Mary E. Clark

Professor of Biology
San Diego State University, California

St. Martin's Press   New York

First published in the United States of America in 1989

Printed in Hong Kong

ISBN 0–312–01580–1 (cloth)
     0–312–01586–0 (paper)

Library of Congress Cataloging-in-Publication Data
Clark, Mary E.
Ariadne's thread : the search for new modes of thinking / Mary E.
Clark.
p. cm.
Bibliography: p.
Includes index.
ISBN 0–312–01580–1 : $39.95 (est.)   ISBN 0–312–01586–0 (pbk.) :
1. Civilization, Modern—1950-   I. Title.
CB428.C58 1989   88–28185
909.82–dc19   CIP

This book is written for Mike, Dave and Bill, and for everyone else who is young and searching

Also by Mary E. Clark

CONTEMPORARY BIOLOGY

Contents

List of Illustrations

List of Tables

Preface

We live in a world undergoing multiple large, rapid and mostly unprecedented changes, in the face of which most of us feel utterly helpless. New technologies that we neither understand nor control are being thrust upon us. Our global population, although its rate of growth is declining, continues to expand in absolute numbers each year – and will do so for at least another generation. Tensions rise; debts mount; the arms race, despite recent token treaties, continues to grow, not decline.

For any thoughtful person – for anyone who cares about the future of the young – this world is a nightmare. Nor do the old ways of thinking about the world help. Concerned citizens fail to find in the rhetoric of civic leaders any hints of new ways of looking at things. And concerned teachers are recognising that the old way of treating history and evolution and political science and religion and physics and anthropology and economics as compartments of knowledge – little pieces of the Western world that could be examined and understood separately – does not work anymore. A coherent picture of reality no longer emerges. In today's world, this kind of reductionism fails – indeed, it is beside the point. Much of what passes for scholarly thinking in universities, which in turn society at large takes to be an improved 'truth', is in fact busy work; it is irrelevant abstraction. The overall theoretical picture being generated no longer corresponds to external reality.

Intuitively, most thoughtful people are aware of this. College students certainly are, and they find the tidy explanations of their world that they are asked to memorise tedious and boring. They cannot articulate what is wrong. How can they, when even their professors appear unaware that there is a problem? What all thoughtful people vaguely sense is an absence of those new modes of thinking Einstein argued for more than two decades ago.

This dawning awareness that our old ways of thinking are insufficient for today's world has been slowly growing and spreading among the more concerned faculty in the universities of Europe and America. About a decade ago, a small group at San Diego State University recognised that what was needed was an integrated vision in our teaching, one that would attempt to focus all disciplines on today's world and connect them together into a meaningful whole. Our goal was to present the multiple aspects of the modern world in a holistic way that allowed the student to begin meaningfully to understand, criticise and evaluate it. From that insight an interdisciplinary course, *Our Global Future*, was begun, with fifteen or sixteen faculty contributing lectures in a connected sequence during the semester. The response from the students has been powerful. Again and again, they tell us that for the first time they see their world in a new way, and are given the skills and opportunity to think critically about it. It becomes the first step toward empowerment.

It was out of this course that *Ariadne's Thread* evolved. The book attempts to provide enough background information to enable the reader critically to evaluate the central ideas presented by each discipline and to interpret those ideas in ways that reflect on the contemporary scene. But it goes a step beyond that function. As it progresses from beginning to conclusion, the book attempts more than we do in the classroom. The commonest student criticism of our course has been that we have laid out all the problems, but provide no solutions. Nor does it satisfy them when we point out that once *they* have become aware then they have moved from being part of the problem to being part of the solution. Students – indeed all citizens, I believe – do not see any satisfactory alternative vision, any new way of thinking, that they can get ahold of, discuss, imagine, criticise, and modify.

For that reason, I have offered in this book the beginnings of an alternative vision, one that seems to me to provide the surest path to survival. It is a path of decentralisation, of local self-sufficiency, of cultural diversity, yet one where a global vision is shared by all. It is a path emphasising cooperation rather than competition, diversity rather than uniformity, social bondedness rather than self-centredness, sacred meaning rather than material consumption. It is, thus, a path that requires the substantially new ways of thinking that Einstein told us we *must* find if we are to survive. Yet it is also a path that draws upon the best of the past. It is this visionary aspect of the book that, I hope, will make it of interest not only to college students, but to all concerned citizens – the 'activists' in environmentalism, in the peace movement, in international affairs, in social work and feminism and minority rights: in short, for all those who care about the future of humankind as a whole.

The central ideas embodied in this book are not mine. Indeed, there is almost nothing to which I can lay claim as 'my own', other than the selection and synthesis itself. It is the juxtaposition, the way the camera lens has been turned onto the subject matter, that is my own contribution. That is what I hope will be read critically and thoughtfully. For my central purpose is not to insist on a particular vision (although of course I feel that the one presented has much merit), but to stimulate widely throughout Western society – and, indeed, all societies – the kind of thoughtful intellectual ferment by which we can gradually yet surely seek our way out of our present impasse. Ariadne's Thread, after all, was a magical ball that unwound spontaneously to lead Theseus safely into and out of the labyrinth housing the Minotaur. If history and science tell us anything, however, it is that humans have always been the generators of their own 'magic', slowly – often unconsciously – creating in their minds new meaning and visions by which they have adapted successfully. The only difference today is that our creative effort must be a *conscious* one, for our own cleverness precludes the past luxury of an imperceptible evolution in our worldviews.

This book, then, is directed not only to the young, but to *all* concerned and caring people. In its pages, much that is supposed to comprise the 'breadth' of a liberal education is summarised, so that any thoughtful person, regardless of prior familiarity, can follow the arguments as they develop. Experienced readers often find it useful to read first the 'wrap-around' chapters of any lengthy book, since these lay out the author's goals and provide a map of how the story unfolds. *Ariadne's Thread* is necessarily long and complex, and I urge the reader to start by reading Chapter 1, 'The Future: A Search for Values' which ends with a description of our pathway through the labyrinth, and then Chapter 16, 'Humankind at the Crossroads', which ties together the many smaller conclusions reached throughout the book, and suggests not only new attitudes for the future but also a means to achieve them through democratic participation. Then the four central questions addressed in the body of the book will more readily fall into place:

What are the limits to growth?
What is human nature?
Whence comes our Western worldview?
Where do we begin?

This book has been in preparation for six years, and during that time I have benefited immeasurably from the generous and supportive criticism of many, many colleagues. First and foremost, I wish to thank my fellow faculty members who have participated in *Our Global Future* semester after semester, and most of whom have offered valuable criticism of this book: Professors Bradley Bartel, William Cheek, Norris Clement, Linda Holler, Henry Janssen,* Albert Johnson, J. William Leasure, Lynn Peters, Stanley Pincetl, Stephen B.W. Roeder,* Alan Sweedler, Eugene Troxell, Sandra Wawrytko, John Weeks, and the Reverend William P. Mahedy. Without their inspirational support and dedication, neither the course nor the book would have been possible.

I am also deeply grateful to the following friends, students and colleagues who have offered invaluable advice: Professors Charles Cooper, Robert Costanza and Herman Daly; Ms Ingrid Dinter;* Professors Lloyd J. Dumas and Johan Galtung; Ms Della Grayson; Professor Helga Guderley;* Ms Lore Henlein;* Professor Maurice Mandelbaum; Dr Peter Manos;* Mr William McConnell; Professor Mary Midgley; Ms Patsy Miller;*

*Indicates persons who have read and commented on entire manuscript.

Mr Terrence Ratigan;* Professors George Somero,* Leften Stavrianos,* Marjorie Turner, Sue Weiler,* and Paul Zedler.

My sincere thanks also go to Mrs Rose Carruthers for her skilful work on several illustrations; to Mr Ed Dixon for dedicated assistance in locating elusive information; to Libby and Laurie Stewart for generous help with the indexing; to Mr Keith Povey for his gently inspired editorial contributions; and to Mr Simon Winder at the publishers for keeping the faith in a dark time.

Despite my deep indebtedness to the thinkers of the past from whom I have heavily drawn and to all those persons mentioned above – and to many others, students in classes, colleagues on various campuses, and friends whose conversations, criticism and comments have generated and nourished new insights – I alone am responsible for the choice of 'facts' that appear in this book, and the interpretations put upon them. Those who disagree with its conclusions will I hope be moved to undertake a parallel journey, putting forth their own visions and recommendations for public scrutiny. To have initiated serious and broadly-based dialogue among Western minds would be the greatest reward any author concerned with the future could possibly have.

San Diego, California   MARY E. CLARK

*Indicates persons who have read and commented on entire manuscript.

Acknowledgements

The author and publishers wish to thank the following who have kindly given permission for the use of copyright material (the following abbreviations have been used: R = reference; T = table; F = figure; OQ = opening quote. Initial numbers refer to chapters. Full citations are given in References and Notes):

Abelson, Philip H. (R. 4–53)
Academic Press (R. 8–6)
Alpert, Peter (R. 1–27)
American Anthropological Association (R. 6–12)
American Association for the Advancement of Science (R. 1–6, 1–9, 1–16, 2–16, 4–1, 4–53, 4–70, 7–11, 12–30, 15–65, 16–10; F. 4–2, 4–3)
American Economics Association (R. 4–1)
American Historical Association (R. 15–66)
American Journal of Physics (R. 2–7; F. 2–5, 2–7; T. 2–1)
American Physical Society (F. 15–1)
Annals of Earth (R. 16–31)
Atheneum Publishers Inc. (OQ–1)
Aussenpolitik Redaktion (R. 14–49)
Aveni, Anthony F. (R. 7–11)
Ballantine Books (R. 6–26)
Baron, Harold (R. 14–59)
Bartlett, Albert A. (R. 2–7; F. 2–5, 2–7; T. 2–1)
Basic Books (R. 12–3, 12–36, 16–29)
Beacon Press (R. 11–44)
Bereiter, Carl (R. 11–37)
Basil Blackwell Ltd (R. 5–1, 9–32, 10–49, 16–8, 16–14)
Bodley, John H. (R. 6–34, 6–35)
Boulding, Elise (Box in Ch. 15, p. 445)
Boulding, Kenneth (OQ–12)
Bowden, Charles (R. 4–12)
Boylan, James (R. 8–42, 8–43, 8–47)
Brookings Institution (R. 14–13)
Brown, Harrison (T. 4–1)
Brown, Lester R. (R. 4–31)
Burton, John W. (OQ–14; R. 14–8, 14–28, 14–50)
California Department of Fish and Game (OQ–4)
Cambridge University Press (R. 4–66, 9–33, 9–34, 9–38, 13–14, 13–31, 13–32, 13–34, 13–54, 13–62)
Jonathan Cape Ltd (R. 14–3)

Capra, Fritjof (OQ–16; R. 16–27, 16–28)
Carey, William D. (R. 16–10)
Cavendish, Richard (R. 7–25, 7–31, 7–34, 8–4)
Century Hutchinson (OQ–8; T. 15–1)
Chargaff, Erwin (R. 16–11, 16–12)
Columbia Journalism Review (R. 8–42, 8–43, 8–47)
Commager, Henry Steele (R. 8–53)
Committee for a Just World Peace (OQ–13)
Common Good Foundation (R. 8–55, 10–9, 10–40, 10–49, 12–1)
Cornell University Press (R. 5–14, 9–54, 15–52)
Costanza, Robert (R. 4–70; F. 4–2, 4–3)
Council on Foreign Relations (R. 7–51, 7–52)
Cousins, Norman (R. 8–40)
Cultural Survival Quarterly (R. 16–2)
Daly, Herman (R. 4-2)
Andre Deutsch (R. 10–38, 10–46, 10–53, 10–54)
Dobzhansky, Theodosius (OQ–5)
Dollars and Sense (R. 12–23, 13–59)
Dorman, William A. (R. 14–37)
Doubleday (R. 3–1, 7–45, 7–49, 8–32, 8–34, 9–49, 14–39, 14–43, 14–67)
Dover Publications (R. 7–28)
Eckholm, Eric P. (R. 4–23)
Edinburgh University Press (R. 8–11)
Educational Foundation for Nuclear Science, Inc. (R. 14–37; OQ–15; R. 15–24, 15–27, 15–29, 15–77)
Encyclopaedia Britannica (R. 9–41, 9–42, 9–43, 10–6)
Esteva, Gustavo (OQ–13)
Faber & Faber Ltd (R. 12–48)
Food and Agriculture Organization of the United Nations (T. 13–2)
Food Monitor (R. 13–41)
Ford, Daniel F. (R. 2–30)
The Free Press (R. 11–1, 11–25, 11–27, 14–5, 14–71)
W. H. Freeman & Co. (R. 2–18, 4–2, 5–21, 11–3, 15–59; F. 3–4B)
Erich Fromm Estate (OQ–7)
Rainer Funk (R. 11–29, 11–30)
Fukuoka, Masanobu (OQ–3)
Gladwin, Thomas (11–33, 13–7, 13–44, 14–33)
Grolier Enterprises (OQ–9)
Guppy, Nicholas (R. 4–3)
Hall, Charles A. S. (R. 2–16)
Harcourt Brace Jovanovich (R. 9–1, 9–7, 9–8, 9–9, 9–14, 9–16, 9–17, 9–18, 9–21, 9–22, 9–24, 9–25, 9–51, 12–25, 12–26)
Harper & Row (R. 1–23, 3–30, 6–34; OQ–10; R. 11–2, 11–38, 11–39, 11–41, 11–42, 14–63, 16–64, 15–35, 15–82)

Harper's Magazine (R. 16–11, 16–12)
Harris, Marvin (R. 3–15, 3–19)
Harvester Press (R. 5–32, 5–33)
Heilbroner, Robert (R. 7–41)
Henderson, Hazel (R. 12–18, 12–24)
HELDREF Publications (R. 15–9)
Holt, Rinehart & Winston (R. 5–41, 6–2, 6–16, 6–17, 6–18, 11–29, 11–30; F. 10–2)
Houghton Mifflin Co. (R. 5–43; OQ–6; R. 6–9, 6–22; T. 15–1)
The Humanist (R. 16–9)
Humanities Press (R. 11–33, 13–7, 13–44, 14–33)
Institut Universitaire d'Etudes du Developpement (R. 14–51, 14–53, 14–54)
Institute for Food and Development Policy (R. 4–6)
Institute for Policy Studies (R. 14–38)
Institute for Public Affairs (R. 14–59)
International Creative Management (R. 16–18)
International Planned Parenthood Federation (R. 13–66)
International Publishers Company, Inc. (R. 14–6, 14–7)
Ithaca New Times (R. 8–49)
Janssen, Henry (F. 14–1)
Johns Hopkins University Press (R. 1–21)
Jolly, Alison (R. 5–17, 5–18, 5–20)
Journal of Counseling and Development (R. 9–2, 9–3, 16–22)
Journal of Religious Ethics (R. 11–53, 11–54)
Journal of the History of Ideas (R. 9–28)
Keen, Benjamin (R. 4–58)
Kennan, George F. (R. 15–41)
King, Jonathan (R. 15–65)
Martin Luther King, Jr. Estate (R. 15–83)
Alfred A. Knopf (R. 8–8, 8–9, 9–11, 10–22)
Kumarian Press (R. 13–76)
Leontief, Wassily (R. 12–30)
Lescher & Lescher (R. 3–60)
Lifton, Robert Jay (R. 15–27)
Lindblom, Charles (R. 12–36)
Little, Brown & Company (R. 8–60)
Lutz, Mark A. (R. 10–4, 12–2)
Mack, John E. (R. 15–24)
Macmillan Press (R. 12–25, 12–26)
Macy, Joanna (R. 13–76)
Malotki, Ekkehart (F. 6–1)
McDougall, Walter A. (R. 15–66)
Meadows, Dennis (R. 1–11; F. 1–2)
Meadows, Donella (R. 7–40)

Mehl, Bernard (R. 8–51)

Methuen & Co. (R. 16–17, 16–18)

Midgley, Mary (R. 5–14, 5–32, 5–33, 9–54, 16–17, 16–18)

MIT Press (R. 10–49, 16–8, 16–14)

Momaday, N. Scott (R. 16–1)

Montagu, Ashley (R. 5–11)

Monthly Review Press (R. 13–4, 13–58)

William Morrow & Company (R. 11–45, 11–55, 13–3, 13–12, 13–16, 13–7, 13–27, 13–28, 13–47, 15–50, 16–15)

Multinational Monitor (R. 14–66)

John Murray, Publishers (R. 9–1, 9–7, 9–8, 9–9, 9–14, 9–16, 9–17, 9–18, 9–21, 9–22, 9–24, 9–25, 9–51)

Nagler, Michael (R. 16–38)

NAL Penguin, Inc. (OQ–5)

The Nation (R. 10–62)

National Council of Teachers of English (R. 16–32)

National Forum (R. 11–10, 11–11, 14–77, 14–78)

National Geographic Society (R. 16–1)

Natural History (R. 6–11)

Natural Resources Defense Council (R. 14–27)

New York Academy of Sciences (R. 1–24, 15–8)

New York Review of Books (R. 15–41)

The New York Times (R. 14–45)

Nietschmann, Bernard (R. 16–2)

Nobel Foundation (R. 5–40)

W. W. Norton (R. 7–41, 12–31, 14–47, 16–4)

OMNI Publications (R. 8–63)

Ontario Institute for Studies in Education (R. 11–37)

Ophuls, William Patrick (R. 16–25, 16–26)

Opie, Iona and Peter (R. 8–21)

Orbis Books (R. 7–32, 7–42, 7–48)

Oxford University Press (R. 4–64, 4–71, 5–12, 5–30, 8–21, 9–5, 9–6, 9–30, 9–31, 9–37, 9–39, 9–40, 16–20)

Peace and Change (Box in Ch. 15, p. 445)

Pendle Hill Publications (R. 15–67)

Penguin Books Ltd (R. 14–49)

Prentice-Hall (T. 13–1; R. 14–11, 14–16)

Princeton University Press (R. 1–18, 1–19)

Public Concern Foundation (R. 11–6, 11–49)

Putnam Publishing Group (R. 14–9)

Random House, Inc. (R. 3–15, 3–19, 11–3, 11–13, 11–28, 11–51, 12–12, 12–38, 14–3, 14–19, 15–7; F. 15–2)

Ratigan, Terence (R. 13–64, 14–55)

Review of Economic Studies (OQ–12)

Review of Social Economy (R. 12–14)
Rodale Press, Inc. (OQ–3)
Royal Society, London (F. 3–3)
Rutgers University Press (R. 4–58)
Seven Locks Press (R. 16–37)
Short, Roger V. (F. 3–3)
Sierra Club, San Diego (R. 4–26)
Sigma Xi (R. 5–17, 5–8, 5–20)
Simon, Julian L. (R. 1–16)
Simon & Schuster (R. 2–30, 6–25, 9–20, 14–35)
Sivard, Ruth Leger (R. 2–19; F. 2–2)
South End Press (R. 14–68)
Soviet Life (R. 15–74)
Sphere Books Ltd. (R. 12–47, 16–13; T. 16–1)
Spretnak, Charlene (R. 16–27, 16–28)
Stavrianos, Leften S. (R. 13–3, 13–12, 13–16, 13–17, 13–28, 13–47, 13–55, 15–50)
Steinhart, John S. (F. 4–1)
Sterling Lord Literistic, Inc. (R. 12–38)
St Martin's Press (R. 5–32, 5–33, 8–59, 14–12, 14–15, 14–17)
Thompson, William Irwin (R. 2–23)
Sweedler, Alan (R. 14–76)
Thompson, William Irwin (R. 16–31)
Threepenny Review (R. 8–50)
Thurow, Lester (R. 12–3)
Time, Inc. (F. 8–1)
Times Books (R. 15–1)
Trever, John (F. 12–1)
Frederick Ungar Publishing Company (R. 15–33)
Union of Concerned Scientists (R. 15–16)
University Books, Inc. (R. 7–3)
University College of North Staffordshire (R. 8–2)
University of British Columbia Press (R. 13–19, 13–20, 13–21, 13–22, 13–23)
University of California Press (R. 11–19, 11–22, 15–15)
University of California Regents (F. 6–2)
University of Chicago Press (R. 3–16, 7–26, 8–11)
University of Illinois Press (R. 10–47, 10–50)
University of Texas Press (R. 4–12)
University of Toronto Press (R. 5–28)
University of Wisconsin Press (R. 7–18)
Unwin Hyman (R. 8–5)
US Catholic Conference (R. 12–28)
VEB Deutscher Verlag der Wissenschaften (F. 7–2A and 7–2C)

Viking Penguin Inc. (Ref. 12–48)
Wadsworth Publishing Co. (F. 13–2)
A. P. Watt Ltd (R. 7–25, 7–31, 7–34, 8–4)
Wawrytko, Sandra (R. 13–8)
Wessel, James (R. 4–6)
Wolin, Sheldon S. (R. 8–55, 10–9, 10–40, 10–49, 12–1)
World Priorities Institute (R. 2–19; F. 2–2)
Worldwatch Institute (R. 2–27, 2–40, 2–42, 2–43, 2–49, 4–23, 4–31)
Yale University Press (R. 5–13)
Yergin, Daniel (R. 2–23)

Every effort has been made to trace all the copyright-holders, but if any have been inadvertently overlooked the publisher will be pleased to make the necessary arrangement at the first opportunity.

Introduction
to the
Labyrinth

1 The Future: A Search for Values

We in the western world have rushed eagerly to embrace the future – and in so doing we have provided that future with a strength it has derived from us and our endeavors. Now, stunned, puzzled and dismayed, we try to withdraw from the embrace, not of a necessary tomorrow, but of that future which we have invited and of which, at last, we have grown perceptibly afraid.

Loren Eiseley, 1962[†]

INTRODUCTION

THE FUTURE: It's our life, and our children's and our grandchildren's. What will it be like? Can all of us look forward to lives more or less like people lead today? Or are there going to be some drastic changes? If major changes are indeed inevitable, can widespread pain and chaos be avoided? Is it true, as M. Giscard d'Estaing, the former President of France, said: 'All modern day curves lead to disaster'[1]. What curves was he talking about? What disasters?

In the Western nations and in many other parts of the world as well there is a touch of unreality about people's lives. It is as though daily life – going to work, going shopping, going to college, watching TV, making love, paying the bills – were a kind of dream from which we know we must sooner or later suddenly awaken. But what will the *real* world be like when we are wrenched out of our reverie? And how much longer can we remain in that twilight zone? How much longer can we stick our heads in the sand, pretending that nothing much will change? How much longer can such struthianism* prevail (Figure 1.1)?

PREDICTING THE FUTURE

Perhaps one reason we are reluctant to remove our heads from the sand, to awaken from our reverie, is that we do not know which of the two sets of soothsayers to believe. Shall we listen to those who predict what has come to be called 'gloom and doom'? Or shall we believe those who predict a rosy future where humans will inevitably march onward and upward, on

Struthios is Latin for ostrich.

Figure 1.1 Struthianism: a contemporary malady

the preordained track of 'progress'? If 'gloom and doom' are inevitable, then by all means, let's keep our heads in the sand as long as possible. But what if suffering and extinction are *avoidable* provided we *act*? In some ways, the prophecies of the 'glad tidings forever' crowd are not as believable as they would have us think. It's a little hard to maintain continuous optimism in a world of ongoing economic, social and political conflict, a world armed to the teeth with life-extinguishing nuclear warheads, a world where men continue to design and build even *more* lethal weapons, and where nobody is making much headway in stopping the whole insane process.

We begin this chapter, then, by identifying and briefly evaluating three sets of 'predictions' about the future. The first prediction is of immediate extinction, the consequence of a nuclear holocaust. The next two, the 'Cassandra' future and the 'Pollyanna' future, are both longer term predictions, into the next century and beyond, premised on our ability to avoid nuclear war.

Nuclear Extinction

Until the end of 1983, it was still possible for people to argue about 'nuclear survival', and for spokesmen of both the Soviet Union and the United States to talk about 'Civilian Defence' and the percentage of survivors after a nuclear war. The nuclear war film, *The Day After*, watched on television by more than half of all Americans, portrayed a goodly number of survivors who, when the radiation dissipated, emerged from their shelters into a bright, sunny, although badly devastated world. Toward the end of the film, the farmers discuss how to prepare the fields for safe sowing; it is clear that remnants of humankind are going to survive.

Then, in November 1983, a diverse group of scientists assembled from NASA and from numerous universities in America and elsewhere issued the prediction that if only a small fraction of the global stockpile of nuclear weapons were detonated, it could cause a 'nuclear winter', plunging the Earth into freezing darkness for anywhere from three months to a year or more. Not only would humankind likely become extinct; so would most of the plants and animals that now inhabit the Earth.[2]

Why did it take so long for the idea of a nuclear winter to be born? After all, there were lots of calculations about the number of people who would be immediately incinerated or blasted to bits, estimates of those who would later die of burns or wounds, of radiation sickness, of epidemics and, much later, of birth defects due to mutations. It was known that animals even at great distances from the blasts would be blinded, and that many species would perhaps be extinguished – the radiation-hardy insects and grasses being exceptions.[3] It was known that virtually all electronically stored information would be erased by the electromagnetic pulse that nuclear detonations generate. It was known that the protective ozone layer in the stratosphere would be largely destroyed, exposing Earth's surface to excess ultraviolet radiation that could cause severe burns, blindness, and skin cancer. Yet no one, it seems, had thought of a nuclear winter.

In retrospect, of course, a nuclear winter seems obvious, as new insights often do. Yet if ever there was a warning about how little we are able to predict in advance all the effects, both good and bad, of our powerful modern technologies, the nuclear winter is such a warning. Our imaginations simply do not generate all possible consequences of our acts. Only when we actually commit those acts – or something very similar to them occurs – do we learn, by trial and error, what later seems so obvious.

The idea of a nuclear winter seems to have come from three directions. First, there were the violent eruptions of Mt St Helens in the state of Washington in 1980 and of El Chichon in Mexico in 1982. Every volcanic eruption spews ash, soot and dust into the upper atmosphere, which have detectable effects on global climate for anywhere from a few months to a year or two. With each such natural experiment, atmospheric scientists are always busy with their latest equipment, trying to measure the global effects. Second, the Mariner explorations suggested the possibility that planetwide dust storms on Mars might have caused certain phenemena observed there, and triggered the thought that parallel phenomena could occur on Earth. The third was the proposal that the mass extinction at the end of the Cretaceous era 65 million years ago of many marine plants and animals – and perhaps of the dinosaurs as well – was caused by the impact of a giant meteorite on Earth. Presumed to be 6 miles in diameter, this huge planetary visitor bombarded the land with such force as to generate a blanket of dust and haze that enveloped the Earth for a number of months. With the Earth shrouded in a dusty darkness, which blocked out life-giving radiation from the sun, the surface temperatures fell dramatically. Without light, the plants died and so did the animals that fed on them. Whatever did not perish first from the cold died later of starvation.[4]

Whether the meteorite story proves true or not, it and the other ideas mentioned caused a group of scientists to realise that no one had calculated the atmospheric consequences of a nuclear exchange. How much dust, soot, and ash would nuclear weapons belch into the atmosphere? How would this alter Earth's climate – in addition to all the other terrible effects of such weapons? Preliminary calculations were done, and the results were horrifying. Based on certain assumptions about the amount of smoke and its global dispersal, it was estimated that *less than one per cent* of the world's nuclear weapons, if detonated over cities where they would cause firestorms and massive amounts of smoke, oily soot, and dust, would virtually blot out the sun in the Northern Hemisphere for at least two weeks, and create subfreezing land temperatures for up to several months. Furthermore, this massive dust cloud would create widespread vertical and horizontal winds, that would carry the dust and fallout to the Southern Hemisphere as well.[5]

The biological consequences would be devastating, with massive destruction of agriculture, forests, and other ecosystems. For human survivors, there would be no drinking water (contaminated and frozen); no safe food; no heat (power plants disrupted; forests destroyed by fire); no medical care. Once the dust clouds cleared, there would still be high doses of ionising and ultraviolet radiation. Here is what some twenty biologists have concluded:

> We emphasize that survivors, at least in the Northern Hemisphere, would face extreme cold, water shortages, lack of food and fuel, heavy burdens of radiation and pollutants, disease, and severe psychological stress – all in twilight or darkness.
>
> The possibility exists that the darkened skies and low temperatures would spread over the entire planet. Should this occur, a severe extinction event could ensue, leaving a highly modified and biologically depauperate Earth . . .
>
> It seems unlikely, however, that even in these circumstances *Homo*

sapiens would be forced to extinction immediately. Whether any people would be able to persist for long in the face of highly modified biological communities; novel climates; high levels of radiation; shattered agricultural, social, and economic systems; extraordinary psychological stresses; and a host of other difficulties is open to question. It is clear that the ecosystem effects *alone* resulting from a large-scale thermonuclear war could be enough to destroy the current civilization in at least the Northern Hemisphere... In any large-scale nuclear exchange between the superpowers, global environmental changes sufficient to cause the extinction of a major fraction of the plant and animals species on the Earth are likely. In that event, the possibility of the extinction of *Homo sapiens* cannot be excluded.[6]

Precisely how large a nuclear war would threaten the extinction of the human race depends on a number of variables besides how many bombs are dropped: whether the world's forests are ignited; whether oil refineries and other smoke-producing targets are hit; whether bombs are aimed at cities or at military installations. Yet as astronomer Carl Sagan points out, 'enough' bombs to set off a global climatic catastrophe will still be a small fraction of the total stockpile.[7]

This bombshell prediction shocked the United States Congress, which called on the Pentagon for an environmental impact statement. Although scientists at the United States national weapons laboratories have been critical of the assumptions used by Sagan and his colleagues, as have some other scientists, everyone agrees that *some* climatic effects will result from a nuclear exchange.[8] As Pentagon advisor Charles Zraket has said:

> Assuming that it withstands additional scrutiny, nuclear winter suggests that it is not possible [to conduct] a protracted war involving large numbers of nuclear weapons ... It will reinforce the existing belief that a first strike makes no sense, because it may be suicidal. And it renders the notion of a real civil defense program, which is already in disrepute, even more disreputable.[9]

Assuming we avoid nuclear extinction and the destruction of civilisation by nuclear war, we still face many other critical problems that demand attention in the coming decades, and so we turn to the other two sets of predictions, the 'Cassandra' future, and the 'Pollyanna' future.

The 'Cassandra' Future

In his epic poem about the fall of the ancient city of Troy, Homer tells how the prophetess Cassandra, the youngest daughter of King Priam, in an anguished frenzy foretold the destruction of their besieged city. Ever since, her name has been linked with prophecies of doom that people refuse to believe.

The best known of the recent forecasts of a potentially disastrous future is the 'prophecy' sponsored by The Club of Rome, known as *The Limits to Growth*.[10] The Club of Rome is an informal international organisation of businessmen, scientists, educators, industrialists and others who, under the leadership of Dr Aurelio Peccei, an Italian entrepreneur involved in several major corporations, undertook to analyse where the world's economy was headed. In the early 1970s it sponsored a study, carried out by an MIT (Massachusetts Institute of Technology)-based team of computer modellers. Their task was to acquire the best information available on global demography, resources, technology, and pollution, and to project the interactions among them to the year 2100.

The outcomes the group came up with were uniformly dismal. By the year 2100, most non-renewable resources would be gone, food would become scarce, industry would fail through lack of energy and raw materials; after a brief population explosion, there would be massive die-offs from famine and pollution (Figure 1.2).

Realising that its data base might be in error, and also that outcomes might be modified through deliberate policy decisions – restriction on births, on resource consumption, or on pollution production – the MIT group generated numerous alternative models. Even the most optimistic of them, however, ended with fewer *per capita* resources available than at present. In short, for the average human being on Earth, material existence in the future would likely be rather less affluent than it is now; for those in today's highly industrialised societies, it would probably be distinctly less luxurious, especially if wealth were more equitably distributed around the globe. No realistic combination of circumstances produced an onwards and upwards spiral of increasing material wealth and affluence.

The MIT model envisions an eventual equilibrium state between numbers of people and their economic throughput on the one hand, and the natural environment on the other. Earth can support only a limited number of people at a given level of affluence. Furthermore, this equilibrium state *will be imposed* – if not by humans, then by Nature. The MIT team's message is that our widely accepted belief in the possibility of indefinite unrestricted growth is in error.

> All the evidence available to us, however, suggests that of the three alternatives – unrestricted growth, a self-imposed limitation to growth, or a nature-imposed limitation to growth – only the last two are actually possible.[11]

Such prediction, as one might expect, met with a flurry of angry denials, especially from a group at the University of Sussex in England.[12]

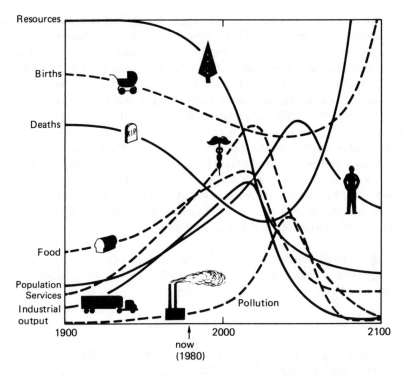

Resources
Births
Deaths
Food
Population
Services
Industrial
output

1900   now (1980)   2000   2100

Source: M. E. Clark, *Contemporary Biology*, 2nd edn (Philadelphia: W. B. Saunders, 1979) Figure 1a.7, p. 7; adapted from D. H. Meadows *et al.*, *The Limits to Growth* (New York: Universe Books, 1972) Figure 35, p. 124.

Figure 1.2 The 'standard' world model.

The Club of Rome modellers developed a set of *assumed* interactions among the variables shown; assumptions were based on interactions in the past (1900–1970), extrapolated without change in feedback parameters to the year 2100.

The unhappy outcome – *if we do not change direction* – is clear enough: as population increases, resources are depleted while pollution rises; deaths increase as the supply of goods and services fails.

Critics charged that the assumptions of the modellers were wrong and that future social and technological adaptations were underrated. Other models, based on more generous assumptions, also failed, however, although they took longer to do so.

Interestingly enough, almost none of the critics of *The Limits to Growth* deny that there indeed *are* limits... somewhere. The major points of disagreement are rather with the numbers used for constructing the curves shown in Figure 1.2, and with the assumptions about how the variables – births, deaths, resources, food – interact. How, for example, would pollution *in fact* affect human life span? How would social and political systems *in fact* respond to new environmental pressures?

The University of Sussex approach looks at recent trends in single segments of the global system and linearly extrapolates these independently into the short-term, only two to three decades ahead, and finds no significant problems. The MIT approach links all these segments together and watches how they interact over a much longer period, a century or more, and finds great problems. But small errors in assumptions become exaggerated the more complex the system is and the further into the future one looks. Such predictions are easy to criticise. Are there any short-term models that signal a different trend than found by the University of Sussex group?

One such study was undertaken by the Carter administration, and in late 1980 the United States government released a comprehensive prediction of the state of the world at the end of the twentieth century. Entitled *The Global 2000 Report to the President* it was the product of thirteen agencies, headed by the Council on Environmental Quality and the Department of State. It summarises its findings as follows:

> The trends reflected in the Global 2000 Study suggest strongly a progressive degradation and impoverishment of the earth's natural resource base. [13]

Among its principal findings are the following:

1. The global population will continue to grow by 200,000 people per day, reaching 6.35 billion in 2000.
2. The gap between the material well-being of people in rich and poor nations will widen.
3. *Per capita* food consumption in the poorer countries of South Asia, the Middle East and Africa will remain inadequate, or decline even further.
4. Improvements in world food production will require increasing fossil-dependent subsidies to raise yields – fertiliser, pesticides, power for irrigation, and fuel for machinery.
5. Oil production will peak in 1990; subsequent price rises will force poor countries to rely on fuelwood, demand for which will exceed supply by 25 per cent.
6. Fossil and nuclear (uranium) fuel reserves are unequally distributed around the world, and often create environmental damage during their exploitation and use.
7. Mineral resources will become more costly, as poorer ores are tapped.
8. Regional water shortages will become more severe due to population demands and the effects of deforestation.
9. The world's forests are disappearing; 40 per cent of remaining forests in less developed countries will be gone.
10. Destruction of the fertility of agricultural land will accelerate.

11. Global environmental degradation will increase owing to dispersal of hazardous wastes, atmospheric carbon dioxide buildup, depletion of the stratospheric ozone layer, and damage from acid rain.
12. Up to 20 per cent of all species on Earth are likely to become extinct, primarily through habitat destruction.

Compared to a number of other global projections, however, (the Latin American World Model; the World 2 Model, the World 3 Model, and the World Integrated Model of the Club of Rome; the United Nations World Model; and the Model of International Relations in Agriculture), the *Global 2000 Report* is optimistic. Yet even it warns:

> If these trends are to be altered and the problems diminished, vigorous, determined new initiatives will be required worldwide to meet human needs while protecting and restoring the earth's capacity to support life.[14]

Later on it states:

> Yet there is reason for hope. It must be emphasized that the Global 2000 Study's projections are based on the assumption that national policies regarding population stabilization, resource conservation, and environmental protection will remain essentially unchanged through the end of the century. But in fact, policies are beginning to change. In some areas, forests are being replanted after cutting. Some nations are taking steps to reduce soil losses and desertification. Interest in energy conservation is growing, and large sums are being invested in exploring alternatives to petroleum dependence. The need for family planning is slowly becoming better understood. Water supplies are being improved and waste treatment systems built. High-yield seeds are widely available and seed banks are being expanded. Some wildlands with their genetic resources are being protected. Natural predators and selective pesticides are being substituted for persistent and destructive pesticides.

> Encouraging as these developments are, they are far from adequate to meet the global challenges projected in this Study. Vigorous, determined new initiatives are needed if worsening poverty and human suffering, environmental degradation, and international tension and conflicts are to be prevented. There are no quick fixes. The only solutions to the problems of population, resources, and environment are complex and long-term. These problems are inextricably linked to some of the most perplexing and persistent problems in the world – poverty, injustice, and social conflict. New and imaginative ideas – and a willingness to act on them – are essential.

The needed changes go far beyond the capability and responsibility of this or any other single nation. An era of unprecedented cooperation and commitment is essential . . . A high priority for this Nation must be a thorough assessment of its foreign and domestic policies relating to population, resources, and environment . . . There are many unfulfilled opportunities to cooperate with other nations on efforts to relieve poverty and hunger, stabilize population, and enhance economic and environmental productivity. Further cooperation among nations is also needed to strengthen international mechanisms for protecting and utilizing the 'global commons' – the oceans and atmosphere . . .

Long lead times are required for effective action. If decisions are delayed until the problems become worse, options for effective action will be severely reduced.[15]

The 'Cassandra' future, then, warns of major catastrophes only IF WE CONTINUE TO ACT AS WE DO, IF WE DO NOT TAKE STEPS TO ALTER COURSE. The prophecy is *not* inevitable. We have it within our power to do something, to act, and the sooner we begin, the less painful will the adjustments be.

The 'Pollyanna' Future

In contrast to the cautionary futures projected by the 'Cassandras', there are the prophets of rosy futures who tell us that all we need is faith in ourselves and who encourage us to continue as we are in order to achieve the best of all possible futures. These cheerful visionaries, like novelist Eleanor Porter's turn-of-the-century heroine, Pollyanna, are possessed of an irrepressible optimism.

Among the leading optimists is economist Julian Simon who, together with the late Herman Kahn, has promulgated the notion that more is better. Simon argues that global population growth, far from being a problem, is a positive blessing. More people are not a *drain* on resources but rather they are *producers* of wealth:

Population growth and productivity increase are not independent forces running a race. Rather, additional persons cause technological advances by inventing, adapting, and diffusing new productive knowledge.[16]

In Simon's view, the more babies born into the world, the better, since in the long run, they will be net *producers* rather than net *consumers*. The more people there are, the larger will be the amount of new knowledge that is created: there will be more inventions, more patents and so forth.

Human imagination can overcome all future shortages: our brains are our 'Ultimate Resource'.[17]

As evidence for this view, Simon looks at the relative costs of consumable resources in the developed countries of the world over the past century or two; in all cases, costs relative to wages or to the Consumer Price Index have decreased.

> The costs of raw materials have fallen sharply over the period of recorded history, no matter which reasonable measure of cost and price one chooses to use ... [T]hese historical trends are the best basis for predicting the trends of future costs, too.[18]

According to Simon, since the price of copper has declined, it must be becoming more, not less abundant; since the price of wheat has declined, it, too, will continue to be more abundant and even cheaper. So too will energy:

> The history of energy economics shows that ... energy has grown progressively less scarce, as shown by long-run falling energy prices.[19]

In Simon's view, then – indeed, in the view of many economists – price is a true measure of scarcity. When the price goes down, something is becoming more abundant.[20] Curiously, this view is held even when logic dictates otherwise. Because more countries today are competing to produce oil does not mean that there is more oil in the world. It means only that what there is is being pumped out faster. Today, oil is abundantly *available* but it is *not* more abundant. In terms of future predictions, this is obviously a critical distinction.

Many economists also argue that whenever a resource grows scarce, its price rises and this makes it 'economic' to exploit lower-grade deposits. Today it 'pays,' for example, to drill much deeper oil wells and to pump oil from depths that were once 'uneconomic'. When the price of oil goes even higher, then other extraction technologies will become 'economic'. When the aluminum-rich ore, bauxite, is depleted, it will become 'economic' to utilise lower grade ores. The difficulty with this simple kind of thinking we shall uncover in the next chapter; it is, however, characteristic of the 'Pollyannas'' approach.

Finally, if resources really become scarce, the 'Pollyannas' predict that we will always find suitable substitutes. Here is the view of economists Harold Barnett and Chandler Morse on the matter of resource limits:

> Nature imposes particular scarcities, not an inescapable general scarcity. Man is therefore able, and free, to choose among an indefinitely large number of alternatives ... Advances in fundamental science have

made it possible to take advantage of the uniformity of energy/matter – a uniformity that makes it feasible, without preassignable limit, to escape the quantitative constraints imposed by the character of the earth's crust... Science, by making the resource base more homogeneous, erases the restrictions once thought to reside in the lack of homogeneity... [T]he particular resources with which one starts increasingly become a matter of indifference. The reservation of particular resources for later use, therefore, may contribute little to the welfare of future generations. The social heritage consists far more of knowledge, equipment, institutions, and far less of natural resources, than it once did. Resource reservation, by limiting output, and thereby research, education, and investment, might even diminish the value of the social heritage.[21]

In this view, resource conservation is wrong; it is by using *up* resources that we can create a brighter future.

Whether one believes the future painted by the 'Cassandras' or the 'Pollyannas' depends on which underlying premise seems more likely to prevail: the constraint placed by the physical limits of the Earth to support humans, or humankind's cleverness at overcoming any such physical limits on what can be consumed. The 'Pollyannas'' future implies a great faith in scientific and technological inventiveness. Like the average person on the street, the optimists look to a continuing supply of technological fixes to get us out of our present and future predicaments. The 'Cassandras', on the other hand, argue that although science and technology certainly have a role to play, major changes are necessary in our attitudes, in our expectations, in our worldview; human beings must adapt to Nature – since the reverse is not possible.

As weather forecasters and other prognosticators have long known, predictions are always in error in some of their details. Hence, neither of the two extremes presented here, neither the 'Cassandras'' nor the 'Pollyannas'' view, is likely to take place exactly as prophesied.

Nevertheless, both sides cannot be equally correct. This book is in general agreement with the interpretation of current global resource trends researched by the Worldwatch Institute and reported in their annual *State of the World* books.[22] It generally accepts the warning of the 'Cassandras'. That does not mean that gloom and doom are inevitable or that the future will be less desirable than the present. It does mean that things will be different – including, as we shall ultimately see, our attitudes, values, and beliefs. There is also a strong possibility that things will in fact be much more pleasant.

To see why this book takes the approach that it does, it will help to look briefly at the limits of science and technology as tools for solving future human problems. Indeed, among our first questions must be: What are the proper uses of science? and How should we value science? Although some of these ideas appear in later chapters, it is useful to introduce them here and so set the stage for what follows.

SCIENCE AND VALUES

In his book about the relationship through the ages between human societies and the soils that ultimately support them, the English author, Edward Hyams, had this to say about modern notions of science:

> For some incomprehensible reason the rise of this kind of detached inquiry into phenomena is, like the rise of a western art and literature, called the Renaissance. But nothing was reborn: not even the most arrogant of the Greeks had quite attained to this notion of inquiring into nature as if man stood god-like outside it. Galileo, Leeuwenhoek, Newton, were, in their methods, expressing the peculiarly Judeo-Christian idea that men were God's principal tenants, the rest of creation the fittings and stock let with the property.[23]

Modern science, with its conscious and formal process of inquiry into Nature, is an utterly new way of understanding the universe, one which tends to place its kind of knowledge above other ways of knowing. It implies that humans are now 'in charge'. When such knowledge is used to manipulate Nature we call it 'technology', and over the past half century this manipulative capacity has grown to the point where we can now annihilate ourselves as a species. Our technological cleverness seems to outweigh our wisdom. Here is how Philip Siekevitz, a professor of cell biology at Rockefeller University, puts it:

> So it seems that if a technology is available, [if] our brains have mastered a bit of a process, we use that technology, hoping that further improvements will rid us of any earthly malfeasances, hoping that in time more information will become available for our brains to use, even if [it is] not really understood.

> But we learn no lesson, give no moral to our children. We go on not having understood what is meant by our technological act or knowing. Some day, in the far, far future, the evolutionary processes will perhaps have gotten rid of this failing attempt at gaining knowledge; either we will be the ancestral progenitors [of a wiser species], or else we will be discarded, an unsuccessful offshoot, a dinosaur of the mind, one of nature's failures.[24]

What Science Is – and Is Not

There is among people living in modern industrialised countries – scientists and non-scientists alike – a certain unquestioned belief that science really *can* explain human existence and that, with enough research, it really *can* find ways to 'fix' our problems. One day, the discoveries of the astronomers and cosmonauts will explain the meaning of time, space, and existence; one day the physicists will harness an inexhaustible supply of energy that will be available in unlimited amounts whenever we need it; one day the chemists will invent a way to extract gold from the ocean 'economically'; one day economists really will discover how to have an expanding economy without using up resources; one day, political scientists will devise a government that makes war impossible . . . and so on. All of the sciences will eventually invent technological fixes for solving both our material and our social problems. People won't have to do anything.

But there is no guarantee that scientific knowledge will automatically produce benefits for human beings and other forms of life. In fact, some would argue that, given our present science-spawned capacity for extinguishing all life, science has so far been more dangerous than beneficial. In addition, there are instrinsic limits to the scientific method that prevent it from creating the hoped-for technological fixes, and it is important to perceive these limits, for they show us that science can reveal only a part of the world, and moreover, this part is inevitably biased. These ideas, outlined briefly here, are explained more fully elsewhere.[25] We begin with a brief summary of how the human mind works, a subject taken up in more detail in later chapters.

How the mind thinks

The human brain is constructed to make mental images of the information it receives from outside, but it does not detect *all* of the information that makes up the world. First of all, our senses are insensitive to much of it. Human eyes, for example, cannot 'see' ultraviolet or infrared radiation. Although we now have instruments to extend our perceptions, these are employed only in restricted circumstances, and by a few people. Moreover, we do not pay attention to all of the information that is actually detected by our senses; our brains filter out and ignore much of it. What enters our consciousness is thus highly selected and inevitably the mental image each person forms about the world differs to some degree from that of every other person.

There are thus frequent gaps in our mental image of the real world, where large pieces seem to be missing, and so our minds *invent* connections that help us to make sense of the whole. We do that every time we propose an hypothesis to explain why something is as it is. We do it less consciously every time we *believe* something to be true. This filling-in process makes a comprehensive whole out of the partial world we perceive. Every child, as it grows up, repeats this process of mental imaging of the world, and many of the unexplained gaps are filled in by myths or beliefs passed down from one generation to the next. This is a central part of the process of cultural transmission.

This mental image we call a *worldview*, and it is this way of seeing and understanding the world that informs our actions; it is the means by which we find our way about in the world and know what to expect of it. Because much of our understanding of the world has been passed down from parents, or has been acquired from our peers, each person's worldview has a strong cultural bias. This of course is essential: if individuals are to interact smoothly in a society they must share many of the same beliefs, values, and biases; to suppose that someone can invent her or his own worldview from scratch is utter nonsense. However, this does not mean that we do not bear a responsibility for the *consequences* of our socially shared worldview. Indeed, that is precisely what this book is about.

Thus, from our partial sampling of true reality, we construct what philosopher Michael Polanyi has called our 'personal' world, and it is full of prejudices and assumptions about those many, many aspects of the world that we cannot personally test. The next question is: How does scientific knowledge affect the 'correctness' of our worldview? To answer this, we must begin with what science is.

The world of science

The reality that science reveals is limited by what it is that science itself defines as valuable scientific knowledge. Polanyi recognises three desirable attributes, of which the first two are inherently 'scientific'. All apply jointly, 'so that deficiency in one is largely compensated for by excellence in the others'.[26]

First, a scientific statement must be *accurate*: How *certain* is this statement? How closely does it reflect reality? The estimation of accuracy almost always requires a measurement, a numerical value. Scientists have constructed the elaborate testing process known as *statistics* to judge the probable truth of a scientific assertion. Science thus studies mainly that which can readily be quantified, ignoring large areas of human experience: pain, pleasure, esthetics, love, honour ... most of the things that really matter. (It is perhaps not surprising that human happiness in a scientific age is measured by countable entities: one's income; the square-footage of one's house; how many friends one has; the number of cars in one's garage.)

A second characteristic of a scientific statement is its *general* applicability, what Polanyi calls its 'systematic relevance'. Unique events, no matter how carefully measured, have no scientific interest unless they illuminate a larger domain of knowledge. Our personal experiences, however important to the individual, are seldom of 'scientific' interest.

The third attribute of a useful scientific statement is that it have intrinsic interest; scientists simply do not seek for explanations of aspects of the world that are not related somehow to their own image of it – *their own worldview*. Only that part of reality will be investigated that is in some way of interest to the individual scientist, to the scientific community, or to society as a whole. Scientific activity, like all human behaviour, is informed by the current worldview.

The limits of science

This leads us to recognise three major limits on what science can tell us: complexity, current scientific visions, and society itself.

The limits of complexity   Those things of most immediate interest to us – economics, politics, social customs – are too complex to be studied quantitatively in ways that allow us to make predictions about them; if we do measure aspects of them, it is frequently the least interesting part of the subject that gets investigated. The limits placed by complexity on scientific understanding are twofold. One has to do with loss of accuracy in the necessity for simplification of a complex problem, the other with the loss of meaning inherent in the emergent properties of complex systems as we reduce them to their component parts.

As Polanyi points out, only relatively simple things are readily accessible to scientific investigation, and when the subject is too complex, we make simplifying assumptions about it in order to study it at all. As an example, we can take a seemingly straightforward study conducted by botanist Peter Alpert, of where mosses grow best on boulders strewn across a hillside. Even to start on such a project, it was immediately necessary to replace the uneven shapes of the real boulders with a simplified geometrical shape – in this case, a sphere.[27] Such a simplification, of course, already distorts the *real* world into a form our minds and measuring devices and mathematical models can handle. Alpert further describes all the other simplifications and assumptions needed to conduct even such an apparently simple study. More complex subjects undergo even more distortion.

In a quantitative, predictive sense, then, we can know much about a container full of single atoms, something about inanimate rocks, less about the development and functioning of living organisms, very little about ecosystems, and practically nothing about culture, economics, politics or the arms race. This does not mean that the latter are not fit subjects for study and understanding. It does mean that we are unlikely ever to be able to make precise predictions from that understanding about future events.

The limits imposed by reductionism on our scientific understanding are even more constraining. For a long time, Descartes's idea of a mechanistic world of cause and effect held out the hope that if only we could understand the behaviour of the smallest parts of something, we should understand the whole; if only we could understand the behaviour of atoms, then we could predict the behaviour of molecules, and from their behaviour we could explain cells, and then organisms, and then whole communities... and so on, until we could comprehend the universe. It turns out, however, that it is not possible to know precisely both the mass and the acceleration of any particle at a given instant, as this theory demands, since our very act of measurement perturbs the system. The notion, first proposed by the French mathematician, Laplace, that complex systems could eventually be understood by reducing them to their substituent parts, is thus physically unattainable. But even if it were possible to make the necessary measurements, the information gained by our minds would be empty – still merely atoms (or possibly quarks and leptons, or whatever are the physicists' latest 'ultimate entities') banging about in space. To conceptualise molecules, or cells, or organisms, or ecosystems, or societies, it is not enough just to know about atoms; we must have in our minds *empirical experience* of those larger systems, of their emergent properties, if we are to give them meaning.

The limits of scientific visions   Scientists cannot think about images of the world that are *not already in their heads*, what Thomas Kuhn called scientific paradigms.[28] Scientists therefore tend to believe the data that seem to 'fit' the current model, and to disbelieve and ignore that which does not. 'Breakthrough' thinking occurs when a scientist has a sudden new vision of 'how the world is', and such insights are often unsupported by any direct evidence, at least initially. Some examples are Copernicus's guess that the planets rotate around the sun; Dalton's concept of indivisible atoms of each of the elements; Darwin's insight into the process of natural selection; de Broglie's invention of the wave theory of particles; and Einstein's vision of relativity.

Such enormous leaps in understanding, however, are few and far between. Most scientists have only mini-visions that scarcely rock the current paradigm. Rather, their work serves to polish it up and expand it slightly, giving the impression of a smooth and orderly sort of scientific 'progress'. In reality, the science of each era has its own limited vision, seeing the world from its own particular angle.

The limits imposed by society   The larger vision of society also influences what aspects of reality science 'sees'. Scientists do not work in a social vacuum, and today little if any important science is done with the aid of well-equipped laboratories and well-trained technicians, both of which require considerable funding to maintain, well beyond the scientist's private means. Not surprisingly, corporate support for science goes for research in which corporations are interested and from which they ultimately expect to make a profit. Non-profit private foundations tend to fund science with specific social goals in mind: the conquest of cancer or lung disease, for example. Governments, too, fund mission-oriented science. Over 60 per cent of United States federal 'R & D' (research and development) funds go to military-related projects, and much of the rest goes to research designed to increase longevity or the level of material consumption. Contemporary society is focused on national and personal security and comfort; these are its tacit goals.

As a result, the great preponderance of scientific inquiry in the United States and other industrial societies has been directed toward an increase in our knowledge and understanding of highly selected aspects of the world in order that we could *augment our power to manipulate them to our own preconceived advantage*. In this our science and the technology that comes from it have been enormously 'successful'. For example, science has caused a population explosion in developing countries, a longevity problem in developed countries, and the amassing of 15,000 megatons of nuclear firepower – an equivalent of three tons of TNT for each man, woman and child alive on the Earth. Science has had many other consequences, but I mention these to show what extraordinary powers we have acquired – and what extraordinary problems we have created – by selectively channelling science along particular paths. We have expended virtually no effort on understanding other aspects of the world that remain even today largely unknown and unexplored. In particular, I refer to our minuscule efforts to understand ecological systems, global climate, and the process of evolution. Exploring these aspects of our world would not necessarily increase our ability to manipulate it, but would certainly give us a better grasp of how Nature functions and of the possible side-effects of our newly acquired manipulative powers. We have essentially ignored the goal of wisdom in our scientific investigation of the world.

Another consequence of this skewed scientific enterprise is the rapid rate of technological change wrought by the highly selected scientific knowledge that we have acquired; it has made the future so uncertain that we almost stop thinking about it. University graduates are being reminded as they enter the 'real' world that what they learned today will be out-of-date tomorrow. Given the enormous power of our present technologies, this kind of struthianism seems to me an extremely dangerous outlook to adopt. For if we fail to take control of our scientific enterprise and redirect it into areas that will provide the scientific wisdom we need, we shall destroy, one way or another, both the environment that supports us and the social institutions that hold us together. As Siekevitz warned, we shall continue to seek more and more technological power without understanding how to use it wisely and safely.

Scientific Values and Other Values

To summarise, science has a number of important social values, and many important limits. Among the values of science is first of all *power*, which Western society places at the top of the list. Science gives us power, in technological terms, to exploit and control our environment and also to threaten, control, and exploit our fellow human beings. But science also has value in illuminating our worldview – in filling in some of the gaps in our understanding of the real world. In doing so, it serves two functions: it fulfills our *innate curiosity* about how things work and, more importantly, it increases our *wisdom* about how to act.

Our need for wisdom cannot be overemphasised. Science, by illuminating for us at least some of the complexities of Nature, can provide us with an ultimate *boundary* for our actions. If we perceive how Nature works we can tell when we are threatening its ability to function in a healthy fashion. By knowing how soils work, we can judge when we are degrading them. By understanding evolutionary processes, we can learn to value living species that at present seem to have no 'economic' qualities. In other words, by understanding how the natural world works, we can better judge our correct place in it, and recognise when we overstep the boundaries. By setting such boundaries, science provides us with the possibility of an indefinitely viable Earth to inhabit. Within that framework, within those boundaries, we are free to create whatever sorts of societies we wish. Science, if we would but let it, can provide us extraordinary wisdom about how to survive.

But science provides no argument whatsoever about whether the human species *ought* to survive. This is not a scientific question. Many thousands of species have gone extinct in the past; why shouldn't *Homo sapiens*? Science can only tell us, *if* we decide we want to survive, what the boundary conditions are, what the 'rules of the survival game are', so to speak.

Nor can science as defined here set out what kind of social systems we *ought* to construct. Scientific empiricism cannot decide, for example, between communism and capitalism. Social systems lie outside the domain of accurate scientific prediction. That does not mean that we are powerless to understand social systems, however, simply that the scientific method is not the appropriate tack to take. For that purpose, we have to utilise other kinds of knowing and understanding that are quite different from those that the quantitative discipline of science has to offer. We need to explore ourselves intuitively, philosophically, historically, religiously, esthetically, morally. We have to ask: What does it *mean* to be human? What is it that gives value and dignity to human life? How should we think of ourselves in the world? How should we act towards others, both friends and strangers? What is the appropriate way for human beings to be?

Such explorations will not guide us to a single utopian society, but they should provide important clues about the essential components of a society compatible with the multiple needs of human nature. By combining scientific and humanistic understandings we may thus begin to remodel our worldviews and so improve the prospects for human survival.

OUR PATHWAY THROUGH THE LABYRINTH

If we agree that the direction in which the world is headed at the moment, determined as it is primarily by the beliefs and values of the Western worldview, leads to catastrophe – if we accept the 'Cassandra' vision – then that set of beliefs and values has to be consciously questioned and changed. That is the function of this book. Our path, as we follow Ariadne's Thread out of the labyrinth, will necessarily follow a tortuous course, for we shall have to pursue many different concepts and explore many different subjects as we try to make sense of the whole. Keeping all these various thoughts in mind at once is a difficult task, and it may help if we briefly chart our course here. We shall address four central themes which are related in the following ways.

Part I of the book lays out the boundaries placed by Nature on humankind. It asks: *What Are the Limits to Growth*? Who is more nearly correct: the 'Cassandras' or the 'Pollyannas'? It asks how much energy and resources there are in the world, which are renewable and which are not, and how long we can expect the non-renewable resources to last. Part I also explores the nature of ecosystems, showing how they maintain themselves in more or less stable states over hundreds of thousands of years, and how, after accidental disturbances, they tend to return to their former conditions. Finally, it reviews the relationships down the ages between human societies and their environments, showing how in fact human economic systems are really only modified ecosystems. This allows us to see the weak points in our modern industrial economies, with their enormous dependence on fossil fuels and other non-renewable resources, and their tendency to destroy even the renewable resource base on which all human societies ultimately depend. We conclude that the 'Cassandras' are, on the whole, right: if we don't change course, we are in for some serious ecological trouble during the coming decades.

Such a dire prophecy naturally leaves many people completely frustrated. They see the present trends in the world, and especially in the 'advanced' industrial countries, as inevitable, driven by an immutable and irredeemable 'human nature'. Part II of the book thus leaves, for the moment, the question of future ecological collapse and delves into the question: *What Is 'Human Nature'*? We quickly discover that our prehuman ancestors were social, sharing, and closely bonded before they became human. They were not, as the Western world has supposed, greedy, selfish, aggressive and individualistic.

From this important insight we turn to an examination of the great variety of cultural value systems that have and do exist in the world, and then to an inquiry into the world's major religions and their differing worldviews. What we discover is that there is not *one* human nature, but many: that we, in the industrialised West, are not genetically predetermined to behave as we do. Finally, by studying how each of us as we grow up acquires our worldview from those around us, we gain insight not only into the cultural nature of our beliefs and assumptions, but also into how a society might consciously set out to change its worldview.

This brings us to Part III, which goes back several hundred years into Western history and asks: *Whence Comes Our Western Worldview?* Whence came the ideas out of which evolved both capitalism and, more recently, Marxist communism? Whence came our notion of possessive individualism, a characteristic shared by few previous societies? How did it lead, step by step, first to the idea of private property, then to the notions of capital and interest, and finally to efficiency of production? As we trace these ideas, we watch how the purpose of an individual human life switched from one of fitting into a social whole, whatever one's rank, and of looking toward an ultimate reward in Paradise, to seeking happiness in material terms, on this Earth, by competing with one's fellow beings.

The ascendancy of economic efficiency and of competition encouraged a profound social revolution. Imperialism and exploitation, and the rise of nation states replaced medieval feudalism and opened the door for industrialisation, led by individual entrepreneurs. Usury became respectable and so, eventually, did unearned profit. Success in a competitive world became a sure sign of godliness. The lone entrepreneur was replaced by a corporate entity, extending control over resources and markets, but the justification remained one of free enterprise of individuals in a free-market economy. The consequences, both ethical and sociological, have been profound. We shall focus particularly on the destruction of social cohesiveness and the increase in social alienation that competitive materialism engenders.

This brings us to Part IV, which asks: *Where Do We Begin?* It starts by examining prevailing assumptions about modern industrial economies: the notions that continuing economic growth is essential and that GNP is a useful measure of well-being. By exposing the fallacies underlying these assumptions, we can see how to change our thinking from a growth-oriented economics that destroys the Earth to a steady-state economics where each society lives within its environmental budget, preserving its natural support system of soil, water, air, and minerals for future generations.

We then turn to the Third World, where 75 per cent of today's people live, whose forebears not only did not participate in the benefits of the Industrial Revolution but were, in fact, as colonies, exploited by it and, indeed, contributed a large part of the wealth now enjoyed by Europe and North America. While the past cannot be undone, there is no moral justification for continuing this exploitation into the future. But it can be halted only by understanding its past and present realities.

In order to bring about a more just world, one that is less exploitative of both natural and human resources, major political changes will be necessary not only in the capitalist countries, but in the Soviet Union, and in most Third World countries. If a worldview is defined as our beliefs and assumptions about how to behave in this world, then politics is either the means for enforcing those beliefs and assumptions, or the means for changing them. We thus explore how political systems work and what is necessary to change the beliefs and assumptions by which they function.

When politics fails, violence ensues. Within a society, we call it revolution; between societies, war. The result is much the same in that after unconscionable suffering a new set of beliefs and assumptions is imposed on a society, which is all too often highly antagonistic to the new ideas. In today's world, rife with nuclear weapons, overt coercion even between non-nuclear antagonists is highly dangerous, since the major global protagonists – the United States and the USSR – are seldom far behind the scenes. The power of nuclear weapons that science has given us – *all of us* – makes escalating violence among nations an historical anachronism, unless we are willing to contemplate the very high possibility of extinction. We are forced, willy-nilly, into seeking non-violent solutions to our conflicts.

It is at this point that we arrive at the last turnings in our struggle out of the labyrinth. How, indeed, can we emerge from this colossal mess? To start with, a few overall 'rules' can be laid down. The first is that indefinite material growth is not possible. The rich must taper off now; the poor require a fairer share of renewable resources, but must not become dependent on non-renewable resources. Ultimately, each ecological region must learn to subsist within its ongoing solar, climatic and geological income, stabilising its population and its *per capita* material consumption to within that level.

This process can occur only if politicoeconomic and cultural boundaries are redefined – as they once were – along natural ecological boundaries. Exchanges of excess wealth between these regions must not endanger the subsistence capabilities of the trading partners, since this encourages exploitation and the selling off of one's ecological capital for a temporary economic gain. Societies can no longer treat their support systems as exploitable commodities, selling off the fertility of their soils (in the form of cash crops) or other crucial resources at the expense of future generations. Only when each society preceives the ecosystem it inhabits as a sacred and inviolable trust to be preserved and passed on to future generations will survival of humankind be assured.

The inviolability of an ecosystem must be defined by scientists, and be recognised and respected not only by the society that lives within its borders, but also by those living in other ecosystems elsewhere. To destroy or annex the support system of another society is to invite the retributive destruction of one's own, given the existence of our destructive weapons. Any society that exceeds the capacity of its ecosystem must take steps to reduce its numbers. Its neighbours, anticipating this possibility, must be trained in the techniques of non-violent defence, thus making exploitation from outside a difficult, indeed, an ultimately impossible adventure. Societies ecologically self-sufficient in their subsistence needs can indeed be made impregnable by non-violent means.

While the existence of numerous, diverse societies seems to invite ongoing conflict, due to the potential misunderstandings that continually arise between those holding diverse worldviews, this difficulty seems to me more solvable than that which would arise if a single worldview – a single world order – were to be imposed, top down, on all of the world's people. While such a monolithic government might temporarily prevent strife, there is no guarantee that *any* of today's worldviews is viable for all time. Any that might be imposed – that of the United States, of the USSR, of China, of India – is full of ecological and social fallibilities that, were it the only system going, could lead to global collapse.

In my view, the best hope of humankind is to maintain as rich a diversity of social types as possible, with the expectation that each of these experiments in the human future will cross-fertilise with others, and thus maintain the vital diversity essential for indefinite survival. If this is to occur, only two universal 'rules' must be agreed upon: first, no society shall exceed its ecological bounds; second, all societies shall seek for the success (within its ecological bounds) of every other society. Competition for ascendancy in world trade, power, or military might are simply empty, meaningless concepts for the future. By encouraging diversity elsewhere, each society ensures a rich source of ideas and techniques for its own future. The goal of a society is not to supersede others, but to perfect its own social skills by interacting and learning from others how to modify itself. From time to time, cultures, and their supporting ecosystems, may coalesce together, or fragment into smaller units, according to what seems reasonable at each future period in history. The opportunities to explore human aesthetic, scientific, and social ingenuity are endless – provided we can agree that not all humans have to be born into this generation, and that not all resources have to be consumed by this generation. Humans do not require absolute security, but they do require hope.

Not everyone will agree with the ideas as they are developed in this book, especially with the goals proposed in Part IV and the Conclusion for a viable future. Indeed, they may not be the most desirable goals, although given my particular mental image of the world, they seem the most self-consistent to me for assuring the indefinite survival of humankind. The function of the ideas proposed here is not to force the reader to a particular conclusion, but rather to offer a possible cohesive model, and so to stimulate the reader's imagination. My short-term goal as author is to arouse the consciousness of as many people as possible and to encourage dialogue, and I invite others to modify, to criticise, and to propose comprehensive alternatives. If that happens, then this book will have more than served its purpose.

REACTIONS TO WHAT IS COMING

The material in this book forces an examination of basic beliefs and assumptions of one's inner worldview, and psychologically that can be a very uncomfortable process. So, too, is facing up to the enormity of the global problems at hand. Students in our course at San Diego State University usually respond in one of three ways. One is disbelief: 'All that can't be true; surely someone is hard at work on the solution. Every problem has to have a technological fix'. A second response is utter selfishness: 'I couldn't care less. So what if they blow up the world? I'm going to live while I can'. The third alternative is to care deeply, yet feel helpless in the face of the enormity of the difficulties ahead. There is a sense of despair and depression: 'I guess I'm glad I know, but I sort of wish I didn't. Why does it have to be *my* future that's ruined?' Similar thoughts must have passed through the minds of millions upon millions of other human beings caught up in the turmoils of history: black Africans chained to the fetid holds of slave ships; Jews behind barbed wire in Hitler's death camps; villagers dying of bubonic plague in Medieval Europe; millions of peasants today – in Bangladesh, Nigeria, Vietnam, Central America – suffering torture, poverty, or starvation. Despair is not new in human existence.

Eventually our students discover a fourth response, and that is to become part of the solution. As long as one refuses to face problems, to care about them, to ponder them, one is simply *part of the problem*. If I perceive a wrong and make no effort to correct it, I condone that wrong. We *are* responsible for each other. Our interrelatedness as human beings is universal and absolute. I owe my very existence to untold thousands of ancestors who established societies and upheld worldviews based on notions of human caring. The push of the prevailing moral thought and argument throughout the ages has been toward caring, and that responsibility lies equally with our own age.

Nor is it only other people toward whom concern has been directed.

Nature, the world around us, has also been an object of human concern throughout prehistory and much of the historical past, as well. In fact, the present age is marked by its singular lack of concern about our surroundings, which shows up not only in environmental degradation, but in a general loss of day-to-day sensibilities. We fail to see the ugliness around us because we close our minds off from our surroundings. The ecological movement, which has been based on sound scientific arguments as well as on aesthetic values, has been treated until recently by the larger society as an irritating red herring. Nature, along with every human artifact, has become merely a commodity for our convenience, something to be bought, sold, exploited and discarded.

This book is thus not merely a catalogue of the world's problems; it is a call to recognise that their origins lie in our own attitudes, in the way we see, think about, and act in the world. It is our worldview that creates our problems. At the precise instant when a person understands and accepts that it is aspects of his or her own viewpoint that have created, contributed to, and helped to sustain the world's problems, then that person suddenly becomes a part of the solution.

The first step is awareness, from which spring innumerable imaginative possibilities for change. Although many of one's ideas may not be practical at the moment, each new or altered personal behaviour pattern, no matter how small, is a contribution, and every contribution matters. The mighty rivers of the world, after all, are simply untold billions of raindrops and snowflakes, each of which fell separately to Earth. Who shall say which raindrops are more important than others?

Part I
Nature's Constraints:
'What are the
Limits to Growth?'

These days the average European or American believes that the only healthy economy is an expanding economy, in which more and more material goods are produced, consumed and discarded. In fact, the well-being of nations is measured in terms of the 'economic growth rates'. If the rate of growth increases, all is fine; if it decreases, it is a sign of trouble; and if growth ceases altogether, it is utter catastrophe. Other measures of well-being are mostly ignored. As long as the economy is growing, it does not matter that school graduates can barely read elementary level books, that violent crime is increasing each year, or that inches of irreplaceable topsoil are being eroded from croplands by 'modern' farming techniques.

Deep down, most people sense that something is wrong, but they are not quite sure what it is, and have not the information at their fingertips to refute the complacent assurances of the big corporations and vocal economists who argue for growth.

Part I of the book examines the way the living world functions, discussing the physical and biological rules that are inherent in Nature, and which constrain all human economic systems. Chapter 2, 'Energy and Exponentials', explains the laws of energy, and describes its available forms, its global distribution, and its role in the exponential economic growth of the twentieth century. Chapter 3, 'The Economics of Spaceship Earth', outlines the basic principles of ecosystem functioning, including energy flow and nutrient recycling, and relates these principles to hunting-gathering, agricultural and industrial economies. Chapter 4, 'Our Environmental Credit Card Account Becomes Due', shows how we have been living off Nature's capital through nutrient depletion, pollution, climate alteration and species extinctions. It advances a new philosophy toward the environment.

2 Energy and Exponentials

Man seems to have no function except that of dissipating or degrading energy...of accelerating the operation of the second law of thermodynamics.

Henry Adams, 1910[†]

INTRODUCTION

The vague presumption that modern technology is overcoming our dependence on the physical universe is sufficiently widespread to cause alarm, for it signals a general ignorance of the laws of matter and energy. This chapter, which relates the principles of energy to modern life, should dispel any such misconceptions.

The material world can be factored into matter – the atoms of the various elements – and energy – the forces that determine how these atoms behave over time. Without energy, matter remains inert and virtually unchanging, as on the lifeless moon. And so our first concern must be with the types and sources of energy available on Earth.

Energy drives all living systems, *including economic systems*. It would be a serious mistake to assume that the present apparent abundance of oil and natural gas, brought on by price deregulation and a sluggish global economy, means that the world's energy needs are now solved. Far from it. The world in the foreseeable future faces a grave shortage of readily accessible energy. It is therefore imperative to understand the forms in which energy exists and the laws that govern its distribution in the universe and its ability to do work. There are certain rules of Nature that human beings can never overcome; our bodies, even our brains, ultimately must obey the laws of matter and energy.

ENERGY: WHAT IS IT?

The presence of energy is evident everywhere. Our five senses are 'energy windows', keeping us in touch with our surroundings through a constant flux of energy. We 'see' radiant energy of those wavelengths known as visible light, and we 'feel' radiant energy of longer wavelengths known as heat. We also can tell, through direct loss or uptake of heat by our fingers, whether we are touching an ice cube or a hot toaster. Our ears 'hear' vibrations of air molecules; our noses and tongues detect certain chemicals. We also sense the presence of energy all about us: tree branches

31

bending in the wind; water under pressure, spurting into the air from a fountain; the feel of our pulse as our heart contracts. Indeed, everything that 'happens' happens because energy is flowing from one place to another.

It is this spontaneous *flow* of energy that permits work to be done: the work of growing up into a tree or a human adult, the work of ploughing soil, or of moving passengers from New York to London, or of removing pollutants from drinking water or industrial chimneystacks or car exhausts. Whatever is not spontaneous requires the input of outside energy in the form of work. Water spontaneously flows downhill. To push it uphill requires the heat of the sun to evaporate it and make rain clouds that return it to the mountains, or mechanical pumps that do the same. Likewise, electrons spontaneously flow, as electric current, from the negative pole of a flashlight battery, through the lightbulb's filament, which they heat up and cause to glow, and finally return to the positive pole of the battery. The two poles of the battery take part in chemical reactions which generate and absorb electrons, thus creating the flow. Eventually, the chemical 'charge' of electrons is used up; it is equally distributed within the battery, which is now 'dead'. To push the electrons back into the negative compartment, to reverse the chemical reactions that have spontaneously occurred, takes work; the battery must be recharged from an outside energy source.

Energy has value because it can do work.

Energy for Work

Energy comes in many different forms. Some are quite familiar: an electric current; kinetic energy (mass in motion, such as a waterfall, or the head of a hammer swung by a carpenter); radiant energy (X-rays, light, radiowaves); and thermal energy (due to the vibrations of atoms and molecules). All these forms of energy can flow from one place to another – from a high voltage to a low voltage, from the top of the waterfall to the bottom, from the sun to the Earth, or from a hot car engine out the exhaust to the cool atmosphere – and because they are flowing, they can do work. The electric motor turns the egg beaters, the waterfall turns a wheel to grind flour, sunlight provides energy for plant growth and to evaporate water from the ocean and make it fresh, and thermal energy causes gas to expand, push against the pistons in a car engine, and so propel half ton of steel and plastic with a person inside. In each case, work has been done.

Stored Energy

Energy can also be stored, in which case it is not so obviously capable of doing work. Of the above forms of energy, electricity is the hardest to store. Small amounts can be stored in cumbersome batteries. But the electricity needed to drive even a small car a couple of hundred miles requires a trunkful of heavy batteries. Much energy is expended just to lug them around with the car. It is because electricity cannot be stored conveniently that electric power utilities like to spread their energy load out as evenly as possible over the 24 hours of each day. It is inefficient to have a huge generating plant working at full capacity only a couple of hours a day and at half capacity the rest of the time.

Hydropower

The most familiar form of stored kinetic energy is the water stored behind the dams on the various rivers of the world. Collected during the spring snowmelt, the water is released gradually over the year through the giant turbines of the hydroelectric plants, pouring electricity into the power grid. While it is stored the water represents *potential* energy; only when it is allowed to flow spontaneously downhill can it do useful work.

Biomass: stored sunlight

Sunlight is stored on Earth in various ways. The constant flow of energy from an enormously hot sun to a relatively cool Earth provides a powerful energy flux for doing a number of jobs. The sun's energy drives the hydrologic cycle, which provides a constant supply of fresh water and the potential for generation of electricity. Before dams were built, erosion caused by this running water laid down a fresh layer of fertile topsoil on the floodplains of the world's great river valleys with each spring flood, insuring continuous agricultural yields.[1] The sun-driven water cycle also accounts for the weather, and especially for the wind, which is a potentially important energy resource in many parts of the world.

All these are consequences of the constant flow of sunlight. But the most immediately important consequence to humans is the sun's direct support of all life on Earth through the process of *photosynthesis*. The chemical bonds by which all living things are synthesised out of non-living air, soils, and water, are formed by the capture of the sun's energy in the leaves of green plants. Some of each day's ration of sunlight is trapped in their energy-rich chemical bonds, and once captured that energy is sequentially transferred from plants, to animals, and finally to the decay organisms in soil and water. Eventually, all the stored energy is used up doing the work of various organisms, and the atoms are rearranged in the original low-energy, non-living configurations. Useful energy is required to create and sustain life because life is a high-energy arrangement of atoms and molecules that would not occur spontaneously.

Fossil Sunlight   The sunlight stored in lettuce is a few weeks old; in wood, from ten to 100 years old; and in fossil fuels, millions of years old. In Earth's geological past, long before humans appeared, there were periods when decay processes did not keep up with the capture and storage of sunlight by plants. Over the several hundred million years of the Carboniferous era, lush fern forests thrived, but before they could decay, the dead trees were buried beneath the Earth's surface where pressure and heat converted them into great beds of coal. In the ancient shallow seas, small aquatic organisms that sank to the bottom also escaped decay, accumulated, and were eventually compressed to form today's oil and gas reserves (and in other places, tar sands and oil shale). The fossil fuels we are burning today are essentially 100 to 500 million-year-old sunlight. Although it took millions of years for them to form, humans will have consumed these fossil fuels during just a few centuries, exploiting their stored chemical energy by burning them.

Nuclear Energy

A quite different form of stored energy is that locked in the nuclei of atoms. It was Einstein who explained the relationship between energy and matter in his famous equation, $E = mc^2$. Under certain conditions, the nuclei of atoms lose a small amount of mass, m, which is converted into radiant energy, E. The conversion ratio is a very big number, c^2, the square of the speed of light. A tiny particle emitted from an atomic nucleus may thus give rise to an enormous amount of radiant energy.

Atomic nuclei are held together by attractive forces known as binding energies. As Figure 2.1 shows, both the very light and the very heavy elements have lower nuclear binding energies than do the atoms of intermediate size. Large nuclei tend to split, or *fission*, into small nuclei. This happens most easily when these nuclei are bombarded by energetic neutrons, which are emitted during spontaneous radioactive decay. When a heavy nucleus fissions, it releases not only radiant energy from its small emitted particle, but also one or more neutrons, which can cause a chain reaction of fissions to occur in neighbouring atoms. If the mass of fissionable atoms (usually uranium or plutonium) is big enough, the system goes 'critical' and a violent explosion results with the sudden release of radiant energy in the form of X-rays, light, and heat – an atomic explosion, as in the bombs dropped over Hiroshima and Nagasaki in 1945. In today's nuclear power plants, fission reactions are kept well below the critical level, and the emitted energy is used to convert water to steam for generating electricity.

At the other end of the atomic scale, light elements tend to fuse into heavier elements, again with the conversion of a small amount of mass into a large amount of energy. Indeed, this fusion process is what provides the

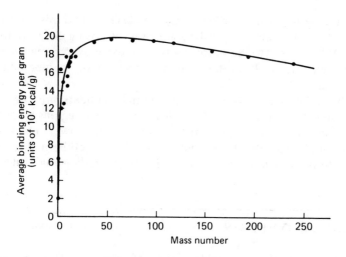

Source: W. L. Masterton and E. J. Slowinski, *Chemical Principles*, 4th edn (Philadelphia: W. B. Saunders, 1977) Figure 24.5, p. 596.

Figure 2.1 Plot of binding energy per gram vs mass number (atomic weight). Very light and very heavy nuclei are relatively unstable.

radiant energy of stars such as our sun. But, as in the sun, fusion occurs only at very high temperatures, which cause nuclei of hydrogen, the lightest element, to collide with such force that they fuse together to form helium or other heavier elements, at the same time emitting energy. This reaction, using heavy hydrogen or deuterium as the 'fuel', is almost thirty times more energetic than a fission reaction burning the same weight of uranium. This explains the extraordinary power of fusion, of 'hydrogen' bombs.

We shall have more to say about the future of nuclear power at the end of this chapter.

Thermal Gradients

One other important form of stored energy is heat itself. Wherever there is a heat gradient, work can be done, usually through the expansion of gases. The Earth has two natural sources of stored heat. One is the ocean, whose upper layers are heated by the sun while the deeper layers remain cool. The principle for making use of this gradient is the reverse of that used in refrigeration. For the latter, the power of an electric motor is used to compress gas in a condenser located outside the space to be cooled since compressing a gas generates heat. When the compressed gas then expands in the coils inside the refrigerator, it takes up heat. A pump returns the heated gas outside, where the cycle is repeated.

Ocean Thermal Energy Conversion (or OTEC) reverses this process. The working gas spontaneously condenses by losing heat to the cool depths. When raised to the warmer surface, it expands, turning a turbine to generate electricity. It is then pumped back to the cooler depths for recondensation.

Earth's other naturally occurring thermal gradients are the familiar geysers and hot springs located in volcanically active regions where steam is available for generating electrical or mechanical power. Unlike ocean thermal gradients, which depend on sunlight, geothermal energy is intrinsic to the Earth itself, arising from ongoing radioactive decay in the Earth's core.

ENERGY CONVERSION

Energy thus comes in a wide variety of forms. Some is immediately usable, but most is stored and must be converted to another form in order to do useful work. We commonly measure these forms of energy in different units: there are calories of food, kilowatt-hours of electricity, and British Thermal Units (BTUs), or therms (10^5 BTUs) of gas. Oil is measured in barrels (42 US gallons per barrel). A convenient energy unit to which all these can be converted is the 'quad', a quadrillion (10^{15}) BTUs, a very large number indeed. Another commonly used large unit is million barrels of oil equivalents per day (mbd), which is about equal numerically to half a quad per year. For instance, in the late 1970s, the total US energy consumption was about 37 mbd of oil equivalents[2] or 74 quads per year.[3] In discussing global energy needs, we shall use the quad.

Although we can express different forms of energy in equivalent units, this does not mean that we can readily convert energy from one form to another and expect to get the same amount of useful work out of it. For example, the mechanical energy of a waterwheel used to grind flour directly can grind more flour than can a mill powered by electricity generated by the same waterwheel. Some energy is lost in converting mechanical energy into electricity and back again. The energy laws known as the Laws of Thermodynamics tell us that these losses cannot be cured by more efficient conversions; they are instrinsic to the conversion process. Thermodynamics tells us the *maximum* efficiency we can obtain from any process.

The Laws of Thermodynamics

Almost everyone has heard that 'energy is neither created nor destroyed'. This statement, known as the First Law of Thermodynamics, simply means that in any process the total amount of energy remains constant, although energy may be converted from one form to another. In a toaster, for example, electrical energy is converted to heat; in a hydroelectric plant, kinetic energy of falling water is converted into electrical energy. (If no turbine is present to intercept the water as it falls, the energy that would have driven the turbine appears at the bottom of the waterfall as heat: the waters of Niagara Falls are a fraction of a degree warmer at the bottom than at the top.)

But if energy is never destroyed, why do we need a constant supply of it? Why does Earth require the continual input of energy from the sun to maintain life? If we have enough energy today, why do we need more tomorrow? For one thing, the Earth quickly loses much of the energy it receives from the sun; it is returned to outer space as reflected light ('earthlight' was observed by the astronauts on the moon), or as radiated heat. In fact, over a year, the Earth loses about as much energy to space as it gains from the sun, thus keeping its average surface temperature nearly constant.

There is another reason though why we constantly need the sun's energy. The concentrated sunlight which arrives at one point on the Earth's surface is soon dispersed as heat over a wide area. Whenever energy becomes uniformly distributed, it cannot do work, which in turn is necessary to maintain all organised systems. Without work to constantly mend, repair, tidy-up and generally maintain them, all organised systems tend to break down. Only energy that *flows* can do work, but no flow can occur when energy is evenly distributed.

The fact that both energy and matter tend to become randomly mixed, to even out levels of high and low potential, to become disordered, has extremely important consequences for both living systems and economic systems. These systems are highly ordered and are not in the least random. Yet anyone who has done such ordinary things as filing reference cards *in alphabetical order* in the library, or folding clothes into *separate* piles of sheets, shirts, pants and towels, or who has separated dust from furniture, dirt from dishes, or weeds from flowerbeds, knows that to maintain *order* requires a constant input of work. In fact, some things tend to become irreversible 'messes'. Anyone who has beaten up a whole egg and then discovered upon re-reading the recipe that the white was first supposed to be separated from the yolk will know precisely what Bertrand Russell meant when he said 'You cannot unscramble an egg'.[4]

Not only does the Earth and everything on it tend to 'run down' and require energy to restore it, but not all of the energy used in the restoration process actually becomes potential energy of the system being restored. By way of a simple example, take two lakes at different elevations and connect them so that all of the water from the upper lake must run down a series of waterwheels. Nowhere is the water falling freely, it is always pushing against a cup-shaped paddle. Now each waterwheel has its axle rigidly connected to a parallel waterwheel whose cups are turned in the *opposite* direction so that they transfer an equal volume of water from the lower lake and return it to the upper lake. For every gallon of water that falls downward, another gallon, in theory, would be lifted upwards. The system would be in perpetual motion, constantly restoring itself without the aid of any outside energy. Although thousands of ingenious attempts have been made, no one has ever succeeded in building such a perpetual motion machine. When models of such machines are displayed at science exhibits, there is always a hidden motor driving them.

This inability to convert 100 per cent of the energy released by one system into stored energy in a second system is just another way of stating the Second Law of Thermodynamics, which says that in every energy transfer, some useful energy is 'lost'; it becomes degraded into a form which cannot do work. The measure of this degradation we call *entropy*. In the systems we have been discussing, the loss is mainly due to friction, and turns up as 'waste' heat in the environment. Whenever we convert energy from one form to another, there is always some frictional loss and the system gets hotter. Thus, when we interposed an electrical generator and motor between the falling water and the mill grinding the flour, the mill could not grind as much as before due to the frictional losses in converting the kinetic energy of the waterwheel *first* into electrical energy and then *back* into mechanical energy. The old waterwheel simply converted the kinetic energy of the water directly into mechanical energy, and so was more efficient.

A commonplace example of an inefficient energy transfer system is the gasoline-powered automobile. Only about 13 to 20 per cent of the energy in gasoline or petrol actually moves the car; the rest ends up heating the engine (which therefore has to be provided with a cooling system), or going out the exhaust in the form of hot gasses. When an engine continues to run while stationary in traffic or waiting at a stoplight, *all* of the energy it burns is being converted into waste heat; it is getting 'zero miles per gallon'.

The Second Law of Thermodynamics predicts what the theoretical maximum efficiency is for various forms of energy conversion. For example, when electricity is generated by steam-driven turbines, the maximum electrical energy obtainable for a given amount of fuel consumed to produce steam is given by the formula:

$$\frac{T_H - T_L}{T_H} \times 100 = \text{per cent efficiency}$$

T_H is the temperature of the steam and T_L that of the cooling system that forms the thermal gradient necessary for work to be done. Both temperatures are measured on the absolute scale, in degrees Kelvin. If the steam is as hot as boiling water (373° K) and the cooling system is ice water (273° K) then:

$$\frac{373 - 273}{373} \times 100 = 27\% \text{ efficiency}$$

Just over one quarter of the energy released from the fuel can be converted to electricity; the rest goes into the cooling water or up the stack. In some power plants, the steam is superheated to about 450° K, and the cooling water is around 290° K, so they function at around 35 per cent efficiency. This is considered excellent efficiency for steam-electric generation.

CURRENT HUMAN ENERGY CONSUMPTION AROUND THE WORLD

The nations of the world fall into one or the other of two broad economic categories, the developed industrial nations and the developing nations. With one or two exceptions, all of the former lie north of a line that connects the Rio Grande with the north coast of the Mediterranean, stretches across the Black Sea and along the southern boundary of Siberia, and finally loops down to include Korea, Japan and the Hawaiian Islands. The only major industrialised countries south of this line are sparsely populated Australia and New Zealand, partly industrialised Taiwan, and the Union of South Africa which economically speaking is really two countries. The world is distinctly divided into North and South, and this is especially true when it comes to energy use.

Figure 2.2 shows the energy consumption *per capita* around the world. Although the world as a whole uses around 250 quads of energy per year, energy use is by no means uniform among all nations. The United States is the most profligate country of all; with only five per cent of the world's population, it consumes about 30 per cent of the world's energy, or five times the *per capita* average. People in most other industrialised countries – Europe, the USSR, and Japan – consume about half as much as Americans. And the 75 per cent of the world's people living in the less developed countries – who will become 80 per cent by the year 2000 when global population exceeds six billion – are consuming 20 to 30 times less energy per person than Americans.

Not only are the amounts of energy consumed widely divergent between the North and the South; so are the forms of energy consumed. Over 80 per cent of the world's energy comes from fossil fuels, but almost all of this is consumed in the industrialised North. The South depends mainly on firewood, the age-old fuel for cooking and heating still used by most

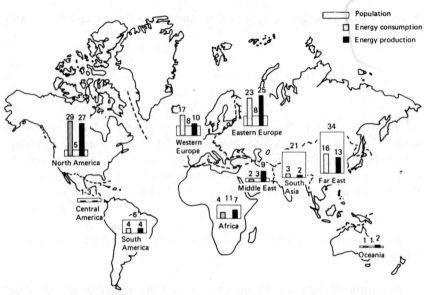

Source: Adapted from R. L. Sivard, *World Energy Survey*, 2nd edn (Leesburg, VA: World Priorities, 1981). Data sources: Population – *World Resources 1986*, a report of The World Resources Institute and The International Institute for Environment and Development (New York: Basic Books, 1986) pp. 236–7; Commercial energy – *International Energy Annual, 1985* (Washington D.C.: Energy Information Administration, US Department of Energy, 1986) pp. x, xi, 24–5, 16, 18, 54, 64, 81; Fuelwood: R. L. Sivard, *World Energy Survey* (1981) pp. 36–41.

Figure 2.2 Distribution of world energy consumption and production and of world population, as percentage of world totals (1984).
(Energy figures include firewood.)

peasant farms and villages; commercial energy (hydroelectric power, fossil fuels, or nuclear power) is almost always restricted to cities.

It is no accident that the industrialised countries consume so much fossil fuel, since most of the world's coal, the first fossil fuel to be exploited, lies under their soil. Japan is the only major industrial country almost totally deficient in fossil fuels. Europe, although it still has considerable coal, is now highly dependent on imports of oil. Both Japan and Europe import large quantities of oil from the Middle East. The United States, once a net exporter of energy, having increased its appetite and depleted its own resources, now imports energy, mostly in the form of oil from Mexico. The Soviet Union, on the other hand, has large amounts of coal and gas, and produces more energy than it consumes. It is the leading exporter of natural gas, mainly to Eastern Europe but also to Western Europe.

A few of the less developed countries have fossil fuel reserves. China has abundant coal fields, and at the moment both China and southeast Asia are energy self-sufficient. Those few other countries in the South with oil reserves – Liberia, Nigeria, Venezuela, Mexico – sell much of the oil they pump to pay the interest on their international debts and keep up with the demands of their growing populations. Except for the Middle Eastern countries who have accumulated petrodollars in America and Europe and so have the option to industrialise, the exploitation of oil by less-developed countries has not opened the door to economic development.

Meanwhile, those underdeveloped countries that have both burgeoning populations and no fossil fuels are rapidly depleting their wood supplies. Much of sub-Saharan Africa and parts of India and Nepal are gravely denuded of wood. In some areas, people spend one or more days a week just gathering wood for cooking. When wood becomes scarce, cattle dung is dried and burned instead. Over-dependence on these forms of biomass for energy can have several dire consequences as populations continue to grow. The removal of trees and shrubs encourages erosion, washing away the fertile topsoil and making the land useless for growing fuel, fodder, or crops. But if the land is replanted in trees for fuel, it cannot be used to grow food, which in some areas is already scarce. (Later on, we look at the ability of each country to feed its people in coming years.) Much woodland in developing countries has already been cleared for agriculture, as happened centuries ago in medieval Europe and central and southern China.[5] Finally, unless the ash from burned wood and dung is returned to the pastures and hillsides, soils become impoverished and less able to support growing populations.

For peasants in poor countries, buying imported fossil fuels to supplement local 'free' energy resources is out of the question. They have no spare cash. What money they earn goes to buy food, or for seed, fertiliser, or a small irrigation pump to ensure enough food for next year.

Energy Resources in Industrialised Countries

In the industrialised North, fossil fuels are the primary energy source. In the United States, which will serve as a model, they are almost 90 per cent of the total energy consumed; nuclear power plants generate around 6 per cent, and hydroelectric, wood and solar power account for 5 per cent. Figure 2.3 is a flow chart of the initial, intermediate and final forms of the energy consumed. Although the diagram appears complex, it really makes four main points:

1. Almost all US energy comes from fossil fuels.
2. About one-quarter of the raw energy consumed is wasted in the process of generating electricity.
3. The transportation system is highly inefficient, getting only a fraction of useful work out of the fuel it consumes. This represents a waste of about one-fifth of the total US energy consumption.
4. Different fuels have distinctly different end uses.

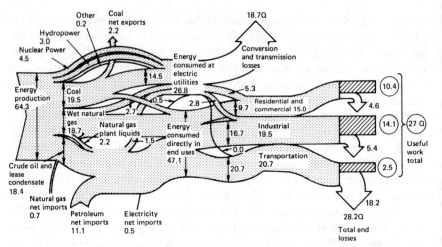

Source: Adapted from *Annual Energy Review 1986* (Washington, DC: Energy Information Administration, 1987) p. 3; end-use efficiencies based on 1979 Brookhaven data.

Note: A quadrillion BTUs, or 1 'quad', = 10^{15} BTUs (British Thermal Units).

Figure 2.3 US energy sources and uses in 1986 (in quadrillions of BTUs). Total consumed = 74 quads, useful work = 27 quads (or 36 per cent 'efficiency'). Energy sources were coal (23%), natural gas (26%), oil (40%), nuclear power (6%), renewables (5%).

In terms of the future, such a picture is both encouraging and discouraging. The enormous amount of wasted energy suggests there may be room for significant further energy conservation without sacrificing any substantial end-use benefits. The discouraging fact is that fuel sources are not directly substitutable. For instance, neither coal nor 'renewable' energy sources (solar, hydroelectric) can *directly* replace diminishing supplies of oil for running cars. Either people will have to depend less on this form of transportation, or else convert other kinds of energy into liquid fuels (synfuels), with the inevitable conversion losses that this entails. For running cars, a quad of coal or nuclear power does *not* equal a quad of oil.

But we are getting ahead of ourselves. Surely all the new oil and gas finds will stave off the need to conserve or develop new sources of energy. What about gas in the North Sea, the Alaska pipeline, oil in Mexico, and the mountain ranges in the Rockies full of oil shale? Won't these 'save' Britain and America for the foreseeable future? Won't Europe and Japan be able to count on the Middle East for a long time ahead? At this point it is necessary to take a long, sober look at the world's fossil fuel reserves – both known and estimated – to see how long they are really likely to last.

The Recent History of Energy Use

If we look at global energy use over the past 200 years, we discover that there has been an ever-increasing consumption of energy, mainly of fossil fuels by industrialised countries. Nor has this increase been constant, year after year; rather, it has grown. In mathematicians' terms, the increase in energy consumption has been not arithmetic, but *logarithmic*. The difference between these two kinds of growth is the difference between simple and compound interest.

Consider a deposit of $100. With seven per cent simple interest, in 10 years the balance will be $170 (100 + [10 × 7]), in 20 years it will be $240 (100 + [20 × 7]) and so forth. Every decade the balance increases by a steady $70. But suppose, instead, compound interest of seven per cent. Each year the interest is calculated on principal *plus* interest from the previous year – interest is earned on interest. At this rate, in 10 years the $100 deposit is doubled; it becomes $200. In 20 years, $400; in 30 years, $800. (With simple interest, the balance would be only $310 in 30 years.) The difference in these two forms of growth is plotted in Figure 2.4

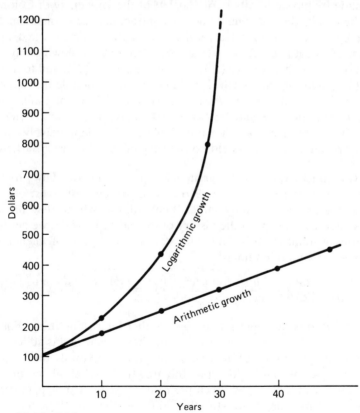

Source: M. E. Clark

Figure 2.4 The difference between arithmetic (simple interest) growth and logarithmic (compound interest) growth when the interest rate is seven percent a year.

This same sort of logarithmic rate of change characterises the rate of *consumption* of fossil fuels. The world has been annually speeding up its depletion of a fixed amount of wealth, its reserves of coal, oil, and natural gas. Since around 1870, the world's use of oil, for example, has doubled every ten years. The same is true of global natural gas and coal consumption (although coal use in the United States remained constant from about 1910 to 1970, when increase in oil consumption was at its height). *It is important to realise that the spectacular economic growth of the industrialised countries has been driven almost entirely by this logarithmic growth in fossil fuel consumption.* The big question is: How long can such growth continue?

The meaning of exponential growth

One of the clearest explanations of the meaning of exponential growth was made by physicist Albert A. Bartlett of the University of Colorado.[6] In it, he recalls the old legend about a contract between a mathematician and a king. As reward for a service, the mathematician asked for the amount of wheat needed to cover a chessboard in the following way: Place one grain on the first square (2^0), two on the second (2^1), and four on the third (2^2), eight on the fourth (2^3), and continue to double the number of grains until the 64th square, which would contain 2^{63} wheat grains. On the whole board, there would be twice that many grains, or 2^{64} (less one grain). These numbers are shown in Table 2.1. The king naïvely assumed that sixty-four doublings would not be very much. But, asks Bartlett:

> How much wheat is 2^{64} grains? Simple arithmetic shows that it is approximately 500 times the 1976 annual worldwide harvest of wheat! This amount is probably larger than all the wheat that has been harvested by humans in the history of the earth! How did we get to this enormous number? It is simple; we started with 1 grain of wheat and we doubled it a mere 63 times!
>
> *Exponential growth is characterised by doubling, and a few doublings can lead quickly to enormous numbers.*[7]

A careful look at Table 2.1 reveals another interesting thing. For each period of doubling (each square), the number of *new* grains added to the board is one grain greater than *all* of the grains previously on the board.

We can use this model to depict the growth in global oil consumption, which for decades averaged seven per cent per year. It so happens that at seven per cent annual growth, the doubling time is 10 years. Let us represent the global consumption of oil during the decade 1950–60 by the square shown in Figure 2.5. By analogy with the chessboard, that amount equalled all the previous oil consumption. During the decade 1960–70,

Table 2.1 Filling the squares on the chessboard

Square no.	Grains on square	Total grains thus far
1	1	1
2	2	3
3	4	7
4	8	15
5	16	31
6	32	63
7	64	127
8	128	255
⋮	⋮	⋮
64	2^{63}	$2^{64} - 1$

Source: A. A. Bartlett, 'Forgotten Fundamentals of the Energy Crisis', *American Journal of Physics*, 46, p. 876, 1978 (adapted).

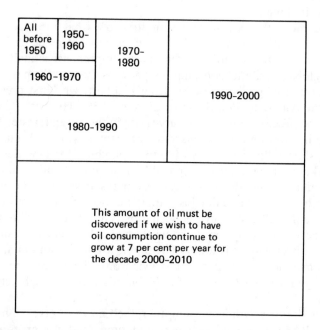

Source: Adapted from A. A. Bartlett, 'Forgotten Fundamentals of the Energy Crisis', *American Journal of Physics*, 46 (1978) p. 880.

Figure 2.5 Pictorial representation of past growth in global oil consumption and of future projections, if the same seven per cent annual growth rate that existed until the mid-1970s were to continue.

Note that each decade uses as much oil in ten years as was used in all of previous history.

however, the amount consumed was equal to all that prior to 1960, and so on. Each decade, the amount of oil needed equals all that was *ever* used before. How long would oil supplies last if global demand were to keep growing at seven per cent a year?

Experts disagree about how large the world's fossil fuel reserves actually are, but one way to put an upper limit on the answer is to make the absurd assumption that the whole world is made of oil (or coal, or natural gas) and ask, how long would such a reserve last at a seven per cent rate of increase in fossil fuel consumption? The answer is somewhere between 350 and 360 years, or about a century longer into the future than coal, for example, has been used in the past. (Smelting iron with coke began in England around 1700.[8]) A decade before the Earth was totally consumed, half of it would still be left, and a mere 50 years before it disappeared, only 1/32 of it would have been used – the reserves still left would appear enormous. By this analysis, we see that smaller errors in the estimates of our actual reserves – and of such 'giant' finds as the North slope oil fields in Alaska or the natural gas and oil finds in the North Sea off Britain – are trivial in the long run.

People, nevertheless, always want to know what the true fossil fuel reserves are. Since the whole Earth obviously is not made of fossil fuels, when would we *really* run out? The most comprehensive analysis of both US and global fossil fuel consumption patterns, both in the past and in the future, is that of geophysicist M. King Hubbert, who first presented his data to the American Petroleum Institute in 1956.[9] Hubbert did not try to identify the precise amount of reserves still in the ground; instead, he gave high (optimistic) and low (conservative) estimates of each kind of fuel. Despite almost three decades of active searching for new deposits, the proven (known) and unproven (possible) reserves still fall within his range of predictions. Using the most recent estimates of reserves, it is possible to calculate how long the major fossil fuels will last under various conditions of use. Table 2.2 gives several scenarios. One assumes zero growth in global energy consumption – the world as a whole does not increase its annual use over what it is today. The others are for various annual growth rates, from a modest one per cent to a vigorous nine per cent per year. The range of years in each case is for Hubbert's low and high estimates of reserves.

Even with no growth in consumption, the known reserves of natural gas and oil will be used up in less than one lifetime. If new reserves are found, more energy will be needed to extract them from the ground and it will be ecologically more damaging to do so. Their *net* worth will be less.[10] Oil and gas are perceptibly finite.

Coal is another story. Today it is used at a fairly modest rate, so the static reserves seem enormous, enough for between two and thirty centuries. But if global use of coal were to increase annually, it too would

Table 2.2 Global fossil fuel reserves

Fuel	Coal	Oil	Natural gas
Global annual consumption (1984) in Quads*	83.1†	120.7	62.4
Reserves (Proven → Possible) in Quads**	25,063–302,260	3863–12,171	3833–9799
Lifetime for different annual growth rates ‡††		Years	
0% (Static)	302–3637	32–101	61–157
1%	139–362	28–70	48–94
3%	77–157	22–46	35–58
5%	56–104	19–36	28–44
7%	44–79	17–30	24–36
9%	37–64	15–26	21–30

Notes:
*Figures for global annual consumption of coal, oil and natural gas are from *International Energy Annual, 1985* (Washington, DC: Energy Information Administration, 1986) pp. 64, 25 and 54, respectively.
**Figures for proven reserves of coal, oil and natural gas are from *Annual Energy Review, 1986* (Washington, DC: Energy Information Administration, 1987) pp. 231, 229 and 229, respectively. Figures of possible reserves are from R. L. Sivard, *World Energy Survey* (Leesburg, VA: World Priorities, 1981) p. 80.
†All energy units have been converted into Quads using the following conversion factors from Sivard, p. 42: 1 short ton coal = 0.907185 metric tons, and 10^9 metric tons = 28 Quads; 10^9 barrels of oil = 5.54 Quads; 10^{12} cu ft natural gas = 1.057 Quads.
‡Lifetimes, T_e, for fractional growth rates per year, k, of a reserve of amount, R, and initial annual consumption, r_0 are calculated from the equation:

$$T_e = 1/k \cdot \ln \frac{kR}{r_0} + 1$$

This is equation (6) in A. A. Bartlett, 'Forgotten Fundamentals of the Energy Crisis', *American Journal of Physics*, 46 (1978) p. 887.
††Note that an annual growth rate of 2 per cent would be needed just to keep up with current rate of increase in global population.

disappear in a century or less, even at the most modest growth rates. Many futurists assume that coal will *replace* gas and oil in the decades ahead,[11] but if this occurs, the proven coal reserves will last only 84 years, *without any growth in energy consumption whatsoever*. The possible coal reserves would last about ten centuries, without any growth in energy consumption, but again even modest growth would make them last only a lifetime or two at most.

It seems unrealistic to suppose, however, that with the world population

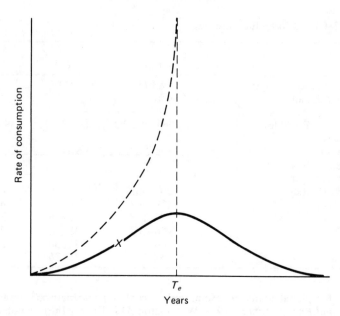

Source: M. E. Clark.

Figure 2.6 Exponential consumption and actual consumption of finite resources.

The dashed line is the curve for continuing exponential (logarithmic) increase in the consumption of a resource; at some time, T_e, the resource is completely used up and consumption abruptly stops. This time corresponds to the lifetime of fossil fuel and mineral reserves given in Tables 2.2 and 2.3.

The solid line is that predicted by Hubbert for the actual consumption of a finite resource. The point X is the inflection point where the rate of annual *growth* in consumption begins to decline; the top of the curve is when the absolute annual consumption begins to decline.

expected to double in a few decades, global energy consumption will remain static. With even a modest two per cent increase in energy consumption (just enough to keep up with population growth) the world's fossil fuels would be gone in about a century.*

The world, of course, will not continue to use up its fossil fuels at an ever-increasing rate; the consumption curve will not simply shoot up to the sky one day and then fall to zero the next, as indicated by the dashed line in Figure 2.6. Instead, economic and political forces will intervene, causing the annual growth rates to taper off, become static, and eventually to turn negative (the solid line in Figure 2.6). Indeed, global oil consumption

---

*This calculation, of course, takes no account of the skewed distributions in population densities and in fossil energy reserves noted earlier. But if there is to be anything like global justice, the annual increments of energy consumption in the South will need greatly to exceed population growth rates, while energy consumption in the North remains static or even declines. A two per cent annual *global* growth rate may thus be reasonable.

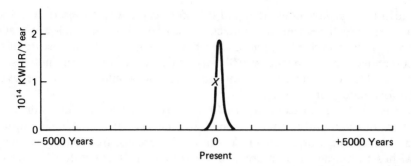

Source: Adapted from A. A. Bartlett, 'Forgotten Fundamentals of the Energy Crisis', *American Journal of Physics*, 46 (1978) p. 885, and Figure 69 in M. King Hubbert, A National Fuels and Energy Policy Study, Serial 93–40 (92–75), Part I (1973) (Washington, DC: US Government Printing Office, 1978).

Figure 2.7 The era of fossil fuel consumption, as projected by Hubbert. Seen in the perspective of human history 5000 years into the past and 5000 years into the future, the time required to consume fossil fuels covers a brief five centuries. Note that at present we are just at the point of inflection, where the percentage annual growth rates begin to decline, although absolute growth still continues upwards.

actually *declined* 14 per cent between 1979 and 1984 because stagnant economies, high prices, conservation, and substitution with other fuels in the major industrial countries all reduced demand; the poor countries naturally could not increase consumption at such prohibitive prices.[12] The precise shape of the curve is thus hard to foresee. Yet nearly three decades ago, Hubbert argued that there would be a fairly symmetrical shape for the consumption curve of any finite resource. The curve he predicted for fossil fuels, shown in Figure 2.7 in the perspective of 10,000 years of human history, is so far right on schedule. We are just beyond the inflection point (*X*) on the rising side of the curve, with declining but still positive annual growth rates in total global energy consumption. The peak is only 50 years away, when the world's population will be over twice what it is now, and fossil fuels will be gone forever sometime within 400 to 500 years. What took around 500 million years to accumulate will be dissipated in less than 1000 years.

Mineral scarcity and energy demand

At the same time that we have been mining fossil fuel reserves at ever-increasing rates, we have also been mining mineral reserves in exponential fashion: gold and silver, mercury, lead, copper, and tungsten.[13] Certain economists (such as Simon and Kahn, mentioned in Chapter 1) try to reassure us that this doesn't matter, since unlike energy, which can only be used once to do work, the atoms that are present on

Earth have always been here and therefore can always be used over and over again. All we need are more clever technologies to utilise lower-grade ores, or to recycle the minerals we have already used and discarded.[14] Somehow human ingenuity will solve this inconvenient but not insurmountable obstacle. It is an idea we all would like to believe. Unfortunately, it is wrong.

Take gold as an example. In Fort Knox, the US Treasury stores an enormous amount of gold – around 260 million ounces – backing up its credibility as a world banker. Yet in all the oceans, free for anyone who wishes to extract it, is more than 8000 times that amount. Why aren't people rushing to the oceans to extract gold? The answer is simple. One would have to process 125 million tons of sea water to recover one ounce of gold! The energy costs of such an undertaking would exceed by thousands of times the value of the gold recovered. *A too-dilute resource is no resource at all*.

In fact, it is the lower energy cost of extracting minerals from rich ores that causes us to continue exploiting them rather than recycling our scrap metals and other wastes; recycling takes energy. But as rich ores are used up, it will require more and more energy to extract the same amount of minerals from ever-larger volumes of low-grade ores. No matter how clever our future technologies, the Second Law of Thermodynamics sets an absolute minimum for the amount of energy necessary to extract a mineral from a given quality resource base, and that minimum increases about 70 per cent every time the mineral content of the ore is halved.[15]

In addition to the direct costs of separating a mineral from its ore, there are also questions of wastes to be disposed of, and of environmental damage for mining and processing ores that must be repaired. Even though iron and aluminum are among the most abundant elements on Earth, they are not as 'inexhaustible' as Kahn and others claim. Low-grade iron ores are more costly to process, and although the centre of the Earth is almost pure iron, there is no feasible way to penetrate hundreds of miles of molten magma to the hot, radioactive core where the iron is. Similarly, most clays contain aluminum, but it takes much more energy to extract it from them than from aluminum-rich bauxite, which is in limited supply.

At the very same time that fossil fuel resources are dwindling, ever-increasing amounts of energy will be required to utilise lower-grade mineral resources. The era of cheap energy and rich ores is coming to a close; soon it may be energetically 'cheaper' to 'mine' old garbage dumps than to utilise untouched but poor quality mineral resources. The facile manner in which we have exploited easily obtainable, concentrated minerals and vast deposits of concentrated energy to build our present industrial society will soon become a thing of the past.

Table 2.3 shows the situation for several major mineral resources. At 1986 consumption rates – that is, without any growth in consumption – the

Table 2.3 Global mineral reserves†

Mineral	1986 Static reserve* (proven → possible) years	Probable average annual growth rate 1983–2000 (%)††	Years of reserve at probable growth rates††
Aluminum[1]	294–950	3.7	67–97
Chromium	682–3273	4.9	72–103
Cobalt	262–342	3.3	69–76
Copper	70–198	2.6	40–70
Gold	30–48	1.7	24–70
Iron	240–932	2.4	80–131
Lead	42–414	1.0	35–164
Manganese	444–?[a]	1.4	141–?
Mercury	40–97	1.2	33–211
Molybdenum	134–237	2.5	59–77
Nickel	140–180	3.1	54–61
Platinum	259–407	3.2	70–83
Silver	26–?[b]	2.0	21–?
Tin	26–?[c]	1.0	23–?
Tungsten	77–?[d]	2.6	42–?
Zinc	45–270	1.9	33–95

†1986 data are from *Mineral Commodity Summaries, 1987* (Washington, DC: Bureau of Mines, US Department of the Interior, 1987).
*Static reserve is the years of this resource at 1986 levels of global consumption; 'proven' are known economic reserves; 'possible' are reserves inferred from geological topography.
††Data are from *Mineral Facts and Problems*, (Washington, DC: Bureau of Mines, US Department of the Interior, 1985). All figures are based on new mineral demands, over and above expected supplies from reclaimed (recyled) minerals.
[1]The data are for bauxite, the most economic aluminum-yielding mineral. At higher energy costs, aluminum can be extracted from clays and other minerals that are in virtually inexhaustible supply:
(a) Uncertain amounts of seabed reserves of manganese exist.
(b) In future, more silver is likely to be obtained as a by-product of mining copper, lead, and zinc, as long as the latter ores last.
(c) 'Sufficient [tin] resources are available to sustain present production rates well into the next century' (comment from *Mineral Commodity Summaries*).
(d) No 'possible' reserve estimates were given for tungsten.

most concentrated ores of mercury, for example, will last only 40 years, while less concentrated but still 'economic' ores will last nearly a century. But these will require more energy to extract. The critical role of energy in expanding our mineral consumption is evidenced by the fact that in the early 1970s, before oil prices shot up, the global demand for most minerals was growing at more than two per cent a year, for a doubling time of less than 35 years. Suddenly, high energy prices created a stagnant global economy, with a consequent levelling off in demand for minerals. If the global economy again grows at its 1974 rate, the remaining years of mineral reserves will shrink sharply. By 2050, 12 of 14 major minerals could be in critically short supply globally. (These figures tell us nothing, of course, about the fact that mineral deposits occur evenly over the Earth's surface, a fact that even today creates international tensions among the 'strategic interests' of various industrial nations.)

It is evident that, in future, recycling of minerals is going to play an increasingly major role in the economies of all nations who wish to remain 'self-sufficient'. For instance, instead of importing 90 per cent or more of its manganese, platinum, bauxite and cobalt each year, the United States could recycle much of these minerals – provided, of course, that their use did not entail irretrievable dispersion into the environment. Uses involving such dispersive waste of materials are clearly to be avoided. A prime example of effective recycling occurs with the silver used in photography, one of silver's main uses. When most camera buffs developed their own film, waste silver was washed down the drain. Today, most films are centrally processed and commercial developers recover almost all the 'waste' silver, recycling it back into new film. The postage pays for the energy needed to collect the silver in one place where it can be conserved. Only if parallel practices are applied to other minerals can the world hope for indefinite supplies.

As rich mineral resources disappear, so too do readily available fossil fuel reserves. The 'possibly recoverable' fossil fuels in Table 2.2 are mainly less accessible, poor quality, hard-to-obtain deposits which will require ever-increasing amounts of energy to extract and refine. The effort required to locate and recover new reserves – especially oil and natural gas – constantly increases. The outlook for future oil finds in the United States is grim:

> Extrapolation of energy costs and gains from petroleum drilling and extraction indicates that drilling for [US] domestic petroleum could cease to be a net source of energy by about 2004.[16]

In other words, the energy cost needed to discover new reserves exceeds the energy to be obtained from exploiting them. While much fossil energy may still lie in the Earth, the effort required to locate and extract it may exceed its energy content.

These limitations make the utilisation of diffuse energy resources decidely less attractive. For example, low-grade Texas panhandle oil reserves require masses of steam or compressed carbon dioxide to force them out of the Earth;[17] both are 'energy-expensive' technologies. Water, already in scarce supply for agriculture, would be pumped hundreds of miles to be squeezed as steam into the Earth to eject oil. At some point, the energy required to import water equals or exceeds the energy in the extracted oil. Likewise, the low-grade coals of the American north-west plains, a semi-desert area, will be useful only if strip-mined and trans-

ported as a water-borne slurry by pipeline to Eastern and Southern manufacturing areas. But the energy cost of importing the necessary water could
consume a large part of the net energy that is exported – not counting the
damage caused to donor water regions from co-opting their water resources, and the local damage through massive strip-mining.

The United States is thus trapped in an enormous energy resource
tangle. Energy, minerals, and water are being 'swapped' over ever longer
distances in order to maintain a level of resource consumption that no one
locality could long sustain. What began a century ago as local sharing of
resources has grown into nationwide – indeed a global – exchange and
competition for ever more scarce supplies of energy and minerals, the
underpinnings of modern industrial economies. America's dilemma is
becoming the world's.

LOOKING TO THE FUTURE

At present, energy conversions for which humans are responsible are
being carried out, in the United States, at a rate more than 100 times
greater than that at which energy could be converted by the unaided
manual labor of the whole population. It is as if each one of us had 100
full-time slaves working for us. Although very few people could afford to
have 100 full-time servants, most Americans can and do command the
equivalent amount of energy.[18]

An era of abundant energy at prices especially favorable to global
expansion of industry and transportation has come to an end ... A
profound change in the world's use and choice of energy resources is
slowly underway ... In the years immediately ahead [however] there is a
risk of a very serious energy crisis unless more effective public action is
taken to conserve supplies and allocate them equitably.[19]

These observations underscore the magnitude of our energy dependence
and the likely severity of the coming crisis. The first quote is from physicist
Robert H. Romer's introductory textbook on energy, and the second
comes from a booklet on world energy by social–economic analyst Ruth
Leger Sivard. It is obvious that the entire industrialised world is built on
'non-renewable' energy resources which are almost half gone. We are not
cleverer than earlier societies in the efficiency with which we utilise
available energy; we have merely discovered and learned how to consume
as fast as possible gigantic stores of fossil sunlight.

Although Americans are vaguely aware that there are energy problems,
where the United States will get its energy from in the future is not a burning topic, even among politicans. This seems particularly strange when we
consider that the gigantic US military machine – which constantly figures in

political discussions – depends mainly on fossil fuels to keep running, not to mention the fossil fuel-powered industrial base on which the military relies. Furthermore, this energy-base is highly centralised, in giant dams, huge power plants, and enormous oil refineries. All are strategically vulnerable. So are the Alaska oil pipeline and the numerous natural gas pipelines. Moreover, America imports a growing proportion of her energy from abroad, without which her present economy would be crippled. Yet in the 1980s almost no government attention has been paid to the country's long-term energy needs. While some industrialised nations have longer range energy plans, it is safe to say that none has sufficiently grappled with the problem.

Although the average person may know little about energy, there has certainly been no shortage of studies. We next highlight the various options suggested for both the short-run and the long-run, noting both social and physical consequences.

Energy Options and Social Options

In a world in which fossil fuels exist in finite amount, there are essentially three eventual 'solutions' to the energy problem. One is to produce electricity from nuclear energy and then convert that into other forms needed for various sorts of work. The second is to make use of 'renewable' energy resources – solar, biomass, wind, tides, hydropower, geothermal energy. And the third is to cut back on energy use, to be less wasteful, to be more efficient, to conserve. These three options are not mutually exclusive, and probably all three will make some contribution in the future. But they embody divergent kinds of social values. This is particularly true between nuclear energy and 'renewable' energy. (Everyone, it seems, is for being less wasteful and more efficient, but the degree of possible energy-saving is hotly argued.)

The long-term energy debate tends to be polarised in the following way. On one side are the so-called 'hard path' proponents. They believe that continued economic growth is essential, that such growth depends on an increasing use of energy, and that the only feasible source of that energy is from nuclear fuels – from fission, breeder and eventually fusion reactors. Nuclear energy is virtually inexhaustible, since a large proportion of the Earth's rocks contain uranium for breeder reactors and the oceans have generous amounts of heavy hydrogen (deuterium) for fusion reactors. This scenario envisions continued centralisation of energy production in giant nuclear power plants, owned and operated by large corporations or by governments. It is essentially an extrapolation of the present institutions of industrialised society into the long-term future, locking in place the social patterns that presently exist. While this group thinks elimination of 'unnecessary' energy waste is important, any significant cut-back in overall energy consumption is considered economically and socially detrimental.

On the other side, are the 'soft path' proponents, most of whom believe that continued expansion of the materialistic economies of the industrialised nations of the North is neither feasible nor desirable. Indeed, they see many of the local and global social problems as arising from the processes of industrial, economic, and political centralisation. For them, a retreat from growth to a more modest, steady-state level of energy and material consumption is the only sustainable route to the future. With less energy demand, it will be possible to supply the basic needs of all human beings from 'renewable' energy resources, provided the world population levels off fairly soon. Since 'renewable' sources of energy are diffuse, spread over wide areas, their use would entail industrial and economic decentralisation. Each locality would harness its own energy supply – whatever it happened to have in abundance: solar (deserts, tropics), wind (Newfoundland, the American Midwest), ocean thermal gradients (Hawaii) and so forth. While such an approach might fit well with the present social institutions of many of the underdeveloped nations of the South, it would entail some major institutional and social changes in the industrialised nations of the North.

Much of the debate, therefore, revolves not only around the feasibility of different energy technologies, but around basic differences in social visions. If in fact we opt for some contribution from *both* 'hard' and 'soft' paths over the next few centuries, the question arises: Can we have overlapping centralised and decentralised social institutions working together harmoniously? While the Laws of Thermodynamics and other laws of physics and chemistry put ultimate constraints on our behaviour, they do not help much in deciding between the options described here; these are social and political matters.

For reasons that will become clear in later chapters, this book urges the adoption of multiple, decentralised energy resources. But that is getting ahead of our story. Here we need to examine critically these two long-term options. Before we do that, however, we must inquire into the short-term future, and particularly into the possibilities for the conservation of energy.

Energy Conservation

No matter which ultimate path one prefers, almost all sides agree that for the near term – the next 20 years or so – in the industrialised North there will continue to be an increase in effective energy demand, although it will not be as steep as before. This demand will be met in three ways: there will continue to be more use of fossil fuels; modest additions from both nuclear and 'renewable' sources will occur; and finally there will be significant improvement in energy conservation.[20] This seemingly gradual change in energy use in the coming decades results from the enormous inertia built into a modern industrialised economy. For one thing, the lead-time required to develop, test and finally adopt new technologies is seldom less than ten years and often spans several decades. For another, the enormously high capital investments in infrastructure – refineries, power plants, high-tension wires, pipelines, oil tankers, highways – and the complex institutions that support them – the large corporations, banks, government bureaux – are highly resistant to change. It seems easier to go on using oil and other fossil fuels, even if we know they are growing scarce, than to tear down the old system and build a new one. It is easier to hope that nuclear power will replace fossil fuels without seriously dislocating these institutions, than to think in terms of renewable energy, which certainly would cause some major changes.

Yet as Figure 2.3 shows, there is considerable room for reducing waste without decreasing end-use consumption. Western Europe has about the same standard of living as the United States but consumes about half as much energy *per capita*. While the American military accounts for some of this difference, the rest implies unnecessary energy waste in the United States. For a time, America could sustain economic growth without increased energy consumption merely by improving the efficiency of energy use, a process already set in motion by the 1970s oil crises.

A group of Swedish theorists has designed a scenario for Sweden's economy that would decrease energy requirements by more than one-third by the year 2010, while permitting a 50 per cent increase in the consumption of goods and services.[21] The scenario would eliminate Sweden's dependence on nuclear power and greatly reduce her need for imported fossil fuels. (Sweden possesses no coal, oil or natural gas of her own.) She would continue to use her abundant hydropower and make up the difference with energy from biomass – wood from her huge forests and from specially cultivated stands of rapidly-growing poplars and willows.

Energy conservation is the key. All three major sectors of the economy – industry, housing, and transportation – would introduce far more energy-efficient technologies. If the most efficient of newly budding technologies were to be implemented throughout Sweden's two major industries, both the paper-pulp and steel industries could reduce their energy consumption by 50 per cent or more. Retrofitting of current buildings could save over one-third of the energy needed to heat and ventilate them, whereas new homes and office buildings could be built that use less than 20 per cent of the energy of existing buildings! In transportation, savings of 50 to 70 per cent are possible with new cars that get double or treble the mileage of today's cars. While industry, it is felt, will automatically respond to the energy economies of new process technologies, the cost of constructing energy-efficient buildings and automobiles is about the same, so that without government regulation or tax incentives these energy savings are less likely to be introduced. There seems to be no reason why similar energy savings should not be possible in virtually all of the industrialised nations; the technologies are there for all to use.

Two further energy-saving schemes already in place in several European cities could be adopted elsewhere if institutional barriers could be overcome.[22] As Figure 2.3 indicates, a huge proportion of energy is wasted during the generation of electricity; about 60 per cent of the heat generated is simply lost to the environment. By channelling this 'waste' heat, usually as steam, from the power plant to nearby buildings – a process known as district heating – energy otherwise used to heat these buildings can be saved. Such schemes obviously work only if the power plant is sited reasonably near a fairly densely populated area.

A second place in Figure 2.3 to 'capture' waste energy occurs in those industries which generate steam as part of their industrial process. By harnessing the steam to turn turbines and generate electric power before it is finally released for processing, the factory produces some of its own electricity, thus decreasing its dependence on outside power plants. This combined process, known as cogeneration, is not restricted to industry; any institution that has large and parallel requirements for heat and electricity, such as shopping malls, hospitals, and universities, can make use of it. As a study from Harvard has pointed out:

> The advantages of cogeneration are substantial – about half as much fuel is used to produce electricity and steam as would be needed to produce the two separately.[23]

Many other energy-saving practices could also be introduced into industrialised societies to curb energy waste without seriously altering social habits. They include: improved insulation of buildings; comprehensive recycling of wastes, which in the United States could reduce annual energy consumption by five per cent;[24] and the introduction of subsidised urban rapid transit systems as well as efficient interurban railways in high density regions or provinces.

While Sweden may be able to live off her own 'renewable' energy resources by adopting more efficient technologies, it is more problematic whether increased efficiency alone will suffice for other industrialised countries with higher population densities and fewer 'renewable' resources than Sweden, in particular France, West Germany and Japan. Even the United States ultimately faces a serious energy shortage. In almost every case, the urge has been to follow the 'hard' energy path. Japan is well on her way to becoming the world leader in nuclear power production.[25]

The 'Hard' Path

Ever since the Second World War, most industrialised countries have poured massive amounts of public funds into researching nuclear power.

(This impetus was fed in several instances by the parallel military applications of nuclear research.) There has also been some research on extracting and processing low-grade domestic fossil fuels such as tarsands and oil shales, in order to maintain energy independence during transition from fossil to nuclear fuels. In the United States, the 'synfuels' push in the late 1970s was a response to the OPEC oil embargo. Since then, however, US funds for energy research have declined sharply and the energy problem at present is virtually ignored.[26]

Before critiquing the increasing use of nuclear power, we should note some major drawbacks of the present tendency to expand consumption of coal to supplement dwindling supplies of oil and natural gas. Aside from its unfortunate effects on those who must prise it from the ground – physical accidents and chronic black-lung disease – coal mining, especially surface strip-mining, tends to destroy vast tracts of land. If the damage is to be healed, a fair proportion of the energy dug from the ground must be used simply to restore the land to its former fertile state.[27] But even land superficially restored is 'sick' underneath, leaching out acids from the subsurface rock into local groundwater and rivers for decades or even centuries.

Combustion of fossil fuels, particularly of coal, has a serious environmental consequence, namely the production of acid-forming gases, NO_2 and SO_2. When these gases, which are mainly released from the chimneystacks of heavy industries and power plants, combine with water droplets in the air, they fall to Earth as 'acid rain'. The gases, however, may be blown hundreds or even thousands of miles, across international borders, before falling on rivers, lakes and forests, where they destroy both fisheries and trees. The mix of pollutants released from both cars and power plants produces the toxic brown smog that damages lungs, forests and many agricultural crops. To prevent this environmental degradation by modernising coal burning plants and putting pollution control devices on cars means using up part of the precious energy being produced, making what remains more costly to the consumer. (Recovery of the industrially useful acids from chimneystacks, however, may sometimes pay for itself.)

Another and ultimately even more severe problem with increasing our global combustion of coal is that carbon dioxide (CO_2), the major product of such combustion, accumulates in the atmosphere, resulting in what is called the 'greenhouse effect'. Carbon dioxide in the atmosphere tends to retain heat at the Earth's surface rather than permitting it to radiate out into space, leading to a rise in the average temperature. Some meteorologists believe that even a modest increase in fossil fuel consumption in the next century would cause Earth's temperatures to 'approach the warmth of the Mesozoic, the age of dinosaurs . . . Thus, CO_2 effects on climate may make full exploitation of coal resources undesirable'.[28] As with nuclear winter, the magnitude of the effects of CO_2

buildup is disputed by scientists, but since it is an experiment no one can carry out ahead of time, caution would seem to be called for. Once CO_2 is released, it cannot be recalled.

The ecological consequences of unleashing in two or three brief centuries the energy that took hundreds of millions of years to accumulate are hardly likely to be trivial, and we shall return to them in later chapters. On the other hand, the problems with the 'hard path' substitute, nuclear power, are quite different but no less difficult. Of the three possible forms of nuclear power, only fission and breeder reactors are now technologically feasible; fusion may or may not prove itself, and we shall not consider it further here.

Like coal mining, the mining of ores containing the nuclear fuels uranium or thorium is hazardous, and the waste tailings (detritus) from the mines and processing mills present a permanent source of radioactive radon gas that escapes into the atmosphere and of radium that leaches into local groundwater, affecting nearby residents.[29] In addition, the question of reactor safety has been repeatedly raised. Although nuclear plants normally release trivial amounts of radiation (less than coal-burning plants), the probability of a serious release during a malfunction was not admitted by the industry until the near meltdown at Three Mile Island in Pennsylvania stimulated some investigative reporting. For decades the US Atomic Energy Commission, which was both promoter and regulator of nuclear power, had covered up and suppressed its own reports of potentially serious safety problems. Daniel Ford, in his book *The Cult of the Atom*, concludes:

> With a hundred [nuclear] plants expected to be in operation in the United States during the next ten years, there is at least a 50 per cent chance during this decade that one of them will have a meltdown or some other form of major accident.[30]

The serious accident at Chernobyl in the Soviet Union, which contaminated milk and other food products across much of Europe as well as the water supplies for Kiev, showed the magnitude of just one accident.[31] Although designs for inherently safe nuclear reactors have been proposed – reactors that would be immune from mechanical or human failure because they would rely on the principles of physics to shut themselves off[32] – they would still suffer the other drawbacks of fission reactors.

The spent fuel rods, still highly radioactive, require either reprocessing or disposal. Both pose major problems, since the rods contain plutonium, one of the most dangerous substances known (only a fraction of a milligram can cause lung cancer), and a key substance for making atomic bombs. Plutonium's radioactive half-life – the time required for half its radiation to decay away – is 24,700 years. It represents a legacy of danger for generations to come, and there is so far no satisfactory way of permanently disposing of it. Meanwhile, seepage of highly active wastes into soil and possibly groundwater has occurred from several 'temporary' US storage sites. Reprocessing fuel rods, a costly and specialised task, occurs at only a few places in the world, none in the United States.[33]

The containment vessels of both nuclear power plants and waste storage sites are constantly bombarded by radiation that makes them both radioactive and brittle. When today's reactors were built, it was assumed they would have a safe productive lifespan of three or four decades, after which the fuel would be removed and the shell encased in concrete for a century while its radioactivity decayed to safe levels. (Over the years, 'dead' power plants would gradually accumulate as concrete pimples on the landscape.) Now we know the rate of ageing of the high-pressure reactor vessels is far faster; they are much more subject to cooling failures and potential meltdowns – or shutdowns in the case of the proposed 'safe' reactors – than first supposed.[34] Furthermore, two long-lived isotopes which no one predicted ahead of time are being produced at high levels in the walls of reactor vessels. Instead of decaying away in decades, these isotopes have half-lives of 80,000 years (nickel-59) and 20,300 years (niobium-94). Instead of encasing old power plants in concrete, it now appears they will have to be disassembled by remote-control cranes and special blowtorches; then the whole lot, power plant, cranes and all, will have to be safely buried, all with commensurate costs to the consumer.[35] At the same time that these new safety problems surfaced, research funds for their study in the United States were cut.[36]

Since the isotope of uranium, ^{235}U, required as fuel for the fission reactors that comprise almost all of today's commercial nuclear power plants is scarce, it is no answer to fossil fuels. One-fourth has already been consumed, mostly for weapons. Even at today's modest consumption rate, global reserves will last for only 50 to 60 years.[37] If the other uranium isotope, ^{238}U, were burned the outlook would become quite different, since this isotope comprises 99 per cent of natural uranium and can be obtained from much lower-grade ores. If an atom of ^{238}U is bombarded by high energy neutrons, it is converted to unstable ^{239}Pu, which then fissions, giving off both heat for generating electricity and further high energy neutrons for converting more uranium to plutonium – hence, the 'breeder' reactor: more fuel is produced than is consumed – until the ^{238}U is gone. Obviously, plutonium 'wastes' from today's spent fuel rods could serve as fuel for this type of reactor, thus reducing somewhat the problem of waste disposal. Breeder reactors are being introduced in Europe and Japan.

In operation, plutonium breeder reactors can be highly dangerous. The amount of plutonium needed for efficient breeding is potentially explosive, unlike the case with fission reactors. Even a small accident could emit significant amounts of plutonium into the air, requiring rapid evacuation of local residents. These reactors must also be cooled with liquid metals, such as sodium, which are themselves explosive when in contact with air or water. Any ruptures or cracks that might occur in reactor vessels could have the most serious consequences. Breeder reactors suffer from the same dangers as fission reactors, but in much higher degree.[38]

The spread of breeder reactors would make the proliferation of nuclear weapons, especially in the hands of terrorists, far easier than it is today. Refining, transporting and storing fuels are all opportunities for capture of plutonium. Even in a politically stable world, the vigilant surveillance of reactor fuels and buried wastes would have to go on uninterrupted for thousands of years. As a group of political scientists has pointed out: 'No social order in history has lasted more than a few centuries'.[39] Does the world – could the world – have the political stability needed to pursue the hard path? Those who strongly favour nuclear power claim only nuclear power can provide that stability. Their argument is that energy use at today's levels or higher is essential, that only nuclear power can provide such energy, and without it, conflict is inevitable.

The 'Soft' Path

Not so, say the 'soft path' proponents. If as much effort had been spent on research into 'renewable' forms of energy as on nuclear energy, we would be well on our way to a stable energy future. As Denis Hayes, the first director of SERI (the now defunct US Solar Energy Research Institute in Colorado) has said:

> Oil and natural gas are our principal means of bridging today and to-morrow, and we are burning our bridges. Twenty years ago, humankind had some flexibility; today the options are more constrained.[40]

In other words, we should be using some of our gas and oil energy to develop 'soft path' alternatives.

We begin with hydropower, today's major contributor of 'renewable' energy in the industrialised world, often copied in the Third World. In general, large dams are counter-productive in the long run. Their reservoirs often inundate and destroy fertile agricultural land or rich natural ecosystems. Then, too, the reservoirs usually fill with silt in a few decades and can no longer store as much irrigation water. Meantime, lands downstream, deprived of this nutrient-rich silt, come to require the application of costly fertilisers to maintain productivity. Reservoirs also evaporate large amounts of water that otherwise might have seeped into the ground to replenish wells in nearby dry areas.

On the other hand, there are many small sites around the world where water tumbles freely over steep drops, and these could be efficiently and relatively harmlessly used for generating power. Only about one-third of these sites are now tapped, and as Hayes points out most of the unused potential lies in the countries that are most devoid of fossil fuel resources: Asia, Africa, and South America.[41] These countries could perhaps garner about 15 quads a year of end-use electric energy from small-scale hydropower.

Wind, too, has been used to do work for millennia, the first windmills having originated in Persia about the time of Christ. Since then, they have been used to grind flour, saw wood, make paper, pump water, and generate electricity. Rural America, in the early part of this century before the national power grid was established, garnered what electricity it had from the wind. Today there are experimental tests of large wind propellors, capable of generating 2000 kilowatts in high winds. But even cheap wind machines, capable of working at low windspeeds, can do useful work, particularly in areas where alternative sources of energy are not available. Physics offers some potential ideas for new wind machines that might be developed.

Since wind blows intermittently and most people prefer energy at their convenience rather than Nature's, the question of storage arises. Three possibilities exist. One is to pump power into a national power grid whenever the wind blows. By act of the US Congress, it is now possible for the owner of a windmill to sell excess power to the local utility; when the wind fails, the consumer draws on the grid, which serves as a backup. Another solution is to use the wind's energy to compress air that is released as needed to generate electricity. Or when the wind is blowing, excess electricity can be used to hydrolyse water, using the hydrogen gas that is produced for cooking, heating and automotive fuel as needed. These second two options require considerable capital outlay, and may be feasible only for small communities rather than individual households.

Hayes estimates that with virtually no environmental damage, 600 quads of wind power 'can be commercially tapped at the choicest sites around the world'.[42] This is about twice the world's present total energy consumption from all sources!

Then there is the sun itself. Every year, it sends 28 000 times more energy to Earth (including the wind and hydropower mentioned above) than all of humankind's commercial uses put together. Whether we use it or not, that energy will continue to flow for several billion years. Drawing on data of 'soft-path' advocate Amory Lovins, Hayes states that 'most of the energy budgets [of Canada and the United States] could easily and economically be met using existing technologies'.[43] Solar collectors for heating hot water, a major domestic energy requirement, are becoming more common in industrialised countries. Solar energy can also be used to heat and air condition buildings, dry crops, cook food, purify water, pump water, and perform many industrial tasks requiring low levels of heat.

It takes a more intensive focusing of sunlight and a more highly capitalised system to obtain higher grade forms of energy from the sun, such as electrical energy. One possibility is solar 'farms' that would contain acres of mirrors all reflecting sunlight onto one central tower where a heated liquid turns a turbine that generates electricity. More feasible, however, is direct photoelectric conversion. New types of photovoltaic cells are being made, at ever lower costs, that efficiently convert light into electrical energy. The energy needed to manufacture such cells, even today, can be 'paid back' in two years, after which time, they are free producers of energy. The Japanese are very optimistic about photovoltaics.[44]

Solar electrical energy is available only during the day unless special storage methods are employed, like those described for wind power. For people living near the oceans, however, there is the 24-hour availability of the sun's heat stored in ocean thermal gradients and obtainable by means of the OTEC device described earlier. Heavy capital outlay means it can be used only by communities, not individuals.

Green plants and wastes from animals are yet another secondary form of sunlight. There are two main strategies for using this resource and one must be careful to distinguish between them. One strategy is to utilise agricultural wastes – inedible portions of crops, wood chips, garbage, cattle dung – for fuel, either by burning it directly for cooking and heating, or by fermenting it into alcohol or methane, using these as fuel and the residue for fertiliser. Today the only cooking fuel for millions of India's poorest people is the dried cakes of cow dung they collect for free. To manufacture biogas (methane) from cow dung requires a modest amount of capital equipment, and the derived gas becomes a saleable item that the poorest cannot afford.[45] (The reader will begin to discern how social patterns are determined by the forms of energy available and the manner in which they are distributed.)

Another strategy for using biomass is to grow crops – sugarcane (for alcohol), trees (for wood and methanol), and guayule, a desert shrub (for petroleum-like oils). Brazil is cutting down large forest acreages and planting sugar cane to produce alcohol for powering automobiles, thus making her less dependent on foreign oil imports. Any number of authors are optimistic about how much sunlight can be harvested to meet the world's energy needs through the husbanding of fields and forests. The difficulty often is that a different set of authors, seeking to solve the world's future food problems, claim much of the same acreage as potential agricultural croplands or grazing pastures. At the same time, a third set of writers see these same 'underexploited' lands as a potential biomass source to substitute for oil feedstocks in the petrochemical industry. Too often, the same limited resource is expected to do more than one job in meeting future crises.[46] Nevertheless, there is certainly scope for reclaiming wastelands – parts of Africa's deserts, of China's and Nepal's deforested hillsides, of the salinated fields of the Indus and Nile valleys – and getting them once again to yield food, fibre *or* energy. With modest help, such reclamation projects could be undertaken, but the United States, among others, has been unwilling to take the lead and give its necessary approval to such self-help programmes for energy production in developing nations.[47]

Lastly, there is geothermal energy. This resource already supplies small amounts of energy in limited localities; it generates electricity in Geyserville, California, and heats homes in parts of Iceland, New Zealand and France. Of more widespread interest is the potential for tapping the hot magma that lies everywhere beneath the Earth's crust; so far no serious explorations of this possibility have been undertaken.[48]

SOME CONCLUDING THOUGHTS

As we shall see later in this book, 'renewable', 'soft' energy, with its diffuse base, is likely to be not only environmentally but politically and socially a better answer to the energy needs of the human population. We can do no better here than close with the words of Denis Hayes, who argues that we already have available 'soft' technologies to provide 40 to 50 per cent of our global energy needs within a generation.

> While no single solar technology can meet humankind's total demand for energy, a combination of solar sources can. The transition to a solar era can be begun today; it would be technically feasible, economically sound, and environmentally attractive. Moreover, the most intriguing aspect of a solar transition might lie in its social and political ramifications.[49]

Most renewable energy sources have a relatively low capital outlay and are therefore suited to individuals or small communities; they do not require centralised institutions to be effective. The ramifications will become apparent in later parts of this book. But first we need a clear understanding of ecosystems and our obligatory relationship with them.

3 The Economics of Spaceship Earth

A snake seizes a frog in its mouth and slips away into the grass. A girl screams. A brave lad bares his feelings of loathing and flings a rock at the snake. The others laugh. I turn to the boy who threw the stone: 'What do you think that's going to accomplish?'

The hawk hunts the snake. The wolf attacks the hawk. A human kills the wolf, and later succumbs to a tuberculosis [germ]. Bacteria breed in the remains of the human, and other animals, grasses, and trees thrive on the nutrients made available by the bacteria's activity. Insects attack the trees, the frog eats the insects.

Masanobu Fukuoka, 1978[†]

INTRODUCTION

There is a delightful cartoon showing two ETs in a spaceship. As they approach Earth, one says to the other: 'You'll like this place – it has atmosphere!' Indeed, Earth is one of the few heavenly bodies with an atmosphere, and no other is like ours. Earth is also unique in having liquid water, lots of it, and dry land covered with a thin, but highly fertile layer of soil. How true is the statement: 'Below that thin layer comprising the delicate organism known as the soil is a planet as lifeless as the moon'.[1]

Earth, this home of ours, is a very special place. Eight thousand miles in diameter, it carries an atmospheric mantle, mainly nitrogen and oxygen, that extends about 150 miles into space, although all of the clouds and weather are contained within the inner 7 miles. From the depths of the oceans to the tops of the highest mountains is a distance of only about ten miles. Yet within this narrow region is found all the life in the universe. Almost all organisms living on land are supported by soil, a wonderfully complex mixture of eroding particles of rock, of decaying organisms, of water and air, and of a host of bacteria, fungi, amoebae, earthworms, nematodes, sow-bugs, millipedes, centipedes, and insects. The soil which gives birth to life is, itself, teeming with life. This miraculous stuff, found nowhere else in the universe, varies in depth from a mere quarter of an inch in the crevices of bare rocks, to several feet in the fertile alluvial plains of the world's great river valleys. In the astronomical distances of the cosmos, however, it is scarcely measurable. Yet on this thin skin of soil, all human beings have been – and are – absolutely dependent.

When ecologists speak of the fragile Earth, they mean this narrow layer of atmosphere, soil, and water – oceans and lakes – that supports all life. They call it the 'biosphere'. On either side, life is impossible. In space, there is no air to breathe; temperatures are extreme; and there is a constant barrage of life-damaging radiation. Beneath the hard surface crust of the Earth, not more than a few miles down, lies a semimolten mass of magma which moves the continents about. From time to time it erupts through volcanic vents to the surface, spewing out lava, cinders, and toxic gases. Deeper still is Earth's extremely hot core, consisting of molten iron, nickel and a few other elements, kept hot by pressure from the outer layers and by the spontaneous decay of radioactive elements trapped in the interior.

No wonder Earth, with its thin life-giving mantle, has been called a 'spaceship'. It is the only place where life could have evolved and survived. It is a finite habitat whose well-being we are now severely threatening. Whether humankind might one day become independent of Earth is beside the point; it certainly is not now. Our only choice is *to maintain and care for our planet as a reliable support system for ourselves and our fellow life forms*.

This chapter deals with the mechanisms and laws, so far as biologists understand them, of how that support system functions. Just as there are Laws of Thermodynamics that describe how matter and energy behave, so are there Laws of Life that describe how living systems work. Understanding them helps us function more appropriately. We begin with a description of Earth and its ecosystems, and the laws they obey. Then we show how human economic systems, from the emergence of human beings over two million years ago until the present, are really modified ecosystems that must obey the same laws as any other ecosystem if they are to persist indefinitely. This is an enormously important point that far too few people have yet understood.

PHYSICAL PROPERTIES OF SPACESHIP EARTH

Three physical factors interact together on Earth to ameliorate the temperature extremes of space. The first is Earth's favourable distance from the sun; the second is the large proportion of nitrogen in the atmosphere, which readily transmits both light and heat; and the third is the large amount of water on the Earth's surface. Water, unlike the compounds common on other planets, has a remarkably high heat capacity and remains liquid over a wide temperature range. The oceans act as an enormous heat sink, absorbing heat in summer and releasing it in winter; they maintain Earth's surface temperature within relatively narrow limits, neatly buffering it against the thermal extremes that occur elsewhere in the planetary system.

Liquid water has another special life-giving property. It is an excellent solvent, and hence forms the perfect medium for the multiple chemical reactions which constitute life's metabolism. In the absence of liquid water – when organisms are frozen or dried out – all life activities cease. It is not surprising that Earth has been called the 'Water Planet'.

The modern Earth also has a large quantity of atmospheric oxygen. This, however, is a direct consequence of living organisms themselves. When Earth formed around five billion years ago, the primordial atmosphere contained no oxygen, being composed mainly of water vapour, carbon dioxide, methane, nitrogen and ammonia. Only after life evolved and photosynthetic algae living in the oceans began to produce molecular oxygen as a waste product did oxygen accumulate in the atmosphere.[2]

This event had two important consequences. It allowed the rapid evolution of oxygen-breathing animals. Before, only anaerobic life had been present. Even so, neither plants nor animals could migrate from the protected seas onto land until enough oxygen had accumulated, about a billion years ago, to create an ozone shield in the outer atmosphere. Ozone, a gaseous molecule composed of three atoms of oxygen (O_3), is formed when ultraviolet light from the sun splits ordinary oxygen (O_2), a process which occurs mainly in the stratosphere. This ozone layer absorbs the sun's most harmful ultraviolet radiations – those that would be destructive for living organisms – and acts as a permanent shield for all terrestrial life. (The accumulation of oxygen in the atmosphere and the development of an ozone shield are an excellent example of how living organisms can significantly alter their surroundings.)

It is evident that the optimistic image of permanent spaceships garnering solar energy and space dust for growing crops and manufacturing products is fraught with enormous dangers: extremes of temperature, the absence of an atmosphere, and the presence of harmful ultraviolet light, not to mention frequent lethal showers of protons irradiated from solar flares.[3] The exploitation of space is not a realistic solution to Earth's contemporary problems. Life is adapted to live on Earth.

ECOLOGY AND ECOSYSTEMS

In recent decades science has 'rediscovered' what native Americans and other primitive peoples intuitively understood: namely, that all living organisms profoundly interact both with one another and with their non-living surroundings. The modern study of this system of myriad interactions is called *ecology*. Earth, as a whole, comprises a giant *ecosystem*. It receives a constant flow of energy from the sun, which powers virtually all of the biological and non-biological phenomena that occur. But aside from an occasional meteorite, our planet receives no outside matter;

virtually all the atoms now present have been here since Earth was formed. Earth is a materially closed system, relying on the sun to energise it.

But Earth is too large and complex to study as a whole. Indeed, for many purposes it is not meaningful to do so, since there are great differences from one place to another in the types of climate, topography, and life forms that are found. Essentially, any circumscribed region can be considered as an ecosystem: a meadow, a valley, a lake. It may be as small as a fish tank or a compost pile; it may be as extensive as a continent. Clearly, boundaries are arbitrary and every ecosystem interacts with all others in varying degrees. Ecologists frequently divide Earth into larger zones, each with its own physical characteristics, always remembering that they are all interconnected by the oceans, the atmosphere, and the migratory species of molecules and organisms that traverse their boundaries. Our focus is mainly on the major terrestrial zones, or biomes, although of course the oceans and fresh waters are also important to human societies.

Ecosystems Are Shaped by Climate

Ecologists distinguish several major biomes, each characterised by a particular set of climatic conditions. These include the sparsely vegetated hot dry deserts, many of them located near the equator; the semi-arid shrublands or chaparral, with wet winters and dry summers; the central continental grasslands of the American plains, the African savannah, and the Asian steppe, characterised by more extreme temperatures and limited year-round precipitation; the deciduous hardwood forests, with similar temperature extremes, but more rainful; the taiga or great coniferous forests, with year-round precipitation and cool, short summers; the arctic and mountainous tundra, comprising small shrubs and dwarfed trees, where the growing season is brief and the soils are often frozen to quite near the surface year-round; and in the wet parts of the equatorial latitudes, the hot tropical rainforests, with tall stands of broad-leaved trees entwined with vines.

Each biome occurs on several continents and has a similar appearance wherever it occurs. Under similar sets of climatic conditions, quite different types of plants and animals, having become adapted over centuries to the same climate, come to look like each other. Although the chaparral shrubs of coastal California and those of the southern mountains of Chile belong to quite different biological lineages, they look very similar; they have small, hardened leaves, and after a brush fire the vegetation in both areas is able to resprout from its roots. Likewise, the euphorbias, a group of African desert plants, look very much like the cacti of American deserts, yet the two are not at all closely related (Figure 3.1). Time and again, quite diverse plants – and animals too – have acquired similar anatomies and life patterns because they have become adapted,

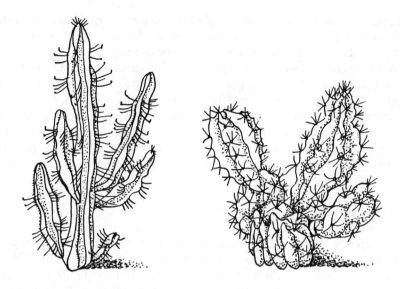

Source: M. E. Clark; art by R. Carruthers

Figure 3.1 Convergent evolution.
An African euphorbia (left) and an American cactus (right) have very similar growth forms – reduced leaves, some of which are thorns, and thick, succulent stems for storing scarce water – which are adaptations to similar arid climates; yet these two plants are botanically not at all closely related.

through the slow processes of natural selection, to living under similar sets of physical conditions, a process biologists call 'convergence'. On land, the primary determinants of ecosystem structure are the annual cycles of temperature and rainfall. Climate, and to a lesser extent soil, topography, and latitude, *powerfully shape the forms and life patterns of the organisms that inhabit an ecosystem. Organisms are genetically, physiologically and behaviourally adapted to fit in with their total surroundings.* Ecosystems are not haphazard assemblages of living things.

Ecosystems Repair Themselves

Occasionally a natural catastrophe occurs – a flood, a fire, or several years of drought – and temporarily destroys part of an ecosystem. When this happens, the damaged area usually undergoes a sequence of stages when first one group of organisms dominates, then another, until it returns to its former condition. After a forest fire, for example, first grasses, then shrubs, and finally trees dominate the landscape. This process, known as *ecological succession*, shows there is a particular stable state to which each ecosystem returns after it is disturbed. Often called an 'equilibrium state', this mature, climax condition constitutes a natural balance among all the species in an ecosystem, and between them and their environment; this natural balance gives the ecosystem its built-in stability.

While the conditions favouring stability in natural ecosystems are far from being understood in detail, one important factor appears to be the degree of species diversity and the complexity of species interactions.[4] Oversimplified communities, such as modern agricultural systems, are inherently unstable and liable to a variety of catastrophes from which they do not spontaneously recover. To see why diversity favours stability, we turn to the complex interactions among living species within an ecosystem.

Ecosystems: Structure and Function

To describe the interrelationships of organisms in an ecosystem, we shall use one particular ecosystem as a model, a decidous forest in the hills of New Hampshire, but the same basic ideas apply to virtually all natural ecosystems. The forested ravine under study is bounded on its upper edges by the tops of the surrounding hills and has a single stream flowing out of it. Its non-living components are the sloping hills, covered by a thin layer of topsoil, in which are found both particles of disintegrated rock and dead organic matter or humus – the remains of once living organisms. Beneath the topsoil lies the less fertile subsoil, and finally the parent rock, which breaks through to the surface on the steeper cliffs. Many little surface rivulets converge to form the central stream, while ground water gradually seeps downhill through the deeper layers of soil. And there is the air, which sometimes brings with it both rain and particles of dust imported from other ecosystems. The stream exports not only water, but silt and dissolved minerals. On average, the water that is lost by transpiration (the evaporation of water from plant leaves) and by runoff is replaced by rainfall, while most of the exported silt and minerals are replaced by gradual erosion of the underlying rock. The forest is thus not totally isolated from its surroundings, being both an importer and an exporter; but in general the level of water and minerals in the soil is fairly constant from one year to the next.

The living organisms, of which there are thousands of kinds, extend from insects above the tops of the trees down to the deepest plant roots and the minute organisms found within the soil. Taken together, all these organisms compose the living *biomass* of the ecosystem and the amount of new biomass formed each year is called the ecosystem's *productivity*.

The photosynthetic *primary producers* upon which all other organisms depend are the green plants: the trees, shrubs, annual flowers and grasses. As shown at the bottom of Figure 3.2, they capture energy from sunlight, combining it with non-living mineral nutrients from the air, soil and water

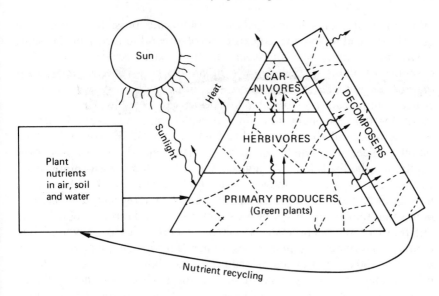

Source: M. E. Clark

Figure 3.2 Trophic levels in an ecosystem (food pyramid).

By the process of photosynthesis, green plants capture sunlight in the bonds of energy-rich organic molecules, derived from low-energy mineral nutrients in the environment. This energy, still in the form of bonds in organic molecules, is transferred from plants to herbivores, and then to carnivores; dead organisms and wastes are finally consumed by decomposers.

At each trophic level, most of the embodied energy (about 90%) is returned to the environment as heat, formed during the oxidation of food to CO_2 and H_2O (the process of respiration). The decomposers extract the last of the original energy, recycling the spent nutrients for further use by plants.

It is thus the continuous capture of the sun's energy – through photosynthesis – that maintains all life. This energy can be used only once, although nutrients are recycled indefinitely.

to generate high-energy plant biomass. Through *photosynthesis*, the sun's energy is stored as chemical energy in living plants.

Plants require some sixteen elements in order to grow. Their largest requirements are for carbon dioxide from the air and water from soil. These are synthesised by the leaves into sugars and other plant constituents; the waste product is oxygen, which is released to the atmosphere. Plants also need substantial amounts of nitrogen, but they can use it only in forms supplied by soil bacteria. Nitrogen in the air (N_2) is unavailable to them. Plants also require generous amounts of phosphorus, as phosphate, that comes both from decaying animal wastes and from eroded rock. In fact, virtually all the nutrients plants need come to them through decay processes *brought about by other forms of life*. The primary production of plants which makes all other life possible is thus itself directly dependent on the co-existence of other life-forms in the same environment, and these organisms in turn exist because of the *prior* existence of plants. *Life is thus a continuously cyclic process; today's life depends on yesterday's, and tomorrow's life depends on today's.* This is a critically important Law of Life. No species can exist alone!

Among the prominent non-plant species living in the New Hampshire ecosystem are all the animals that feed on plants: the squirrels, deer, rabbits, mice, hummingbirds, finches, snails, aphids, honeybees, bark beetles . . . the list is long. As a group they are called *herbivores*, or 'plant eaters'. As Figure 3.2 shows, their food provides both the nutrient molecules and atoms they need to grow and the energy stored in chemical bonds that they need for synthesis, maintenance, and reproduction. About one-tenth of the plant material eaten by herbivores is converted into biomass; the remaining nine-tenths is burned for its energy or is excreted. For every 100 calories of acorns a squirrel eats, about 10 calories' worth are incorporated into the growing tissues of the squirrel, and nearly 90 are combined with oxygen in the process known as *respiration* to produce the waste products, carbon dioxide and water, and to release the energy the squirrel needs to run around, crack nuts, digest food, make urine, mate, raise young, and so forth. All of the energy the squirrel expends eventually finds its way back into the environment as heat. Unused calories are excreted in urine and faeces.

Besides herbivores there are *carnivores*, animals that eat other animals: frogs, mink, wolves, owls, hawks, flycatchers, spiders and the rest. They in turn utilise about one-tenth of the flesh they eat for their own growth, converting the rest to energy (which again, ultimately becomes waste heat), carbon dioxide and water, and a small volume of excreta. There may be two or even three levels of carnivores, the carnivores who feed on other carnivores being called 'top carnivores'.

This sequence of feeding levels, or *trophic levels*, in an ecosystem, is commonly called a *food chain*. At each succeeding step in such a food chain – from plant → herbivore → carnivore → top carnivore – only about 10 per cent of the food eaten is converted into biomass of the next trophic level; the rest eventually appears in the environment as waste heat and recycled carbon dioxide and water. The energy has been used to maintain the organism's life processes.

The rapid decrement in available energy at each feeding level means that very few chains have more than four or five trophic levels. It takes ten times as much sunlight captured by photosynthesis to produce a rodent-eating bald eagle as it does a grain-eating wild turkey which weighs about the same, and it takes ten times as much sunlight to produce a strictly meat-eating human being (say, an Eskimo) as a strictly vegetarian human being (say, a Hindu mystic, like Mohandas Gandhi). This is another critically important Law of Life, since it relates the amount of energy available to the number of organisms – including human organisms – an ecosystem can support.

All organisms eventually die and, together with the excreta they gave off in life, they form the food source for yet another group of organisms, the *decomposers*. The species involved in decay processes are the most varied of all, and include such diverse organisms as bacteria, single-celled protozoa, mushrooms and puffballs, dung beetles, termites, sow-bugs, earthworms, centipedes, millipedes and many others. These myriad soil organisms extract the last of the energy from dead organic matter, converting them back into water, carbon dioxide, phosphate, ammonia or nitrates, and the other mineral nutrients that plants require to grow. The most important decay organisms are the microbes that complete the last steps in recycling these basic nutrients, but their function is greatly impaired if the small soil animals, the soil invertebrates, are destroyed, as often occurs when farmers use heavy doses of pesticides on crops.

The critical idea illustrated in Figure 3.2 is that *the mineral nutrients in an ecosystem are recycled over and over again, while energy from the sun flows through only once.* Captured in the energy-rich chemical bonds of photosynthetic plants, this energy is gradually dissipated along the food chain, the last of it finally being extracted by the decomposers as they complete the nutrient recycling process. It is this constantly renewed flux of sunlight onto the leaves of green plants that drives all the living processes in an entire ecosystem.

Factors that Influence Stability

Two further points will elucidate the ecological weaknesses of modern agricultural ecosystems on which our lives depend. One is that the continuous presence of living plants is essential for retention of mineral nutrients in an ecosystem. In a valley in New Hampshire, next to the one described earlier, foresters experimentally cut down all the vegetation but left it lying in place. Over the following months they discovered an enormous increase in the mineral nutrients being lost out of the ecosystem via the stream.[5] While the decomposers continued to release nutrients into the soil, there were no longer any living plant roots to taken them up again; they simply washed away. Furthermore, water that would have been retained in the soil by the roots or transpired by the vegetation, flowed out of the valley instead, increasing the amount of nutrient leaching from the soil. Losses of major nutrients from the ecosystem ranged from ten to 40 times higher than before the vegetation was cut. All this occurred long before the cut plants decayed and released their own stored nutrients.

Similarly, when humans remove plants, even temporarily, from ploughed fields or cleared forest lands, soils can become depleted of their mineral nutrients and lose their fertility. Often the topsoil itself is eroded away when exposed to heavy rainfall. The new practice of 'no-tillage' farming, which dispenses with ploughing to reduce erosion, is being introduced in some areas, but it requires weed control through heavy application of chemicals which unfortunately destroy the natural recycling in the soil.[6] As we shall see, many attempts to increase farm yields are eventually destructive of the basic ability of the ecosystem to function. A temporary increase in productivity today can lead to impaired productivity tomorrow.

The other factor that causes agricultural systems to be unstable is their *oversimplification*. In Figure 3.2, the diversity of species within each trophic level is indicated by dotted lines. In the New Hampshire valley, there are perhaps a hundred or so plant species, several hundred herbivores (mostly insects), perhaps fifty carnivores (again, mostly insects), and perhaps a thousand species of decomposers. Instead of a single food chain, the ecosystem is in fact a complex food web, with some species feeding at more than one trophic level, and each organism having many predators. It is an intricate network of checks and balances, of which only a few are easily measurable by ecologists. Not only do the different species interact as members of a food web; they also provide one another with habitats, mating grounds, hiding places, nesting materials, look-out posts, water reservoirs, and such services as pollination, seed dispersal, grooming (of tree bark by insect-eating creepers, for example), and so on. It is this complex web of interlocking functions that creates the stability of natural ecosystems.

In human-managed ecosystems, on the other hand, the modern tendency has been to simplify – to plant vast areas to a single crop, which lies exposed to attack by any explosively reproducing pest that comes along. Most crop pests, such as insects or fungi, undergo several generations in a single season; their entire life-cycle may require only a few days or weeks. In a diversified natural ecosystem, there are only a limited number of host plants for each pest's offspring to feed on and a large variety of potential predators to keep them in check; the number of pest offspring in each generation stays small. With monoculture, however, every plant is a suitable host, and there are few predators to control the pests. Almost every egg laid grows up and reproduces, and since female insects can lay dozens of eggs, the laws of exponential growth quickly take over. The entire crop may be threatened with destruction in a few weeks. The usual response has been the massive application of pesticides, which besides being costly, are generally toxic to people, to soil organisms, and to nearby ecosystems.

In conclusion, the complex biotic structure of a natural ecosystem helps to ensure its continued stability and productivity. The presence of many complexly interacting species seems to maintain an 'equilibrium'; the ecosystem is buffered against sudden swings in populations, such as sudden outbreaks of 'pests'. If these do occur, it eventually recovers back to its 'equilibrium' state, much as it would after any other catastrophe. The retention of nutrients within an ecosystem is also coupled to its biotic diversity. Highly complex, mature ecosystems lose very few nutrients; they are constantly recycled. Consequently they are 'sustainable' – they remain fertile and productive indefinitely. Moreover, under parallel conditions mature ecosystems maintain a greater amount of biomass than do less complex systems. They are considerable more efficient than modern agricultural systems at utilising sunlight and nutrients![7]

The Concept of Carrying Capacity

Within a stable ecosystem the population of each species remains fairly constant year after year. Occasionally there are dramatic fluctuations – an outbreak of tent caterpillars, an excessive die-off of mature pine trees due to drought, a sudden outmigration of lemmings in the arctic tundra. Many insect species, such as aphids, and many rodent species, such as fieldmice, have annual population cycles, increasing enormously in summer and dying back in winter. In some years, populations of wildflowers are much greater than in other years, giving rise to memorable displays on hillsides and meadows. Yet overall, the populations in a given ecosystem vary about a mean value, year after year. Each species seems to exist within a range of tolerance, seldom growing so large that it takes over, and seldom growing so small that it disappears altogether. For each species, the maximum population that an ecosystem can support is known as its *carrying capacity*.

The carrying capacity for a species is a balance between its own abilities to reproduce, find nutrients and escape predation and disease, and the opportunities and limitations existing in its environment. Does the ecosystem provide the right temperature range, and enough sunlight, water and other amenities? (Neither trees nor fieldmice, for example, can live where soils are not deep enough for roots or for burrowing.) Are required nutrients and nest sites or germinating sites abundant or scarce? Are there other organisms competing for the same resources? Are there refuges to escape from predators? If the species is a parasite, are the proper hosts present? (Most parasites are restricted to one or a few host species.) Whether a particular species is present in an ecosystem and the size of its population therein thus depends upon the sum total of a multitude of factors.

What happens when a new species colonises a mature ecosystem? How are the populations of other species affected? What happens when we introduce another herbivore into Figure 3.2, as happened for instance when our ancestors began to migrate out of Africa and across the globe? Will the carrying capacities of other species be raised or lowered? Will their populations go up or down?

Much depends on how the newcomers make a living. If they are able to utilise aspects of the environment that were previously underutilised, then the carrying capacity for other herbivores may be relatively unaffected. Such is probably the case for the !Kung bushmen in the Kalahari desert of Africa, whose main diet is the mongongo nut, a highly proteinaceous food source that is present in adundance; there is more than enough for other herbivores. In fact, in that ecosystem, herbivores are scarce not from lack of food, but from lack of water. The success of the bushmen lies in their ability to obtain water in ways that other species cannot – by digging up water-containing tubers, for example, and carrying water in ostrich eggshells.

If however the newcomers are competing with others for scarce food resources, or nest sites, or other necessities, then there exists a 'zero-sum' game, where the population of the new species grows at the expense of the older ones. On the other hand, the advent of a new species may increase the carrying capacities for species in other trophic levels. If they are animals, their nutrient-rich excreta, deposited in new places, may create fertile habitats for seeds to sprout; predators may have a new food resource to exploit; and parasites may acquire a new host to infest.

Whatever the net effect of the new species, the ecosystem sooner or later tends toward a new 'equilibrium' state. Or that was the case until human populations began to have a significant impact on Earth's ecosystems. We turn now to the history of human population growth.

THE GROWTH OF THE HUMAN POPULATION

A change in the size of a population occurs when the birthrate and deathrate are out of balance. The birthrate depends on the ability – and in the case of humans, the inclination – of females to reproduce. The deathrate, on the other hand, is almost always determined by environmental factors: nutrient availability, the prevalence of diseases, and the number of predators, although among humans intentional killing by others can also be important.

As with all species, the reproductive potential of humans is greater than what is needed just to maintain a constant population. If the species is not to go extinct, it is necessary that populations be able to build up again following catastrophes such as mass starvation or epidemics. During her reproductive years, a human female may ovulate 400 times, and the maximum number of children she might give birth to, if she nursed each child for less than a year, is around twenty. Yet such large families are seldom met, even in societies where people are well fed and use no contraceptives. In one such society, the Hutterites of Canada, the average number of children per female is eleven, births being spaced at

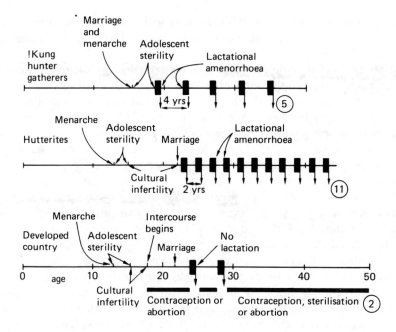

Source: Adapted from R. V. Short, 'The Evolution of Human Reproduction', *Proceedings of the Royal Society B*, 195 (No. 1118) p. 1–224, *Contraceptives of the Future* (1976) Figure 5.

Figure 3.3 Patterns of human birth spacing.

The !Kung hunter-gatherers use no contraceptives, but have only a few children, relative to each woman's total potential, because of late menarche (due to nutritional levels) and adolescent sterility, and to wide birth spacing, due to prolonged lactation which, perhaps together with a low-calorie diet, prevents ovulation. Infant and childhood mortality result in two to three children who survive to adulthood.

The Canadian Hutterites, although becoming fertile at an earlier age, postpone marriage until their mid-twenties. (Premarital intercourse does not occur in this deeply religious sect.) No contraceptives are used and they wean their babies early, so that lactational amenorrhoea is reduced and spaces births at only two-year intervals. The completed family size averages eleven children, most of whom live to adulthood.

In modern industrial societies, a high calorie diet permits menarche and fertility to occur at younger ages than in hunter-gatherers, but although pre-marital intercourse is common and women also remain fertile until their late forties, the average number of children is two. Breast feeding is little practised and low reproductive levels are attained by abortion or contraception (and by sterilisation in later years).

approximately two-year intervals. In most societies throughout history the number of births per female has been far lower. The factors bringing this about are summarised in Figure 3.3.

Two important biological factors limit human reproductive potential. One is undernutrition; poorly nourished women may fail to ovulate.[8] In low-calorie societies, both the age of menarche (onset of menstrual periods) and of fertility (beginning of ovulation) are often postponed. Secondly, when a woman nurses her child, especially if she does so at frequent intervals throughout the day as in primitive societies, her cycles are suppressed and she neither ovulates nor menstruates. This effect is enhanced in thin women, since lactation prevents accumulation of the body fat necessary for ovulation. Even today, this phenomenon, known as 'lactational amenorrhea', prevents more births worldwide than all other forms of contraception combined.[9] The recent spread of bottle feeding without concurrent introduction of contraceptives in developing countries could thus have grave impacts on population growth. Lactation has probably been the major factor curbing full realisation of reproductive potential through most of human existence. As Figure 3.3 shows, contemporary primitive peoples, such as the !Kung women in Africa, have an average of five children, two or three of whom survive childhood diseases and grow to reproductive age.

Even this low fertility rate results in a net reproductive potential greater than needed to replace the parents, leading slowly but surely to significant population growth. (As Chapter 2 showed, even a small percentage annual increase quickly grows to a very large number.) Evidence exists that almost all past societies practised some form of cultural control over births. An obvious strategy is to postpone adolescent intercourse for several years, creating a period of cultural infertility. (This is why Hutterite women average only eleven rather than fifteen or sixteen children.) Another is to place ritual taboos on intercourse; for months after the birth of a child a woman may be 'unclean' until ritually purified. Ancient societies also attempted various forms of contraception: coitus interruptus (early withdrawal of the penis); cervical caps made of leaves, halves of lemons, or fibre sponges soaked in solutions with spermicidal properties. Primitive tribes in Ecuador used the root of two species of sedges to induce temporary sterility with remarkable success, and elsewhere in the Amazon basin other plants – vines and herbs – are used as oral contraceptives, abortifacients, and self-sterilisers.[10] Other societies attempted to induce abortions, often endangering the pregnant woman's life. These methods, however, probably did not effectively control population size over the long run.

A more drastic measure for holding down population growth was the widespread practice of infanticide.[11] Deformed infants were automatically killed at birth; so too was one of a set of twins, since the mother simply

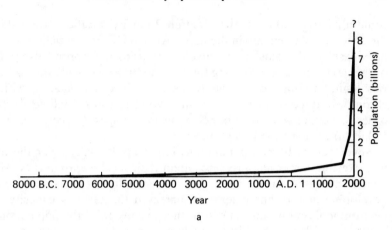

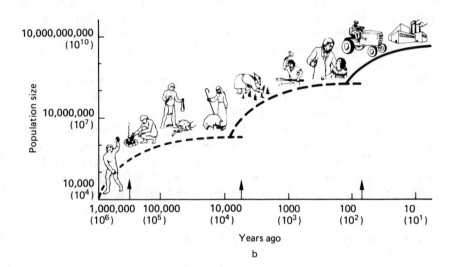

Source: M. E. Clark, *Contemporary Biology*, 2nd edn (Philadelphia: W. B. Saunders, 1979) Figure 1.1, p. 6 (part b of Figure).

Figure 3.4 History of global population.

Human population growth over the past million years may be expressed in two ways:

a: Linear scale, in which population prior to 8000 BC disappears in the thickness of the line.

b: Log–log scale, in which three 'spurts' of population growth can be seen; one must, of course, remember that to reach the first plateau required 900,000 years, and the second 'spurt' occurred over perhaps 9000 years. Today's 'spurt', however, is occurring in decades, not millennia.

could not carry and nurse two children. In many societies – the Eskimo in the artic, the Yanomamö in the upper Amazon basin – female infants were heavily selected against,[12] effectively decreasing the reproductive potential of the next generation. Among the Eskimo, male mortality during hunting was high, evening out the adult sex ratio. The Yanomamö, with their chronic shortage of women, recently evolved institutionalised warfare in order to abduct women from neighbouring villages; the consequence was to kill off some of the excess males.[13]

Nevertheless, human populations imperceptibly grew over the millennia, exerting increasing pressures upon their local environments. Mostly, people migrated or died off, but occasionally they avoided crisis by developing new technologies that increased the carrying capacity of the environment and permitted still further expansion of the population.

When we depict the history of human population growth on a linear scale, these early spurts disappear in the thickness of the line used to draw the graph; only in the last 10,000 years does the line begin to move noticeably upward, shooting up almost vertically in recent decades (Figure 3.4a). Such a graph starkly portrays the dramatic increase in population during the last two centuries.

Using a log-log scale, however, we can simultaneously depict both the small, gradual changes in the past and the larger, more rapid changes that are presently occurring (Figure 3.4b). We see that there have been three 'bursts' of global population growth, each of the first two being followed by an almost level plateaus when population growth rate was extremely low. These three 'bursts' are associated with the widespread adoption of three innovative economies: tool- and fire-making hunting–gathering economies; traditional agricultural economies; and industrial economies.

ECONOMIC SYSTEMS AS ECOSYSTEMS

Hunting–Gathering Economies

If we somewhat arbitrarily set the 'origin' of the human species at one million years ago, then for 990,000 years our direct ancestors were hunter-gatherers. They lived as any other omnivorous species does, by collecting plants – roots, fruits, leaves, nuts – and by catching lizards, rodents, and fish. There is good evidence from the formation and wear of fossil teeth that our earliest ancestors had a diverse diet with considerable animal as well as plant foods. With the passage of millennia, the skills of human hunting bands gradually improved, allowing them to bring down larger species of game, of which there was an abundance. While the men (and sometimes the women without children) hunted, perhaps two or three days a week, the children and their mothers, carrying the infants with them, daily gathered plants and the eggs of birds and reptiles, and snared small animals.

The time spent collecting food each day averaged around three or four hours.[14] The rest of the time was spent cooking, resting, gossiping, visiting others, fashioning and decorating tools, and preparing for sacred rituals. People were apparently quite healthy once they had passed through the dangerous childhood years. Aside from arthritis, their fossilised bones show little sign of the common diseases that later plagued agricultural societies, and surviving hunter–gatherers are almost free of such common modern afflictions as cancer, heart disease, high blood pressure and diabetes.

The image of our ancestors struggling long hours to eke out a meagre living in a harsh environment, with neither pleasures nor comforts, is thus simply not accurate. Both archaeological finds and the lives of extant pre-agricultural peoples (prior to their disruption by contact with Western cultures) give evidence that hunter–gatherers seldom suffered extreme hardship, and certainly were not the crude, unskilled 'savages' pictured by many nineteenth and early twentieth century writers. As anthropologist Marvin Harris reminds us:

> [T]hey have aptly been called the 'master stoneworkers of all times'. Their remarkably thin, finely chipped 'laurel leaf' knives, eleven inches long but only four-tenths of an inch thick, cannot be duplicated by modern industrial techniques.[15]

The primeval 'Garden of Eden' is apparently less mythological than many of us have been taught to believe.

The knowledge and skills of our earlier ancestors were, of course, of a different sort from our own, but it is doubtful that, at least since 150,000 or so years ago, one could say our ancestors were less intelligent than we are. Thus, for at least 90 per cent of the existence of modern *Homo sapiens sapiens*, intelligent people, genetically indistinguishable from ourselves, lived in closely knit bands, usually of less than 200 people, with complex social interactions, and in close harmony with their surroundings.

Figure 3.5 shows a schematised ecosystem containing such a primitive hunting–gathering tribe. (The word 'primitive' is used here in its original sense meaning *earliest* or *original*; the pejorative sense of being *inferior* is a wholly derived meaning.) The size of the *Homo sapiens* compartment is small and bridges both the herbivore and carnivore trophic levels. There were seldom more than one or two persons per square mile and often far fewer in less luxuriant regions. In order to live, it was necessary for these people to move every few weeks or so to a new campsite and exploit a new part of the environment.

Hunting–gathering people expended virtually no energy on the

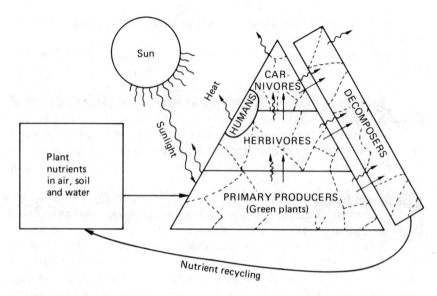

Source: M. E. Clark

Figure 3.5 The ecology of a hunting–gathering economy.
   The human population, which is small relative to the total number of animal species in the ecosystem, displaces a proportion of competing herbivores and carnivores. Usually, this meant only lowered populations of such species but sometimes it involved species extinction. In general, however, the ecosystem remains in equilibrium, and plant nutrients continue to be recycled in the usual way.

maintenance of their surroundings, depending like any other species on solar input and natural recycling processes in the ecosystem to keep up a constant supply of life's necessities: food, fibre, medicinal herbs, firewood, skins, and the building materials needed to make temporary shelters. They expended about 0.3 calories or less of energy for each calorie of food harvested from their naturally renewable surroundings. This is not to say, however, that such people were without any effect on their environments. On the contrary, they occasionally had a large impact, as one or two examples will indicate.

   As noted above, the presence of humans in an ecosystem automatically suppresses, however slightly, the populations of all competing species – the herbivores and carnivores that share the same food supplies and other environmental resources. To the extent that humans thrived, these other species did not. But since in most areas populations of hunter–gatherers were sparse, it is unlikely that *competition* from humans was a major factor in altering ecosystem 'equilibria'. Indeed, the human brain and human dexterity often permitted exploitation of previously unused parts of an ecosystem, so competition with other species was minimal. Early human impact came in other ways.

Some 500,000 or so years ago our ancestors learned to make fire, which became, at first by accident and gradually by intent, a tool for manipulating the environment to their advantage. Here is geographer Carl Sauer's assessment:

> Even to Paleolithic man, occupant of the Earth for all but the last 1 or 2 percent of human time,* must be conceded gradual deformation of vegetation by fire. His fuel needs were supplied by dead wood, drifted or fallen, and also by the stripping of bark and bast that caused trees to die and become available as fuel supply. The setting or [accidental] escape of fire about camp sites cleared away small and young growth, stimulated annual plants, aided in collecting, and became elaborated in time into the fire drive, a formally organized procedure among the cultures of the Upper Paleolithic *grande chasse* and of their New World counterpart . . . Burning as a practice facilitating collecting and hunting, by insensible stages became a device to improve the yield of desired plants and animals.[16]

The late Paleolithic American plains Indians, for example, burned off large areas of forest and shrub – to drive game, as a strategy in warfare, or simply to keep the forest open for travel.[17] The fire drives noted by Sauer were used to stampede herds of large mammals over a steep cliff, or to force them into a box canyon where they could be easily killed. Paleontologist Paul Martin has suggested that the extinction during the last Ice Age of many large mammals – woolly mammoth, giant sloth, sabre-toothed tiger, long-horned bison – was due to such fire drives. He believes that, as modern humans migrated from Africa and western Asia south into Australia and north into Europe and Siberia, thence over the land bridge across the Bering Sea into the New World, and later found their way into New Zealand and Madagascar, they drove the big mammals to extinction in one area after another. Martin refers to this as 'Pleistocene overkill', since, he argues, far more animals were killed than could be consumed.[18] Not surprisingly, his explanation of the Pleistocene extinctions has aroused much controversy and is by no means generally accepted. Yet there can be no doubt that most hunting–gathering peoples, especially as the quality of their stone and flint weapons became more and more sophisticated and their populations gradually grew and used more and more resources, began to have significant impact on their surroundings.

---

*The length of 'human time' on Earth and hence the fraction of it – before the onset of agriculture around 10,000 years ago – when only hunting–gathering occurred, depends on when we decide our ancestors became 'human'. If only *Homo sapiens sapiens* is 'human', then our kind goes back around 150,000 years; if we consider *Homo erectus* and *Homo habilis* as 'human', then our species is around 3 million years old. The details of human evolution are taken up in Chapter 5.

Despite these impacts, however, hunting–gathering peoples, with their low population densities (their maximum numbers around the world barely reached ten million persons) had only minuscule effects when compared to the agriculturalists who largely supplanted them.

Agricultural Economies

Since the dawn of archaeology the myth has been propagated that agriculture was a clever 'discovery' – a marvellous breakthrough invention, a grand successor to fire – that permitted the human species at last to settle down, to have leisure time, and to increase its population, which had hitherto been kept in check by environmental limitations. Unfortunately, almost every part of this assumption is probably wrong, or at best only half true.

As already noted, hunting–gathering peoples controlled their populations by practising both infanticide and prolonged lactation. Through such culturally reinforced behaviour patterns, these societies limited their impact on the environment. There is also good evidence that they had long understood the relations between seeds and plants and sometimes even gave care to their favourite wild plants. Here is what Marvin Harris says:

> The Shoshoni and Paiute [hunting–gathering Indians] of Nevada and California returned year after year to the same stands of wild grains and tubers, carefully refrained from stripping them bare, and sometimes even weeded and watered them.[19]

Nor did people have a 'yearning' to settle down, as shown by the persistent urge of today's Eskimos and Australian aborigines to go on 'walkabouts' over their old hunting–gathering areas, where they lived until contact with whites constrained their lives. Finally, there is evidence from the Middle East and the upper Nile that hunter–gatherers some 18,000 years ago used sickles to harvest wild grains, had grinding stones to make flour, and storage pits to preserve excess food from one year to the next.[20] People knew the basics of agriculture long before they became settled farmers.

Why, if people long knew about raising plants, did they not become full-time farmers until around 10,000 years ago? What triggered this change, which by 2000 years ago had affected all humankind except for a few hunting–gathering societies remaining in marginal niches of the world? Recent archaeological evidence suggests that the first agriculturalists had not an easier, but a far harder life than their hunting–gathering predecessors. Skeletons associated with the earliest settlements indicate both a high infant mortality and a high incidence of disease among adults. Average life expectancy at birth fell from around 30 years for hunting–

gathering peoples (the same as it is today in many underdeveloped countries) to under 20 years. Furthermore, the proportion of skeletons showing evidence of violence suddenly rises at this time, suggesting increasing strife and competition.[21]

Archaeologist Mark Cohen argues from these rather unsavoury data that people did not prefer agriculture as a way of life; rather, it was forced on them as a rising population pressure finally made survival by hunting and gathering impossible.[22] According to his hypothesis, the increasing population density led first to the disappearance of large game; then people spread to lands not previously inhabited, utilising less preferred foods, including less game and more fish and plant foods. Among the latter were wild grains, which as we mentioned were harvested in the upper Nile Valley some 18,000 years ago. An increase of grain in the diet is evidenced in fossil remains by severe wearing down of molars; flour, which is ground by rubbing stones together, contains far more sand and grit than other foods. Eventually, permanent settlements arose which depended almost entirely on cultivated plants for sustaining both humans and their now domesticated livestock. Archaeological evidence shows that by the end of the last Ice Age, about 10,000 years ago, hunting–gathering peoples had migrated into almost every habitable land mass. The Earth was, quite literally, 'full',[23] and agriculture was beginning in earnest at many sites around the world.

Recent evidence from southwestern Egypt paints a picture of how the gradual agricultural transition occurred in that region. Now a formidable desert, the area contained seasonal lakes and rivers during slightly wetter times around 7000 to 8000 years ago. At first the wild grasses growing on the muddy shorelines of the lakes were visited by nomadic hunting–gathering peoples in the wet seasons to graze their partially domesticated cattle, which yielded them a sure supply of milk and blood to supplement their normal wild diet. But just 100 years later, the same areas were inhabited by settled peoples who not only herded cattle, but also planted barley. Not long after, the archaeological excavations show, the same area was settled permanently for 1500 years, by people who first kept cattle and later, as the climate grew drier, sheep and goats. They also planted primitive forms of wheat and barley in the drying lake beds, following each rainy season.[24] The transitional stages seem clear. Nomadic hunter–gatherers, under the slow but relentless pressure of increasing population and a less lush environment, first domesticated animals and relied on them for food in periods of stress; then, while still hunting small game and gathering wild plants, they began to plant grain during their seasonal migrations; and finally, they settled in one place and came to depend both on what they cultivated as well as on wild forage for their stock and small game for themselves.

The obvious advantage of the agricultural transition was that the land could support more people. The carrying capacity for humans was increased, as people focused more of their own energy back onto the ecosystem (Figure 3.6). This transition, however, also brought severe penalities. The human diet switched from high protein to high carbohydrate, with a tendency to protein-deficiency and increased body-fat storage. Although people's general health declined, the pregnancy rate climbed. In fatter women, lactational amenorrhea may have lost some of its contraceptive effectiveness; more importantly, with softer foods available, infants could be weaned years earlier. Birth spacing decreased.

But the increased birth rate did not immediately bring an increase in population. Crowded sedentary existence, in close association with domestic animals, brought a sudden surge in transmissible diseases, and the infant mortality rate soared as measles, malaria, anthrax and other sickness passed among the populations. Moreover, early agriculture was probably hard work. With low productivity seeds and only primitive hand tools for preparing the soil and harvesting crops, yields must have been poor. Every able person, including children, spent long hours in the fields. The plough, the wheel, and animal traction were still to come. Only gradually did improvements in agricultural technology lead, on the one hand, to a longer life expectancy (back to 30 years or so), and on the other, to a rise in population and its two concomitants: a hierarchical society and 'civilised' warfare.

The ecological consequences of agriculture

As Figure 3.6 shows, agriculturalists expand the carrying capacity of their environment by consciously manipulating it – by expending their own energy to make the environment more suitable for their own survival. It is a development which has had – and continues to have – profound ecological effects. The kinds of agriculture practised, as well as its ecological consequences, have varied according to local climate, soil, and the types of domesticable species available.

In agricultural terms, there are basically four kinds of soil-climate systems, each of which has been exploited – and often over-exploited – in different ways. These are the rich alluvial flood plains of the world's great rivers; the wind-formed loess soils of central Asia and North America; the temperate forest lands; and the tropical rainforests. But first, a few more words about soil, this 'stuff' out of which all human-kind ultimately makes its living.

Soil Most of us regard soil as just 'dirt', yet it is far more than the pulverised particles of bedrock that slowly release mineral nutrients needed by plants. Soil is a living entity, where plant roots, multitudes of bacteria, yeasts, and larger fungi, and a wide variety of protozoa and

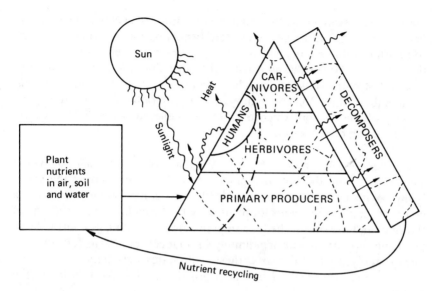

Source: M. E. Clark

Figure 3.6 The ecology of an agricultural economy.

The human population now occupies a much larger proportion of the ecosystem, and domesticated animals (mainly herbivores) and plants push aside the natural species. The ecosystem is no longer in equilibrium, and requires the input of human energy, especially in the husbanding of domesticated plants, to maintain itself. It is, however, still entirely powered by the sun.

Despite their lack of stability, such ecosystems could be maintained for long periods, provided that humans did not interfere with the normal recycling processes. Such interference did of course sometimes occur, resulting in erosion, floods and other ecological catastrophes.

invertebrate animals all interact closely, making a living off of one another and off the decaying wastes of animals and plants. In this microcosm the last of the energy produced by photosynthesis is finally extracted. This living activity not only recycles the nutrients needed by plants; the humus it generates helps retain moisture, preventing the soil from drying out; the soil animals, as they move up and down, ensure aeration and proper drainage, as well as redistributing nutrients to different depths. (Anyone who has ever watched a dead leaf disappear into an earthworm's burrow will immediately know what this means.) Without these activities, plants would not grow. Yet as we saw, plants too play an important role by preventing the loss of soil nutrients. Their roots form 'freeways' along which soil organisms move, and the deep roots help to break down the bedrock beneath the soil, thus replenishing what naturally erodes away at the surface.

Soil, being a living entity, is susceptible to damage or destruction by humans. As described by the English writer Edward Hyams, when humans first became agriculturalists, they were benign parasites, living off natural soil fertility without damaging it. Often, however, this led to over-exploitation; humans became (and mostly still are) disease-causing organisms, slowly destroying soils. Ultimately, of course, we must become contributing members of the soil community, building up rather than destroying the fertility of the soil on which our very existence depends.[25] But we are getting ahead of our story.

Agriculture on alluvial flood plains   All great rivers carry a natural load of silt, eroded from the uplands, much of which is deposited on the plains below. With each flood season, a fresh layer is left behind, providing a constant supply of nutrients for plant growth. Depending on the silt-load of the river and the size of the human population exploiting it, a fertile flood plain can support humans indefinitely without concern for its fertility. Not unexpectedly, four of the six earliest civilisations were built on the flood plains of great rivers: the Nile; the Tigris–Euphrates; the Indus; and the Hwang Ho.

Of these, the Nile, with its heavy annual load of silt eroded from the bare mountains of Ethiopia, was the most successful in resisting human exploitation, having maintained the fertility of its flood plain for some 10,000 years. For centuries, simple but effective means were used by the Egyptian *fellahīn* for retaining the flood waters long enough to irrigate their crops each year. (Since the Aswan High Dam was built, however, this fertility has been lost and has to be replaced by imported fertilisers. A gain in irrigation water has meant a loss of soil nutrients. It remains to be seen whether this altered ecosystem can maintain the expanded Egyptian population indefinitely. At the moment, it is running on an external supply of energy.) Until recently, humans were non-destructive parasites of the Nile flood plain.

Other alluvial-based civilisations took a toll of their ecosystems much sooner than did the Egyptians, becoming disease organisms rather than mere parasites. In most cases, collapse came when the civilisation expanded its cultivated acreage beyond the natural flood plain. First the people built extensive irrigation canals, such as criss-crossed the lands between the Tigris and Euphrates rivers of ancient Mesopotamia, where the new acreage supported the famed cities of Ur, Sumer, and Babylon. Then they expanded into upland areas, where soils were thinner and less fertile. These originally wooded areas, lying above the Fertile Crescent in the Near East and above the Indus Valley in present-day Pakistan, were gradually cleared. The wood was used as fuel for making bricks or smelting metals, or for shipbuilding, and the land was cultivated for crops. At this point, the ecosystem began to collapse, eventually destroying the civilisation that had exploited it.

The culprit was erosion, aided by a gradually drying climate. Without vegetation to break the force of the rain, the hillsides soon eroded. Instead of seeping into the loose, fertile soil and replenishing the ground water, storm waters rushed over the surface of the land, carrying away the top soil and digging deep gullies. The uplands became barren wastelands covered by coarse shrubs, while in the plains below the excessive amounts of silt, instead of being a benefit, became a nuisance. Silt clogged the irrigation ditches and canals and filled the channels of river beds, causing catastrophic floods and often forcing rivers to change course. The ancient civilisation of Mesopotamia, which at its zenith numbered around 20 million people, collapsed, unable to keep open its water supplies.[26] The same area in the 1960s supported but 7 million people. Only with petrodollars can Iraq today import the crop subsidies and food needed to feed the 14 million people who now live there.

In the Sind district of Pakistan, the ancient removal of upland trees to clear the land for farming also had a devastating effect, but of another sort. When the soil lost its fertility for agriculture, the original trees failed to return, and the tamarisk and scrub that replaced them slowly but permanently altered the climate, switching it from moderately wet to semi-arid.[27] Although the lower Indus Valley is still fertile today, its surrounding highlands and parts of Baluchistan, which once boasted an irrigated agriculture, are now 'a region of desolation, of icy winter winds, and burning summer heats, and of sporadic, unreliable rainfalls incapable of supporting more than a nomadic agriculture practised by a meagre population of migratory peasants'.[28]

The Hwang Ho, or Yellow River, in China is yet another major alluvial system which, however, has had a somewhat different history, since the civilisation of northern China did not begin on its flood plain but on the rich loess soils of the western highlands. To understand China, we must understand the limits for agricultural exploitation of this quite different type of ecosystem.

Agriculture on loess soils Like alluvial flood plains, loess soils are often rich and fertile, but they are formed not by silt-bearing water but by wind. They are thus characteristic of the semi-arid grasslands and parklands of continental interiors. Fertile soil particles eroded from the mountains by the wind are deposited on the plains of the steppe, where they are trapped by the grassy vegetation. Such soils may range from a few inches to 30 or more feet in depth.

It was on the rich yellow loess soils of western China that Chinese civilisation arose several thousand years ago, with frequent disastrous results for the later populations that settled on the eastern flood plains. Denuded of their natural grass by the neolithic farmers and exposed to wind and rain, these soils began to erode. Because of their great depth, however, it was several millennia before they lost their fertility, or until great clefts were cut into them, creating miniature Grand Canyons that made much of the land inaccessible. Meantime, as the neolithic Chinese population grew it expanded into and eventually filled the alluvial flood plain below.

This enormous upland erosion added greatly to the natural silt load of the Hwang Ho, giving it its yellow colour. Over the centuries, its silt repeatedly destroyed irrigation systems and, by filling the river's channel, forced 'China's Sorrow' to overflow its banks innumerable times and seek new routes to the sea. In 1852, despite heroic efforts by thousands of peasants to reinforce its banks, the river burst its retaining walls, inundating thousands of square miles of farmlands and killing millions of people – and found a new outlet to the sea 400 miles to the north! (A similar fate is predicted for the Mississippi River delta, since that river, despite the efforts of the US Army Corps of Engineers, threatens to seek a new route to the sea, leaving the cities of Baton Rouge and New Orleans high and dry.[29])

The black loess soils of central and western Asia have largely escaped the same fate, not because they are less susceptible to erosion but because nomadic herding peoples – known throughout history as the Golden Horde – used their land less destructively by grazing animals on its permanent grasses; they regularly drove off the encroaching farmers. Of this area, Edward Hyams writes:

> While these two forces held each other in balance, the soils in question were protected from abuse: the peasants checked the overstocking of grasslands by making war on herdsmen; and the herdsmen protected the grasslands against destruction by the plough.[30]

Loess lands elsewhere have been less fortunate. Parts of Chile and Argentina are being converted to desert from over-use of semi-arid grasslands, as are enormous regions of Africa, including not only the sub-Saharan region known as the Sahele, but large parts of Kenya, Tanzania, Botswana and Somalia. In North America, pioneer farmers needed only 50 years to turn Oklahoma grassland into a dust bowl, and erosion throughout the American midwest still greatly exceeds the rate of soil formation. Just 100 years ago, much of today's mesquite scrublands in New Mexico, Arizona and the Mexican state of Sonora were covered with grass and oak parkland, but overgrazing by livestock, abetted by a long dry spell, gradually and irreversibly changed the landscape over hundreds of square miles.[31] Even the Siberian steppe, so long protected, is now experiencing the early stages of sheet and gulley erosion since it has been 'opened up' by the Soviets to modern agriculture.

Agriculture on forest lands with seasonal climates The world's forests fall into two large categories: those, mainly in the tropics, that are wet year round, and those, mainly in temperate regions, that experience annual wet and dry seasons. Of the latter, many have already disappeared from the Earth. Some, as in Baluchistan and Iraq, have been replaced by scrub vegetation, which is all that will now grow there. Similar events occurred around the hilly shores of the Mediterranean, from the once wooded Italian and Greek peninsulas to the formerly tree-covered slopes of Mt Lebanon, home of the famed 'cedars of Lebanon'. Over the centuries, these majestic trees were gradually stripped away, until today only tiny, carefully protected clumps remain. The rest of the giant mountain is 'as barren as the mountains of the Sahara'.[32] With regard to the ecological changes in ancient Greece, here is Plato's fourth century BC description of what had already happened to the Attic peninsula:

> [W]hat now remains compared with what then existed is like the skeleton of a sick man, all the fat and soft earth having wasted away, and only the bare framework of the land being left.[33]

In some regions major attempts, some of them fairly successful, have been made to preserve the remaining soils. Large areas of the Mediterranean lands have been planted to olive and vine, both capable of eking out a living on steep, semi-arid hillsides because of their deep root systems. In other areas extensive agricultural terraces were constructed and some have been maintained for centuries. Both practices sharply decrease the rate of erosion.

After the last Ice Age receded from northern Europe the tundra and grasslands left behind, where hunting–gathering peoples found abundant game, were gradually replaced by dense forests, and it was into these that the first agriculturalists ventured in small groups. They would settle for a few years on natural patches of grassland or on small clearings they made in the forest, but as soon as the crop yields declined they moved on to a new site, allowing the old one to regain its fertility.

Although temperate forests, left to themselves, create rich soils, when the trees are removed and farming commences, the soils begin to age; without the deep roots of trees to extract minerals from the bed rock, soil nutrients are gradually lost. As the European population increased it would eventually have depleted the forests and exhausted its resource base, but this did not occur owing to the introduction of husbandry – the careful recycling of crop and animal wastes back into the soil to maintain its texture and fertility. European farmers were perhaps the first to discover the need to nurture and maintain soils.

Even with the practice of husbandry, however, it is possible to force too much productivity from soils, especially when population pressures build. The threatened over-exploitation of European soils has been averted three times in history. The pressure placed by the demands of the Roman

Empire on the soils of Europe was relieved by the fall of Rome, which was followed by a thousands years of feudalism and a fairly stable population that husbanded the soil. When the population again grew around AD 1500, mercantilism had begun to flourish, and much farmland was enclosed (especially in England) in order to raise sheep and cattle for export. Although the peasants suffered gravely, the land was saved since grazing is less damaging to soil than tilling. Finally, the growing population of the Industrial Revolution, which again threatened the land, was mainly fed by imported grain from North America and meat from South America. The soils of Europe were temporarily preserved from over-exploitation at the expense of those of the New World.[34] Today the threat has returned; 'modern' farming has made Europe food self-sufficient, but is again threatening its soils.[35]

Agriculture in tropical rain forests   The shifting forest agriculture practised in ancient times in Europe and until just a century or so ago in parts of Scandinavia and Siberia, is most commonly found today in the wet tropics. Trees in a small area of the dense forest are felled, allowed to dry, and then set alight to clear the land. The ash forms a nutrient fertiliser for the crop plants, which are planted in holes made by a dibble stick in the soft Earth between the stumps. While they are growing, the crops are weeded and cared for. In the tropics of the ancient Old World, commonly grown plants were cassava (manioc), yams and other tubers, pulses (beans, peas and their relatives), and bananas; upland rice was grown in some areas. It was in the New World that maize (corn), tomatoes, chillies, peanuts, and sweet potatoes originated, and these were grown with beans and sometimes manioc. Within a few decades after the European 'discovery' of the New World, however, tropical plants from both hemispheres were grown around the world, regardless of where they originated.

Depending on the quality of the soil, the gardens are used for two, five, or occasionally ten years, and then, when the soil is exhausted, they are abandoned. The villagers prepare a new site, transplanting cuttings and tubers to it; the tropical jungle gradually overgrows the abandoned clearing, the deep roots of the trees drawing up nutrients from the bed rock, gradually regenerating the soil. This is a case of a natural ecosystem that has been disturbed undergoing the process of ecological succession, finally restoring its former 'equilibrium' state.

Shifting agriculture was probably first used as a food supplement by hunting–gathering peoples to augment their diets when game or wild fruits became scarce owing to population pressure. Gradually, however, it became the mainstay of existence in many areas, as populations increased. But, in general, shifting agriculture cannot support a large population, since the same areas cannot be used except at long intervals. In the tropics of the Old World, as the populations continued to rise, agriculturalists began to burn over their abandoned gardens to encourage the growth of grasses on which to graze domesticated animals. (This was not done in the New World as no grazing animals suitable for domestication existed there.) Through this process, large tracts of tropical forests were permanently replaced by less productive grasslands in parts of Sumatra, the Philippines, Africa, and Madagascar. Once again, early agriculture tended to destroy its own resource base when population pressures rose.[36]

A case of tropical agriculture pushed to, and beyond, its limit may have been the Mayan civilisation on Mexico's Yucatan peninsula. Emerging about AD 300 from a shifting agricultural community, the Mayan peoples reached dramatic population densities of 750 persons per square mile by AD 800 by planting permanent orchards of a staple tree, the breadnut, and by year-round irrigation of gardens. About that time, however, the civilisation began to decay, not through conquest but apparently from inner collapse. As orchard yields declined from overuse, they were converted to less productive grassland, and the resulting erosion led to a loss of soil quality and a silting up of the irrigation channels.[37] It was the soil that failed.

Compared to shifting agriculture, permanent agricultural sites in the wet tropics have been far less common except in those areas of southeast Asia where sufficient water has been available for year-round irrigation of terraced rice paddies. Such terraces are remarkable for having retained their fertility for many centuries. This is due partly to the use of human and animal fertiliser, partly to the occurrence in these paddies of a water fern which harbours nitrogen-fixing blue-green algae that supply nitrogen to the rice plants, and perhaps most importantly, to the rich content of dissolved nutrients, leached from volcanic soils of the highlands, that the irrigation water constantly supplies.

Elsewhere in the wet tropics, however, the continuous cultivation of cereals and similar crops is difficult since nutrients are not stored in the soil; the abundant rains leach them away. In such ecosystems, free nutrients never find their way into the soil; upon being released from decaying matter by fungi associated symbiotically with the forest root systems, they are immediately taken back up into the living trees. When tropical forests are felled and burned, the nutrients thus released remain in the soil only a year or so before being washed away. When modern cash crops such as sugar cane are grown, their sale abroad earns the money necessary to import the fertilisers needed to continue to grow them. Just like the post-Aswan Dam soils of Egypt, the soils of the wet tropics, when permanently put to annual crops, become nutrient-importing ecosystems; they are no longer self-sustaining.

A further complication in some areas of the tropics is the existence of a type of soil which, when exposed to air and sunlight for a few years, turns into a cement-like substance known as laterite. This makes marvellous building stone (the famous Cambodian temple of Angkor Wat was constructed of laterite) but it is impossible to plough or dig it; one might as well try to farm a slab of concrete.

Thus, despite the high productivity of undisturbed tropical forests, their agricultural potential is limited. We shall consider the proposed exploitation of the world's remaining tropical forests by modern industrial agricultural techniques in the next chapter. First, though, we must examine the ecological relationships of industrial economies generally.

Industrial Economies

The economic systems we have discussed so far were powered entirely by their annual income of sunlight: work was done by human power, by animal power, or by wind or water power. The major contrast between industrial economies and all previous ones is their huge dependence on fossil sunlight in the form of oil, coal, and natural gas. By the addition of extra energy to an ecosystem, over and above that derivable from today's (or last year's) sunlight, not only has the yield of agricultural crops been increased, but also the production of manufactured items. Through the addition of massive amounts of energy, it has been possible to increase both the population fed by a given amount of land and the *per capita* consumption of material goods as well. This effect is shown schematically in Figure 3.7.

This figure tells us a number of important facts about industrial economies. First of all, *a modern industrial economy is really just a highly modified ecosystem, subject to the same ultimate laws of thermodynamics and biology as is any other ecosystem.* Once this is clearly seen, the significance of the environmental problems of our modern world is better understood. Second, the 'nutrients' of an industrial economy are no longer just those chemicals that green plants need in order to grow; they also include such raw materials as metals and other strategic minerals that factories need to turn out manufactured goods – things like iron, copper, sand, asbestos, talc, and petroleum as a chemical feedstock. Third, as already noted, in addition to sunlight captured via photosynthesis, a great deal of *extra* energy is needed to keep the economy going.

Another important fact is that in industrial economies, almost the entire ecosystem is given over either to supporting humans and their domesticated plants and animals or to making room for the industrial infrastructure: cities, freeways, airports, harbours, oil refineries, military bases, garbage dumps and so on. The space allotted within the industrialised ecosystem to 'non-economic' species becomes smaller and smaller as industrial societies grow. The rate of extinctions today is many times higher than ever before in Earth's history!

Finally, Figure 3.7 points out the enormous quantities of pollutants

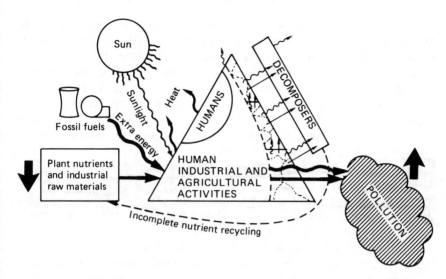

Source: M. E. Clark

Figure 3.7 The ecology of an industrial economy.

The human population is larger, and the fraction of the ecosystem given over to supporting that human population is extremely large, leaving only a small number of plants, animals and decomposers to carry on natural recycling processes.

Industrial economies depend on the addition of fossil fuel energy in amounts far greater than the solar energy fixed by the plants themselves. Much of that energy is used to extract raw materials, the 'nutrients' of the industrial system, but little of it is used to recycle industrial wastes, which pile up as pollutants in the ecosystem. Agricultural nutrients also are not completely recycled, requiring larger and larger inputs of fossil energy into the food cycle.

Industrial economies today thus cause ecological degradation both by depleting the resource base of both agriculture and industry, and by poisoning the system with pollutants. In addition, genetic reservoirs of wild species are being driven to extinction.

generated by modern industrial economies, which in turn destroy the health of the entire system. While pre-industrial agricultural economies degraded their own support systems through greatly accelerating the processes of erosion, industrial economies are capable not only of causing erosion (sometimes on a grand scale), but also of releasing highly toxic substances into the environment: radioactivity, poisons, acids, asbestos fibres and so on. Furthermore, they produce prodigious amounts of less dangerous wastes, such as scrap metals, paper, plastics, glass, garbage and sewage, all of which accumulate somewhere in the ecosystem as general pollutants.

In natural ecosystems, as seen in Figure 3.2, there are decomposers that recycle *all* the nutrients. There simply are no decomposers, however, to recycle many of the wastes in industrialised ecosystems – the non-biodegradable metals, plastics and glass. Even those wastes that are biodegradable are usually produced in such quantities they swamp the capacity of natural decay processes. The consequence is that industrial economies – as already noted in Chapter 2 – are using up their 'nutrients', their natural resources, at a rapid rate, and at the same time are accumulating a growing load of unrecycled wastes. The faster such economies 'grow', the faster both processes occur. It is a situation that obviously cannot continue indefinitely.

Indeed, as resources grow scarcer, pollution will also have to decline. There will be a growing incentive for people to put more energy and effort into repairing rather than replacing their material goods, and into recycling spent resources and recovering valuable 'wastes'. The so-called durable products of an industrial economy will be much more durable than they are today; things will be built to last for decades, even centuries. Throwaway items will be seen as a burden to society, not as they are now, a blessing to economic growth. The production of pollutants in any form will be seen as profligacy – not just as an unsightly inconvenience, but as economically wasteful.

But we are jumping ahead. To change our ways will require not only economic pressure, but new ways of thinking altogether. How we came to hold the worldview which is responsible for today's techno-industrial society we shall explore further in Part III. But first we need to take a closer look at just where we stand in relation to our environment.

As we have seen, the human population is growing explosively. By the year 2000, today's five billion will have grown to well over six billion; by 2050 there will be ten billion, twice as many as now. Population growth, like the wheat grains on the chessboard, leads to astronomical numbers.

Clearly we need to ask some serious questions. First, when will it stop? For reasons discussed in Chapter 13, the human population is expected to level off at 10 to 12 billion around the middle of the next century. Second, can Earth support so many? What about the damage the *present* population is doing? Could Earth provide a decent living even to all its present inhabitants? If so, could we further double everything – food, houses, clothes, cars, roads, hospitals, cemeteries, schools, libraries – without totally destroying our ecosystem? By examining current stresses on our support system, we can see what changes are most critical and which can perhaps be postponed.

SOME CONCLUDING THOUGHTS

To sum up, looking at the way human societies have accommodated to population pressures in the past leaves us with grave cause for concern. In each case the tendency has been to use up the natural reserves that Earth formerly possessed. Hunter–gatherers slowly but surely used up easily accessible plants and animals; traditional agriculturalists, slightly more rapidly, used up the Earth's soils, and industrial societies are busily consuming fossil fuels while destroying even more species and more soils. Time after time, instead of learning to maintain – to nurture and care for – Earth's resources, humans have dipped into Nature's non-renewable reserves: her living species, her soil, her water, her minerals and finally, her stored fossil fuels – consuming each in turn. For 2 million years we have been living beyond our means, charging our overdraft to Nature's environmental 'credit card account', but the amount owing is now growing exponentially. It's time to begin paying the bill. Ours is the first generation to have the global perspective to perceive, if we choose to, the folly of our ways, and so deliberately to change them.

4 Our Environmental 'Credit Card' Account Becomes Due

This we know: The earth does not belong to man; man belongs to the earth. This we know: All things are connected like the blood which unites one family. All things are connected...

Whatever befalls the earth befalls the sons of the earth. Man did not weave the web of life: he is merely a strand in it. Whatever he does to the web, he does to himself.

Attributed to Chief Seattle, 1854[†]

INTRODUCTION

Man has probably always worried about his environment because he was once totally dependent on it.[1]

Thus did two economists begin an article on the relationship between the environment and the economy. Their words imply that human beings today are somehow exempt from any serious dependence on their natural surroundings; it is a premise that is surprisingly widely held. Most people living in industrialised societies presume that their lives are somehow uncoupled from Nature. Food no longer comes from the soil, it no longer depends on the weather, and it no longer has to be carefully stored over winter; it comes, instead, all year round, from the local supermarket, frozen, canned or bottled, or shipped in fresh from some more benevolent climate. So, too, do all of the accoutrements of life – clothes, housing, automobiles, entertainment – seem to emerge effortlessly from nowhere; to the 90 per cent who live in cities, they seem disconnected from any sort of natural processes. Nature is one thing; the economy is quite another. As long as one contributes to the 'system', one can demand and get back from it all one needs. No one asks what ultimately supports the 'system'.

We take it for granted that our modern industrial economy represents a major advance in the relationship between human beings and their environment. We have better control over our surroundings; we seldom go hungry; we live in well-heated homes, with plenty of hot water; we have access to medical treatment. Not only do we not suffer; we also have positive benefits – instant entertainment, rapid personal transportation, and some form of financial security for our old age. Never before have humans had such wealth, comfort, and convenience at their fingertips. Surely we have 'earned' all this by our scientific and technological cleverness?

With only a superficial understanding of how the world works, it is easy to believe we have conquered Nature. We live in a secure cocoon, spun by our industrial economy. Occasionally our complacent existence is jarred by some natural phenomenon – a volcanic eruption, a flood, an ice storm – but from these it quickly recovers. Only when we begin to understand the implications raised in the last two chapters does our belief in 'progress' start to waver.

Herman Daly, one of a small but growing group of economists who understand how the Second Law of Thermodynamics sets limits on our material growth, makes our environmental dependence clear by his response to the economists' quotation with which we began:

The implication is that man is no longer totally dependent on his environment, or at least that he has become less dependent. Presumably, technology has made man increasingly independent of his environment. But, in fact, technology has merely substituted nonrenewable resources for renewables, which is more an increase than a decrease in dependence . . . Man is an open system [constantly exchanging matter and energy with his surroundings]. What was man three months ago is now environment; what was environment yesterday is man today. Man and environment are so totally interdependent it is hard to say where one begins and the other ends. *This total interdependence has not diminished and will not in the future, regardless of tecnology.*[2] (emphasis added)

Daly also recognises that almost all our modern wealth has been gleaned at the expense of Earth's non-renewable resources. We have been – and are – living on a one-time 'bank account' of fossil energy and mineral deposits both formed over eons of geologic time. To have become as dependent on them as we now are is singularly imprudent. Not only are we using up these one-time resources at a galloping rate; we are also living far beyond our day-to-day income of potentially renewable resources – of soil, of water, and of other living species. Instead of husbanding them so they will last forever, we exploit them. We are borrowing from the future.

We are not paying our way at all. We are living off past and future natural capital that we have suddenly learned to exploit. Like irresponsible credit card users, we live high today by borrowing from tomorrow. The rate at which we are running up our 'Environmental Credit Card' balance is increasing exponentially. Our credit is already way over-extended. In her own way, Mother Nature – the ultimate giver of all resources – is calling us to account!

ENERGY AND ENVIRONMENTAL EXPLOITATION

There are two ways in which human populations bring about the destruction of their local environments and exceed their permanent local carrying capacities: by increase in sheer numbers and by increase in affluence – *per capita* consumption of Earth's wealth.

As we saw in Chapter 3, global population is burgeoning, mostly in the underdeveloped and developing countries of the world. Many of these people live at or barely above subsistence level, but even so populations are in places exceeding what the local environments can tolerate. The spreading over-exploitation of forest and scrub lands for fuelwood, for grazing, and for agriculture is causing an inexorable decrease in the carrying capacity of these lands. But often (as we shall see later) the problems faced by these peoples are due to their forced exclusion from underutilised lands rather than to local environmental limits.

The rate of destruction caused by the sheer numerical pressure of poor and powerless people is as nothing, however, compared to the destruction caused when massive, energy-consuming technologies are introduced into the world's ecosystems. As shown in Figure 3.7, the enormous power possessed by industrial economies permits environmental destruction on a grand scale. Nor is the power of the wealthy to wreak destruction restricted to their own borders; it extends by various forms of economic intervention into all those ecosystems where natural wealth is to be had for the taking – if you have the energy and technology to do so.

Although the relationships between rich and poor countries are explored in detail later, it is important here to see the close connections between energy, wealth, and environmental destruction. Consider the explanation Nicholas Guppy, an authority on tropical rain forests, gives for their destruction:

> Discovering the reasons for the sudden dramatic acceleration in rain forest destruction is like slicing into a multilayered cake. On the surface, obvious factors include population growth; the increased need for land for agriculture, stock-raising and settlement; the wish to raise capital for development; and a rapidly growing demand for timber, fuelwood, and other forest products.

> Yet when we look carefully we find that these are not necessarily decisive, or even important causes. Underneath are other layers – of social mores, of political expedients, of national and global economics, and of ideological conflict. Indeed today, almost everywhere, destroying rain forest is a means of *avoiding* tackling real problems by pursuing chimeras: a 'license to print money' which yields quick cash at the cost of ultimate catastrophe. Each time we read of floods, landslides, starvation,

and loss of life in India, Brazil, the Phillipines, Haiti, East Africa, and elsewhere in the tropics, we are reading about problems caused or exacerbated by environmental destruction, often of rain forest . . .

Visiting such [logged] areas it is hard to view without emotion the miles of devastated trees, of felled, broken and burned trunks, of branches, mud, and bark crisscrossed with tractor trails – especially when one realizes that in most cases nothing of comparable value will grow again on the area. Such sights are reminiscent of photographs of Hiroshima, and Brazil and Indonesia might be regarded as waging the equivalent of thermonuclear war upon their own territories.[3]

Wherever powerful technologies are being unleashed to extract immediate wealth and profit from an ecosystem, environmental destruction is seldom far behind. Future carrying capacity is being forfeited to momentary gain.

Table 4.1, which summarises the patterns of global energy consumption over the last million years, explains why the present impact of humans on the environment is so enormous. It is the product of two exponentials: one of population growth, the other of *per capita* energy consumption.

Before humans acquired the use of fire some 500,000 years ago, the daily energy consumed by our hunting–gathering ancestors consisted mainly of the food they ate, around 3000 calories per day. The simple addition of cooking food almost tripled *per capita* energy consumption; this was doubled again by the advent of agriculture, and had increased somewhat further in Europe by the Middle Ages. Thus, while the human population increased 500 times between 1 million and 1 thousand years ago, its global energy demand increased nearly 3000 times! Throughout all that time, the energy consumed to maintain human economies came entirely from primary and secondary solar sources: plants, animals, water, winds and human labour. No wonder these economies occasionally destroyed the ecosystems that supported them. They withdrew the energy needed by the ecosystem to maintain its own stability, especially the vegetation necessary to stabilise the soils and whose nutrient content should have been recycled for renewed plant growth. Only economies that did not extract more energy than the ecosystem could tolerate continued to thrive.

Today there are essentially two kinds of economies: industrial and pre-industrial. As Table 4.1 shows, 75 per cent of the global population lives in developing countries where the majority of people still rely almost entirely on their solar income; these are subsistence economies where the *per capita* energy consumption is still quite low. But in the past 100 years, the number of people living at that level has increased eightfold.

The other 25 per cent of the world's people live in industrial economies. On average, the peoples of Japan and of West and East Europe (including the USSR) use about half as much energy *per capita* as do North

Table 4.1 *Per capita* and global energy consumption for different types of human economies

Economic system	Years ago	Maximum global population* (approx.)	Daily calories/person†	Global daily calories consumed by human population
Hunting–gathering (before cooking)	1,000,000 to 500,000	1 million	3000	3×10^9
Hunting–gathering (after cooking)	500,000 to 10,000	10 million	8000	8×10^{10}
Early agriculture	10,000 to 2000	300 million	15,000	4.5×10^{12}
Middle Ages	1000	500 million Europe 10% Rest 90%	23,000 15,000	$\sim 8 \times 10^{12}$
Today North America Europe, USSR, Japan Third World	0	5000 million 5% 18% 77%	314,000 157,000 15,000+	$\sim 2.8 \times 10^{14}$

Notes:
*R. Leakey and R. Lewin, *Origins* (New York: E. P. Dutton, 1977) p. 143; J. Weeks, *Population: An Introduction to Concepts and Issues*, 2nd edn (Belmont, CA: Wadsworth, 1981) p. 46.
†Harrison Brown, *The Human Future Revisited* (New York: W. W. Norton, 1978) pp. 30–3, with *per capita* figures for industrialised nations upgraded from 1970 to 1980 levels.

Americans (Figure 2.2) but have only a slightly lower living standard. Together the industrial economies account for 80 per cent of the world's total energy consumption, most of which comes from fossil fuels; they are largely responsible for today's human global energy consumption being 90,000 times what it was a million years ago!

One might suppose that in the process of consuming such prodigious amounts of energy the industrial economies would be using some of it to husband their environments: to protect their air and water supplies, to recycle nutrients, to preserve and protect the soil, the forests, the fisheries. But despite some conservation efforts, the *net* impact of industrial economies continues to be in the opposite direction. On a *per capita* basis, they are far more destructive of the environment than are pre-industrial economies. Their access to vast amounts of energy permits them to exploit the environment in ways that non-industrial economies can manage only with difficulty, at much higher population densities, and over much longer periods of time.

Oddly enough, industrial societies treat the protection of the environment and the husbanding of resources as a 'luxury' affordable only during periods of economic growth and increasing affluence. During slumps and recessions, regulations designed to protect the environment are eased and its destruction is considered an inevitable, if regrettable, necessity – 'to keep the economy growing'. The consequence of this short-sighted approach is further to degrade the environment, so even *more* energy will be needed in the future just to maintain today's level of economic throughput of goods and services. It is a clear prescription for disaster.

Throughout this chapter, it is important to keep in mind that the massive energy consumption of modern industrial societies is largely *destructive* of the environment. Little energy goes into nourishing, maintaining, preserving, and recycling; most of it goes into extracting, exploiting, manipulating, and consuming. We place ourselves in quadruple jeopardy. Not only do we deplete our resources; we also create pollutants, alter climates and cause extinctions – each of which potentially lowers the carrying capacity of Earth for the human species. We now turn to these.

NUTRIENT DEPLETION

The ecosystems of modern industrial economies require two sorts of nutrients: mineral nutrients for manufacturing and plant nutrients for growing forests, fields and pastures. A few words about the former are needed before discussing the loss of plant nutrients from our ecosystems.

Industrial Nutrients

As for fossil fuels, demand for the mineral nutrients of our manufacturing system has been growing exponentially. Two scientists from the Oak

Ridge National Laboratory in Tennessee recently analysed future US mineral supplies – to the year 2100.[4] Intending to be optimistic, they predict that 33 of the 65 elements needed by industrial economies will still be available in 'unlimited' supply *if* three basic assumptions are met: (1) a lower rate of global population growth than the United Nations and most demographers foresee, together with only modest demands from developing countries where 90 per cent of people may then be living; (2) new technological breakthroughs in the recovery of minerals from lower-grade ores; and (3) an increasing rate of energy consumption – eventually reaching eight times what is used globally today* – to generate the capital needed to extract and refine these lower grade ores. Their entire projection assumes a continuing expansion of energy consumption, without explaining where that energy will come from. Finally, they predict that technological breakthroughs will devise new synthetic materials from elements still in 'infinite' supply to substitute for the remaining 32 elements which will then be exhausted.

This is a good example of the 'technological fix' approach that focuses on one problem (in this case, mineral resources) and treats all other problems as marginal (population growth) or non-existent (future energy resources). The social and political implications of these sweeping changes in the global industrial resource base are never mentioned. The only human element allowed to enter the calculation is the confident presumption of human inventiveness.

This kind of prophesying finds ready acceptance. Scientists who make the predictions like to think that science is crucial to the future of humankind, and non-scientists feel relieved from any responsibility to change their lifestyle and attitudes. While some aspects of such a narrowly conceived projection may prove valuable in planning the future, it is a mistake to assume that it will be fulfilled *in toto*. Indeed, if global energy consumption does increase significantly during the next century, most of it is likely to go to producing food and fuels for human subsistence rather than to prop up a failing industrial system. To perceive this, we turn to the status of the nutrients required for agriculture.

Agricultural Nutrients

Of the 800 calories a day needed by hunting–gathering peoples, 3000 calories represented their intake of food. But they expended only 600 or so calories a day to obtain this food, for a ratio of calories expended per calorie eaten of 0.2 to 0.25. By the time traditional agriculturalists, after

---

*The authors predict a growth rate of global 'gross national product' of 2 per cent per year for 120 years, or 3.43 doubling times. Since GNP is tightly linked to energy consumption, this predicts a global energy consumption by 2100 of at least eight times what it is today. As we saw in Chapter 2, this is highly unlikely.

several millennia, had selected for highly productive cultivars, they needed to expend even less energy to obtain a calorie of food: 0.02 to 0.05.[5] Today, with the help of tractors, irrigation pumps, fertilisers and other chemicals, the average American farmer expends 2.5 calories of fossil fuel energy-equivalents to produce one calorie of food, or 50 to 125 times as much as his pre-industrial predecessors. This is the real basis for the spectacular productivity of modern industrial farming.

> Between 1970 and 1979, farm machinery use [in the United States] increased by 29 percent; use of fertilizer and irrigation rose by almost 30 percent; and the overall application of agricultural chemicals jumped by almost 60 percent... [O]n-farm energy use [rose] to 7.1 percent of our nation's total energy use by 1981... [T]o keep our farm economy chugging away required the equivalent of *half* of net energy imports for the entire nation.[6]

Furthermore, in industrial societies food is processed, packaged, transported, stored in warehouses, and distributed via supermarkets before it is consumed. This requires another five calories per calorie of edible food. Finally, home storage in energy-gobbling refrigerators and cooking take another 2.5 calories. The grand total is *10 calories* of fossil fuels expended for every calorie of food consumed on the American dinner table.[7]

Figure 4.1 summarises the energy required for various types of food production. Shifting agriculture and cereal farming take only one-fifth the energy of hunting and gathering. All forms of modern agriculture, on the other hand, have far higher energy requirements per net calorie of food. One can readily see why fish and beef prices are high; both have high fossil fuel requirements. Since much of this energy is expended on irrigation and on the fertilisers needed to boost the productivity of tired soils, we shall concentrate on these two resources: water and soil.

Water

A major energy input into farming has been the addition of water. Pre-industrial farmers relied on rain; on water diverted from rivers, sometimes stored behind small dams or in underground cisterns; on spontaneously flowing wells or springs; or on water drawn by human, animal, or wind power from wells. The advent of fossil-fuel powered machinery made it possible not only to extract quantities of water from great depths, but also to transport rivers of water over mountain ranges from one drainage basin to another. Both have had far-reaching effects on the environment, as the following few examples demonstrate.

At some depth beneath the surface, the soil of most areas is saturated

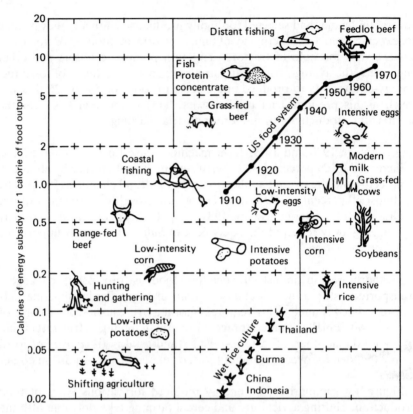

Source: Adapted from J. S. Steinhart and C. E. Steinhart, 'Energy Use in the U.S. Food System', *Science*, 184 (1974) Figure 5, pp. 307–16.

Figure 4.1 Energy and food production.
  The energy needed to obtain each calorie of various kinds of food sources varies widely. The solid line indicates the energy history of the US food system since the early part of the twentieth century. (Note the logarithmic scale.)

with water. The upper level of the underground reservoir, the 'water table', is the depth to which wells must be dug. Most ground water flows slowly to the sea, being replenished by rain water seeping down from the surface. Flow rates vary from five feet per day in rainy areas to five feet per year in deserts. If the rate of water removal by wells exceeds the rate of replenishment, the water table falls, and wells have to be sunk deeper and deeper. Such 'water overdrafts' are occurring in numerous places around the globe.[8]

  The American Southwest is particularly impoverished with respect to water. The Santa Cruz basin near Tucson, Arizona, is overdrafted annually by a half-million acre-feet (an acre-foot being the water needed to cover an acre to a depth of 12 inches). The water table in some areas has dropped

110 feet in 10 years. In California's rich San Joaquin Valley, which produces around $5 billion worth of farm products annually, the overdraft is 1.5 million acre-feet each year. As the water table falls and fuel prices rise, it costs more and more to bring water to the surface. In 1970, it cost Texas farmers $1.50 per acre-foot to pump water; in 1980, the cost was $60 per acre-foot.[9] Since the water is not replaced, the soil over it gradually subsides; in the San Joaquin, the ground has sunk nearly 30 feet in some places. In central Arizona, deep cracks and fissures are disrupting drainage and irrigation canals, endangering homes, and damaging wells and highways; they also hasten erosion.[10]

Underneath part of the American plains lies a pre-Ice Age fossil lake, the Ogallala Aquifer, that stretches from Nebraska to Texas and is responsible for the thousands of acres of irrigated fields so visible from the air. This great reservoir, isolated from replenishment by surface seepage, is being rapidly depleted, having perhaps only 40 years of useful life remaining – less in some places, such as west Texas. Overall, the United States is drawing from the ground 21 billion gallons per day more water than is being returned to it by rainfall and seepage from rivers.[11]

The social consequences of scarce water are far-reaching. As writer Charles Bowden puts it:

> [I]n arid lands [water] rearranges humans and human ways and human appetites around its flow . . . Humans build their societies around consumption of fossil water long buried in the earth, and these societies, being based on a temporary resource, face the problem of being temporary themselves.[12]

Not only is ground water being mined. It is also being contaminated with pollutants that may take hundreds of years to disappear.[13] Along coastlines, wherever too much fresh water is removed from the soil, sea water seeps in, spoiling the ground water for domestic or agricultural use. Finally, as if these problems were not enough, non-agricultural demands for water are also increasing. As lower-grade mineral ores are mined, it takes more and more water to recover the metals.[14] In parts of the Texas panhandle, scarce water is being pumped from the Ogallala and injected into inactive oil wells, forcing their residual oil to the surface – which in turn can be used to pump more water . . . ![15] And if low-grade coal in the northwestern Plains States is indeed strip-mined and shipped, as proposed, by slurry pipeline to the factories of the East and South, then an enormous new demand for water will develop in this already semi-arid region.

When populations settle in arid areas and deplete their local water supply, they generate political and economic pressure on other areas to supply them with water. Almost the entire American West faces this problem. In the 1920s, the city of Los Angeles bought the water rights of farmers in the Owens Valley, east of the Sierra Nevada mountains, turning once profitable cattle ranches into a sagebrush desert. In the 1930s, it heavily tapped the Colorado River. In the 1960s, it turned to northern California. Today it is diverting fresh water that normally flows into Mono Lake, an inland saltwater sink lying far to the north. The lake is a refuge and nesting ground for millions of migratory waterfowl, but as Los Angeles has diverted water from streams that normally feed the lake, its level has fallen dramatically, creating a giant alkali dust bowl around the perimeter and greatly decreasing the area available for wildlife.

While farmers, miners, cities, and even nations quarrel over water rights to rivers, the use of water for all purposes remains largely underregulated and underpriced. In fact, American farmers have been given government depletion allowances for using up their underground water. Just like oil companies, they are being rewarded for consuming rather than conserving.[16]

As irrigation costs rise, more land is being converted back to dry farming with greatly reduced yields. A farmer in the Texas panhandle, for example, can produce 373 pounds of cotton per acre if he irrigates his land, but only 25 pounds if he does not. As water disappears and farmers expand their acreage to maintain profits, dust bowl conditions will reappear, since non-irrigated lands are highly susceptible to wind erosion.

Irrigation often creates a new problem: salination. As irrigation waters evaporate they leave behind a deposit of salt that accumulates in the soil and eventually becomes toxic for plant growth. To flush away this excess salt, large excesses of water are required – far more than the plants actually need. To prevent water logging, this extra water must be drained from the fields through an extensive system of underground porous pipes (known as 'tile drains'), and an elaborate network of pumps and drainage ditches. The problem becomes even more difficult – as in the case of the San Joaquin Valley – if there is no convenient place to dump the salt-laden water. Either local lowlands must be turned into salt marshes, or the water must be pumped over mountains to reach an outlet to the sea. Salination today is creating problems around the world: in the American West, in the Nile Delta of Egypt, in the Indus Valley of Pakistan, to name but a few.

Despite these multiple hazards there have been some remarkable advances in the conservation and use of water in arid lands. The global pioneer is Israel, which now utilises 95 per cent of its 25 inches of annual rainfall. The Israelis have developed salt-tolerant plants, drip-irrigation techniques to apply water only to plant roots, more efficient means to trap rain water, and industrial processes that recycle their own waste water. In Israel, water is no longer a privately owned commodity. 'To [try to] get a license to dig a private well is *via dolorosa*' says Israel's water commissioner.[17] In the Indus River valley of Pakistan, a new irrigation technology based on the intentional depletion and annual recharging of ground water reservoirs may be able to solve the problem of salination in that region, at least for the next century or so.[18]

Finally, widespread reclamation of sewage and other domestic waste water could not only conserve water, but also recycle its rich nutrients. Ideally, sewage is first treated anaerobically to digest its organic matter to simpler compounds. The solids are then composted and used on field and forest. The water, which still contains useful plant nutrients, can either be used directly to fertilise fields, or it can be used to grow aquatic plants. These in turn are harvested and used for fertiliser, and the now purified water is ready for reuse.[19] More and more communities are experimenting with such sewage reclamation systems.

With new public attitudes, new regulations, and new institutions, future global water problems might be made less critical than they now appear – but it will likely require significant changes in our attitudes toward water.

Soil

The annual erosion from lands covered by natural vegetation averages 40 to 60 tons of topsoil per square mile, an amount easily replenished by the processes of soil formation.[20] In some agricultural areas, however, erosion rates of 40 to 60 tons of topsoil *per acre* are common. (There are 640 acres in a square mile.) In western Nebraska, wind erosion alone removes 186 tons of soil per acre per year.[21] These are 400 to 2000 times the natural erosion rate. The Mississippi River dumps 10 tons of mid-western topsoil into the Gulf of Mexico every second, or 300 million tons a year; the Hwang Ho and Ganges each carry off 1.5 billion tons a year of topsoil from Chinese and Indian farmlands.[22] Erosion and leaching of soil nutrients are, together with the failure of water supplies, the major factors threatening agricultural productivity worldwide. Erik Eckholm, an ecologist with the Worldwatch Institute, writes:

> A Somali nomad builds his herd to record size, but the grass land is overgrazed, his cattle grow thin, and sand dunes bury pastures. A farmer in northern Pakistan clears trees from a mountain slope to plant his wheat; soon after, fields downstream are devastated by severe floods. In Indonesia, a peasant burns away luxurious hillside vegetation to plant his seeds; below, rice production drops as soil washed down the mountain chokes irrigation canals.[23]

Erosion first merely reduces soil fertility, causing farmers who can afford it to apply ever greater amounts of expensive fertilisers to maintain yields, but if prolonged it turns productive land into wasteland: eroded hillsides in China, Nepal and Pakistan, and deserts in Africa, the Middle East, the American Southwest and Latin America. The naturally arid areas of the world (more than a third of the land surface) have recently grown by nearly 20 per cent due to human actions: overgrazing of livestock, clearing of forests, and indiscriminate use of fire.[24] Every year around the globe, 20 million acres of cropland – an area the size of Maine – are converted to desert-like wastelands.[25]

While population pressure is an important cause of desertification in Africa and Asia, in places like the United States it is economic pressure. In the United States, the government's Agriculture Stabilization and Conservation Service has played contradictory roles: aiding soil conservation while subsidising crop prices. But farmers got little advice or financial incentive to preserve soils; instead, the government paid them guaranteed prices for crops they otherwise would have lost money on because they were grown at high cost on poor quality soils. Says California farmer Ian McMillan.

> The problem is ecological, and our agricultural system is not designed to deal with ecological problems. The driving force of modern agriculture is not to conserve the soil resource; it is to exploit that resource in every possible way to produce maximum, immediate economic gain. The result is a fiercely competitive industry – a form of economic cannibalism in which even the slightest concern for the resource is a weakness to be penalized.[26]

Today many small American farmers are in an insurmountable bind. Each year they borrow thousands of dollars to purchase seed, pump water, and buy fertilisers, pesticides, gasoline, and equipment, the energy subsidies needed for 'modern' farming. Hence they must get the highest yields to pay back the loans, plus interest, and have enough left to live on. But each year their overworked land is less fertile, so to get the necessary yields to pay back the loans, ever greater energy subsidies are needed. It is a vicious circle. As Texan district conservationist Walter Bertsch comments:

> The great majority of farmers know what has to be done to stop the soil from blowing [away], but they can't afford to do it. In the short run, they've got a bank loan to meet.[27]

This bind began in the 1970s when the US government gave incentives to farmers to convert marginal, erosion-prone lands to crops; food for export was part of the nation's foreign policy.[28] But when global grain production improved and prices dropped, the government withdrew its crop subsidies and guaranteed bank loans, leaving the farmers deeply in debt while claiming they were 'inefficient'. Thousands have gone bankrupt. Although the government began to reverse its policy in 1985, giving small subsidies for land returned to pasture and forest, for many family farmers this came too late.

While nearly half a million acres of American croplands are being lost to erosion each year, three-quarters of a million more prime acres are lost to urbanisation; they are paved over for subdivisions, airports, and industrial parks.[29] The growing industrialised population is encroaching on its own life support system. A small group, the American Farmlands Trust, is trying to offset this process by finding ways to pay farmers the difference between their land's agricultural value and its development value in the hands of speculators, thus ensuring that the land is kept forever as farmland. The Trust also seeks permanent zoning changes that would forbid developments on arable lands, as well as other measures to prevent the continuing attrition of fertile farmlands to non-agricultural uses.[30] So far, the battle has been won in only a few scattered localities: the prevailing notion that land, like water, is not a publicly shared trust but a matter of private ownership where the owner has the right of total exploitation, makes any serious attempt at land preservation almost impossible.

If the loss of cropland in America with all its wealth and opportunity to preserve its soils is serious, soil loss globally verges on the catastrophic. Worldwatch scientists Lester Brown and Edward Wolf make an important point about how much arable land is available:

> Projections of world food production always incorporate estimates of future cropland area, but what has been lacking thus far has been any effort to project changes in inherent productivity of the projected crop-land area.[31]

They estimate that due to population growth there will be a 19 per cent global decline in cropland area per person by the year 2000, but the *topsoil* per person will decline 32 per cent. In very few countries are steps being taken to reverse this erosion, since most governments prefer to invest in weapons rather than sound agricultural practices.[32] Moreover, much of the available cropland in the poorer countries is being used to grow cash crops for export rather than the food and fuel so desperately needed by local people. Luxury foods – coffee, bananas, pineapples, cocoa – are grown for profits which seldom benefit either the lives of the poor or the land itself. It is part of the global exploitation process identified at the beginning of the chapter.

The Tropics

As the arid and temperate lands show increasing signs of stress under heavy use, the world is turning expectantly to the wet tropics – the vast natural rainforests of the Amazon basin, central Africa, Indonesia and

Malaysia. Can they be converted to farmlands to feed a burgeoning population? Can they be used to grow sugar cane for making alcohol – a
substitute for gasoline? Can they be used to raise cattle, rubber, timber and
other money-making crops? In short, can industrialised agriculture be successfully transplanted to tropical rainforests?

The answer to these questions is 'No'. Except in a few special areas,
tropical rainforest cannot be permanently exploited in these ways. When
the trees are removed, the entire ecosystem collapses: its climate changes,
and its soil rapidly loses fertility. What was once rich and productive, in a
few years becomes useless wasteland.[33]

The reasons for this are found in the peculiar nature of tropical forests,
which are among the most richly diverse and productive ecosystems
known. Forty to fifty per cent – perhaps even more – of all living species
are found there, the majority of them not yet known. They live in great
profusion, interacting in an enormously complex and stable network.
Despite poor soils, they are remarkably efficient at conserving nutrients.
Recycling efficiency in the Amazon basin is judged to be better than 99 per
cent, due to the close association between soil decomposers and plant roots
described earlier.

The forests also create a remarkably stable climate for themselves,
maintaining a narrow temperature range and constant high humidity. They
accomplish this by manipulating water. The heavy forest canopy intercepts
much of the rain before it hits the Earth; some evaporates and the rest falls
to the ground so gently it causes little erosion. The water quickly percolates
into the soil where much is taken up by the tree roots and evaporated back
into the sky via the leaves. Called 'transpiration', this process returns 50
per cent – sometimes up to 80 per cent – of the rain water into the
atmosphere. There it soon condenses to form clouds, which reflect the
sun's heat and often shed their water again on another part of the rain
forest, or else on a neighbouring dry region. Throughout the Amazon the
average raindrop falls at least twice before flowing into the Atlantic.

When rainforests are logged and cleared for agriculture, the entire
system breaks down. Because the trees are no longer there to transpire,
rainfall decreases by a third, and the rain that does fall hits the Earth hard,
erodes it, and then runs off in torrents instead of being reabsorbed by the
soil and the vegetation. The resulting flood waters, loaded with silt and
leached nutrients, do enormous damage downstream, flooding villages and
rice fields, destroying fish populations in the rivers, causing landslides, and
silting up dams. The soils quickly become impoverished and some, when
exposed to air and sunlight, form rocklike laterite.

Over the past couple of centuries almost half the world's tropical forests
have disappeared, but today this one-time-only exploitation is accelerating. If it continues, the rainforests will be gone before the middle of the
next century. The most aggressive programmes are in Indonesia and

Malaysia, with Brazil and the Philippines not far behind. On the 'reclaimed' lands, costly fertiliser must be applied to maintain production; without it, rice yields, for example, may fall by 50 to 75 per cent in just two or three years.[34] Unable to afford the necessary fertiliser, many of the Brazilian peasants sent to colonise the Amazon have failed. They now work as labourers for wealthy cattle ranchers who were able to buy up large tracts of land cheaply. Even these ranches, however, have an average life of but two to seven years before weeds, erosion, or soil exhaustion force abandonment.[35] Similar events are occurring in southeast Asia.

Many tropical countries today must import what they once exported. Nigeria, Ghana, Thailand and several other countries, their forests now depleted, must import timber, using precious cash earnings to do so.[36] Brazil imports two-thirds of her rubber while permitting destruction of her native rubber trees; the fish in her rivers, now being destroyed by floods, would produce more protein per acre than cattle; and wild Brazil nuts would net more cash earnings than the settlers' crops.[37] Finally, large tracts of forest are being cleared to grow sugar cane for alcohol. It is Brazil's attempt at energy self-sufficiency; she now imports almost all the oil needed for her automobiles. The country's leading environmentalist believes the energy obtained from the alcohol may actually be *less* than all the energy subsidies needed to grow and process the cane.[38] The social havoc wrought by all these misguided efforts matches the environmental destruction. But more of that in later chapters, when we explore the forces driving all this activity.

If the destruction of the Amazon continues unchecked, two climatologists, Eneas Salati and Peter Vose, predict significant changes in the local weather. Flooding and erosion will continue, though at a decreasing rate as the annual rainfall decreases by one-third or more. Since the trees will no longer be pumping water into the sky, the humidity will fall, resulting in forest and brush fires, now virtually unknown in rainforests. The lack of almost continuous cloud cover will lead to climatic extremes typical of continental interiors elsewhere: hot summers and cold winters. Particularly hard hit in the Amazon will be the neighbouring central and southern Brazilian highlands where such tropical cash crops as bananas, sugar cane, citrus fruits and coffee are grown. Their present mild climate and predictable rainfall are due to the propinquity of the rainforest; when it disappears, these rich agricultural regions will become arid and experience temperature extremes, including frosts that the lucrative tropical crops cannot tolerate.[39]

The Oceans

We now turn briefly to what some consider the resource of last resort, the oceanic fisheries. As Figure 4.1 indicates, modern fisheries consume a

great deal of fossil energy. With such power at their disposal, today's
fishermen easily exceed the 'maximum sustainable yield' of many of the
world's fisheries: that is, they catch more fish than the resource can
continue to supply year after year. The profits are needed to repay the
huge capital investments in ships and processing plants. Such overfishing
has contributed to the sudden collapse of fisheries in several places around
the world: the sardine fishery off the northern California coast in the 1950s;
the North Sea herring fishery in 1969; the southwest African/Namibian
pilchard fishery in 1970; and the Peruvian anchoveta fishery in 1972.[40] The
only way to increase the global fish catch is to expend even more energy
and exploit the distant Antarctic waters. Yet even this vast fishery, which
at its most productive is unlikely to more than double our global fish take,
appears to be a highly uncertain resource, varying greatly from one year to
the next for no apparent reason.[41] Our only recourse is to aquaculture, the
cultivation of fish in lakes, ponds, and coastal bays. Many developing
countries are already successfully experimenting with these techniques.

In summary, our recent efforts to increase our well-being at the expense
of the environment have yielded only brief, one-time gains; resource
depletion is not a healthy survival tactic.

POLLUTION

Figure 3.7 showed how unrecycled wastes from our industrialised
ecosystem build up as pollutants in the environment, further contributing
to its degradation as a life support system. In her book *Silent Spring*,
Rachael Carson warned back in 1962 that manmade environmental toxins
were destroying wildlife and possibly harming people. Use of the pesticide
DDT, her main target, was banned in the United States ten years later
(although US manufacturers continue to sell large amounts of it abroad),
and many of the bird species it had threatened with extinction are making a
recovery – including the bald eagle, America's national emblem.[42]
Meantime, hundreds of new chemicals are being introduced each year into
agriculture and industry, a great many of which are proving toxic for
humans and their environment.

A particularly noxious substance is dioxin, a contaminant of herbicides
and defoliants widely used in agriculture. It was present in Agent Orange,
the defoliant sprayed on the jungles of Vietnam. Dioxin not only causes
acute illness, but is also linked with birth defects and with the high
incidence of cancer among Vietnam War veterans.[43] In an episode that
occurred near St Louis, Missouri, in 1971, dioxin-contaminated oil was
widely sprayed on roads, schoolyards, and horse paddocks, causing severe
illness in many people, the deaths of dozens of horses, chickens and pets,
and the widespread contamination of farms and streams. In one town, over

800 families were affected. Although dioxin was identified as the culprit in 1974, it took the US government nearly a decade to respond to the problem. Meantime local people were exposed to 2000 times the legal levels of this poison. There still has been no satisfactory solution; to collect and incinerate the 500,000 cubic yards of contaminated soil, as has been proposed, would cost a half billion dollars![44]

This is only one small example of incidents that are turning up with increasing regularity around the globe. In 1973 large quantities of the toxic fire retardant, polybrominated biphenyl (PBB), contaminated cattle feed in Michigan, killing thousands of farm animals; it is still present in the bodies of virtually all Michigan's 9.2 million residents.[45] Several years after the 1979 release of radiation from the overheated nuclear power reactor at Three Mile Island in Pennsylvania, excess deaths were being reported among the elderly living downwind.[46] In 1982 the pesticide heptachlor, a known carcinogen that is regularly sprayed on pineapple plants in Hawaii to stop root damage from mealybugs, found its way into the local milk supplies; cows had been given sprayed leaves for fodder. The contaminated milk was marketed and drunk by Hawaii's babies and children for two months before the test results revealing the contamination were released to the public.[47] Massive addition of antibiotics to cattle feed in South Dakota resulted in a 1983 outbreak of antibiotic-resistant *Salmonella* infections in humans in four states. Indeed, most of the increasingly common outbreaks of gastroenteritis caused by these dangerous drug-resistant bacteria have now been traced to animal food sources.[48] In December 1984, 45 tons of methyl isocyanate, a chemical with the properties of cyanide gas, leaked from the Union Carbide plant in Bhopal, India, killing at least 2600 and making over 200,000 people ill.[49] And the total effects of the Soviet nuclear disaster at Chernobyl will not be known for several years.

Besides these local crisis situations, people in many areas are daily bombarded by low levels of pollutants present in air, water, food, clothing, houses and soils. Ozone, a component of urban smog, accelerates the development of emphysema. Lead, still present in the environment, not only causes widespread mental retardation among urban children; it has now been linked to high blood pressure, heart attacks and strokes.[50] Asbestos particles, long known to cause lung cancer, are released from automobile brake linings, from pavements in schoolyards, and from synthetic building boards. (Although some uses of these substances have been outlawed, those already in the environment, especially asbestos, will continue to remain dangerous.) Groundwaters in the United States are becoming contaminated with what some have called an 'alphabet soup' of toxic substances: TCE, EDB, DBCP, dioxin, alpha-BHC, toluene – as many as three dozen manmade organic chemicals. A US government survey of 954 cities showed that 30 per cent have contaminated underground water supplies.[51] Rural water supplies are often contaminated with pesticides and synthetic fertilisers; the latter can cause fatal methemoglobinemia in babies – a massive destruction of the blood's ability to carry oxygen.

Perhaps the greatest long-term threats to human health come from improper disposal of industrial wastes. No one paid much attention until the Love Canal scandal of 1978. The 'canal', an abandoned excavation site in the city of Niagara Falls, New York, was used for over a decade as a dump for toxic chemicals. In the mid-1950s, the site was sold to a local school board, which built a school on it; dozens of homes were constructed nearby. When toxic substances began oozing to the surface in basements and backyards around the neighbourhood, various levels of government shied away from responsibility for rehousing the 700 families affected or indemnifying their losses. Eventually people were temporarily evacuated until steps could be taken to reduce further seepage, but the problem is not fully resolved.[52]

Yet Love Canal is only the tip of the iceberg. In 1980 the US Congress established a five-year 'Superfund' of $1.6 billion to clean up an estimated 2000 hazardous sites around the nation. Little has happened, however, except to move wastes from one place to another without achieving permanent solutions. Moreover the problem is far, far worse than first thought. A 1985 report to Congress judged that there are 100,000 *priority* sites in urgent need of cleaning up. Philip Abelson, an editor of *Science* magazine, summarises the report's conclusions:

> Experience during the last 5 years indicates that the costs of cleanup will be enormous. The OTA [Office of Technology Assessment] estimates that it may be necessary to spend several hundred billion dollars in an effort requiring as long as 50 years.[53]

The report recommended that hazardous waste sites be immediately isolated and stabilised to prevent further contamination of the environment and damage to human health. Progress so far is virtually nil.

This huge cost clearly represents a payment now due on our environmental 'Credit Card' Account, part of the hitherto unpaid overhead of a modern industrial economy. It is becoming apparent just how much energy and effort must go into closing the nutrient cycles in our industrial economy – into converting the dashed line in Figure 3.7 into a solid line.

Chronic and acute hazards to human health are not our only pollution problems, however. Some pollutants directly damage the environment, decreasing its productivity. Fertilisers and sewage bring about excessive 'enrichment' of lakes, rivers, estuaries and bays, causing an overgrowth of algae, which soon die; the ensuing decay deoxygenates the water, destroying animal life. Many areas of Chesapeake Bay, for example, are so depleted of oxygen that the famed blue crab no longer inhabits the deeper waters, the striped bass fishery has declined 75 per cent in 20 years, and oysters and clams are dead.[54] Toxic wastes spawned by industry are poisoning the environment. Sediments along 200 miles of the Hudson River in New York are so heavily contaminated with polychlorinated biphenyls (PCBs), poured into the river for decades by a General Electric plant, that fish from the river are inedible. It is not certain whether dredging up the contaminated sediments will make the problem better or worse.[55] The ozone in urban smog drifts over neighbouring farms and forests, killing sensitive crops and trees. Ozone attacks the chloroplasts of susceptible plants, causing a yellow mottling of the leaves. The ponderosa pine forests bordering California's San Joaquin Valley, as well as those near the great conurbations of Los Angeles and San Diego, are seriously affected; in some areas virtually every pine tree is dead or dying.[56]

The most widespread, and widely known, environmental pollutant today is acid rain. All high-pressure combustion processes, whether in power generating plants, manufacturing industries, or automobile engines, generate nitrogen oxide gases. The burning of fossil fuels, especially low-grade coals, also releases sulphur oxides. All these gases combine with water molecules in the atmosphere to produce two of the strongest acids known – nitric acid and suphuric acid. Eventually they fall to Earth as acid rain, which is sometimes as acidic as vinegar or lemon juice.

When acid rain falls on buildings or monuments made of limestone or marble, it etches them; the lovely friezes of the ancient Parthenon in Athens, Greece, show the ravages of air-borne acids in recent years. The biological effects are even more widespread. The major culprit is toxic forms of aluminum which the acids release from the soil into the inter-stitial water, whence they make their way into lakes and streams. These aluminum complexes severely damage the rootlets of trees, such as spruce and beech, and in lakes and streams they poison young fish, particularly trout and salmon fry, and also crayfish.

The problem has become international. Decades ago, in order to prevent local air pollution, industries built tall chimneystacks to disperse their toxic effluents. Consequently, acid-forming gases and fly ash were – and continue to be – jettisoned high into the upper air stream and blown hundreds of miles. Scandinavia receives its acid rain mostly from Britain. In Sweden, 18,000 lakes have become sterile, and in Norway, all the fish have died in 1000 out of 4800 lakes surveyed. In central Europe, Germany's famous Schwarzwald is severely damaged, and 247,000 acres of Czechoslovakia's Erzgebirge forests are dying. Not even Siberia escapes.

In North America, some of the acid rain that falls on New England and southeastern Canada may be locally produced, but most blows in from the American Midwest. On the west coast, forests and lakes in the Sierra Nevada, the Cascades, and the Rocky Mountains are expected soon to show damage beyond what they already experience from ozone. Besides the pollutants from west coast cities, there is the import of toxic emissions from newly installed Mexican copper smelters just south of the border.[57]

Again, the environmental 'Credit Card' Account is now due and payable. Technical solutions are already available: tighter controls on exhaust emissions from cars; more efficient scrubbers on industrial chimneystacks; fluidised bed systems for burning coal, where a bed of crushed limestone is used to precipitate out sulphur oxides before they can escape to the atmosphere. It is simply a matter of adopting them.

CLIMATE ALTERATION

Besides pollutants, our industrial economies also seem to be generating long-term effects on climate. Even 500 years ago, Christopher Columbus, so his son Ferdinand records, recognised the impact people have on local climates:

> [O]n Tuesday, July 22d [1494], he departed for Jamaica ... the sky, air, and climate were just the same as in other places; every afternoon there was a rain squall that lasted for about an hour. The admiral writes that he attributes this to the great forests of that land; he knew from experience that formerly this also occurred in the Canary, Madeira, and Azore Islands, but since the removal of forests that once covered those islands they do not have so much mist and rain as before.[58]

We have already seen how rainfall in the Amazon Basin is generated by the presence of widespread tropical forests. Recently, two meteorologists have made models of the rates of evaporation, convection, cloud formation and precipitation for different levels of worldwide vegetation cover. During the summer, transpiration by plants has the double effect of increasing the humidity over vegetated areas of land and also of cooling the air so that clouds tend to form and the water is returned to the soil. In a well vegetated world, rainfall over land is maximal. But if continental areas were to become largely denuded of their permanent vegetation – as is now happening – the top layers of soil would lose their water, the air over the centres of the continents would heat up to 40 or 50 °C (105–120 °F) during the summer, and there would be almost no rainfall in most of Europe, Asia and northwestern North America – the very places in the northern hemisphere that now receive the most rainfall. Such a denuded Earth would also be prone to atmospheric pressure extremes and violent wind storms. The authors conclude that the Earth's vegetation cover and the way that it is modified by humans have a significant impact on the weather: the more vegetation, the milder the climate and the greater the rainfall.[59]

Humans are also affecting Earth's stratospheric ozone shield. While the ozone that is generated in smog is toxic for living organisms, the stratospheric layer of ozone, as noted in Chapter 3, is crucial for life since it screens out harmful ultraviolet (UV) radiation. This protective shield is being depleted by gaseous chemicals emitted mainly by modern industrial societies. They diffuse into the stratosphere where they remain for long periods, gradually breaking down the ozone molecules. Probable culprits identified so far are chlorofluorocarbons (CFCs), which have numerous industrial uses, and nitrogen oxides which, as we have seen, are products of combustion emitted by airplanes, automobiles, and industrial chimney-stacks; a further contribution comes from nitrogen oxides emitted from the soil by nitrogen recycling bacteria.

Measuring stratospheric ozone is a difficult task, but every means of doing so indicates a recent decline of half a per cent per year. Which of the pollutants are most to blame is not clear,[60] but since each one per cent depletion of ozone increases harmful UV light reaching the Earth by two per cent, there is no room for complacency. In view of the global doubling in the use of nitrogen fertilisers every ten years or so, a panel of the US National Research Council recommended, back in 1976, that a close eye be kept on the potential impact of fertilisers on the global ozone shield:

> The present assessment of the problem appears to be that there is no excuse for complacency – and equally no cause for immediate alarm.[61]

We cannot go on indefinitely increasing the amounts of nitrogen added to the soil in order to increase our crop production without endangering Earth's protective ozone shield which makes terrestrial life possible.

A more widely publicised concern is the accumulation of carbon dioxide (CO_2) in the atmosphere, which will cause a global warming effect. On average, the amount of sunlight Earth receives each year is balanced by the radiation to outer space of an equal amount of energy, mostly heat. Over the past century, however, the combustion of fossil fuels and the clearing of forests have increased the atmosphere's content of CO_2 by more than 20 per cent, a trend that is expected to continue for several more decades at least. Since CO_2 readily transmits incoming sunlight but absorbs some of the radiated heat, its increase will cause an atmospheric warming known as the 'greenhouse effect'.

It is not certain just how much the Earth will heat up, or what the ecological consequences will be. In the geologic past, when the world has warmed up (for unknown reasons), rainfall patterns have been different than they are today and so has climate, with northern regions being much more clement than at present. Under similar climatic conditions in the future, a warmer and wetter Canada and Siberia would be the world's major grain producers, while Europe, the United States and China become semi-arid. The Antarctic ice sheet would begin to melt, causing a gradual rise in sea level of 15 or 20 feet or more, inundating low-lying urban and agricultural areas. How rapidly these changes might occur and their precise extent are uncertain. Nevertheless, the US Environmental Protection Agency predicts that the first effects will be evident before the year 2000:

> Our findings support the conclusion that a global greenhouse warming is neither trivial nor just a long-range problem.[62]

Here is yet one more reason for caution in trying to solve our global economic problems by destroying forests and increasing our consumption of fossil fuels. Devegetation and CO_2 buildup are both likely to affect global climate in similar ways. Ultimately the effects, even if slower in coming, may be as severe in their impact as those following a nuclear holocaust. It is one more item on our environmental 'Credit Card' Account.

EXTINCTIONS AND GENE LOSSES

The extinction rate of living species today is the highest since life began, and it is due directly or indirectly to human activities. As shown in Figure 3.7, modern industrial societies have squeezed non-domesticated wild species out of the picture. In terms of extinctions, only a nuclear holocaust could be a greater biological catastrophe. Except for a handful of biologists and environmentalists, this fact seems unimportant to most people, who neither know nor care whether some obscure species exists or not. Why should a giant dam be stopped because of a tiny fish called the snail-darter? Why should a major housing development or industrial park be blocked because a few insignificant wildflowers are endangered? Do we really need whales in the ocean? Why shouldn't humans simply stamp out 'uneconomic' forms of life?

A little reflection shows that there *are* reasons for being concerned over the loss of so many species, over and beyond our moral sense of community with our fellow living things. For one thing, other species have much to teach us. Scientists simply do not know how the diverse organisms living together in various natural ecosystems have sustained themselves as viable communities for hundreds of thousands of years. If we destroy them we shall have no models from which to learn how indefinitely to maintain our manmade ecosystems, upon which we now utterly depend. We shall have lost our most important educational resource, Mother Nature, and if this chapter says anything, it is that modern societies are far from indefinitely sustainable.

Along with natural ecosystems, industrial economies are also eliminating the few remaining primitive cultures and their ancient folk wisdom regarding the husbanding and use of wild plants and animals. Traditional knowledge accumulated over thousands of years will soon be forever lost in the course of 'progress'.

Wild species also have potential uses in the future. Yet of all the organisms living today, sicentists estimate that less than half are known. The unknown species mostly reside in those tropical jungles where destruction is going ahead the fastest. Among these unknown species – and indeed, amongst many that are known but little studied – there are certainly many with important future uses for humans as raw materials, medicines, food resources; there are potential microorganisms for industrial processes; and there are potential agents for control of human or crop pests. To destroy all this without knowing what it is we destroy seems the height of imprudence – like bulldozing a library housing books whose language we have not yet understood.

But we need to preserve wild species for an even more direct purpose, namely, as a source of genes for constantly improving our domesticated plants and animals. Improvements in the yields of rice, wheat and maize have depended on the availability of wild relatives of these cultivars, or on genetically diverse domesticated types grown by peasant farmers in various parts of the globe. Both these sources of 'wild' genes are rapidly disappearing, leaving humankind with a limited genetic reservoir from which to develop new disease-resistant and pest-resistant strains in the future. We are not now clever enough – and we may never be clever enough – to invent the necessary genes to keep our basic plant and animal food resources thriving in the destabilised environments we have created. Rice, for example, maintains its resistance to new diseases by crosses with the wild rice species, *Oryza nivara*. The ongoing need for new genes never ceases, since the pest organisms that attack crop plants evolve as fast as scientists develop new strains. We have only to think of parallel cases in human diseases: of antibiotic-resistant bacteria causing the rise in *Salmonella* infections, and of the global resurgence in malaria with the evolution of DDT-resistant mosquitoes.

A few of the world's nations, recognising this problem, have established gene banks – cold storage centres for preserving seeds of wild plants in case they are later needed for developing new hybrid crops.[63] Gene banks for the indefinite storage of animal eggs and sperm are scarcely thought of, since the storage problem is far more difficult. Existing seed banks are far from comprehensive, however, and important species are going extinct before their seeds can be preserved. Already scarce funds barely maintain banks at present levels. In any case, gene banks are a poor – indeed, a probably futile – substitute for natural ecosystems where evolutionary changes can continue to occur.

Instead of recognising the vast potential value of the rich natural diversity we have inherited from literally hundreds of millions of years of evolution, we are destroying that information at an ever faster rate. Again, we are borrowing from the past and the future to live in the present. Species extinctions are the last in our list of environmental overdrafts now due and payable.

TOWARD A NEW WORLDVIEW

Like it or not, we shall have to face up to Nature's demands regarding the overdraft on our environmental 'Credit Card' Account. The question is whether we will be painfully driven to it as our support system crumbles away beneath us, or whether we will take action now and avoid the worst. The latter, however, demands a new attitude toward Earth. As the eloquent ecologist Aldo Leopold put it, we need a land ethic, a sense of community with the soil and the plants and animals that grow from it:

> All ethics so far evolved rest upon a single premise: that the individual is a member of a community of interdependent parts. His instincts prompt him to compete for his place in that community, but his ethics prompt him also to co-operate (perhaps in order that there may be a place to compete for).

> The land ethic simply enlarges the boundaries of the community to include soils, waters, plants, and animals, or collectively: the land . . . In short, a land ethic changes the role of *Homo sapiens* from conqueror of the land-community to plain member and citizen of it. It implies respect for his fellow-members, and also respect for the community as such.[64]

A Philosophy of Community

Popular television programmes about Nature tend to present the myriad interactions that go on among species from a single point of view – that of struggle, competition, and exploitation. A lion chases down a young zebra, an osprey dives after fish, a snake winds its way up a tree to eat a bird's egg, the tongue of a motionless frog suddenly snaps up a passing insect. In each case, the suspense builds, often to sinister music; sympathy is generated for the 'hapless' prey; and the narrator, after the 'grisly' act is over, inevitably reminds us that 'this is Nature's way'. No wonder viewers imagine Nature as brutal, an enemy to be tamed and subdued.

This attitude, as we shall see later, emerges from our Western world-view, and may even subconsciously affect the way ecologists 'see' the world they study. For instance, modern ecology frequently employs the language of capitalist economics, using phrases such as resource allocation, productive efficiency, competitive strategy, nutrient capital, energy budgets,

parental investment in offspring. Ecologists, geneticists, evolutionists –
almost all of them regard the organism as an individual struggling against a
hostile environment in order to eke out a perilous and ephemeral existence. Now all this is true. Every living thing is mortal, and more often than
not meets its end at the hands (so to speak) of another organism. The films
and the research data are accurate – so far as they go. But what they report
is only part of the story; they present but one view of the world.

An alternative is to focus on *relatedness*, the pervasive interdependencies among individuals and species. There are instances of symbiotic
mutualism where two distinct species have an absolute need for each other:
the algal and fungal cells that live in close association to produce lichens,
the first colonisers of barren rocks and boulders; the micro-organisms in a
cow's stomach that live on the cellulose the cow eats, and so provide it with
food it could not otherwise digest; the orchid plants which are pollinated by
only one species of bee, that seeks no other flower. In each case, two
species have coevolved into total interdependence.

Most cases of interdependence are of a less specific, more general sort.
There are species that interact because of their reciprocal functions –
predator/prey, flowering plant/pollinator – but where one *particular*
species is not essential. Most flowers have more than one pollinator; most
predators have more than one prey. As long as there is *a* pollinator species
or *a* prey species available in an ecosystem, then flowering plants and
predatory animals can survive in it.

There are thus 'guilds' of species – groups of organisms that perform the
same function in an ecosystem and make their living in essentially the same
way.[65] Just as in a medieval village there had to be *a* farmer, *a* blacksmith,
and *a* carpenter, so in an ecosystem there must at least be *a* producer
species (green plant), *a* pollinator, *a* predator, and a minimal diversity of
decomposer species to carry out each recycling process. This view sees
ecosystems as *non-random* assemblages of interacting species performing
multiple essential functions. *No species exists in isolation*: each organism is
both the *supporter of* some species and is *supported by* others. The idea
that organisms not only adapt to their environments but actively create
them is admirably stated by evolutionist R.C. Lewontin:

> Organisms alter the external world as it becomes part of their
> environments. All organisms consume resources by taking up minerals,
> by eating. But they may also create the resources for their own
> consumption, as when ants make fungus farms, or trees spread out
> leaves to catch sunlight . . .

> It might be objected that the notion of organisms constructing their
> own environments leads to absurd results. After all, hares do not sit
> around constructing lynxes! But in the most important sense they do . . .

[T]he biological properties of lynxes are presumably in part a consequence of selection for catching prey of a certain size and speed, i.e. hare . . .

[This] metaphor of construction rather than adaptation leads to a different formulation of natural selection and evolution . . . Organisms . . . both make and are made by their environment in the course of phylogenetic change.[66]

In discussing evolution, scientists still commonly speak of the 'forces of natural selection' acting on an individual organism. What is largely ignored by converting the 'forces of natural selection' into an abstract, intellectualised concept, is that these 'forces' have a reciprocal interdependence with what is acted upon; they are both *actors* and *acted* on. The organism being selected 'for' or 'against' is in its turn a selective force within the ecosystem. In Lewontin's example the existence of a hare 'selects' *for* the presence of lynxes (the hare provides a food source); but a hare's propensity to escape 'selects' *against* the presence of lynxes, especially if the hare turns out to be quicker than the lynx. On the other hand, hares that are too easily caught would soon become extinct, threatening the extinction of lynxes as well. In such a situation, who is 'selecting' what? What 'forces' are at work? From this view, it is the 'fitness' of the multiple interactions in the entire ecosystem that is selected for in the course of evolutionary time, not the fitness of this or that particular species in an otherwise static system. 'Fitness' of individuals is a matter of 'fitting in' with a functioning whole.

From this point of view, humans are but part of a biotic community, shaping it and being shaped by it. If we continue to overexploit and destroy it, then we shall go extinct just as surely as would the lynx if it too cleverly consumed all its resource base.

A New 'Economics'

As Figure 3.7 showed, a modern industrial economy is simply a modified ecosystem maintained by massive infusions of fossil sunlight. It also showed that our current wealth is borrowed from the future; through failure to recycle our wastes, we deplete resources and pollute the environment. To complete this recycling process requires a level of energy input that has so far been considered 'uneconomic'.

In energy terms, industrial societies have never lived within their incomes, and as discussed in Chapter 2, their dependence on non-renewable fossil fuels has grown greater with time. When the United States gross national product (GNP), a measure of economic output, is plotted against annual energy consumption, we find an extraordinary correlation (Figure 4.2).

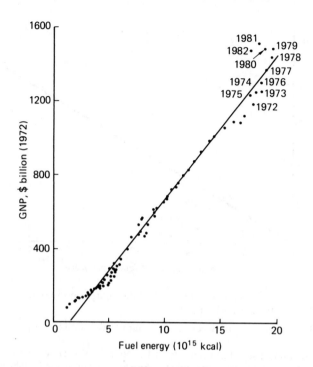

Source: Adapted from C. Cleveland *et al.*, 'Energy and the U.S. Economy: A Biophysical Perspective', *Science*, 225 (1984) Figure 1.C, p. 891.

Figure 4.2 Relationship between GNP and energy consumption
The linear relationship between fuel consumption and Gross National Product (GNP) in the United States for the years 1890–1982 (calculated in constant 1972 dollars).

During periods of boom, energy consumption increases; during depressions, it decreases.[67] If we count wealth as the annual *consumption* of goods and services, then wealth depends directly on energy.

Ecologist Robert Costanza has shown that the market prices for most goods and services correlate well with the total energy embodied in them, including the energy inputs from labour, from the sun, and from government services (education, regulation, and so forth). When the dollar value contributed to the GNP by each sector of the economy is plotted against its energy content, all but a few sectors fall along a straight line (Figure 4.3). Those sectors falling below the line are 'underpriced': agriculture, fisheries, and primary energy sources such as coal, gas, oil and wood.[68] Each of these sectors is exploiting Nature in ways that are being charged up to our environmental 'Credit Card' Account!

If we begin to conserve energy – to heat our homes more efficiently – the future curve in Figure 4.2 will shift upwards: more GNP output per unit of

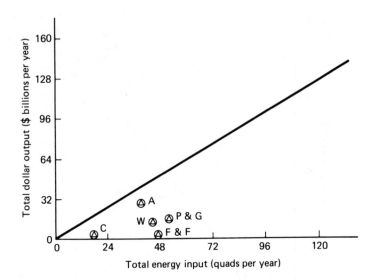

Source: Adapted from R. Costanza, 'Embodied Energy and Economic Valuation', *Science*, 210 (1980) Figure 4D, p. 1223.

Figure 4.3 Relationship between energy input and dollar output in the US economy

The linear relationship between output (in dollars) of various sectors of the US economy and the total energy, including indirect energy inputs from labour, government and solar energy, embodied in each sector (based on 1967 econometric data).

Note that the sectors that are 'undervalued' (lying well below the regression line) are those that exploit Nature in non-sustainable ways: energy, agriculture, forestry and fisheries.

C = coal mining; A = agricultural production; W = wood production; F&F = forestry and fisheries; P&G = crude petroleum and natural gas.

energy input. Conservation following the rise in fuel prices in the late 1970s resulted in the energy savings shown in the early 1980s.

On the other hand, we will have to begin paying for the unpaid portion of our present wealth, the part now charged to our environmental 'Credit Card' account. This will include paying higher prices to meet the real costs of agriculture, fisheries, coal mining and deforestation. It will require expending additional energy to clean up pollutants, recycle wastes, and care for and nurture the environment. Since these are not 'economic' goods and services, the slope in Figure 4.2 will drop downwards. Yet the energy expended for clean up will generate jobs and create a pleasanter environment. Indeed, as shown later on, GNP is a poor yardstick of human well-being; it is not how much we *consume*, but what amenities we *have* that constitute wealth.

Energy is clearly crucial to an economic system. Nobel prize-winning physicist Frederick Soddy, who became fascinated with economics, said in 1926:

If we have available energy, we may maintain life and produce every material requisite necessary. That is why the flow of energy should be the primary concern of economics.[69]

What we need is an accounting system for our economy based on the principles of ecosystems, which includes the true energy costs of its maintenance. Costanza writes:

The required perspective is an ecological or 'systems' view that considers humans to be part of, and not apart from their environment. A few economists have already taken this perspective, and the implications for a new ecological economics that links the natural and social sciences are great.[70]

This notion – that economies are really ecosystems, and that they are driven in precisely the same way by their energy throughputs – will recur throughout the book.

SOME CONCLUDING THOUGHTS

While this chapter has catalogued our global environmental problems, it has neglected to mention the numerous small, and occasionally large, efforts being made at many levels and in various countries, to solve them. Some of these 'experiments-in-the-future' will be met later, in Part IV. They do exist and they are a hopeful vanguard that concerned readers can join. But there is still a large task ahead in raising awareness – and changing world outlooks – among the vast majority of the world's citizens, especially those of the powerful, dominant West. Solutions lie less in the possibility of 'breakthrough' technologies than in the ability of people to bring about social, institutional and value changes that will allow them to adopt the more appropriate technologies that already exist.

We can do no better than to close this part of the book with a suggestion of the direction in which our next steps should lie – the final words of Aldo Leopold:

By and large, our present problem is one of attitudes and implements. We are remodeling the Alhambra with a steamshovel, and we are proud of our yardage. We shall hardly relinquish the shovel, which after all has many good points, but we are in need of gentler and more objective criteria for its successful use.[71]

It is all a matter of opening up the Western mind to new modes of thinking.

Part II
The Human Animal:
'What is
"Human Nature"?'

Whenever anyone suggests changing society by appealing to the 'better instincts' of human beings, someone else always kills the idea by saying 'You can't change human nature. People are born greedy, competitive, and aggressive'. This has given rise to the idea of 'social Darwinism' – namely that human behaviour is selected for on the same basis as are the inherited traits of other organisms. But do we really understand the forces of selection as they act in Nature? Are we genetically programmed to behave as we do? Part II of this book supports the idea that although socialisation *is* part of our genetic inheritance, the type of society we establish and the values we hold are purely learned.

Part II begins with human origins. Chapter 5, 'The Emergence of Human Nature', explains evolutionary theory and then explores recent thinking on the biology of bonded primate societies and the rise of human cultures. It argues for the oneness of our emotional and reasoning selves. Chapter 6, 'The Cultural Spectrum', looks as anthropologists do at the diversity of human social systems and worldviews, and at their adaptiveness. It considers cultural change and cultural destruction in the face of too rapid change. Chapter 7, 'Religions and Worldviews', extends the recognition of human diversity in beliefs and values to the world's major religions which arose concomitantly with civilisation. It further contrasts the roles of science and religion in formulating today's Western worldview.

Finally, if values differ among cultures, then they must be culturally acquired. This is the subject of Chapter 8, 'On Acquiring a Worldview', which explores how our brains learn to 'see' a culturally selected world, a world that is transmitted to us in various ways. This insight is the key to undertaking a deliberate change in our thinking.

5 The Emergence of Human Nature

The relevance of biology and anthropology is evident enough. In his pride, man hopes to become a demigod. But he still is, and probably will remain, in goodly part a biological species. His past, all his antecedents, are biological. To understand himself he must know whence he came and what guided him on his way. To plan his future, both as an individual and much more so as a species, he must know his potentialities and his limitations.

Theodosius Dobzhansky, 1967[†]

INTRODUCTION

Hereby it is manifest, that during the time men live without a common power to keep them all in awe, they are in that condition which is called war; and such a war, as is of every man, against every man...

In such condition, there is no place for industry; because the fruit thereof is uncertain: and consequently no culture of the earth; no navigation, nor use of the commodities that may be imported by sea; no commodius building;... no knowledge of the face of the earth; no account of time; no arts; no letters; no society; and which is worst of all, continual fear and danger of violent death; and the life of man, solitary, poore, nasty, brutish, and short.[1]

Such was Thomas Hobbes's seventeenth century view of the life of our primeval ancestors. Before the advent of civilisation and the rule of written law, men and women, driven by their inner natures, were condemned to unending violence. For Hobbes, the evils of humankind were inherent, a view still held by those who have never studied the human past. It implies a fundamental nastiness in human nature which only civilisation could possibly overcome. When asked about solving the problems of war, greed and selfishness, people simply shrug their shoulders and say: 'Well, you can't change human nature, can you?'. This image has been reinforced by such popular authors as Robert Ardrey (*African Genesis* and *The Territorial Imperative*) and to a lesser extent by Desmond Morris (*The Naked Ape*) and Konrad Lorenz (*On Aggression*).[2] But what do we *really* know about the kind of beings our ancestors were? What is this raw material – this human nature – out of which we are shaped?

131

This chapter examines what is known about the biological origins and constitution of humankind, in which human nature must be grounded. Yet even among scholars there is confusion and disagreement, since the very existence of culture makes it difficult to see the true nature of either ourselves or our ancestors. Culture biases us in ways we are scarcely aware of.

It is easiest to describe the two extreme positions regarding human nature, while recognising that not every person involved in the controversy holds totally to one or other viewpoint. Stated most simply, it is the old 'nature-versus-nurture' dispute. There are those who interpret human nature, including human culture, strictly in terms of the laws of heredity that seem to govern the natures of most other living species, and there are those who view newborn babies as essentially 'blank paper' (*tabula rasa*) to be inscribed upon by whatever they experience. Both positions have social and political ramifications that others view with alarm, and the entire question of the biological basis of human nature often raises emotions to fever pitch.

As a group, 'sociobiologists' are regarded as determinists who picture human nature as fixed and unalterable. According to certain of them, such behavioural traits as male dominance and aggressiveness are inherent, inescapable 'givens' anchored in our genes. The implied conclusion is that *what is* in a particular society, being inherited, cannot be changed, at least not without great difficulty. Such a view can be – and often has been – used by chauvinists, racists and others, to justify social and political injustices.[3] Perhaps the best known scientist proposing a relatively large genetic component to human behaviour is E.O. Wilson, author of *Sociobiology*, a giant treatise on the evolution of animal societies.[4] He and his followers form the 'pro-nature' side of the 'nature–nurture' controversy, although in fairness they do acknowledge a place for learning in human behaviour. Nevertheless, by the time the ideas of sociobiologists have been simplified by the popular press, the public's image of their message is well expressed by the cover of *Time* magazine's issue on this subject; it showed a couple of puppets making love, their strings pulled from above by strands of DNA.

The 'pro-nurture' side is espoused by many social scientists, especially those of a more 'liberal' bent. For them, the environment is the major determinant of behaviour. But just as the sociobiologists, in their extreme moments, think of genes and DNA as *agents* of human behaviour, so the anthropologists see culture as an *agent* – something external to people, which manipulates them.[5] One behavioural psychologist, B.F. Skinner, heads a school which carries this notion to an extreme. Any innate tendencies that people might have are considered intractable and therefore uninteresting. According to this school, behaviour is almost totally a result of learned expectations of rewards and punishments. We are conditioned by our culture; we are externally manipulated.[6] It is not our genes that are pulling the puppet strings; rather, it is our environment, especially other human beings.

Both extremes sound silly when described so bluntly – and they are. Neither one leaves room for such common human experiences as imagination, curiosity or free will. (We take up the question of free will and related matters in later chapters.) One reason such extreme schools arise and persist is the academics' tendency to dissect problems into smaller and smaller parts in pursuit of ultimate causes. Experts forget that what they are cutting away and discarding *is part of the whole*; they fondly imagine that their discipline explains the world and resent encroachments by others. Thus when Wilson predicted the cannibalising of the social sciences – and perhaps the humanities as well – by the twin sciences of sociobiology and neurobiology, not only did he raise many academic hackles; he also distorted the essential properties of his subject matter, the nature of human beings.

A preferred route is to discern human nature through examining the evidence of our own evolution: early human fossils and artifacts; the behaviours of surviving hunting–gathering peoples, as well as of our modern cousins, the other living primates; the nature of the human brain and its evolution; and last but not least, our own 'common sense' understanding of what it means to be a human being. (Too often what everyone knows intuitively but science has not yet described quantitatively tends to be dismissed as 'mere subjectivity'.) But before we can discuss this evidence, we need to look briefly at evolution – at what it *is*, and is *not*.

THE NATURE OF LIFE

Had the world been designed by a disembodied twentieth century physicist, it is doubtful life would have been invented. In the physical universe, things tend to run downhill. Orderliness decreases and disorder or randomness increases over time. Salt and water spontaneously mix to form oceans; mountains spontaneously erode to form level plains. To reverse either process requires input from an external energy source – the sun or the heat of Earth's core. To achieve order in one place there must be growing disorder – an increase in entropy – somewhere else in the universe. To physicists, the neoformation of ordered systems is a rare event.

Compared to non-living systems, even the simplest bacterial cell is highly ordered – and hence, highly improbable; it cannot occur spontaneously. Organisms thus require external energy to create and maintain their orderliness – food for animals and sunlight for plants. But they require something more, namely *information*, which they get from their parents. Life requires blueprints in order to assemble itself.

The Blueprints

DNA – the parental information, the genes – has become something of a household word. Everyone knows vaguely what genes are, but like television sets and automobiles, most people are not sure how they work. To understand evolution, including our own appearance on Earth, requires a thumbnail sketch of how DNA directs the development of each living organism.

DNA is something like Morse code, where combinations of simple basic signals stand for a letter; the letters, together, form words; and the words form meaningful messages, from a simple SOS to a Shakespeare play. (In themselves, the dots and dashes of Morse code are meaningless and we could never predict *Hamlet* from knowing the physical basis of the electronic impulses emitted when a telegrapher taps his key.) Like dots and dashes, the four basic symbols in DNA – the nucleic acid bases A, T, G and C – when read as linear triplet sequences, code for amino acid 'letters'; these in turn form protein 'words', and when the variously shaped proteins interact together they build a complete product – a living organism.

What is meaningful about DNA, then, just as it is for the strings of letters of the alphabet that make first the words, then the sentences, next the paragraphs, and finally the books of our written language, is the fact that encoded in it are *hierarchies* of order: amino acid sequences constitute proteins; proteins fold up and assemble together to make living cells; cells multiply and aggregate to form organisms; and organisms interact to form biological communities. The components of this organisational hierarchy ultimately are all coded by information-rich arrangements of extremely long threads of DNA. Just as a Morse-code message is a long string of dots and dashes, so this DNA message is a long string of the basic units of inheritance – A, T, G and C – a chain that is over a billion units long for most mammals. Even though the human mind – itself a physical product of DNA sequences – may never be able to explain precisely each and every step by which this living hierarchy arises, it may still grasp intuitively its existence.

Although the laws of physics and chemistry cannot explain how such highly specific DNA molecules could have evolved by chance, information theory can.[7] The clue lies in the initial random formation of a self-duplicating molecule, the DNA of a primordial gene.[8] Once that event had taken place, then any accidental change in the DNA message which favoured its ability to reproduce itself would be retained. Such changes we call 'mutations'. It is as if this year's car models contained computerised robots that could gather the necessary raw materials together and follow an internal programme to duplicate themselves completely – car, robot and all. Errors in following the programme (mutations) would create novel models which would be selected either for or against in the marketplace. We are talking here about *evolution*.

THE PROCESS OF EVOLUTION

Once an idea becomes accepted – buoyed up by masses of irrefutable evidence – we wonder how people could ever have been so blind as not to have 'seen' it. Today we wonder how people could ever have thought the Earth flat or believed it to be the centre of the solar system. Yet even today, despite its powerful explanatory power, the theory of evolution is not accepted by a small but vocal religious minority. But real as their objections may seem to them, each can easily be rebutted by well-substantiated data.[9] We assume here that evolution has occurred and continues to occur.

In 1859, after more than twenty years of pondering, Charles Darwin published his book, *The Origin of Species by Means of Natural Selection or the Preservation of Favored Races in the Struggle for Life*. Knowing that his ideas about evolution flew in the face of popular beliefs, he marshalled every shred of evidence then available to support them. His thesis, in essence, ran thus:

1. Parents of all living species produce more offspring than the environment can sustain.
2. These offspring are inherently variable.
3. Environmental forces select among these offspring the best-adapted individuals.

These tenets form the basis of his theory of natural selection. In Darwin's argument, a particular new variant is not caused by the environment; rather, the environment merely acts on whatever chance variants are already present in a population. In this way, a gradual change in the inherited traits of a population occurs – based on chance mutations in DNA – that eventually leads to the formation of new species.

Recognising the millions of years of geologic time needed for evolution to occur, Darwin knew that in one human lifetime one could not expect to see noticeable changes in oak trees, squirrels or earthworms. (It was, of course, the observable selection by humans for desirable traits when breeding pigeons, plants, cows and dogs, that alerted Darwin to the idea of similar, if slower, selective forces in Nature.) After decades of studying the fossil record, we now know that the rate of evolutionary change is probably not constant; it is quite slow during periods of climatic stability, and speeds up greatly during periods of climatic change.[10]

The 'Ethics' of Evolution

It is unfortunate for our understanding of ourselves that two popular misconceptions have grown out of evolutionary theory. These have to do with notions of 'competition' and 'progress'. Because they are often used in moral arguments about human nature, it is important to clear them up.

The idea that success in the game of life means outcompeting others and caring only about yourself, so your genes will predominate in the next generation, is deeply ingrained in the modern Western worldview. It is unfortunate that Darwin used such loaded words as 'struggle for life' and 'favored races' in the title of his book. But as anthropologist Ashley Montagu has pointed out, it was natural for him to draw a parallel between the competitive war of commerce going on around him and what he saw as the war of Nature in which individuals struggle with each other to survive; it was the same mistake Thomas Hobbes had made two centuries earlier.

> Man sees the world according to the kingdom that is within him; it should therefore not be surprising that nature should have been interpreted [by Darwin] in terms of nineteenth century human relations.[11]

As already noted, the use by ecologists of phrases like 'parental investment in offspring', 'resource allocation', and 'time-and-energy budgets' shows how easily we interpret Nature in terms of capitalist economic activities. It is a dangerous practice, leading us unconsciously to think of Nature as calculated and intentional, and blinding us to the complementary, cooperative, and interdependent aspects of natural systems.

Because genes code for heritable traits, the idea of a 'struggle' in Nature actually led sociobiologist Richard Dawkins to write about 'the selfish gene' and 'the morality of the gene', as though genes had foresight and the power to make moral choices:

> We are survival machines – robot vehicles blindly programmed to preserve the selfish molecules known as genes.[12]

This reductionist view, which relegates all the living world to manipulation by long strings of DNA molecules, is simply not true biologically. A gene is not an isolated entity acting independently; each works in concert with thousands of other genes in an organism, creating a functioning whole. Moreover, the activities of those genes can be deeply affected by the environment in which an organism finds itself. Natural selection does not act on single genes, but on whole organisms.

As far as behaviour is concerned, there are few cases in Nature of heritable behaviours ascribable to one or even a few genes. There are no genes for 'love' or 'selfishness' or 'altruism'. Nor is 'reproductive fitness' a necessarily competitive act. Leaving behind more offspring than somebody else does not necessarily imply that an individual organism has won some sort of zero-sum, 'I win – you lose' game. In fact, it seldom implies this (although getting more to eat and beating out a rival for a mate or a nesting area, because such behaviours are so easily filmed, do tend to be shown a lot on televised nature films!). Being a survivor far more often means evading predation, resisting disease, or being able to utilise a piece of the environment that no one else is using. Aggressive competition between individuals occurs in relatively few species.

Similarly, when biologists talk about allocation of scarce resources among, say, several species of seed-eating rodents, they do not mean the animals are fighting over seeds; instead, each species has seed preferences as to size and distribution that limit the amount of overlap among them and inhibit direct competition. Of course, total resources in an ecosystem are limited, and in that sense species and individuals do compete. But most competition is resolved by limiting the area of the environment that an individual or a species uses, not by direct, aggressive confrontation. The latter tends to be restricted to scavengers such as magpies and hyenas, and of course to some groups of modern humans.

'Fitness' also implies far more that just getting 'more of the pie'. As discussed in Chapter 4, it means fitting better into the general ecosystem, the whole community of organisms. Nature is replete with interdependencies and mutual interactions that show that 'cooperation' or 'fitting in' with the overall ecosystem is just as important to evolutionary 'success' as is 'competition'. (All these words are in quotes because they imply intentions on the part of organisms or their genes which are not at all part of the evolutionary process. There is neither direction nor morality in the workings of natural selection.)

These thoughts are well summarized in the words of evolutionist George Gaylord Simpson:

> To generalize that natural selection is over-all and even in a figurative sense the outcome of struggle is quite unjustified ... Struggle is sometimes involved, but it usually is not, and when it is, it may even work against rather than toward natural selection. Advantage in ... reproduction is usually a peaceful process in which the concept of struggle is really irrelevant. It more often involves such things as better integration into the ecological situation ... more efficient utilization of food, better care of the young, elimination of intra-group discords ... that might hamper reproduction, [and] exploitation of environmental possibilities that are not the objects of competition or are less effectively exploited by others.[13]

Species that survive, do so because they *fit in* with their surroundings: failure to do so leads to extinction. In Nature, 'conquering' one's environment is a meaningless concept.

The other popular misconception about evolutionary theory stems from the fact that over time there has been an increasing complexity of life forms. This lends itself to a vision of 'progress' in evolution, where more sophisticated forms replace earlier forms. The whole of evolution is regarded as a ladder or a pyramid, where 'higher' forms achieve 'success' at the expense of 'lower' forms. Humans, being the most recent and complex, are seen as deservedly at 'the top'.

The words in quotes, of course, are the problem. How shall we define 'success'? By numbers of individuals? There are far more bacteria in the world than people and they leave behind far more progeny. By length of time on Earth? Again, bacteria have been around a thousand times longer than humans. Even the dinosaurs, whose demise we love to point out as an example of overspecialisation, were around for well over 100 million years and dominated the landscape for 50 million years. Can we really say that ancient species, like bacteria, amoebae, and blue-green algae, have 'failed' because they continue to exist without becoming complex? Is not the idea of 'improvement' purely a human one, and not a part of the natural scheme? As philosopher Mary Midgley puts it:

> If man wants to set up a contest in resembling himself and award himself the prize, no one will quarrel with him. But what does it mean? [14]

Indeed, human existence absolutely depends on the continued existence of those very bacteria and other life forms that have been around far, far longer than we. Instead of an evolutionary ladder, we might usefully think of an evolutionary bush, with many equally important branches all stemming from a single rootstock; some branches have changed greatly over time, while others have remained apparently unchanged. Yet all share the same genetic code and have had common ancestors at some time in the past. [15]

Significantly, the great majority of human societies in the past valued and held sacred other living species with whom they felt a close affinity and interdependence. The peculiar notion that humans are 'superior' beings with a 'right' to dominate the Earth comes, at least in part, from Judeo–Christian cosmology, as we shall see later. The view that 'competition' and 'progress' are natural phenomena is of more recent origin. Its greatest apologist was the nineteenth century social Darwinist, Herbert Spencer. By asserting that Nature is inevitably 'red in tooth and claw', he justified the cut-throat commercial competition of modern industrial society, including the right to destroy other species and to dominate more 'primitive' peoples.

HUMAN EVOLUTION

Humans and the great apes shared common ancestors. The amino acid sequences in the proteins of chimpanzees and humans are more than 99 per cent the same. [16] This close relationship between apes and people suggests that we may learn something of the heritable behavioural natures of our early ancestors by looking at the behaviours of modern primates. Behaviour is at least as important as anatomy in ensuring the survival of organisms; in fact, the two are closely related.

Primate Bondedness

All species make a living by becoming more or less specialised – by being particularly good at something. Birds fly; moles dig; deer run very fast – and primates think. As mammalogist Alison Jolly puts it:

> If there is an essence of being a primate, it is the progressive evolution of intelligence as a way of life.[17]

Moreover, the evolving primate path combined intelligence with increasing social bondedness. Feelings and intellect run in parallel!

Primates arose in tropical rainforests as arboreal creatures making a living mainly from fruits, but they also ate leaves, gum, insects, flowers, and sometimes small animals. To move safely about in the forest canopy a hundred feet and more above the ground required binocular vision coupled with grasping appendages. These two attributes selected for a large brain, which in turn permitted the development of manual dexterity and learned behaviour.

In rainforests, food is patchy, and this may explain the prevalence of social behaviour among primates. In tropical fruit trees, especially figs which are a preferred food, the fruit on a single tree tends to ripen all at once, but at irregular intervals. The sharing of a ripe tree by numerous individuals thus requires a certain degree of social tolerance. As Jolly says:

> If early primates traveled together from one fig tree to another, they had already begun to fulfill a condition of longer-term social bonding.[18]

The earliest and strongest primate bonding no doubt arose between mother and offspring. Stephen Jay Gould, an evolutionary zoologist at Harvard, notes that primate evolution is marked by progressively retarded development to accommodate the evolution of an enlarging brain.[19] Because their ever larger heads could not squeeze through the mother's birth canal, infants were born in a more and more embryonic state, with most brain enlargement occurring after birth rather than before as is true for most other mammals. This caused a great prolonged period of infant dependence – five or more years in the case of the great apes – and lifelong bonds developed between females and their offspring. This extended contact also permitted the transfer of learned skills between generations.

We now know that social learning is widespread among primates, much more than once thought. For example, one small population of blackcapped cebus monkeys in Colombia has discovered how to open the hardest black palm nuts, a trick unknown to other populations of these monkeys. Among baboons, hunting is not innate, but isolated troops have learned to kill other animals for food. In Japan, there was the famous case of Imo, the juvenile macaque monkey, who discovered how to wash grit off sweet potatoes and separate rice grains from sand by floating them on water. After a few years, her discoveries had spread throughout the troop. Wild chimpanzees have been observed to teach one another to use sticks to fish for termites, and they regularly communicate by vocal calls and bodily cues the location of food, the presence of snakes, and other pieces of environmental information. Recently Washoe, the tame chimpanzee taught American Sign Language since her infancy, has been spontaneously teaching ASL signs to her adopted infant son. As Jolly says:

> Local traditions in foods, in tool-using, and in defense against predators are dependent on the ability of a small, more or less cohesive group of primates to learn from one another.[20]

Our nearest living relatives are the chimpanzees whose social structure somewhat resembles that of human hunting–gathering societies. Groups of around 50 share a territory, fissioning into subgroups when food is scarce, and recongregating through loud calls when a rich supply is located. Fission prevents competition during times of scarcity, but chimpanzees prefer being together, spending long periods in mutual grooming and other social activities.

Ethologist Michael Ghiglieri notes that such behaviour refutes the expectation of competition among individuals that Darwinian theory seems to demand:

> Currently accepted theories of evolution hold that natural selection operates through the individual rather than through the group or the species... The reproductive strategy of the males of most mammalian species is based on defending food and excluding other males from females that are in estrus. Yet male chimpanzees in the wild do just the opposite: males at Ngogo [in Uganda] display cooperative behaviour in both feeding and mating.[21]

In order to explain how such cooperative behaviour might have evolved in various species, scientists developed the theory of kin-selection, an elaborate calculus erected to show that if an individual is generous towards close relatives (e.g., siblings and cousins) that naturally share the same genes, then on the whole, that individual's genes will be preserved in future generations. Self-sacrifice and altruism are merely disguised selfishness.[22]

A difficulty in placing this argument on a scientific footing when applying it to highly complex and flexible primate social behaviours lies in having to assign numerical benefits and risks to subparts of an overall behaviour pattern. Surely there is not one set of genes for food sharing and another for rescuing a relative from the jaws of a predator? More likely is a genetic basis for a strong, generalised bonding that then results in both activities. On average, the risks of starvation from sharing one's food or of dying in trying to protect a relative from an attacking leopard are outweighed by the benefits of all the group's social activity. Once the propensity for social attachments evolves, it creates an entirely new basis for selection, the social group.[23]

Our prehuman ancestors almost certainly lived in small cohesive groups that ranged over a familiar home territory; used simple tools and taught each other new, innovative tricks; and could modify their behaviour to suit new conditions.

The First Humans

The first animals on the human branch of primate evolution arose in Africa. The search for fossils occurs mostly along the great Rift valley in eastern Africa, where a giant fault in the Earth's crust has exposed layers of ancient sediments containing bones and artifacts.[24] Early finds were often misinterpreted according to the preferred ideas about human evolution at the time, leading to the presumption that our earliest ancestors, the Australopithecines (southern apes), were weapon-using killers. More careful observations show that their supposed 'weapons' were in fact left-over bones from hyena banquets, and that the skulls of their 'murdered' comrades had been crushed centuries after death by rockfalls in the caves where they had died. More cautious interpretations of fossil finds lead to a more benign image of our earliest forebears.

The two characteristics distinguishing us from our nearest ape relatives are an upright gait and a large brain, and we now know that bipedal locomotion came first. Our large gluteus maximus (buttocks) muscle permits the erect posture that distinguishes the hominid line of evolution. Walking upright freed the hands of our early ancestors – not, initially, for carrying tools or weapons, but for carrying food. The primary basis of the earliest hominid societies was the propensity to *carry* food back to a home base and to *share* it with others.[25] Such a strategy, which greatly increases the chances of survival of the entire group, is still absolute in hunting–gathering societies and remains in ritual form even in modern industrial societies which celebrate shared family feasts such as Thanksgiving and Christmas.

The earliest hominid fossils are 3.5 million years old, their ancestors having branched off from the great apes from 5 to 7 million years ago. That they all possessed a perfectly developed upright gait means that bipedalism developed in a couple of million years, a very short time indeed, evolutionarily speaking. There is evidence that these earliest hominids branched into several contemporary species. Two or three types of Australopithecenes and an early form of humans known as *Homo habilis* lived simultaneously on the borders between the receding forests and the open savanna grasslands of eastern Africa. *Homo habilis* attained a population of perhaps one or two million, and began to spread into Asia. Their brains were only about 800 cubic centimetres (cc) volume, about twice that of chimpanzees and half that of modern humans.

About a million years ago a new form emerged in Africa, presumably a descendent of *Homo habilis*, with a brain of 1200 cc. *Homo erectus* also spread across the Old World, including Java and the caves outside the capital of China where the famous Peking man fossils were found. They learned to use fire and make crude shelters, permitting them to move into colder regions such as Europe and northern China. Around 250,000 years ago, the first *Homo sapiens* appeared, again in Africa. Equipped with slightly larger brains and somewhat more robust bodies than modern humans, they are known as Neanderthal men because the first skeletons were found in the Neander River valley in Germany. Unjustly classified as bumbling mental defectives, the Neanderthalers left evidence of personal awareness and a capacity for abstract thinking. They shaped tools and, at least by 60,000 years ago, were burying their dead, surrounding them with flowers, stone axes and wreaths of antlers.

About 40,000 years ago, at the height of the last Ice Age, Neanderthalers finally disappeared having given way to modern humans, *Homo sapiens sapiens*. Arising in Africa about 150,000 years ago, this new form moved to the Near East and then across Asia and Europe.[26] Among them, art began to flourish: cave paintings, engravings on ivory and bone, and clay figurines. These artifacts seem to have been associated with sacred rituals, for they are found in the depths of labyrinthine caves, far from where daily activities would occur. Although the grave-building Neanderthalers must have been capable of fairly sophisticated communication of symbolic thoughts, only *Homo sapiens sapiens* possessed the structures of throat and larynx (voice box) that permitted full-fledged language to evolve. When this occurred is not known, but from about 50,000 years ago until the present, human evolution has been mainly cultural. The only skeletal difference between people living then and now is size, which may simply reflect better nutrition today. The superficial differences among modern so-called races are genetically trivial.

The Human Brain

Human nature – by which we mostly mean human thoughts, feelings and behaviour – depends on that remarkable organ, the human brain, about whose structure and function there are several popular misconceptions. Figure 5.1 identifies the various regions of the human brain and their major functions.

The base of the brain, or medulla, is an extension of the spinal cord that conducts signals into and out of the brain; it also regulates breathing, heart rate and other vital functions that continue even when one is deeply unconscious. Just above the medulla is a convoluted outcropping, the cerebellum, which coordinates muscular activity. The incoming and outgoing nerve tracts from the medulla continue through the midbrain to terminate in the thalamus, a relay centre to the cerebral hemispheres of the forebrain; along the way, information is sorted out and filtered so that only 'important' signals reach the cortex. The two large hemispheres, the hallmark of the human brain, are connected by a huge bundle of nerve fibres that keep the two halves in touch with each other. Underneath the convoluted outer region which forms the cerebral cortex is a diffuse limbic system which is interwoven with the basal ganglia. (We shall speak more of these regions in a moment.) Under the thalamus and in close touch with the limbic system lies the hypothalamus, where the endocrine system interfaces with the brain. Many of our basic drives – thirst, hunger, sex – are initiated in this region.

There are two main ways to study the functions of the human brain. One is to observe humans with various brain lesions; the other is to study animal behaviour after experimentally either exciting or destroying specific regions of the brain. We begin with the first of these.

Damage to circumscribed areas of the cortex can interfere with vision, hearing, smell, motor activity, or sensory inputs from the skin. There are also two areas located only in the left cortex that have to do with speech. Broca's area, in the front part, has to do with forming sounds in the larynx and mouth; Wernicke's area is further back and has to do with the organisation of language. Lesions in these specific areas cause the impairments that sometimes follow strokes.

Other, more diffusely located brain functions are also rather asymmetrically distributed between the two hemispheres. In addition to language skills, the left hemisphere is important in sequential processing and analysis of information – it is where 'logical thought' occurs. The right hemisphere perceives and processes information in a spatial and holistic fashion, being more important in such creative activities as insight, painting and musical skills. Normally the two hemispheres work together, exchanging information between them via a huge bundle of connecting fibres, the *corpus callosum* (Figure 5.1). Interesting experiments have been done on 'split-brain' patients, in which the superficial connections between the two hemispheres are almost completely severed. (This operation often relieves severe epileptic seizures without seriously affecting mental

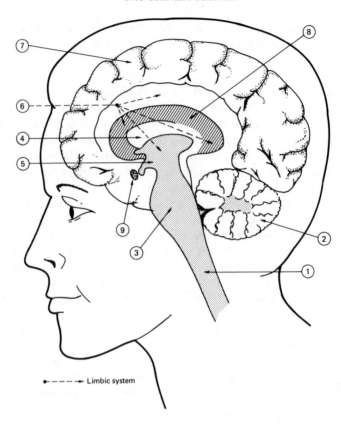

Source: M. E. Clark.

Figure 5.1 The human brain.

The following main functional regions of the human brain are seen in midsection:

1. *Medulla*: regulates vital functions
2. *Cerebellum*: coordinates muscle movements
3. *Midbrain*: conducts, and filters, incoming and outgoing signals
4. *Thalamus*: relay station from midbrain to cortex; many nerve connections here
5. *Hypothalamus*: seat of basic drives: interfaces with endocrine system, the chief chemical regulator of the body
6. *Limbic system*: a diffuse peripheral area (indicated by branched lines) connecting the cortex, thalamus and hypothalamus; plays a role in memory and sense of 'self'
7. *Neocortex*: the primate region of the brain which is the most highly developed in humans; comprises the paired central hemispheres – the 'right' and 'left' brains. Besides specific sensory and motor functions, the cortex carries out thought and memory functions
8. *Corpus callosum*: bundles of nerve fibres connecting right and left halves of the neocortex
9. *Pituitary gland*: the master endocrine gland, that connects and coordinates brain and hormonal activities

function.) Such a patient, when viewing an object only with the left eye, cannot say what the object is. This is because information from each eye, passing across a much deeper connection in the brain, projects onto (is 'seen' by) the opposite cortex. Hence, left eye information reaches the right cortex, which however, lacks the speech centres to 'tell' what it saw, and in split-brain patients the information cannot be relayed, as it normally would be, to the left cortex so it can 'speak'.[27]

Beneath the cortex lie the limbic system and basal ganglia of the forebrain, the focus of widespread misconceptions about 'human nature'. Because in animals stimulation through electrodes inserted into specific areas of the limbic system leads to passivity, or to penile erection, or to aggressive behaviours, some have assumed that these are the specific and normal functions of these regions not only in the animals studied but also in similar parts of the human brain. Furthermore, the basal ganglia of the human brain are claimed to be homologous in function to the entire reptilian forebrain. It is thus popularly held that human brains are an evolutionary accretion of layers – like extra sweaters added on a cold day – which act almost independently: the basal ganglia where our 'reptilian' selves lie; the limbic system, where we have the emotions of lower mammals; and the outer cortex where our truly human selves reside. Thus neuroscientist Paul MacLean says:

> It cannot be over-emphasized that these three basic brains show great differences in structure and chemistry, yet all three must intermesh and function together as a *triune* brain ... In popular terminology, the three sub-brains might be regarded as biological computers, each with its own special form of subjectivity, intelligence, time measuring, memory, motor, and other functions ... The wonder is that nature was able to hook them up and establish any kind of communication among them.[28]

This interpretation of evolutionary homologies is unfortunately misleading. Naturally the human forebrain has specialised areas of function, as do many other organs of the body such as the kidneys, spleen and gonads. But organ systems do not evolve as independent parts vying with one another for supremacy: they evolve as integrated wholes. It is misleading to interpret repetitiveness in human behaviour or our tendency to cling to established moral and legal codes as 'evolutionary memories' of the repetitive movements characteristic of reptiles, or an erection as a subhuman act because other mammals have similar behaviour controlled by similar regions of the brain. Snake-like and shrew-like propensities are not lurking in the deeper recesses of human minds. The fact that homologous structures appear among evolutionarily related species does not mean that they remain static in their function. Quite the opposite. Evolutionary biologists constantly discover that old structures are turned to new uses. The lobe-fins of certain fishes became the legs of their amphibious descendants, while part of the second gill arch of those fishes became the three little bones of the middle ear of land animals. Surely the brain evolved with time, too. Indeed, other information regarding our supposedly 'mammalian' limbic system gives it a very broad function in humans, modulating many aspects of our behaviour and giving us a clear sense of self – hardly an animal-like trait.

Much has also been made of the fact that there are areas in the rat brain – and probably our own if we could study it – known as pleasure and pain centres. A rat with an electrode implanted in one of its hypothalamic pleasure centres will repeatedly push a lever to stimulate itself, forgetting to eat and drink for hours.[29] But it should not surprise us to learn that there are localised regions in the brains of rats or people which are the internal receptor sites of 'good' and 'bad' sensations. What would surprise us is if a rat with its cortex removed (which does not immediately kill the animal) could continue to give itself hypothalamic pleasure shocks. It cannot. Our 'lower' centres require our 'higher' centres to make sense of them. We function by means of our *entire* brains. Thinking and feeling, cognition and emotion, are inseparable and simultaneous activities that our minds undertake. In his book on the human brain, English neurobiologist J.Z. Young quotes the following observation by two child psychologists:

> [A] human being . . . is not partly cognitive and partly emotional but in being and substance is both of these all of the time and indivisibly.[30]

Finally, big-brained mammals are not a unique pinnacle in an evolutionary hierarchy of living beings. The big predatory fishes of the oceans, the great white shark and the large tunas, each independently evolved well-developed cerebral cortexes, as large and complex as those of sophisticated hunting mammals.[31] Without the underwater equipment needed to track these animals, which swim faster in the ocean than most animals can run on land, we do not yet know much about how they use their large brains. But we mammals are not unique; big brains evolved three separate times during vertebrate evolution.

HUMAN NATURE

If on a clear, moonless night one gazes intently skyward at the Pleiades, it is hard to tell how many stars make up the 'seven sisters'. Yet if one looks slightly aside, the whole constellation suddenly appears much sharper and more distinct. This trick of the eyes has a perfectly simple explanation. The central part of the retina is less sensitive to dim light than are the more peripheral regions. Something similar seems to happen when we concentrate too hard on a complex subject. As we limit our focus to this or that small aspect, trying to explain it in ever greater detail, we lose sight of the whole. Nowhere is this more apparent than when we discuss that most controversial subject of all – human nature. Listen to the English philosopher, Mary Midgley:

> The roots of [our] difficulties are not new. People's understanding of themselves has always been fragmentary ... What is new in this century, however, is the contribution of academic specialization to the splitting process. Mind and body, scepticism and stoicism, god and beast, are now topics belonging to different disciplines. Each is supposed to be discussed in its own appropriate terms.[32]

Today, each discipline contributes its own specialised view of human nature: neuroanatomy, neurophysiology, endocrinology, pharmacology, psychology, ethology, anthropology, genetics, evolutionary biology, history – and last, but scarcely least, philosophy. Indeed, philosophy, which once arched over academic thought, providing a unifying framework, has seen parts of itself pre-empted by various specialties. The result is detailed confusion. Human nature is totally determined; is totally free of any and all constraints; is self-centred; is social; is shaped from within; is shaped from without, and so on. Modern reductionist thought is making no headway at all. Human behaviour and human goals cannot be explained in terms of DNA, or brain parts, or sequential conditioning, or any other single agent. Says Midgley:

> The only remedy for this fragmentation is to stand back and take a wider view of the key concepts as parts of a whole.[33]

Emotion and Instinct

A good place to begin is with the age-old problem of 'emotions'. Ever since Plato, emotions and reason have been treated as two independent warring factions.[34] Within the Western tradition emotions tend to be treated as 'the beast within', lowly, animal-like, genetically imposed, and generally 'bad'. Reason, by contrast, is uniquely human and therefore 'superior'; being the source of the intellect, of rational thought and of freedom of choice, it is 'good'. But if this dichotomy is true, what about the unfeeling genius who skilfully plots and deftly carries out a heinous murder without a single feeling of remorse, anger or other passion? We rightly call such a person *inhuman* and have far less sympathy for her or him than for the wife or husband who shoots an unfaithful spouse in a fit of jealous rage. If emotions make us human, why do we belittle them and fear them?

Emotions are human. They are feelings – our inner sense of what pleases and displeases; they create tastes and provide motives. They are, in fact, instincts,* inborn tendencies to behave in ways appropriate to the survival of the human animal. We readily accept the idea that instincts largely direct the behaviour of dogs, cats, birds or ants; they do not have to be acting 'emotionally' for us to agree that they are behaving according to certain inner drives or needs. Why should this not be true of humans as well? Is not human behaviour also motivated by inner needs – our instincts? Extreme displays of emotionalism are not necessary for us to recognise in ourselves and others longings for food, parental care, companionship, sex, affection and social acceptance, and to understand feelings of anger and frustration when they are not met and of contentment when they are. To convince ourselves beyond doubt that emotions really are 'human', we need only remember that two of our most common emotional displays – laughing and crying – are unique to the human species!

In both animals and humans, exaggerated emotional displays occur only when one drive temporarily supersedes the others. One becomes oblivious to all else when in the midst of an orgasm, or when one's car is skidding out of control on a wet freeway. More commonly, we experience simultaneous sensations, such as hunger, loneliness and drowsiness, which often cannot all be gratified at once. We must choose whichever one at the time seems most important, a choice that often gives rise to inner conflict and discomfort. Ethologists have described the same kind of conflicting drives in animal behaviours.[35] It is the *manner in which conflict is resolved* that differentiates humans from other animals; the human ability to imagine the probable consequences of a particular behavioural choice makes the difference.

The Role of Reason

Most animals avoid extremes of any one instinctive behaviour because of countering instincts or inhibitions, which thus adequately regulate its actions. The regulation is not perfect, of course; in battles over social precedence for mates or food, a wolf sometimes kills another wolf, a lion another lion, or a gorilla another gorilla. But *normally*, the losing individual makes a submissive sign which appeases its antagonist. Although such submissive acts are also a part of human behaviour – smiles, for example, or bowing one's head to expose the vulnerable back of the neck – they apparently are not enough. We seem to need in every culture a behavioural code to restrain aggression and regulate our various instinctive behaviours.[36] Why should this be?

*This broad definition of instinct as a heritable propensity to act in certain ways should not be misconstrued as inflexible or predetermined behaviour comparable, for instance, to the building of a spider's web. For clarification, see Chapter 1 of Midgley's *Beast and Man*.

Like other animals, human beings have evolved a whole range of genetically determined instincts – curiosity, love, appeasement, aggression, sex drive – which have adaptive survival value *when actuated in the right proportions*. The function of rational thought is not to suppress or eliminate these innate propensitites – one cannot imagine a human being without them – but to select, temper and modulate conflicting drives to meet the needs both of the moment and of the future. It is precisely this ability consciously to prioritise conflicting desires that gives humans a behavioural flexibility not available to other animals. *This is what makes us human*. Because we can predict the outcome of alternative actions – not just in the short run but also in the long run – we can deliberately choose amongst them. This potential, which arises from our large cerebral cortex, allows us to adapt our behaviours to widely diverse situations, and was no doubt a major selective force in the evolution of human thought.

This ability to choose among alternative behaviours gives us our sense of freedom, with its inescapable concomitants, the burden of moral choice and the responsibility for one's actions. In *Homo sapiens*, cognitive reflection is designed to be the integrator of inner, conflicting drives. It is thus Midgley's argument – and I find her conclusion inescapable – that our moral obligations are shaped in large part by our natural feelings.[37] We are not incorrigible egoists because those feelings bind us inextricably to our surroundings and especially to other people. Only when we actively suppress this natural bondedness – as our modern Western worldview so strongly encourages us to do – do we get into trouble. Most of our current difficulties can be traced to denial of both our relatedness to Nature and our need for strong social bondedness. This thread runs throughout the book.

As Midgley concludes, it is wrong to say that emotions are a weakness and only intellect counts: 'Normal emotions are as necessary for morality as thought is'.[38] No matter how intelligent, a person without sensitivity for others is an emotional cripple – a psychopath – not a strong, silent hero. Our emotional capacities have a purpose and cannot be safely suppressed. Indeed, our conscience is merely our conscious awareness of the requirements of our own natures. And one absolute requirement of human nature is for culture.

Culture Is 'Natural'

One of the most powerful needs of human beings is to be closely bonded with others of one's species. Virtually all primates have this requirement to some degree. For example, when baby rhesus monkeys are placed alone in cages, they crouch fearfully in a corner, ignore their playthings and become deranged. But if a furry cloth 'mother' with painted features is placed in the cage, the baby explores and plays normally, running occasionally to hug this surrogate parent for reassurance.[39] Such experiments, of course, have not been performed on human infants, but it is clear that one factor in the development of autism in children is too little interpersonal contact. As ethologist and Nobel laureate Nicolaas Tinbergen observed:

> Those therapies [for autism] that aim at the reduction of anxiety and at a restarting of proper socialization seem to be far more effective than, for instance, speech therapy *per se* and enforced social instruction, which seem to be at best symptom treatments, and to have only limited success. Time and again treatment at the emotional level has produced an explosive emergence of speech and other skills.[40]

The power of peer pressure is another indicator of the importance of social intercourse and acceptance to the human psyche. Being black-balled or exiled from one's society is a shameful fate. The Greeks invented ostracism. The English call it 'sending someone to Coventry' when a group refuses to speak to, or even acknowledge the presence of, a fellow member who has offended it. Reduction to the status of non-person is the worst punishment a society can impose on one of its own.

It is often claimed that the basis for strong human ties originated in permanent pair-bonding based on sexual activity. Sigmund Freud insisted that sex is the strongest human instinct, an idea that gained support from the biological fact of continuous sexual receptivity in human females. But Freud was reacting against the heavy taboos on sexuality that prevailed in nineteenth century Europe. Moreover, women are not as uniformly receptive as once thought, but like other cyclical mammals have heightened receptivity around the time of ovulation. In addition, in other primates in which pair bonding is strong, such as the siamang and the gibbon, copulation is very infrequent. There is thus no reason to believe that sexual activity is the main basis for the strength and persistence of human pair bonds.[41] As the behaviourist, Eibl-Eibesfeldt, has pointed out, Freud had it all backwards; the roots of love are not in sexuality, but in parental care:

> From what has been said so far, it should be clear that in point of fact, many behavior patterns which are regarded as typically sexual, such as kissing and caressing, are in origin actually actions of parental care. We remind the reader of this because Sigmund Freud, in a strikingly topsy-turvy interpretation, once observed that a mother would certainly be shocked if she realized how she was lavishing sexual behavior patterns on her child. In this case Freud has got things back to front. A mother looks after her children with the actions of parental care; these she also uses to woo her husband.[42]

Important as human contact is, however, we need culture for more than just its psychic social rewards. Consider for a moment an imaginary experiment with three hunting–gathering cultures, the Eskimo in the Arctic, the Auca in the Amazon jungle, and the !Kung-San in the desert bushlands of southern Africa. If a baby at birth were taken from each culture and transported to one of the others, it very likely would thrive. Had it been, say, an Eskimo male child by birth, it would grow up to have not the seal-hunting habits of an Eskimo, but would instead be skilled at aiming curare-tipped darts from a blowgun at howler monkeys high up in a tree. It would learn the skills necessary to survive in its new environment. But instead of transporting just one infant, suppose we transposed entire populations from one area to another. Overnight, all Eskimos find themselves in the jungle, the Auca are in a dry desert, and the !Kung in a cold, white world of snow and ice. None of these cultures would survive. The human brain, for all its cleverness, cannot invent out of thin air a whole new set of survival skills – technologies and social institutions, let alone psychological attitudes – appropriate to a new environment. (Something akin to this would happen after a nuclear holocaust, although few seem to have appreciated the fact.)

This absolute requirement for cultural transmission of survival skills to meet the needs of each new environment has resulted in the evolution of an enormous diversity of human cultural patterns around the world. Even in the same environment, neighbouring societies may evolve in quite different yet equally successful adaptive patterns – a point taken up in Chapter 6. So great, then, is the potential diversity of human action that even if it were somehow possible for a person to grow up in isolation, the mind boggles at how such an individual could formulate a meaningful pattern of life activities. As anthropologist Ruth Benedict has said:

> It is in cultural life as it is in speech; selection is the prime necessity. The numbers of sounds that can be produced by our vocal cords and our oral and nasal cavities are practically unlimited ... But each language must make its selection and abide by it on pain of not being intelligible at all ... In culture too we must imagine a great arc on which are ranged the possible interests provided either by the human age-cycle or by the environment or by man's various activities. A culture that capitalized even a considerable proportion of these would be as unintelligible as a language that used all the clicks, all the glottal stops, all the labials, dentals, sibilants, and gutterals ... Every human society everywhere has made such selection in its cultural institutions. [43]

It is culture, emerging out of our innate instincts, that provides a mental framework – a worldview – that makes sense of the world. Culture gives meaning to our lives; it tells us who we are; it creates our self-awareness. Culture provides the necessary context so that we *can* function and *within which we are free to act*. An absence of culture, of a social framework in which to participate, does not lead to greater freedom – it leads to chaos and inaction. And cultural plasticity allows humans to live in almost every environment on Earth. Genetic specialisation leading to a limited repertoire of fixed behaviours suited to a single environment is not the human way. Cultural adaption is faster and more flexible.

Culture and Violence

Any discussion of human nature would be incomplete without an inquiry into the sources of social violence. There is almost universal despair that war can ever be ended; it is 'too deeply embedded in our nature'. I think this view is unnecessarily pessimistic, but unless we analyse its origins it could become a self-fulfilling prophecy. If people believe war will end humankind, it very probably will. What follows is a brief outline for such an analysis.

Human beings *above all else* are bonded social beings. Our ancestors were social before they were human: belonging, affection, attachment, physical contact, cooperation, love – all are hominid hallmarks. Social violence results when this bondedness is threatened, when social attachments are endangered, and it finally becomes institutionalised when social bonds are weak or absent. A few examples will illuminate this premise.

Consider first one-to-one personal violence, which among both human and non-human primates almost always results from the threat of a rival to destroy existing bonds: the rival lover threatens sexual bonds; the new sibling threatens child–parent bonds; the strange interloper threatens group bonds. In human societies, the rapist and the thief often have experienced prior social ostracism themselves. Violence in the family is almost always a response to real or imagined rejection, and both cause and response are frequently passed from one generation to the next. A society that has 'a place in its heart for everyone' is a non-violent society.

When primate societies come under pressure, however, either through population growth or through external forces constraining them, their social structures often change, creating situations that tend toward violence. For example, among the populations of langur monkeys in India, the pressure of humans makes a great difference. In the north, where humans are sparse and do not compete much with the monkeys, the troops are small, egalitarian and free of internal strife. Further south, where the human population is much more dense, the monkeys live in large, rigidly structured, hierarchical groups marked by frequent internal squabbling. These distinct social patterns are environmentally, not genetically determined.[44]

A parallel situation occurs for primitive human societies. The archeological record does not provide us with any sound evidence of warlike behaviour prior to the emergence of agriculturally-based civilisations at the end of the Neolithic era. In two contemporary cases, it can be shown that organised warfare among shifting agricultural peoples started in part through outside pressures. Among the much-cited 'fierce people', the Yanomamö, a primitive Indian group living in the Venezualan rainforest, there has been rapid population growth during the last century, a consequence of the introduction of steel axes and machetes. This led to the development of inter-village fighting and the spreading out of the population, so that it now also experiences pressure from outside peoples as well. But not all villages participate in fighting; only those in the most crowded areas are highly warlike. In remote areas, where pressures are less, the Indians are peaceable.[45] A similar case occurred in New Guinea, where fighting among villages sharply increased in frequency and ferocity after Western steel axes were introduced in the 1930s. Axes and other forms of European wealth disrupted the normal patterns of trading among rival clans; brides and pigs became scarce in some areas and tensions rose. Moreover, the new axes gave men more free time from subsistence labour to engage in prolonged and widespread warfare.[46]

Small primate societies where individuals know one another seem to remain non-violent, but when strangers enter a territory they are often threatened or attacked. In the Gombe game reserve in Tanzania, for example, male chimpanzees regularly patrol their territorial boundaries, attacking and even killing strange males who wander into the area.[47] (Fatal violence *within* the group occurred only after human disturbance of the society.) And when baboons brought from different parts of the world were suddenly thrown together in an enclosure at the London zoo, a terrible free-for-all battle ensued that lasted several days. Many individuals of both sexes were killed.[48] Social cohesion apparently can exist only among individuals who recognise and are familiar with one another. Strangers are a threat to a society's internal bonds.

This problem of 'strangeness' became acute for human societies when, at the end of the Neolithic era around 5000 years ago, food production had increased enough to permit not only permanent settlements but also large societies. During the earlier stages of the agricultural transition, productivity was low, the villages were small, and all were engaged equally and virtually full time in the communal effort to survive. But as agricultural skills improved, populations grew and villages of 100 or 200 became cities of thousands. People became strangers, and the excess of food permitted specialisation of jobs. The social structure changed from egalitarian to hierarchical. As societies grew in complexity, they required a managerial elite: law-givers (kings and priests), book-keepers (scribes and accountants), and protectors (a professional soldiery). This non-producing bureaucracy held the productive unit together, and also organised the construction of infrastructure: canals, roads, terraced fields.

The elite class established its authority to do all this through myth and religion, creating a worldview for society in which the old one-to-one bondedness was replaced with emotional ties to a powerful god-like king. Allegiance to a central figure of authority superseded allegiance to the small, lifelong kinship group in people's lives. The abstract state took the place in human hearts once reserved for the personally known members of the tribe and the clan.

A second complementary factor was the establishment of ritualised warfare in the form of blood revenge. Deaths among one's own warriors arouses communal wrath that has to be avenged upon the enemy. Again, the rupturing of bonds is the basis of violence – in this case a violence that becomes mutually reinforcing over time and is finally institutionalised in elaborate, quasi-religious rites. (The massive American anger aroused when Iranian and Lebanese terrorists seized and held United States citizens hostage in the 1980s exemplifies how easily threats to 'our own' can escalate to an emotional level disproportionate to the cause.) The ritualisation of tribal warfare has many examples, among which are the infamous Huron Indians who tortured their captives for days before killing and eating them. Yet archeological evidence shows that such cannibalism arose only around AD 1300 among these Indians, at about the same time that ritual killing and eating of prisoners of war and other victims reached its height among the civilisations of Mesoamerica.[49] Increasing hemispheric population pressure may be the common thread.

The origins of ritualised warfare thus seem to depend upon two apparently incongruous causes. There has to be *enough* surplus wealth to provide the luxury of organised battle: a population that must gather food daily cannot afford prolonged warfare. But there must also be some initiator of ritualised violence, either population pressure or a perceived absence of wealth or power needed by the elite to maintain social cohesion. Once an external threat to social bonds is identified, however, social cohesion is easier to maintain. Ultimately the social hierarchy – later called 'the state' – became the means for extracting the excess wealth generated by the society as a whole and, after investing some of it in infrastructure (roads, aqueducts) or great monuments (pyramids, cathedrals), funnelling the remainder to the most powerful. Such a social arrangement is possible for a protracted period only in a large, anonymous society where cohesion is maintained either through ideological charisma – the well-known 'love of country' – or through internal coercion. In either case, those being exploited are powerless to create a unified force against the privileged few unless they can generate a strong new ideology that is capable of *bonding a mass of strangers together* in a single cooperative movement.

In this view of history, originally propounded by Lewis Mumford, 'civilised' warfare is a special consequence of highly organised megasocieties – or, as he called them 'megamachines'.[50] To maintain cohesion in such giant societies, men and women bonded their hearts and minds to grand ideologies instead of to kinsmen and ancestors. It is still necessary today, and when leaders find the allegiance of the masses slipping, they have only to insert the fear of a foreign 'enemy' ideology into peoples' minds to re-establish cohesion. Modern warfare, then, seems to arise less out of the competition for scarce resources that evolutionary theory would demand than out of a relentless drive by distant leaders to retain the loyalty of a loosely bound mass of followers. As shown in Part IV, it is the need for an 'enemy', an 'evil empire', by both modern superpowers that has fuelled the arms race, even as it destroys the economies of each and threatens the extinction of all humankind. We begin to glimpse some of the roots of our contemporary predicament.

SOME CONCLUDING THOUGHTS

Evidence from paleontology, from ethology, from extant hunting–gathering peoples, and from modern psychology all indicate that the common 'human nature' we share is one of strong interpersonal bondedness arising from our long period of infant dependency. As social animals *par excellence* we need 'to belong' above all else, and in the course of belonging, we come to share common beliefs and values that give meaning to our lives. At the same time, we also acquire culturally transmitted survival skills. These two cultural phenomena create the patterns of our lives, our worldviews, and so passionately are we attached to them that external threats are met by violence. Our human dilemma, then – how to retain our passionate attachment to cultural meaning without extinguishing ourselves in a global holocaust – will be resolved only if we become conscious of the profound emotional role our worldviews play in our lives.

Much of the rest of this book is taken up with examining the origins and implications of the major ideologies, belief systems and worldviews of the modern world. The next chapters explore the great range of human cultures that have existed and do exist; whatever our innate nature is, there is enormous room for choosing our values and our life-styles. We cannot claim that some tyrant genes are forcing us to overpopulate the world, to use up its resources, or to threaten each other with nuclear extinction. These are institutions and rituals that we have learned, and, with the help of insight combined with dialogue and determination, we are free to change them as we pass them along to our children.

6   The Cultural Spectrum

One day, without transition, Ramon broke in upon his descriptions of grinding mesquite and preparing acorn soup. 'In the beginning', he said 'God gave to every people a cup, a cup of clay, and from this cup they drank their life... They all dipped in the water, but their cups were different. Our cup is broken now. It has passed away'... The old man was still vigorous and a leader in relationships with the whites. He did not mean that there was any question of the extinction of his people. But he had in mind the loss of something that had value equal to that of life itself, the whole fabric of his people's standards and beliefs. There were other cups of living left, and they held perhaps the same water, but the loss was irreparable.

Ruth Benedict, 1934[†]

INTRODUCTION

Culture! Westerners often use this word to mean a taste for the fine arts, music and other aesthetic matters. But it has a much broader meaning, namely, the shared values, language and traditions that define a particular group of people, be they Australian aborigines, black Americans, or the ancient Greeks. Culture is learned as a child, and as children we each learned from those around us a particular set of rules, beliefs, priorities and expectations that moulded our world into a meaningful whole. *That is our culture.* It tells us what is correct, expected, normal and right. It explains the world for us. It gives meaning and purpose to our lives. Culture is the socially determined mental framework in which we live. It is our *Weltanschauung*, our worldview, our abstract conception of reality.

Without culture, we would be adrift, unanchored, with no way to fathom the behaviour of those around us and no way of knowing how to behave towards them. We humans have an almost unlimited potential for action. Our instincts are flexible tendencies, not rigid behaviour patterns: 'they deal in ends, not means'.[1]

Not surprisingly, we find in all cultures institutions that serve our shared instinctual needs: the selection of mates, the provision of child-care, the obtaining of food, the curing of illness. Yet the specifics of these institutions and how people feel about them differ widely from one society to the next. When rainfall is needed, Hopi Indians hold rain dances while non-native Americans push for irrigation projects or even cloud-seeding. When two Russian men walk arm-in-arm, they are sharing friendship; in Britain they might be suspected of homosexuality. The Japanese find it

156

unnecessary to mention the obvious and consider the Western habit of small talk as so much babbling. And so it goes. What is normal and expected in one culture is peculiar, amusing, weird or morally wrong in another. No wonder the French word for foreigner is *étranger* – someone who is strange.

Whatever culture we belong to, we naturally consider its worldview 'normal' and use it as the standard by which we judge others. Our ethnocentrism is unavoidable; each of us is raised immersed in a particular culture. And that culture's vision is both the individual's way of knowing who he or she is in the universe, and the society's way of maintaining its cohesiveness. Unfortunately, our universal habit of ethnocentrism too often gives rise to misunderstandings – and to wars – when two cultures interface. *We have now reached a stage in human evolution when it is absolutely necessary for all cultures to be aware of the existence of ethnocentrism and to school themselves to accept a world with a diversity of value systems.*

Why we should strive to retain our cultural diversity is addressed later. At the moment we focus on our absolute requirement for culture. This chapter samples the great spectrum of cultures that humans have created, showing how no *one* set of values or *single* way of behaving is imposed on us by 'human nature'. It shows too that when environments change, so must cultures. Those that fail to adapt, die out. Its ultimate lesson is that the Western worldview which drives us to destroy our environment is no longer adaptive: it lacks appropriate values and goals; it has grown obsolete. By examining a wide variety of other societies, we see that other possibilities exist and that we *can* redesign our society; we also may discover what qualities to seek as we evolve new social institutions.

Because they are easier to understand, this chapter focuses mostly on hunting–gathering and early agricultural societies. By studying them, one can more easily envision the two essentials that any culture, including our own, must provide if it is to continue: (1) successful adaptation to the environment and (2) social cohesion. At the end of the chapter these ideas are applied to cultural change and the functioning of modern pluralistic societies.

WHAT ANTHROPOLOGISTS DO

Unlike other social scientists who study one aspect of human behaviour in depth, anthropologists try to assimilate, as a child would, the entire value system – the total worldview – of a society. Since no single society can make use of all possible human behaviours, it must *select* and *choose*. Each culture must meet basic individual needs, but they are met in quite different ways. It is the mental framework – one might almost say the 'logical rules' – by which each tradition defines proper behaviour, that the anthropologist tries to discover. Although value systems and the moral codes derived from them are widely divergent, that does not at all mean they are dispensible. As historian Will Durant observed:

> A little knowledge of history stresses the variability of moral codes and concludes that they are negligible because they differ in time and place, and sometimes contradict each other. A larger knowledge stresses the universality of moral codes, and concludes to their necessity.[2]

A value system serves to fuse individuals into a socially cohesive whole, which then survives and evolves as a unit. Once such a social system develops, it tends to be conserved, changing only gradually over time. Its values are maintained by customs, rituals and beliefs, that both guide the individual and give meaning to her or his life. Unlike last year's clothes, culture cannot be discarded for whatever new fashion comes along; it is a deepseated part of one's thinking and valuing processes. Most members of a society accept its norms, its rules and regulations, not because they are coerced but because 'that is how things *ought* to be'. The major constraints on behaviour are internal to the individual, not external. It is this underlying value system that anthropologists seek to discover when they study another culture.

The Elements of Culture

Figure 6.1 shows the three interacting primary components of a culture. The *material culture* which depends directly on the ecosystem, includes food, clothing, shelter, and other artifacts. The *social culture* comprises kinship and family groupings and patterns of political and economic arrangements. Sometimes these are coincidental, but other times they are not. Economic and political institutions may be separate from each other, especially if political power entails magical or religious rites.

This leads to the third component, the sacred or *ideological culture* – the belief system and myths of a society. These establish the various ceremonies which mark the changing seasons and the stages of human life. The seven sacraments of Christianity, for example, celebrate stages in an individual's religious development. Many modern holidays have their origins in ancient rituals: Hallowe'en is the old Celtic new year, and May Day, now celebrated in Europe as Labour Day, was the Celtic fertility festival. Almost every society has a midsummer and midwinter festival of some kind. Recurring rituals that reinforce social wisdom help keep society 'on the right track' and guide its major decisions. They are a cultural memory bank from the past, an accumulation of wisdom whose origins are long forgotten.

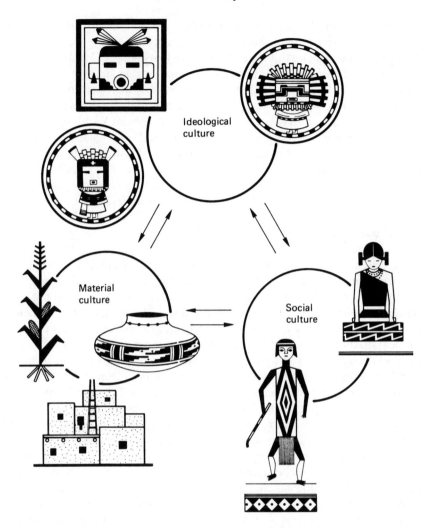

Source: Basic idea from A. Pacey, *The Culture of Technology* (Cambridge, MA: MIT Press, 1983) Figure 1, p. 6; Hopi motifs by Anne-Marie Malotki are in E. Malotki, *Hopitutuwutsi: Hopi Tales. A Bilingual Collection of Hopi Indian Stories* (Flagstaff: Museum of Northern Arizona Press, 1978) various pages.

Figure 6.1 Elements of culture.

Material culture depends on environmentally available resources; social culture comprises the kinship and political patterns; ideological culture embodies the belief system and myths. None of these, however, is independent of the others; all three elements interact to create a meaningful worldview.

Preliterate peoples have an extraordinary oral memory: the Iroquois peoples are able to recite their ancestors back for seven generations and Australian aborigines exhibit a computerlike knowledge of kinship relations out to their fifteenth cousins, including everyone living in a radius of a hundred miles. The Book of Genesis contains what was once an oral record of a long succession of Judaic princes. And, before they were written down in the thirteenth-century *Red Book of Hergest*, ancient Welsh epic poems, dating from the Bronze Age around 2000 BC, had been passed orally from minstrel to minstrel, recording the people's sacred history.[3] Human beings have always spent much time memorising the intricate details of their folklore and myth, and it continues even today as children learn moral values from their parents and their nation's history in school.

'Emic' and 'Etic' Aspects of Culture

By watching, asking, even sometimes doing what others are doing, an anthropologist gleans a partial understanding of the world as seen by another culture. Often the observer realises that these poeple do not 'see' things the same way that she or he does. Westerners, for instance, emphasise products, while native Americans are concerned with process; we deal in 'things', they, in 'doings'. Such differences affect every aspect of life and thought in the two cultures. Westerners also think of time as linear and finite – there *was* a beginning and there *will be* an end. Yet many other people see time as cyclic, as constantly repeating itself. Similarly, Westerners organise thought in terms of conflicting dichotomies: black versus white, success versus failure, right versus wrong, up versus down, forward versus backward. The East sees two divisions, also, but theirs are complementary, not conflicting categories. For native Americans, things come in fours. There are, for example, four directions, each with its sacred function, it psychological characteristics, and its own special colour.

We humans are seldom aware of the subtle ways our minds divide up, order, and arrange the world, unconsciously constructing our thought patterns for us as we grow up. These become our reality, our inner world; yet we are hardly aware of it and cannot explain it. Thus when an anthropologist studying the Havasupai Indians along the Grand Canyon asked 'But why? Why do you do that?' his informant patiently replied, for the dozenth time, ever so slowly: 'That's–the–way–we–do'. That's the way we cure disease, that's the way we plant corn, that's the way we calculate who our kin are.[4] There were no rational explanations to be given. The Havasupi act out of tradition – that is how it has always been done; there is no other explanation. Surprising as it may seem, much of Western reality is equally inexplicable and can be just as incomprehensible to those of another culture as theirs is to us.

The assessment that a people makes of its own behaviour is sometimes called the *emic* or internal explanation. In the Havasupai case, the behaviour is simply a given. At other times, mythological or religious explanations exist. For instance, in the Old Testament book of Leviticus, Chapter 11, pigs, shellfish and all insects (except locusts and grasshoppers) are proscribed as food for the Israelites because they are 'unclean'. Similar religious explanations are found among other cultures for their behaviours – but these reasons are still emic; they are the reasons perceived by the people themselves for what they do.

Sometimes an anthropologist can guess at some practical reason for a particular behaviour. Such an *etic* or external explanation suggests how a traditional behaviour has an adaptive function *via-à-vis* the environment or how it helps bind the society together and reinforce its social patterns. It may do both.

Take as an example the foods proscribed in Leviticus. It can be argued that utilising them would have been ecologically unfavourable for people living in pre-Biblical times when these 'laws' first evolved. Because pigs eat the same food as humans they compete directly for resources. When food is scarce, it makes more sense to eat the food oneself rather than feed it to pigs and then eat them. Likewise, in the Middle East the energy expended gathering small, scarce organisms like shrimps, other shellfish, and insects (except the all-too-abundant swarms of locusts) could result in a negative net energy gain. By proscribing these foods the Bible ensured that valuable time and effort were not expended on non-productive food resources. Describing a food as 'unclean' puts a great prohibition upon it; in Leviticus, all proscribed foods are so labelled. The presumption that pork was included because it can be infested with disease-producing trichina parasites is probably incorrect. The parasites are easily killed by adequate cooking which Biblical peoples almost certainly employed. Besides, dietarily permitted meats such as beef can also contain harmful parasites. The label 'unclean' is inconsistent if argued solely on the ability of a food source to cause disease.

CULTURAL ADAPTIONS

Each society thus has a specific behavioural tradition for successfully maintaining both internal social cohesion and adaptiveness toward its environment. This tradition is reinforced by habits and customs that often exist as elaborate rituals. What aspects of life are chosen as the most important focal points varies greatly from culture to culture; for some it is puberty, for others, heroic deeds, or harvesting food, or an afterlife. In most traditions, seasonal rituals are important, marking the appropriate times for hunting, migrating to winter quarters, planting crops and so forth. Frequently, the function of a ritual given by the people, its emic function, is quite different from its etic or adaptive function. The forms a ritual takes may thus vary widely from one culture to the next, even when it serves the same purpose.

The Diversity of Ritual

Rites of passage are one of the rituals found in almost every culture, including our own. Besides religious ceremonies such as bar mitzvah or confirmation, the advance through puberty to adulthood in Western societies is marked by a number of trials and rewards. Ten or twelve years of formal schooling capped by graduation from high school or gymnasium is the most obvious, but there is also learning to drive and perhaps getting the keys to one's own car; the attainment of the legal drinking age; of the right to vote; of the responsibility to serve in the armed forces. There is the first date, the first job, the first bank account, the first credit card. But although these steps in 'growing up' occur during the period of physiological puberty they are not directly coordinated with it. The rites of passage in other cultures are often more severe and more closely associated with sexual maturation. By briefly glimpsing some of these rituals, we can discover how wide the cultural spectrum can be.[5]

In some societies both boys and girls go through special puberty training. Among the Masai tribe of East Africa the rites include surgical manipulation of the penis or clitoris, which the candidates must accept without a sign of pain. (Such operations may have the etic function of postponing fertility.) In other societies, females are ignored and only males undergo special rites. For latter-day American plains Indians, preparation for warfare was the great goal and was achieved by extensive self-mutilation; candidates cut off strips of skin or even parts of their fingers. Among Australian aborigines, various forms of mutilation were inflicted on the initiates by the older men, ranging from something as relatively painless as perforating the nasal septum or tattooing the face and body, to knocking out the front teeth, circumcision and even subincision – an extremely painful slitting of the entire length of the penis with a flint knife. The rites lasted several days and included many other trials as well.[6]

In other cultures, it is the girls who undergo special rites and the boys who are more or less ignored. Among the Carrier Indians of British Columbia, the onset of menstruation is treated as an untouchable state and young girls must live for three or four years 'buried alive', as they phrase it – living in utter seclusion in the wilderness. And the Xingu Indians of the Matto Grosso region of Brazil sequester pubescent girls in a hut for two years.[7] But not all cultures that pay attention to menarch react negatively. Apache girls experiencing their first menses are believed to have supernatural powers to cure illness and dispense blessings, and are sought out and made much of.

These sometimes drastic rites serve to bond the new generation strongly to society and to inculcate a powerful sense of social responsibility; they are taken very seriously by both young and old. Often important survival information and religious secrets are also imparted, whose importance is underscored by the painful procedures employed. But although rites of passage are treated in widely different ways, they all serve the same crucial functions of social cohesion and continuity. While every human being must pass through puberty, what she or he experiences during that time depends not on the physiology of the process but upon what cultural norms that society most prizes.

The Adaptiveness of Ritual

Examples of traditional ritual practices can provide an understanding of their etic roles in maintaining cultural cohesion and adaption to the environment, and encourage reflection on what rituals may perform similar functions in Western society.

The Tiwi of Australia

In Arnhem Land, an isolated area on the north coast of Australia that long remained unaffected by white settlers, live the Tiwi aborigines, a hunting–gathering people whose lives have changed little over 10,000 years. Their staple food is wild yams which they dig wherever they find them. Yams are supplemented by other plants and by game. About once every 20 years people congregate in huge numbers for a special fertility ceremony that lasts three days. During this time older women teach the younger exactly where to dig and how to prepare an unpalatable – in fact, a highly poisonous – species of yam that they normally ignore. It is repeatedly boiled, leached and roasted to extract the poison. None of the Tiwi can say why this elaborate ceremony is performed – yet it is, once every generation.

As it happens, this drought-resistant plant is a potential life-saving food during protracted dry spells. When asked by an anthropologist, a few Tiwi did recall hearing that their grandparents had eaten this yam years ago, at a time when Western meteorological data record a long dry period. Cultural memory, preserved through ritual, thus provides insurance against intermittent drought, even though people themselves are unaware of the connection.[8]

Potlatch in the Pacific Northwest

The totem pole makers, the Salish, Kwakiutl, Haida, Tlingit and other Indians that lived in the narrow strip of coastal land running from Puget Sound to Alaska were basically hunting–gathering peoples. But so abundant were their resources from the sea – salmon, cod, halibut and seal – that they led a settled village existence, living in much larger groups than other hunter–gatherers. Also unique was their ownership by families of fishing and gathering grounds: a piece of the river, a section of oyster-bearing mudflats, an area of forest for berries and roots. Whenever people settle permanently in large communities, they tend to develop social hierarchies and accumulate personal possessions. This social arrangement, so different from the sharing typical of nomadic hunter–gatherers, led to the acquisition of individual wealth and position by a most unusual process, the potlatch.

Potlatches were great parties to which the local chief invited neighbouring chiefs and all their relations in order to give away his wealth: salmon, blankets, canoes, and especially, engraved sheets of copper whose symbolic value derived from the prestige of previous owners. A copper once owned by several important chiefs might be worth four or five thousand blankets. The more a chief gave away, the higher his social status became. The recipients of his generosity naturally came under an obligation to him, and in turn would invite him to a party and, if they could, shower him with more than they had received. The goal was to outdo each other in generosity. Alternatively, to gain more prestige a chief could destroy his property, putting his rival under an obligation to do the same if he could afford it. As American anthropologist Ruth Benedict observed:

> Manipulation of wealth in this culture had gone far beyond any realistic transcription of economic needs and the filling of those needs. It involved ideas of capital, of interest, and of conspicuous waste.[9]

On the face of it, such a social system seems quite impractical. If one wants status, why not keep one's wealth rather than give it away or destroy it? A chief's whole clan had spent months gathering together, trading for or making the gifts. What function could the give-away potlatch serve? A glance at the map in Figure 6.2 suggests one. Wealth came from the sea, yet many people lived 40 or 50 miles inland without ready access to the coast. Potlatches, to which they too were invited (although they never had a hope of becoming important chiefs) provided a way of redistributing the wealth. It was an early form of 'trickle-down' economics, similar to the famed Moku parties in New Guinea. The potlatch assured the poorer inland clans a share of the coastal wealth. The etic function of redistribution broke down, however, when the emic function of status led to wanton destruction of wealth, a tendency not unknown in modern Western societies.

The !Kung in Southern Africa

A quite opposite form of cultural adaptation characterises the people living in the desert areas where Botswana, Namibia and southern Angola

ALASKA

B R I T I S H

C O L U M B I A

Tlingit

P A C I F I C

Tsimshian

Haida

Bella Coola

Kwakiutl

Nootka

Coast Salish

WASHINGTON

O C E A N

⊢━━━━━━━━┥
← 250 miles →

Source: Catalogue of an exhibition, *Art of the Northwest Coast*, Robert H. Lowie Museum of Anthropology, University of California, Berkeley, 26 March–17 October 1965, by M. J. Harner and A. B. Elsasser; map from p. 7. Sketch of wood carving of salmon carved by Dennis E. Matilpi, Kwakiutl. Art by R. Carruthers.

Figure 6.2 Map of Northwest coast Indian tribes.
  Map of the coast of western North America, showing extent of various Northwest Coast Indian tribes, all of whom engaged in potlatch ceremonies that helped to redistribute wealth from the sea to the villagers living further inland.

converge. The Kalahari San are peaceable hunter–gatherers who live a nomadic existence, consciously sharing with one another. The most studied group among them is the !Kung.[10]

During the wet season these people live in small bands of twenty or thirty, moving every week or two to new areas before they exhaust their plant and animal resources. Over 60 per cent of the diet is plants – nuts, tubers, melons, onions and gourds – that are gathered by the women. Their staple is the abundant mongongo nut. The men use poisoned arrows to hunt game and they set traps for birds like guinea fowl. At nightfall all food is returned to camp and shared generously. In the dry summer, which coincides with our Christmas, the small bands aggregate around permanent water holes for several weeks. The following story takes place at this time.[11]

Anthropologist Richard Lee, during a prolonged study of the !Kung, acquired a reputation for miserliness. Not wanting to alter the way of life he was trying to study, he scrupulously refrained from sharing his canned foods with the !Kung, giving them only occasional tobacco and medicines as a sign of friendship. Finally, at Christmas – a festival they had learned about third hand – he felt able to make an offering of food, the ox that had become traditional with them.

Lee went to a neighbouring Tswana herdsman and purchased the greatest, fattest ox he could find and proudly invited everyone to share it with him. But to his chagrin, during the ten days until Christmas and even during the two days of continuous and joyous feasting, his !Kung friends did nothing but complain loudly about the poor quality of his gift. 'Do you expect us to eat that bag of bones?' said old Benia. When butchering the giant beast, /gau remarked, 'Fat? You call that fat? This wreck is thin, sick, dead!' Lee, who was soon to return to America, grew gloomier as the comments continued. Finally, a Tswana who had married a !Kung woman explained it to him. Among the !Kung, no one ever brags of his accomplishments. It just isn't done. A hunter who makes a good kill must bemoan how useless it is – 'it's hardly worth hauling the miserable carcass home' – and his companions all agree. 'But why this false modesty?' poor Lee asked a renowned !Kung hunter, Tomazo.

> Arrogance! . . . when a young man kills much meat he comes to think of himself as a chief or a big man, and he thinks of the rest of us as his servants or inferiors. We can't accept this. We refuse one who boasts, for someday his pride will make him kill somebody. So we always speak of his meat as worthless. This way we cool his heart and make him gentle.

The !Kung store no food. Each day everyone returns with her or his contribution to share among the entire group. Such sharing and equality are essential for this form of livelihood, and the ritualised behaviour described here insures that cooperation continues unimpaired. A good hunter cannot become a hero, nor demand more food for himself than others receive.

CULTURAL CHARACTER

Just as individual humans have distinct personalities, so too does each culture seem to possess character traits. We speak of the dour Scots, the fiery Latins, the punctual Germans, and the polite Japanese. This does not mean that Scots never laugh, or that Germans are never late for appointments. Nor does it mean that all members of a particular culture have the same personalities. Rather, it is the common observation that every culture seems to have its own attitude, its own mood, its own aura. One culture is friendly, another reserved, a third openly hostile; in one culture, laughter is frequent, in another it is not. Some prize industriousness, others, leisure; some are competitive, others cooperative. Whatever each person's individual personality, her or his behaviour is still largely fashioned by the attitudes and values imposed by society and its institutions. This was well understood by Benedict:

> As a result of his immersion in one particular culture an individual will have a particular character structure . . . [12]

It is of course possible for a society's cultural character to change over time. One of the most recent, purposeful cultural changes occurred in China following the communist revolution. After a century of foreign imperial control, visitors described the Chinese character just prior to 1949 as marked by suspicion, despotism, sloth, corruption, venality, and the shirking of one's responsibilities to society. After several decades of political indoctrination (including loud-speakers in fields and factories), of mass participation in Revolutionary Committees, and of widespread education, the Chinese personality has more recently been described by outsiders as hard-working, thrifty, honest, cooperative, productive, and nationalistic. [13] Obviously, there are deeper aspects of the Chinese character that neither imperialism nor Maoism could alter, particularly an optimistic outlook, an appreciation of Nature, and a strong sense of family; yet it is clear that cultural attitudes and patterns do change over time.

Two cultures of quite different character may be equally well adapted, although in a different fashion, to the same environment. Mere chance – or perhaps an unremembered event deep in the past – fixes one or another 'personality' on a given culture. But once fixed, the cultural pattern provides cohesion for the group. Its members share the same meaning of life; they approve and disapprove of the same things. In short, individual personalities flow along with that of the culture. A few examples will illustrate this point.

The Dobu

On several of the volcanic islands of the D'Entrecasteaux archipelago east of New Guinea lives a Melanesian tribe known as the Dobu. The soil is poor and living is not as lush as on some other islands. Dobu culture is marked by an extreme lack of cooperation, by suspicion, and by treachery. Power comes from malignant sorcery which one either inherits or buys. The object of life is to advance oneself, either by magic or sheer cleverness, *at the expense of others*. To retain a semblance of social cohesiveness in the midst of so much selfishness there are elaborate kinship laws and taboos forcing people to support their relatives. When a marriage occurs, the villages of the prospective husband and wife, which are always hostile since no two villages are friendly, are obligated to exchange formal gifts. Children of a marriage belong to their mother's village, and should the mother die, they must return to her village, from which their father is forever banned. [14]

Some consequences of this cultural pattern seem bizzare. For instance, although husband and wife share a garden, each grows separate plots of yams (the staple food) using seed plants from their home village, and employs private magic to make the yams grow. Naturally a husband's magic will not work on his wife's yams and vice versa. So suspicious are people of one another that the yield of the family garden is kept secret, for it is believed that one person's gain can only be had at the expense of another. People jealously put poisonous curses on their fruit trees to prevent thefts. They can of course lift the curse so as to pick their own fruit, unless an unfriendly neighbour has added his own curse to the same tree. All sickness and death are blamed on the sorcery of another; when a spouse falls ill and dies, the surviving husband or wife is sure to be suspected. Poison is another great threat. Women never leave cooking pots unattended for fear of poison, and upon receiving a gift, a Dobu's phrase for 'Thank you' is 'If you poison me, how should I repay you?'. Even in trading, while the Dobu pretends honesty, friendship and cooperation, he plans treachery. Happiness and laughter are disparaged; dourness prevails. Sexual passion is high, yet people are prudish. Marriages are rocky and divorce is common. The good Dobu excels at suspicion, cruelty and sorcery, for these are the virtues upon which life is built.

The Yanomamö

If the Dobu are aggressive by subterfuge, certain villages among the Yanomamö are openly aggressive, as already noted. Living in villages of 50 to 200, these short, stocky peoples are one of several related groups inhabiting the upper reaches of the Amazon and Orinoco river basins on the border of Brazil and Venezuela. They subsist on a mixed economy of cultivated crops – mostly plantains, bananas, manioc, and yams – and wild food, such as palm fruits and game. Their gardens are of the temporary slash and burn type, and provide 80 per cent of their food. Yet theirs is a strongly male-oriented society, in which hunting and especially warfare dominate the thoughts of everyday life.[15]

Warfare – a surprise raid on another village – is undertaken to avenge the death of a fellow villager and to abduct nubile females. Among the polygynous Yanomamö, women are constantly in short supply, a problem enhanced by preferential female infanticide. Men dominate women; wife beating and inflicting burns on a wayward wife with glowing embers are not uncommon events. The men constantly practise warlike activities, and they use hallucinogenic drugs frequently, especially before a raid. Inter-village alliances are critical and involve 'war games' and contests of strength. All villages in this geographic area are barricaded against surprise attacks and an unannounced stranger is met by a phalanx of drawn bows.

Although their chronic warfare appears to be of recent origin, their explanation is cosmological. They are decendants of Periboriwa, the Spirit of the Moon; their male ancestors were formed from drops of Periboriwa's blood which fell to Earth after he was shot by an arrow. Anthropologist Napoleon Chagnon, who lived among these people, interprets: 'Because [the Yanomamö] have their origin in blood, they are fierce and are continuously making war on each other.'[16] (This seems to be an instance of the unconscious creation of a myth to explain a recent behavioural pattern.) A raid however seldom involves the death of more than one man, and there are elaborate rituals – chest pounding duels, side slapping, and club fights – by which personal feuds can be resolved without any killing at all: 'Thus, Yanomamö culture calls forth aggressive behavior, but at the same time provides a regulated system in which the expressions of violence can be controlled.'[17]

Despite their aggressiveness, Yanomamö are hardworking. Both men and women tend the gardens and gather wild fruits, although it is the women who must also gather and carry home the heavy loads of firewood each evening. Meat is shared, and a hunter is expected to give away most of the game he kills. Says Chagnon: 'Men genuinely abhor hearing their children cry for meat; this calls into question their abilities as hunters and marksmen, both of which are associated with prestige'.[18] The whole pattern of life thus revolves around the *ideas* of warfare, revenge and prowess at arms. Virtue is equated with ferocity. Yet there is also love, affection and sharing.

The Zuñi

In the southwestern United States live the remnants of once numerous city-dwellers – the builders of the great mesa and valley pueblos that flourished until about 600 or 700 years ago. Among these are the Zuñi, whose culture was described some 50 years ago by Ruth Benedict.[19]

A ceremonious people who valued sobriety and inoffensiveness above all other virtues, their entire life was directed towards cooperation, towards relating the individual self to the good of the whole. The compactly built pueblos, some inhabited for centuries, contrast starkly with the widely scattered hogans of the surrounding Navajo, a nomadic sheep-tending people. Despite their apparently desolate environment, the Zuñi relative to most other native tribes were wealthy, having gardens, orchards, sheep, and silver and turquoise. Until recently, they lived according to their traditional patterns, untouched by the Western culture that had so cruelly impinged on other tribes. They shunned alcohol and never used peyote or sought hallucinatory visions through physical ordeals or intoxication. 'The Pueblos distrust and reject those experiences which take the individual in any way out of bounds and forfeit his sobriety', commented Benedict.[20] (Today unfortunately alcohol and drugs, which proved so destructive to other native societies, have found their way into some of the Pueblos.[21])

Pueblo economic life was based on gardens which it was the men's responsibility to tend. Houses and their contents, however, belonged to women and passed from mother to daughters. Although reciprocal family obligations made first marriages economically important, divorce and subsequent marriages were culturally uneventful occasions, involving only the participants themselves. To 'divorce' her husband, a woman simply put his belongings on the doorstep before he came home in the evening; he took them and returned to his mother's house to live. But this seldom happened. Marriages were generally happy; sex was considered an important and pleasurable activity. Until quite recently, homicide, suicide, theft, and even fighting were almost unknown among the Zuñi, and strong emotions were seldom displayed. Warfare was considered an unfortunate event and was not sought. If a man killed an enemy at war, he had to be ceremonially cleansed, and the scalp of the 'worthless enemy' was brought back and ritually adopted by the tribe to redeem the soul of its owner.

The main characteristic of Pueblo life was the total absorption in community ritual. Continuous religious observances insured that the rains would come and bring successful harvest and an abundant life. Most tribal activities centred on the next round of ceremonies. Each village had two or three circular kivas, sacred underground temples, where ancestral spirits were invoked and much of the ritual life occurred.

This dedication to community ceremony meant that selfishness and pride were almost unknown. Throughout all aspects of society, cooperation was uppermost. A man did not seek priestly functions; rather, they were thrust upon him. No one was 'in charge'. Despite individual differences in economic well-being, there was no social hierarchy, either economic or religious. Benedict described the Zuñi thus:

> The lack of opportunities for the exercise of authority, both in religious and in domestic situations, is knit up with another fundamental trait: the insistence upon sinking the individual in the group. In Zuñi, responsibility and power are always distributed and the group is made the functioning unit ... Neither in religion nor in economics is the individual autonomous. [22]

(The boxed text below describes a personal encounter by the author with group egalitarianism in another Pueblo village.)

A VISIT TO OLD ORAIBI

On a tour in 1960 of the American Southwest, I stopped at Old Oraibi, a Hopi village continuously inhabited since the tenth or eleventh century. Perched on the edge of Third Mesa in northern Arizona, the stone and adobe houses, some fallen into decay, others fairly new, look southward across the semi-arid plains below. That day all the adults except one old grandfather had gone to a snake dance in another village. About ten or eleven children, aged twelve to five, agreed to show me their village. They pointed out houses, answered questions, and chattered among themselves. The boys constantly threw sticks, hitting distant rocks with amazing accuracy – and no doubt rabbits whenever possible. When I asked if the people still used their kivas, a chorus replied, as though it was a foolish question, 'Why yes, of course!'

After the tour, I took the children's picture and offered, as is the custom with Indian children, to pay them for posing. But my pieces of change were of all sizes. There were not ten of anything: nickles, dimes or quarters. One older girl, who had enterprisingly tried to sell me a toy bow and arrows painted in gaudy colours, suggested that the biggest ones be given quarters, the next dimes, and the littlest, the nickles. But the eldest boy hung back, pouting and frowning. 'The mothers won't like it', he muttered. 'Everyone has to be treated the same'. The problem was solved by taking a couple of dollar bills to the grandfather and asking him to see that each child eventually received an equal share. He readily agreed, with a gentle twinkle in his eye. There were no more arguments, and as I drove away from Old Oraibi all the children waved goodbye.

The emphasis placed on ritual and cooperation long held the Pueblo tribes intact. Even today the Hopi and others are struggling to retain their cultural character and value systems against repeated attacks by the encroaching industrial world that surrounds them.

CULTURAL DESTRUCTION

All the cultures discussed thus far are under external pressure from intruding cultures. The established social patterns of the Kwakiutl, the Dobu, the !Kung, the Yanomamö and the Zuñi are all threatened by outside forces they have not the institutions to deal with. The Dobu, even as Ruth Benedict was writing about them, were suffering from cultural disintegration. She noted that they were far poorer than their neighbours and had recently suffered severe population decline. It seems likely that their habit of gift-exchanges, so inconsistent with their general selfishness, was a remnant behaviour from a more generous past. Likewise, the warring Yanomamö villages studied by Chagnon are not typical of all Yanomamö, being limited to the regions where the Indians have been forced into constricted areas of the jungle by outside pressures. Most of the preliterate societies that still exist are remnants of cultures pushed into the most marginal habitats – the arctic, the deserts, the jungle niches and other isolated sites. The cultural characters anthropologists observe in many such places, although believed to be ancient by the peoples themselves, may in fact be recently acquired in response to external pressures; they are cultural artifacts of traditional societies caught in a rapidly changing world.

Acts of encroachment, whether caused by missionaries, oil companies, or invading armies, if they destroy the central meaning of a culture also destroy its entire social structure. The Dobu deprived of their treachery and sorcery would have no meaning in their lives, no way of making sense of their world. To deny the Zuñi their ritual beliefs would reduce their lives to a mere routine. In each case, people without their cultural visions would be doomed to lead empty, repetitious lives. Similarly, affluent Westerners abruptly shorn of their freedom to pursue material goals might well find life had become a blank, meaningless existence.

The central meaning of human life thus comes not from our genes nor our instincts, nor even the inventions of our individual imaginations, but from our culture. Western society, with its vaunted 'freedom' and 'independence', supposes itself to have outgrown such cultural constraints. Yet in Part III it becomes clear that we are excessively dependent on a culture in which mechanisation and specialisation are obligatory. Westerners are no more free of cultural constraints than are any of the peoples described in this chapter, nor is our culture any more secure in the long run than theirs. Indeed, there is growing evidence that Western materialism itself is proving an inadequate cultural vision.

When the environment changes drastically, a culture – like any overspecialised organism – may be unable to adapt, and Western culture is no exception to this rule. When that happens, the society concerned loses its cohesiveness and its members gradually lose those qualities we recognise as human. Two examples at different levels of disintegration illustrate this process.

The Yir Yoront of Australia

The Yir Yoront are Australian aborigines, different from the Tiwi discussed earlier, who inhabit the southern Cape York Peninsula that forms the eastern boundary of the Gulf of Carpentaria. For untold centuries this tribe's most important tool was the stone axe, an implement of prestige. Only men had axes, and while women and children often used them, they always had to be borrowed from and promptly returned to their male owners. Axe heads were fashioned from a special sort of stone only found 400 miles to the south. Stones were obtained by trade, mainly for spears tipped with the barbs of stingrays fished from the gulf. A Yir Yoront man thus had to have trading partners to the south and the north in order to obtain stones for axe heads and spears to be traded for them. These items were exchanged during great annual festivals when villagers from an extensive area would congregate for trade, initiation ceremonies, exchange of ideas, relaxation, and a general strengthening of social bonds. These occasions were highlights of the aboriginal year, as Christmas is of ours.

Sometime after 1900, however, local Christian missionaries began giving away steel axes; these lasted considerably longer than the old stone axes and were much prized. But instead of improving life, the steel axes ended by disrupting it.[23] Men who once controlled the making of their own stone axes now found themselves beholden to white men, having to do their bidding to obtain steel axes. Even worse, the missionaries, ignorant of social custom, gave out axes indiscriminately, thus destroying the male hierarchy upon which social organisation depended. Women and children now owned axes and behaved independently; the old social arrangements fell apart. Finally, the central reason for the annual trading festivals was now gone. People still met for a while but there was no enthusiasm. Stone axes had held the larger society together. Steel axes destroyed its texture. A simple technological change ramified throughout the whole social structure and led to cultural confusion – an event utterly unseen by the missionaries and not understood by the aborigines.

The Ik of Uganda

The political colonisation of the African continent by Europeans lasted only about 100 years, yet the havoc left in its wake is enormous. By carving up the map of Africa with arbitrary straight lines, European hegemony forced unrelated tribes into unwanted political alliances and imposed major changes in the territorial boundaries of various tribes. One victim was the Ik (pronounced 'Eek'). Once nomadic hunter–gatherers wandering widely over parts of what are now the Sudan, Uganda and Kenya, the Ik are now confined, owing to 'national' boundaries, to a small mountainous area of northern Uganda which embodies only a fraction of their former lands. It is a poor country, unable to supply these nomadic people with enough food to survive. They try with little success to grow gardens in a land where droughts are common and rain may fall for only a few hours a year. Today the Ik are literally starving slowly to death.[24] (Following the long drought of the 1980s they may be totally gone.)

The effect on Ik society was total. In fact, to all intents and purposes there is no society. Unlike the selfish Dobu who at least owes allegiance to a village and family and has multiple obligations to others, an Ik lives in virtual isolation. Each person must fend for himself or herself. Personal survival is *all* that matters. There is no culture. There is not even love among family members. Sex for its pleasurable aspects is regarded as about equal to defecation; it requires too much energy, and there is no tenderness associated with it. Children are thrown out of the parental house at the age of three to survive if they can. Siblings do not regret the death of a brother or sister from starvation, pointing out that sharing food would only have resulted in two deaths instead of one. Stealing food from the mouths of the weak or the old before they can swallow it is clever practice, so most people eat on the run, gobbling food like hungry chickens. Stealing, cheating and prostitution are ways to get food, water and tobacco. Liaisons, mainly with outsiders who can provide benefits, are affectionless.

A few older people retain the cultural memory of an earlier life when love and family ties existed. Some still carve gourds with ritual designs that are hidden in secret family caves, but they seem to have lost their meaning. Aside from having food in one's belly, life has lost its purpose: it is boring; there is nothing to feel. Useless activities are aimlessly undertaken, such as sweeping around the living compounds, but seldom inside. Shade trees are cut down and the few bushes chopped up for no apparent reason: 'They're in the way.' Bark in far greater quantities than needed to make the cord the Ik use is stripped from trees, causing them to die. It is wanton environmental destruction – a people's statement of utter hopelessness. People laugh only at the misery of another. After a good rainy season when everyone had enough to eat for a time, the bumper crops were neither tended nor stored but left to rot in the fields. Any affluence might have cut off the government's famine relief, that is doled out to one or two people from each village to carry back to the others. But Ugandan officials were ignorant of the Ik's selfishness, and these village representatives, as soon as they were out of sight, gorged themselves on the rations meant for others. Says anthropologist Colin Turnbull, who has studied these people:

So those who were in no real need [the young and fit] who could come to the government stations got the relief intended for themselves as well as that intended for all their truly needy kin. They grew fat on stupidity, while others died.[25]

Thus unsupervised, famine relief hindered any small chance for the development of communal self-reliance among the Ik.

These examples show how easily a society confronted with uncontrollable change can lose its structure and become chaotic – how its people suffer mental and emotional disruption and even death. They also show how well-meaning assistance or technical innovations introduced from outside can destroy rather than benefit a society. Since both direct aid and technological change are in progress around the globe, it is important to identify the most vital factors that should be taken into account if true benefits are to accrue to a changing society. They will appear again in Part IV.

CULTURAL CHANGE

Today about two-thirds of the world's people live at low levels of technology, often in conditions of physical deprivation and even starvation, owing to a combination of factors. These include overpopulation, environmental deterioration, a history of colonialism that has destroyed old political and cultural units and thrown unrelated peoples together in newly emerging nations, and external or internal economic exploitation by powerful elites. Stable traditional societies almost everywhere have been severely disrupted and their newly emerging social patterns are still undergoing great flux. In trying to provide assistance, agencies of technologically advanced nations and non-governmental organisations are faced with multiple considerations, of which the two critical ones from the point of view of social stability are cultural congruence and local autonomy.

Cultural Congruence

Introduction of a new, seemingly beneficial technology can, like the steel axes among the Yir Yoront, destroy the structure of a society, creating major dislocations. An example was the widespread adoption of the 'miracle' grains of the Green Revolution. These genetically engineered seeds, when properly watered, weeded and fertilised, did indeed increase *per capita* food outputs of many countries. But the extra food did not always find its way into the mouths of the poorest, nor did the excess wealth from the sale of exported grains enrich them. Instead, peasants without capital to buy the energy inputs needed to grow the new seeds became landless labourers in country after country, deprived even of the right to obtain a small share of the harvest. Frances Lappé, from the Food First Institute, explains:

> So the number of landless mounts while the number of rural jobs shrinks. Traditionally in many countries even the poorest landless peasant had access to part of the harvest. In India, Bangladesh, Pakistan, and Indonesia the large landowner once felt obligated to permit all who wished to participate in the harvest to retain one sixth of what they harvested. Even the most impoverished were assured of work for a few bags of grain. Now, with increased likelihood of profitable sales, the new agricultural entrepreneurs are rejecting the traditional obligations of the landowner to the poor. It is now common for landowners to sell the standing crop to an outside contractor before harvest. The outsider, with no legal obligations, can seek the cheapest labor, even bringing in workers from neighboring areas.[26]

Any number of other examples might be given of how the introduction of not only agricultural, but also child care, public health, industrial and educational innovations into various cultures has proved socially disruptive, sometimes causing the innovation to be rejected out-of-hand.[27] More often, new technologies have been imposed by central authorities, whatever their social toll. As a result, a subclass of unemployed, both rural and urban, has rapidly accumulated around the world. Like the Ik, these people face an increasingly hopeless situation, deprived of the supports of their original culture yet denied the possibility for creating a meaningful replacement. The world's problems cannot be solved by simple macroeconomics! New technologies must evolve in concert with *each* society's cultural institutions.

Local Autonomy

A second consideration in providing assistance to developing countries is that the change should be *selected and integrated by the recipient culture*, not imposed from outside. In this way, the first requirement of cultural congruence is more likely to be met. Pursuit of this principle, however, can be extremely difficult. For one thing, in those many societies now in the midst of cultural flux, it is hard to determine who speaks for the society as a whole. Those in decision-making positions often represent a small, elite power group. If so, how are the best interests of the people to be discovered and met? Next, how are purveyors of cultural change to be restrained in their dealings with recipient cultures? Many of them are profit-making multinational corporations, who sell such things as pesticides, soft drinks and baby foods to markets 'created' by their own advertising. Each of these products has already had severe negative impacts in various cultures.[28] Finally, who is to restrain the financially pressed leaders of struggling nations from inviting in foreign industries who will exploit the labour of their impoverished subjects, paying them bare subsistence wages to assemble products they can never afford to buy? In today's morass of global indebtedness it is hard for leaders of developing nations not to clutch at such culturally self-destructive straws. Only a universal awareness and sensitivity toward the need to maintain – and often reestablish – the cultural integrity in developing countries can stave off the sort of massive social disintegration of which the Yir Yoront and Ik are only small examples. The global economic calculus almost never takes into account the critical linchpin of cultural integrity (and the social stability it creates) as *underlying preconditions for a productive economy*.

One last homily flows from this. It is often imagined that human beings are infinitely flexible, able to adapt to every new situation. On a geologic timescale this may be true, but not on the scale of a human lifespan. One example of a culture that tried to move out of the stone age into the twentieth century in a single generation is the Manus, a Melanesian tribe living in the Admiralty Islands near New Guinea. Their attempt was described in rather glowing terms by Margaret Mead,[29] who first visited them in 1928, before they had had any significant contact with the outside world. In the 1930s, Catholic missionaries arrived and disoriented their traditional culture, and then the Second World War brought them into contact with American soldiers, whom they much admired. At a propitious moment, when the tribe, hoping for God's miracle to arrive on the next cargo ship and save them, had destroyed all its material wealth, a visionary leader talked them into adopting Western ways: new clothes, houses, schools, economic institutions, and a 'centralised' government over all the villages to deal with the outside world. Within a few years, the people had completely changed their lives.

Mead observed this 'conversion' during a return visit to the Manus in 1953, and her enthusiastic descriptions set this small tribe up as a model for self-determined cultural change. But in 1979, Mead's anthropologist daughter, Mary Catherine Bateson, visited the island and found that 'The New Way' had turned to disillusionment. The new social institutions, once their novelty had worn off, lacked the deeper meaning and ritual of the old, traditional ways. Life had become superficial, flat and boring, the 'modern' economy uncertain, and the future precarious.[30] Conscious cultural change evidently requires more gradual stages with deeper connections to the past if it is to succeed, depending less on charismatic leadership and more on social discussion and the communal creation of a new worldview, sanctified by ritual and tradition.

PLURALISTIC SOCIETIES

A few words are needed regarding those giant nations which are made up of many peoples with a variety of religions, ethnic backgrounds, social classes, and dialects. Most readers of this book will be members of such a pluralistic society. They will recognise that although they share certain legal, economic and material patterns with everyone else, they do *not* share identical value systems or priorities, the same expectations or goals in life. Indeed, many individuals move daily between two or more subcultures which may have quite different social visions. Even so, there are observable distinctions between the polyglot cultures of, say, London, Paris and New York. Somehow, the national stamp impresses itself on a subculture, so that one can readily distinguish between a West Indian from Queensway in London and a black American from Detroit, or an Irish Catholic raised in Dublin and one raised in New York.

Those people living in pluralistic societies who move freely between subcultures adopt two distinct cultural attitudes, often reflected in different forms of speech. In England, for example, the author found that laboratory technicians at the university spoke a local but understandable dialect to her and other 'educated' persons, but among themselves they spoke a working-class dialect incomprehensible to the 'elite'. The educated, of course, were proud of their impeccable 'Queen's English', free from the taint of any local dialect. In America, urban blacks also have a speech for communicating with whites that can be quite different from what they use at home and on the streets of the ghetto.[31] A distinctive way of talking is a powerful means for isolating a subculture from its surroundings and reinforcing the bonds within it.

Accommodation of diverse subcultures within a larger social system is an art that is far from perfected. Such a situation breeds conflict at the same time that it confers richness. Fostering of tolerance in one's own society can be regarded as practice for the greater task of creating global cultural tolerance. Yet overly diverse societies can become culturally dilute and lacking in cohesiveness, a threat both superpowers constantly face. If an individual's subculture – the church, the ghetto, the union, the company – fails to provide significance and meaning for his or her life, a cultural vacuum may develop, to be filled by whatever whims, fads or cults may come along. This thread appears again when we discuss the cultural consequences of Western industrial society on its own people.

SOME CONCLUDING THOUGHTS

We can perhaps draw two important points from this cultural survey to guide us as we plan for the future. One is the recognition, emphasised yet once more because it is so important to grasp, that *no culture, including our present one, is an inevitable outcome of an underlying, biologically determined human nature.* Since culture is not inevitable, it can be changed. Change is difficult; as we have seen, it can easily bring chaos if not thoughtfully undertaken. Yet is *is* possible. The second point, which has to do with cultural temperament and social survival, is more subtle and requires some elaboration.

It was Ruth Benedict's wisdom to recognise that the different levels of cooperativeness in a culture – she called it 'synergy' – are not due to a special goodness, a special generosity, or a special willingness to make personal sacrifices on the part of individuals in order to benefit the whole. Conscious altruism is not necessary.[32] The difference between high synergy and low synergy cultures lies rather in their institutions. By looking at, say, their economic systems (whether wealth is funnelled to the richest or redistributed throughout) or their religious and ideological beliefs (whether secret or magical powers belong to the few or the many and whether they are used for personal gain or general benefit), one can tell the level of synergy in a society. In a highly synergic society, *the good of the individual coincides with the good of the whole because the social institutions create this situation.* Helping others, far from being a personal sacrifice, is a perceived personal benefit. This is the genius of synergy! Hobbesian man – greedy, selfish and aggressive – is a product of human institutions and of culturally evolved worldviews, and so is generous, cooperative and peaceable man. Tomorrow's world probably has but one choice if humankind is to survive.

Where do such institutions and worldviews come from, and how are they altered in the course of cultural change? In the past, with a few exceptions such as the recent cases of the Manus and the Chinese, internal cultural change has probably been accidental, often occurring in response to uncontrollable external pressures. Throughout past millennia, people may have had no concept of free will or any belief that they could fashion their own futures. They acted according to rules laid down by their ancestors, or by the gods.[33] In small hunting–gathering societies and even in the earliest agricultural settlements, it is easy to imagine that this was the case. Since everyone believed in the same myths that legitimised the shared cultural values there was never any reason for questioning tradition. It was simply: 'That's the way we do'. Cultural change came imperceptibly, and myths and rituals were unconsciously altered to explain and legitimise the new worldview. No wonder such peoples still living today cannot deal with rapid change.

With the advent of civilisation, however, change became more rapid. As people grew conscious of it there was a tendency to resist new institutions. To hold the growing societies together, new, more powerful belief systems were required, from which emerged the major religions (the subject of our next chapter). Worldviews could be changed only by changing religious doctrine, a conscious process that has given rise to modern philosophy and its subdisciplines of theology and ethics.

In today's overstressed world, grown small through technology and threatened by nuclear destruction, the values of synergy, of global cooperation, may be necessary for human survival. To the extent that this is true, then we need to study carefully the institutions of synergic societies of the past and compare them with our own. It is the recognition of the *necessity for purposefully changing our institutions* that makes the study of other cultures so vitally important. For the first time in human history it appears that we shall have *deliberately* to create new cultures and new worldviews on a massive scale. Yet it is perhaps not so much the developing countries that are most in need of cultural change, but the developed ones. Anthropologist John Bodley borrows economist Robert Heilbroner's prescription for the developing countries, but applies it instead to ourselves:

> Nothing short of a pervasive social transformation will suffice: a wholesale metamorphosis of habits, a wrenching reorientation of values concerning time, status, money, work; an unweaving and reweaving of the fabric of daily existence itself.[34]

It is Bodley's contention that what is needed is a conscious reconstruction of our worldview, and that much is to be learned from the attitudes and worldviews of earlier societies, including those primitive societies with ecologically sound and egalitarian cultures. It is a mistake to assume that primitive cultures were static: they did change, but in environmentally cautious ways. Moreover, their lives were not at all 'solitary, poore, nasty, brutish, and short', as Thomas Hobbes imagined; rather, they spent relatively little time making a living, averaging four or five hours a day obtaining food and other necessities. One anthropologist has dubbed hunter–gatherers as the original 'affluent society'. And what is even more telling, their way of life, which lasted for perhaps 99 per cent of human cultural existence, is 'the most successful and persistent adaptation man has ever achieved'. As Bodley explains:

> [I]n comparison, newly arrived industrial civilization [is] in a precarious situation with the 'exceedingly complex and unstable ecological conditions' it [has] created ... [W]e should study why hunters were so successful ... [O]ur civilization might actually learn something from them.[35]

It is indeed high time that we critically examine what such words as 'primitive' and 'backward,' and 'civilised' and 'modern' really imply. Our entire conception of 'progress' needs to be dissected and evaluated. These tasks are pursued in later chapters.

7 Religions and Worldviews

It is the function of the prophet to show reality, to show alternatives and to protest; it is his function to call loudly, to awake man from his customary half-slumber. It is the historical situation which makes prophets, not the wish of some men to be prophets.

Erich Fromm, 1981[†]

INTRODUCTION

Philosopher William Hazlitt once observed that 'Without the aid of prejudice and custom I should not be able to find my way across the room'.[1] That simple phrase in fact says something quite profound about the human mind, and hence about human behaviour. Walking across a room was his metaphor for living through the events of everyday life. He was talking about the inner mental map each of us constructs out of our familiar surroundings and in which we live: the synthetic worldview which is a biased, shorthand reconstruction of 'out there'. What exactly this worldview is and how each individual acquires it is the subject of Chapter 8. Here it is sufficient to reflect on Hazlitt's words, 'prejudice' and 'custom'.

Prejudice, bias, whatever we call it, is inescapable. People see and take in what they need to know. We are surrounded by far too much information to notice it all. We quickly learn to ignore the unimportant – and so are able 'to walk across the room' undistracted by the masses of extraneous information bombarding our senses. And it is custom that tells us what biases to adopt, what we should notice and what we can safely ignore. Prejudice, at least in the subconscious everyday routine, probably affects all humans about equally, but the particular biases we have are largely a matter of culture. We are inevitably 'ethnocentric'. In this chapter, we examine a major cultural influence, namely, religion.

Religion today is not popular among many educated persons; I have heard it denounced as the greatest plague of mankind. Marx, of course, condemned it as the opiate of the people. The implication is that if religion were outlawed, most of the world's problems would be solved. But it is not that simple. Religious views inherited from the past deeply affect cultural biases and hence shape our world vision, whether we think of ourselves as religious or not. Entrenched atheists may deny this, but reflection usually uncovers the religious tenets underlying our basic assumptions about the world. This chapter begins that task of reflection. It also shows that the psychic and social functions served by religion cannot be eliminated simply by denying them. Like eating, breathing, laughing and crying, religious beliefs are intrinsic to human nature. The problems created by religious differences will be solved not by suppressing our beliefs but by understanding their origins and their functions.

This chapter begins with the changing social functions of religions during the long history of humankind. It then examines today's major religions, particularly in relation to our global problems. Finally it turns to religious changes currently underway in the world.

RELIGION AND SOCIETY

In the West the word 'religion' conjures up the worship of one or more invisible beings who created and control the universe. 'Faith' implies absolute trust in this mysterious power, who rewards the faithful with everlasting life. In fact, *religio*, the original Latin word, derives from either the word 'to constrain' or 'to recite'.[2] It first referred to the popular Roman mystery religions, access to which demanded secret initiation rites. Among them were the Eleusinian mysteries of Demeter that took place in ancient Attica, and the Persian Mithraic mysteries which the Roman legions carried across Europe. As the inscription at Eleusis reveals, their central promise was immortality:

> Beautiful indeed is the Mystery given us by the blessed gods: death is for mortals no longer an evil, but a blessing.[3]

While Western religions, especially Christianity and Islam, tend to focus on the mysteries of the next life, those of the East concentrate on how to live in this world, to which one is forever tied through many incarnations. There is no word equivalent to 'religion' in Sanskrit, Pali or classical Chinese, the ancient languages of Hinduism, Buddhism and Confucianism; the nearest are 'dharma' and 'tao'. The Sanskrit *dharma* cannot be translated by a single idea in Western thought for it embodies the whole meaning of the universe and of one's appropriate relationship with it. Likewise, the Chinese *tao* means literally 'the way', or how to approach life through understanding the Order of Nature. 'What by nature is, is good'.[4] Meaning, power and authority are not revealed by supernatural beings (although such may exist), but by living in such a way as to make the true meaning of life evident. Holiness comes not by the grace of God but by personally seeking oneness with the universe. There is a deep sensitivity toward Nature, which finds its highest aesthetic development in East Asia.

In the East, religions are both less authoritarian and more pervasive of everyday consciousness than in the West, imparting a quite different cultural vision of the place of humans in the universe. In the West, salvation comes not from seeking understanding about the universe in order to unify the self with it, but from faith and trust in an almighty god who is unquestioningly obeyed. These distinctions have been overly simplified to underscore the deep impact that religious viewpoints have on the 'characteristics' or 'personalities' of different cultures. How sacred beliefs define the world deeply affects the social structure and behaviour of an entire people, binding them together into a single unit. *Sacred meaning is one of the deepest bonding forces societies possess, and in its absence, they tend to disintegrate.* This theme recurs throughout the book.

Whatever in a culture we identify as 'religion', it generally comprises two elements: cosmology and ethics. The second largely derives from the first.

Cosmology

Self-consciousness demands meaning, an answer to the question 'Why?' asked by every child since the dawn of awareness. 'Why are we here?', 'What is life for?', 'Where did we come from, where are we now, and where are we going?' Purpose and meaning are at once our joy and our curse. 'The humanities' – religion, history, philosophy, art – are our attempts to answer these ultimate questions.

Every society has thus invented a *cosmology*: an explanation of how the universe came into being and why humans came to be in it. Such creation myths vary greatly in form. All reflect a sense of time, but time may be linear and short as in the Biblical story of Genesis, or recur in cycles of hundreds of trillions of years as Hindus believe. Generally a Divine Being, whether a heavenly spirit or a sacred seabird, created sky, water, dry land and, ultimately, the founding ancestor of humans: Adam of Judeo-Christian belief; Spider Woman of the Indians of the American southwest. The descendants are therefore the Creator's 'chosen ones'.

Sacred myths also relate the history of a people, recounting migrations, crises and calamities. The flood, droughts and locust plagues recorded in the Bible are familiar to most, but the Hopi, too, tell of past worlds destroyed by fire, ice and flood; they expect the present Fourth World will also succumb, for people are failing to live up to their virtue.[5]

Each cosmology thus provides a distinctive vision about who people are, where they stand in the universe, what their responsibilities are, and how they ought to behave. In this way, cosmology leads to a code of *ethics*, which may subconsciously persist in secular societies long after the creation story has been dismissed.

Ethics

In the secular West, our ethical concerns are limited mainly to relations between people. Most are unaware that in many societies, past and present, behavioural codes include obligations to Nature as well. This extension of intrinsic value, and hence of rights, to things non-human grows out of those cosmologies that regard humans as *embedded within* the universe rather than as *standing outside* or above it. In the former, we belong to the universe; in the latter, the universe belongs to us. As we shall see, the Western utilitarian, 'ownership' vision seems to have emerged from the Judeo-Christian cosmology, and Part III describes how this sense of ownership evolved to its present extreme. The Western ethic can potentially cope with inequities between human beings; it is almost impossible for it to cope with the exploitation of Nature.

In stark contrast are ethical visions embodying a sense of unity with Nature. These are particularly common among hunting–gathering peoples and early agriculturalists for whom the spirit world and Nature are one. When an Eskimo kills a seal, he asks forgiveness of the animal's spirit so it will return in a new body and replenish the food supply. Tree fellers hold propitiary rites for the souls of fallen trees.[6] The Zuñi and Hopi dance to encourage the cloud spirits to release rain. Human beings, as active participants in the unfolding of natural events, have duties and obligations that bond them deeply to the natural world. Heaven is the Father, Earth is the Mother and everything in the universe is their child.

The relations between humans and Nature in the East are less direct and more subtle, yet there, too, one finds a deep respect for the sanctity of life – *all* life, not just human life. An aesthetic sense of Nature, where humans stand as tiny observers within a vast and beautiful living landscape, permeates people's vision (Figure 7.1). This contrast between Eastern reverence and Western domination of Nature is central to this chapter.

SOME THREADS OF HISTORY

From the diverse but always Nature-oriented spirit worlds of hunting–gathering peoples there emerged during the second and first millennia BC the major religious traditions of the modern world – Hinduism, Buddhism, Confucianism, Taoism, and Judaism (from which Christianity and later Islam arose). Their emergence was influenced by three important developments that accompanied the gradual spread of agriculture between 10,000 and 2000 BC, with its concomitant surge in human populations. These were the invention of calendars, the evolution of writing, and the advent of large, impersonal, hierarchical societies.

The Telling of Time

A Westerner deprived of a clock feels uncomfortable. Being without a calendar is even more unsettling. Prisoners mark the passing days and

Source: Interpretation of a watercolour painting, *An Autumn Scene* (1729) by Hua Yen, at the Freer Gallery, Washington, DC. Art by R. Carruthers.

Figure 7.1 A typical Chinese landscape.

In the Eastern vision, people and their dwellings are always small and embedded in a vast natural panorama.

weeks on their cell walls; a person waking from a coma feels disoriented in time. We constantly locate ourselves in a continuum of time. Keeping exact track of time is relatively new in human existence. Hunter–gatherers are well attuned to seasonal cycles, using the moon as their chronometer, but it was the early agriculturalists, their existence dependent upon precisely anticipating the seasons, who developed far more exact means for tracking time. They turned to the movements of the sun and stars.

The peoples of Oceania who used the night sky for navigation between distant islands[7] also planted their crops according to the position of certain constellations.[8] But a more popular guide, especially in temperate latitudes, was the seasonal shift along the horizon of the rising and setting sun. Among the best known solar timekeepers are the great henge monuments of northern Europe and the British Isles. Built between 2500 and 1500 BC, these giant stone circles, such as England's famous Stonehenge, were almost certainly precise sacred calendars where important rituals were observed. Solar observatories of ancient American Indians are also known, including the angular corner windows at Pueblo Bonito in Arizona[9] and the curious sun-dagger calendar where vertical spikes of sunlight pass across concentric circles carved by the ancient Anasazi Indians on the stone cliffs of Chaco Canyon in New Mexico.[10] Both devices precisely identify annual solstices and equinoxes. The Incas in Peru had a series of horizontal sitings marking sunrise throughout the year, as noted in 1653 by the Spaniard, Bernarbe Cobo:

> Suanca was a hill where the irrigational canal of Chinchero passes. On it were two pillars or monuments to signal that when the sun arrived there it was time to begin to plant maize. The sacrifice made there was addressed to the sun and they asked that it would arrive on time so that they would have good reason to plant.[11]

These early rituals served a double function. The 'etic' role, of course, was to synchronise crops with seasons. The 'emic' role was religious; through ritual action designed to insure the cooperation of sun and rain clouds, a society unconsciously bound itself together. For social survival, they are equally important.

To acquire the power of astrological magic that would flow from accurate, long-term predictions of eclipses and conjunctions of the planets, people turned to the night sky. Astronomical observations flourished in Central and South America, in Persia and China. These exact long-term calculations required the invention of number systems. The Mayas learned to reckon to well beyond a billion.[12] Astronomy thus gave rise to mathematics as well as a new concept of time counted on a scale far longer than the lifetime of a single human being or even of several generations.

Along with calendars came clocks. By 3500 BC, the Sumerians had subdivided each year into four-minute units.[13] Not long after, the Egyptians had shadow clocks (sundials), which with water clocks, candle clocks and sand clocks (hour glasses) spread throughout the civilised world. Keeping track of time profoundly changed daily life; it made synchronisation of large populations possible.

The Keeping of Records

Graphic art among preliterate people was of two kinds. There was magical or religious art, such as the intriguing Neolithic cave paintings discovered around the Mediterranean, and there were carved objects used for communication, such as trail maskers, tally-sticks for keeping track of debts, and message sticks (Figure 7.2a).

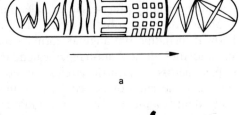

a

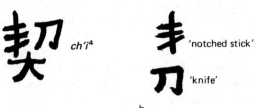

ch'i[4]   'notched stick'

'knife'

b

Source: a from Figure 4, p. 26, and c from various illustrations, pp. 58–64, of H. Jensen, *Sign, Symbol and Script* (© VEB Deutscher Verlag der Wissenschaften, 1958); b courtesy of Dr. S. Wawrytko.

Figure 7.2 Early forms of written communication.

   a: Message-stick. Wooden baton of about a forearm's length, inscribed with a route map for finding one's way to a gathering. This is an Australian aborigine message-stick; read from left to right it gives the path to a tribal meeting place that passes over sand-dunes, a silted stream, several 'territories' and a river.

   b: Chinese ideograph for 'contract' (*ch'i*[4]), made up from the ideographs for 'a notched stick' and 'knife'. (The superscript '4' refers to the pronunciation of this *ch'i*, distinguishing it from other *ch'i*'s.)

   c: Egyptian hieroglyphics. These exist at three levels – words, syllables and letters.
     A few of the original ideograms retain only their original meanings, while others serve also as syllables. (Thus the ideogram for 'swallow' also means 'big' because both spoken words, in Egyptian, contain the same consonants, *w* and *r*. The vowel is never indicated in Egyptian writing, hence *w–r*.)
     Some ideograms became syllable sounds used in many words ('jar', or *n–w*), and others lost their original meanings almost completely, retaining only the initial consonant sounds and so becoming letters of an alphabet.

Sign	Word meaning	Syllable sound	Letter sound	Remarks
⌐⌐	sky			
⌐x⌐	night			
𓃂	(mammal)			Determinative (not pronounced)
🐦	swallow	w–r		w–r also means 'big'
▱	flute	m–ꝫ–		m–ꝫ– also means 'true'
🏺	jar	n–w		n–w is a consonant pair used in many words
𓇌	(reed)		i	⎫
∿∿∿	(water)		n	Original meanings mostly lost; only initial alphabetical (consonant) sounds retained
🐍	(snake)		ds	
☐	(seat)		p	
🐦	(quail)		w	⎭

☐ 🏺 🐦 𓃂
∿∿∿ 🐦 ▱   (p + n + n–w + w + 'mammal') = mouse (p–n–w)

c

As trade increased between agricultural settlements, permanent records of contracts, deeds and tax payments became necessary. Before 3000 BC, clay tablets were being used in the Middle East to keep track of grain rations for people and animals, of how much seed to sow in a field, and of deliveries of wheat, bread and beer. Clay and tokens inscribed with numbers, dating back to 8000 BC, may have had similar commercial uses.[14] A clue to the origins of writing comes from the Chinese ideograph (word-picture sign) for 'deed' or 'contract' ($ch'i^4$) in which one part originally meant 'notched stick' while another means 'knife' (see Figure 7.2b). Indeed, the word 'to write' in most languages comes from words meaning to scratch, scrape or carve, or from names of writing implements such as awl or knife.[15]

The first written prose was simply a picture-record of important events, but gradually stereotyped pictures or hieroglyphics became symbols for objects, actions or ideas. Eventually some symbols became associated with pure sounds – a process known as *phoneticisation*. (It is as though, in English, we combined symbols for 'bee' and 'leaf' to make a new symbol for the spoken word 'belief'.) The final step, which in Egypt occurred sometime before 3500 BC, was going from syllable signs to single letter signs – a true alphabet. Ancient Egyptian writing retained all three kinds of symbols – word pictures, syllable pictures and consonant pictures (Figure 7.2c).[16]

Most written languages developed alphabets in a similar fashion, finally dropping word pictures altogether. Chinese is an exception, remaining primarily a word-picture language with over 40,000 different characters. Japanese writing, which began very late, around AD 400, borrowed only the sounds of Chinese symbols, adapting them to fit the spoken language; Japanese is thus a syllable script.

Writing made large, hierarchical societies possible. Literacy, a highly specialised and often secret skill of priests and scribes, played a major role in the spread of religious and political authority over wide areas. It permitted the bureaucratic book-keeping necessary to organise large numbers of people; it permitted the writing down of laws; and it permitted the recording of historical events and their rewriting when necessary to legitimise a new regime.

The Advent of 'Civilised Megamachines'

The steps leading to the emergence of organised civilisations that time-keeping and book-keeping made possible can be surmised from archaeological and linguistic clues about the migrations of peoples and political events that must have accompanied them. Everywhere a similar pattern appears of strife between two kinds of husbanding peoples, the nomadic pastoralists and the settled agriculturalists[17] – Abel and Cain of Biblical metaphor.

Farmers raised abundant crops on the rich silt deposited annually in the floodplains of the Nile, Euphrates, Indus and Hwang-Ho. They learned to store excess food and survive lean years. On the less fertile slopes of the outlying hills and steppe lived more mobile farmers who also tended flocks of sheep, goats or camels. Trade of their surplus meat, milk and hides for surplus grain, fruits and oils, facilitated by tally-systems and a common language for bartering, suited both groups *until* pressures arose: droughts, plagues or population growth. Pressures were generally greatest in the marginal uplands, causing nomads to migrate into more fertile regions, perhaps swooping down on the farm villages, breaking into grain stores and attacking those who tried to defend them. Archaeologist Kaj Birket-Smith describes their raids:

> If a drought scorches the steppe, the nomads' locust swarms descend on the surrounding farmlands.[18]

As societies grew larger, there arose a need for political management. Small egalitarian societies gave way to hierarchies; chiefs appeared, then kings. At first chiefs were simply 'big men' with no special claims to power beyond their own persuasiveness; they were self-made leaders among equals, and were easily superseded. But as societies grew, chiefs began to claim divine personal powers and so became hereditary god-kings.[19] Political power in the earliest civilisations was thus legitimated by a claim to inherited religious authority with its presumption of magical and mystical potency. It was a claim that would last thousands of years.

The archeological record gives the first evidence of organised warfare at this time. Societies were now large enough and affluent enough to engage in extended combat. Towns were fortified and armed cohorts retained to protect them. During the third and second millennia BC, nomadic warriors from the steppe of central Asia repeatedly invaded China, India and the Near East. These mobile herdsmen, latterly mounted on horseback, regularly overran the growing cities, only to merge with those they had conquered.

Over time, the small, diverse egalitarian agricultural communities were fused into severely stratified societies. The conquerors, as rulers and priests, formed the new aristocracy, traders and artisans lived in the towns and cities; at the bottom was a large peasant class, the serfs who were tied to the land and the slaves who provided labour for ancient engineering projects. Wealth flowed upwards. It was a period of hierarchical feudalism and warring states led by competing princes and petty kings.[20]

During this period a major shift occurred in the social purpose of religious power. In prehistory, the sacred mysteries of the universe were shared within small egalitarian societies and the same ethical constraints applied to everyone. But as hierarchies formed, religious knowledge became the secret property of an ever smaller priesthood from whom flowed sacred authority and political legitimacy. Religion acquired the potential for social control and human exploitation, and usurpation of political power required religious legitimation. Historian Lewis Mumford

THE GOLDEN RULE

Chung Kung asked about true manhood, and Confucius replied, 'When the true man appears abroad, he feels as if he were receiving distinguished people, and when ruling over the people, he feels as if he were worshipping God. What he does not want done to himself, he does not do unto others. And so both in the state and in the home, people are satisfied.

> *The Analects*, p. 168,
> (ed. and trans.) Lin Yutang,
> *The Wisdom of Confucius*
> Illustrated Modern Library
> (New York: Random House, 1943).

That which is hateful unto thee do not do unto thy neighbor. This is the whole of the Torah.

> Hillel (1st century BC)
> R. Cavendish, *The Great Religions*
> (New York: Arco) p. 134.

So whatever you wish that men would do to you, do so to them; for this is the law and the prophets.

> Jesus Christ,
> *The Holy Bible*
> Matthew vii, 12

contends that these highly stratified civilisations – he calls them 'mega-machines' – became rife with injustice, and that the major religions of the world arose as a moral counterforce against the intolerable abuses heaped upon the populace by their 'divine' despotic rulers.[21] It is certainly true that all the great religious leaders – the Jewish prophets, Lao Tsŭ, Confucius, Buddha, Jesus, Muhammad – admonished kings to act righteously and preached that justice was a virtue far superior to mere power. Each preached his own version of the Golden Rule (see boxed text above).

The major religions however did not emerge *de novo*; they built on the diverse beliefs of prehistoric peoples. By looking at the spiritual world they lived in we gain perspective on the later religions that still influence us today.

ANIMISM

The earliest men and women walked in a world populated by spirits, endowed like humans with a reason, intelligence and volition. They dwelt not only in things that moved – animals, water and the wind – but also in trees, in rocks, and in the sky. In the world of primitive peoples, human beings are only one of many kinds of spirits who must all interact harmoniously if things are to go well. Other spirits care what people do and keep an eye on them. Humans are by no means in control. In 1854 Chief Seattle, in a letter said to have been written to the American president Franklin Pierce in response to the government's proposal to buy the Indians' land and move the tribe onto a reservation, eloquently explains this animistic worldview:

> Every part of this earth is sacred to my people. Every shining pine needle, every sandy shore, every mist in the dark woods, every clearing and humming insect is holy in the memory and experience of my people. The sap which courses through the trees carries the memories of the red man ... We are part of the earth and it is part of us. The perfumed flowers are our sisters; the deer, the horse, the great eagle, these are our brothers. The rocky crests, the juices in the meadows, the body heat of the pony, and man – all belong to the same family ...

> The air is precious to the red man, for all things share the same breath – the beast, the tree, the man, they all share the same breath ... The wind that gave our grandfather his first breath also receives his last sigh ...

> This we know: the earth does not belong to man; man belongs to the earth. This we know. All things are connected like the blood which unites one family. All things are connected.

> Whatever befalls the earth befalls the sons of the earth. Man did not weave the web of life; he is merely a strand in it. Whatever he does to the web, he does to himself.[22]

It was also believed that the human soul, like other spirits, was immortal and had potential power after death. Ancestor worship was deeply important. Graves of departed spirits were carefully tended and ritual sacrifices made in their honour to ensure prosperity. This strong focus on sacred authority from past generations guaranteed cultural integrity over long periods of time. (The worship of ancestors in fact lingers to this day, not only among traditional agriculturalists such as the Hopi with their kivas, but among rural Chinese who steadfastly refuse to plough over the graves of their ancestors to make way for much-needed rice. America's idolisation of its 'Founding Fathers' – men of a larger stature than any alive today – is a contemporary example of this ancient tendency to seek authority in the past.)

When primitive societies were overrun and absorbed by burgeoning civilisations religious legitimacy had to be given to political conquests.

Often the goddesses of neolithic agriculturalists were superseded by more powerful male gods. Sometimes the sacred places and practices of the old religion were simply absorbed into the conquerors' belief system. There are numerous traces of animism in the Old Testament, for example: the oracular 'oak of Moreh' (Genesis xii, 6–9) indicates a sacred tree, and 'En-mishpat', the spring of decision, locates an oracular well (Genesis xiv, 7). Other passages can also be taken as a 'hallowing in the name of Yahweh of places already recognized as sacred in the pre-mosaic religion, which was thoroughly animistic'.[23]

There are abundant examples of animistic carryovers into the modern era. Among converts to Islam in Malaysia, it is not the Muslim mullah but the local *pawang* or witchdoctor who is called to exorcise evil spirits.[24] In Brazil, more than a quarter of the nominal 'Catholics' now follow the animistic Umbanda religion of Nigeria. And sacred amulets, such as the St Christopher's medal worn by Catholics to fend off danger while travelling, are likely derived from the stone-carved fetishes of earlier animistic societies.

Despite these carryovers, however, animism is disappearing from the world and with it a central religious tenet that informed our ancestors for untold millennia: human beings, far from being all-powerful, are but one among many kinds of sacred spirits and must be a harmonious part of, not lords over, Nature.

THE WORLD'S RELIGIONS

Owing to the forces of electronic communication and mass destruction now at our disposal, the power of religion to influence global history may be greater today than ever in the past. Never before has it been so critical to understand the diverse worldviews embodied in the major religions that underlie the cultural visions of today's restless societies. We begin with the religions of the East, starting with those of India which are the most ancient.

Hinduism: Enlightenment Through Renunciation

Hinduism arose about 1500 BC following the conquest of the agricultural peoples of the Indus River valley by nomadic Aryan herdsmen from the north. Since then it has undergone continuous change and evolution, branching and diversifying, so that today it is a whole family of religions that nevertheless share a similar worldview.

To a Westerner, Hinduism hardly seems a religion: it has no creed, no church, no set faith, nor does it attempt to make converts. Yet for those who follow Hindu teachings, sacred ritual permeates every aspect of life, from arising in the morning, to matters of washing, eating, drinking, sex, childbirth and all else. It is a personal, private religion, a matter for individual conscience. Hinduism is thus one of the world's most tolerant religions. The nineteenth century Hindu saint and mystic, Ramakrishna, deciding to seek God in as many ways as possible, experienced the divine through a dozen different Hindu approaches as well as Islam and Christianity. All religions were simply different ways to God. Despite its great intellectual freedom, Hinduism has generated remarkable social stability, for its overarching vision is of a giant social organism of which the individual is a living part.

Hinduism interweaves two images: an incomprehensibly long cosmic time and the existence in each individual of an eternal divine self, the Âtman. The present universe is but one of many ruled over by the universal spirit Brahman, that which is everything for all time. During the lifetime of a universe Brahman takes on three forms: Brahmā the creator, Vishnu the preserver, and Shiva the destroyer. Brahmā creates each universe in complete harmony; life is idyllic and people live long lives. But with time, things gradually fall into discord and decay, despite Vishnu's attempts to preserve them. Finally Shiva destroys the universe and the entire cycle begins again. The present universe is neither the first nor the last. Moreover, there are many coexisting universes, each with its own Brahmā and in its own stage of the cycle.

The time allotted to each universe is 100 years in the life of Brahmā. Now one *day* in the life of Brahmā is a very long time, notably eight billion solar years. (Geologists estimate Earth as about five billion years old.) Hindus poetically express this period as the time it would take a bird, flying with a feather in its beak and gently brushing the Himalayas with it once every hundred years, to erode the mountains way. A rough calculation puts the lifetime of Brahmā, and hence of one universe, at 300 trillion solar years. Cosmic time has neither beginning nor end, but infinitely recurs in giant cycles in which a human lifetime is less than the wink of an eye.

The ultimate Hindu goal is to unite one's eternal Âtman with the Absolute Essence of which it is part, Brahman, but this happens only when through acts of renunciation *moksha* or perfect beinghood is achieved. If the liberation of *moksha* has not been reached on dying – as is the case for almost everyone – the self is reborn to live again and again, each new life being at a higher or lower level, according to one's present level of beinghood or *karma*.

Since the Absolute Essence pervades everything – oneself, all life, all matter, and all time – there is an overwhelming sense of unity, of personal identity with the entire universe akin to animism. Striving for oneness with all things leads naturally to *ahimsa*, literally 'not harming other life', which also may derive from earlier animistic beliefs. It was *ahimsa* that guided Gandhi's path to non-violent political action in liberating India from British imperialism.

The Hindu expectation of multiple reincarnations has helped maintain India's rigid caste system. Since caste at birth is predetermined by the level of *karma* in a previous life, people tend passively to accept their present circumstances. But it may also engender a concern for Earth's future that Western expectations of an unearthly heaven fail to elicit. Finally, the striving toward *moksha*, the final liberation of the soul, through ascetism, purifying the body, rejecting personal goals and above all living in perfect harmony with all Nature and for the good of all life give Hinduism a deeply spiritual quality. The practice of vegetarianism, the struggle for *ahimsa*, and the overall gentleness, peaceableness and tolerance of devout Hindus all become readily understandable.

Buddhism: Suffering and Release from Suffering

While Hinduism had no founder and possesses, albeit in many forms, a universal deity, Buddhism started as an atheistic faith established by an historical person. Prince Siddhartha Gautama was born about 563 BC in Nepal, the son of a local chieftain. At 35, dissatisfied with the teachings of the local Hindu sect, he began meditating, sitting cross-legged and motionless for many days and nights under what came to be known as the sacred Bodhi-Tree, the tree of wisdom. Finally one night near dawn he achieved complete enlightenment, without the help of god, and entered *nirvana* – life in an ideal and eternal state free of passions, delusion and suffering. As religious scholar Richard Cavendish explains:

> It is no good asking what nirvana is, the Buddhists say, or whether someone who has attained it survives death or not. The meaning of nirvana cannot be conveyed in words, but can only be experienced, as the Buddha himself experienced it. The Buddha is quoted describing what it is not, rather than what it is.[25]

Instead of leaving Earth at that moment the Buddha – or Awakened One – out of his great compassion chose to stay in the mortal sphere to teach what he had learned. For 45 years he remained in the central Ganges plain, teaching and gathering followers.

Although later Buddhist beliefs came to share with Hinduism the concepts of extremely long, recycling periods of time and the reincarnation of the soul, the Buddha was less concerned with cosmology than with the means for eliminating suffering in this life: 'We are in truth without soul or self; there is no substance in us; all individuality is illusion. By rejecting the self, the ego, we become one with all things.' From the Buddhist point of view, I *am* my ancestors, I *am* my offspring, I *am* the lion club, I *am* all things. Buddha repeatedly offered to lay down his life for wild creatures. Thus, like Hinduism but in a slightly different way, Buddhism stresses both a total immersion of the self in Nature and a connectedness with infinite time, but the individual person is but a tiny part of the total scheme of things. Gentle, kindly, meditative and harmonious living is the paramount quality of aspirants to nirvana. (On a visit to China the author was struck by what can only be called the softness of the Buddhist monks.)

Both Hinduism and Buddhism share a deep sense of unity with Nature and lay down goals for the conduct of individual life. But aside from the general notions of harmony and *ahimsa* they are silent regarding the rules of social organisation. For this we turn to Confucianism.

Confucianism and Taoism

We move now to China where, in the middle of the first millennium BC, two great belief systems grew up side by side and simultaneously influenced the culture around them: Confucianism and Taoism. Confucianism is not a cosmology or religion at all, but a secular moral code; it is deeply involved with social relations. Taosim, on the other hand, resembles the Indian religions in its rejection of practical affairs and its search, through a personal union with Nature, for the eternal Tao. It does not, however, carry the concept of reincarnation, which did not spread throughout China until the fourth century AD when Buddhism fused with Taoism as the popular religion.

Both Confucianism and Taoism grew out of the unrest at the end of the Chou reign. From time immemorial China experienced invasions by nomads from the north and west, and the Chou clan, who overran the earlier Shang warriors about 1100 BC, set up a series of city-states that extended south to the Yangtze River, and established the worship of their own ancestors. Claimed the conquering Duke of Chou:

> Heaven . . . sought among your many regions . . . for one who might be attentive to its commands, but there was none able to do so. There was, however, our Chou king, who treated the multitudes well and was virtuous, and presided carefully over the sacrifices to the spirits and to heaven. Heaven therefore instructed us to avail ourselves of its favor; it chose and gave us the decree of Shang, to rule over your many regions.[26]

Thus did the victorious Duke legitimise his conquest. But he devised a just code of governance to guide the widely scattered feudal lords whom his nephew, the young king, appointed as local rulers. Thus began a dual precedent in Chinese thought that survived into this century: the ideas that the system of governance comes as a 'decree from Heaven' and that rulers have a sacred trust toward the people, which includes upholding ancient virtues such as family ties, filial piety and reverence for ancestors.

Over the centuries the appointed fiefdoms became hereditary; as the

Chou kings lost control, the feudal lords fell to squabbling among themselves, and general moral decay set in. It was during this period of corruption that K'ung Fu-tzŭ lived. Confucius, as he is known in the West, was born in 551 BC, probably the dispossessed younger son of a nobleman. Literate, intelligent and morally uncompromising, he felt deep concern for the suffering of the common people. He never gained rank or position, but spent his life teaching the principles of just government to his equally disinherited colleagues – the accumulated younger sons of the nobility – and thus formed a cadre of disciples trained to rule effectively, a kind of morally qualified civil service.

The teachings of Confucius are entirely secular. He believed that a governing class of 'gentlemen' was necessary, but that anyone might aspire to it if his conduct were noble, unselfish, just and kind; one's status by birth was immaterial. He stressed the study of six classics: poetry, history, music, philosophy, prose, and especially *li*, the principle of social order. It is in the explanations of the unfolding of *li* that Confucianism gains its moral – its almost religious – strength (see the boxed text below). Confucianism stresses propriety and ritual not to appease gods, but as a conscious means of assuring social harmony – as in a great orchestra where all the instruments are disciplined to play harmoniously together. Ritual, custom and manners provide a smooth rhythm to social life. Confucianism seeks the meaning of life not by rejecting society (as with Hinduism, Buddhism and Taoism) but through correct participation in it. In his writings Confucius constantly calls for balance in all things, including harmony between society and Nature, and for purity of motive. Confucianism has had a powerful influence on Chinese moral and social philosophy down to the present day. Its drawback from the modern socialist–

THE PRINCIPLE OF *LI* – 'THE WAY IT IS DONE'*

li is the 'correct' way, the 'courteous' way
li is our moral duty
li creates justice
li is the embodiment of the laws of heaven (Nature)
li is the history of continuous social order
li is the means by which a country is properly governed
li is the crystallization of what is right
li is therefore the basis of social harmony

---

*Extracted from Lin Yutang (ed. and trans) *The Wisdom of Confucius*, Illustrated Modern Library, (New York: Random House, 1943); and from H. G. Creel, *Chinese Thought from Confucius to Mao Tsê-tung* (Chicago: University of Chicago Press, 1953).

communist point of view is that it accepts the necessity of a social hierarchy – albeit a non-exploitative one – governed by an educated elite. As such, it was the ethical tradition from prerevolutionary China most abhorrent to Chairman Mao.

The other great Chinese tradition is Taoism. Although some have claimed that the *Tao Tê Ching* – 'The Way and Its Virtue', the classic exposition of Taoist mysticism – was written at the time of Confucius, it probably first appeared a century or two later.[27] Yet some of its concepts must have been familiar in Confucius's time; certainly he used the idea of a correct 'way', although not in the metaphysical sense of Taoism, and he also adopted the necessity of complementary opposites arising out of the facts of Nature.

It is these two ideas – a correct way and complementary opposites – which the Tao raises to a compelling and inescapable paradox. All of Nature is seen and revered as expressing the absolute harmony of opposites: the Yin and Yang. Yang is the male, active and positive force; the sun and fire are Yang. Yin is female, passive, the negative principle; the moon and water are Yin. It is the great paradox of Tao that what is, also is what is not. 'Do nothing and there is nothing that is not done'. The great sage, Chuang Tzǔ, explains:

> The operations of Heaven and Earth proceed with the most admirable order, yet they never speak. The four seasons observe clear laws, but they do not discuss them. All of nature is regulated by exact principles, but it never explains them. The sage penetrates the mystery of the order of Heaven and Earth, and comprehends the principles of nature. Thus the perfect man does nothing, and the great sage originates nothing; that is to say, they merely contemplate the universe.[28]

The cosmology of the Tao thus eschews any causal explanation for existence, seeking understanding instead through the harmonious balancing of opposites. 'The Way' is not to be found through either struggle or renunciation, but through contemplation. Given that Confucianism, Taoism and Buddhism all stress harmony, it is not surprising that the Chinese have managed over the centuries to fuse them, often venerating all three in the same temple.

If one were to assign differences, Eastern religions might be described as Yin: passive, gentle, meditative, personal, while those of the West to which we now turn, are Yang: hierarchical, authoritative, proselytising, activist.

The Western Religions

The greatest distinctions of Western cosmologies compared to the East are the direct role of God in the affairs of humans and the linearity of time.

God created the universe for humans who, far from being small and insignificant, stand close to God himself. Both God and man act in history, which has a distinct beginning and end. There are no cycles, no reincarnations, no 'second chances'.

The central theme of Western religions is a dialogue between human-kind and the forces of good and evil. Zoroastrianism, which arose along with Judaism among the Aryan tribes of ancient Persia, has two gods, one of light, one of darkness: humans must choose which to serve. In Judaism there is but one omnipotent God; the origins of evil lie within humans them-selves who knowingly commit sin. Both religions describe an originally perfect world in which humankind fell from grace and can be redeemed only at the ultimate end of the world when goodness shall again prevail. Since the faithful have a duty to help bring about this state, the West has a history of missionary fervour and religious militancy.

Judaism: the religion of the Covenant

In relation to their numbers, the Jews have had an extraordinary impact on the world, particularly through Judaism's daughter religions, Christianity and Islam. There are historial reasons why Jerusalem is holy to all three religions.

Judaism, which may date back before the second millennium BC,[29] reveals its ancient tribal origins in being a covenant between God and His chosen people, the semi-nomadic Israelites. God will protect them so long as they live according to the sacred articles of His Law. Judaism, perhaps more than any other religion, is one of nationalism.

The word of God, revealed through prophets and recorded in the Bible, especially in the Torah, the first five books of the Old Testament, describes the creation of the universe, the fall of man, and the history of the Jewish people. It prescribes detailed rules for daily life within this righteous community, whose purpose is to prepare the world for the Messiah and the coming of God's kingdom on Earth. Yahweh, a god of gigantic power, enters actively into history. He floods the Earth; He scourges Egypt with terrible plagues; He parts the Red Sea. He is also a protective Father; He goes to battle with His children against their enemies. He is monstrously jealous, raining down terror on those who worship pagan gods or false idols. He is a male God without concubine or queen, implacable, inflexi-ble, intolerant. Unlike the East, which accepts many paths to truth, the God of the West allows only one. The unbending character of Yahweh has deeply influenced the cultural worldviews of the entire Western world.

Regarding the place of humans in the universe, the eighth Psalm makes it clear that God did not create humankind *within* Nature, but as master *over* it:

[W]hat is man that thou art mindful of him,
    and the son of man that thou dost care for him?
Yet thou hast made him little less than God
    and dost crown him with glory and honor
Thou hast given him dominion over the works of thy hands;
    thou hast put all things under his feet,
all sheep and oxen,
    and also the beasts of the field,
the birds of the air, and the fish of the sea,
    whatever passes along the paths of the sea.
O Lord, our Lord,
    how majestic is thy name in all the earth![30]

Instead of the interactive community envisioned by animistic and Eastern religions, Judaism erects a hierarchy, a 'holy order'. Almighty God is seated in Heaven at the pinnacle of the universe, supported by archangels and angels. Beneath Him is man – male man – who clearly dominates woman. Eve is not the original mother of all people but was created almost as an afterthought from Adam's rib. Next come children, then animals, plants, soils and water. Each exists to serve and obey those higher up. This Western image of a hierarchical universe dominated by a God who exists 'out there' contrasts strongly with the contemplative Eastern search for harmony with the universal Absolute Essence.

Yet at each level of the hierarchy, the God of Judaism demands fairness and justice. Among God's children, man may not dominate man. The rules of social order also include love, compassion and responsibility. It was this important aspect of Judaism that the prophet Jesus preached about.

Christianity: incarnation and redemption

Repeated upheavals in the Middle East in the centuries before Christ strained tribal coherence among Jews. There was increasing division and a growing concern for personal salvation. As Cavendish states:

The older Jewish tradition had been concerned with the destiny and ultimate triumph of the nation as a whole, not with the salvation of individual men and women. The purpose of living was to serve God and multiply his holy nation by having children. No hope was generally held out for the dead, whose usefulness to God and the nation was over ... Now, however, this old tribal concept was challenged by beliefs about a life after death and rewards and punishment based on individual conduct on earth ... The righteous would enter God's kingdom, in heaven or the perfected earth, and the wicked would be consigned to hell for ever.[31]

By the last century BC, crisis was believed to be at hand. Ascetic sects of Jews such as the Essenes were preaching the imminent coming of the Messiah, a saviour proclaiming the coming of God's kingdom on Earth. Into such a world was born Jesus of Nazareth.

Throughout his life Jesus saw himself as a prophet of Israel, one of a long line of reformers who fearlessly spoke God's truth; he may have believed he was the Messiah. He certainly believed the Day of Judgement was near at hand when the righteous would enter the kingdom of God and achieve everlasting life. And the righteous by Jesus's interpretation of ancient Judaic law were to be found among the outcasts of society – the poor, the crippled, the lepers, the sinners. These a compassionate God would forgive far more readily than the wealthy and sanctimonious.

Such preaching, calling into question the legitimacy of the entire social order, had powerful repercussions. Theologian Richard Shaull explains:

> Jesus was condemned to death by the religious and political authorities of his day as a criminal, an enemy of the established order. [He] was killed for political reasons, yet we have so spiritualized the story of his life that this fact escapes our comprehension.[32]

Jesus died an outcast, forsaken by all but his small band of loyal followers. His own claims to divinity were ambiguous. The phrases 'I and the Father are one' and 'I am the way, the truth and the life' appear only in the gospel of St John, written many years after Jesus's death. It was St Paul, himself a convert to Christ some years after the crucifixion, who finally severed the ties with Judaism by refusing to force his own converts to accept the daily rituals prescribed for Jews. And it was he who transformed Jesus from a prophet of Israel into the Son of God and instituted the celebration of the Lord's Supper – the partaking of Christ's body and blood in the Holy Eucharist. Christianity thus was born with a double theme. Its moral message was justice, love, compassion and forgiveness; its mystical message, added after Jesus's death, was redemption through faith in God incarnate. The two have not fared equally.

The monotheistic Jews of Palestine rejected incarnation, but the new religion, following Paul's footsteps, gained ground in the cities of the Roman empire, especially among the poor. There followed a prolonged period of persecution of the enthusiastic flock, known as the 'Church of the Catacombs' because it met secretly in undergound burial galleries. The persecution ceased only when the Roman emperor Constantine adopted Christianity in AD 323, but by then the emphasis placed by Jesus the prophet on love and compassion for those suffering in *this* world was all but forgotten in concern for the personal salvation promised by Christ the Saviour in the *next*. The ethical teachings of Jesus took second place to the metaphysics of the Trinity, the virgin birth and the origins of sin in the concerns of the Church.

Ultimately Christianity spread clear across Europe, absorbing earlier deities as it went. The cult of the Virgin, for example, grew out of the worship of several pagan goddesses.[33] Holy Mother Church, through its hold on the immortal souls of princes and kings, became a unifying political force in medieval Europe. It protected its power by conducting Holy Crusades against the Muslim 'infidels' and Inquisitions against heretics within. Although the rise of Protestantism broke the political monopoly of the Church, as we shall see later it did not swing the emphasis of Christianity back toward compassion and justice. It too was concerned with personal redemption. This was also the message the Christian missionaries took to the heathen around the world.

By focusing on eternity, Christianity like Hinduism has become a preserver and legitimiser of a hierarchical *status quo*. Although Christianity no doubt ameliorated the suffering of Europe's poor, particularly during the Middle Ages, the just society of ancient Judaism implicit in the teachings of Jesus the prophet has largely been forgotten for almost 2000 years.

Islam: submission to the will of Allah

During the first centuries of Christianity, the nomadic peoples of the Arabian peninsula still worshipped nature-gods, but the ideas of Judaism and Christianity were also in the air. Through the life of one man, the most powerful of the local deities, Allah, was to become identified with the Jews' Yahweh who had become the Christians' Jehovah. All three Western religions worship the same God.

The city of Mecca had long attracted pilgrims to its Ka'ba shrine and the sacred well of ZamZam, and it was in Mecca that Muhammad was born in AD 570. At 40 he began to experience divine revelations which were written down as the Qur'an (Koran), the sacred and infallible text of Islam. The Muslim faith that emerged is utterly monotheistic: 'There is no god but Allah.' It accepts Biblical history but utterly rejects the Christian Trinity, treating both Jesus and the Old Testament prophets merely as forerunners of Muhammad, God's final and supreme messenger. Muslims share with Christians the same expectations of a Day of Judgement and an afterlife to be spent in heaven or hell, but their paradise is full of sensual delights awaiting the faithful. Faith, prayer, charity, fasting and a pilgrimage to Mecca are the five pillars of salvation.

The social and political consequences of Islam – even up to the present – arose out of the place and times when Muhammad lived. The wealthy merchants of Mecca had grown arrogant and corrupt while widows and orphans starved. Muhammad's earliest revelations condemned such greed and inequality. Like Jesus, the Prophet and his followers were threatened by the politically powerful. But instead of 'turning the other cheek', they took up arms, defeated their enemies and established a godlike Islamic kingdom on Earth. It is more than a church; it is a complete social order where the most religiously pure are qualified to govern. Muslim law is religious law, interpreted from the Qur'an and from written tradition by the ulamā, a body of learned men charged with interpreting God's will.

Although openly enjoying such earthly pleasures as food and family life, a devout Muslim seeks a quiet existence, abjuring all that creates disharmony. Alcohol, gambling and extramarital entanglements are forbidden. There is a strong sense of honour and of social responsibility, particularly toward the less fortunate. Religious devotion is coupled with a duty to spread the faith both by persuasion and the sword. Within 100 years of Muhammad's death the holy wars of Islam had established an empire greater than that of Rome, spreading from the Middle East across North Africa and into Spain. In the eighth century Muslims overran the Indus plain, establishing Islam under the great Mughal emperors of India.

Today nearly a billion Muslims live around the world. In large areas of Asia and Africa they are a majority of the population. The close historical relationship between politics and religion in the Islamic world still exists, particularly in the Middle East. Not surprisingly, the present interface between Islam and modern materialism is the basis of much of the current unrest in the Arab world. As Cavendish says:

> No other religion has in fact remained so close to its original inspiration, and Islam has preserved its traditional identity and withstood the pressure of materialism and religious skepticism more effectively than any of its rivals.[34]

RELIGION IN A SECULAR AGE

Throughout history, politics and religion have mutually shaped one another. Yet despite apparently similar political histories at the dawn of civilisation, the religions that emerged in the East and West were highly divergent. We cannot yet identify the cultural differences or the accidental charismatic personalities that could account for this divergence. But once in place, each belief system acted to further develop cultural views and shape future history.

With certain important exceptions, ours is commonly seen as an age of waning religiosity. Protestant bishops have even asked such questions as 'Is God dead?'[35] Yet although the overt practice of religion may lapse and details of cosmic belief change, the cultural character generated over the centuries persists in powerful and often subconscious ways. In the People's Republic of China, for example, public self-criticism is expected of officials despite Mao's banning of Confucian teachings and among developing countries China is the most sensitive to the natural environment. Eastern peoples also retain their extended sense of time, contributing to both India's slow progress in eliminating the Hindu caste system and Japanese businessmen's ability to plan 100 years ahead.

The West by contrast thinks in weeks and months, taking the daily pulse of the stock market as its guide; five or ten years is a 'long time' ahead. Its passionate belief in 'upward and onward progress' comes from its sense of linear time. And according to historian Lynn White, the West still takes too much to heart Biblical injunctions to dominate the Earth.[36] Species extinction, environmental destruction, and unbridled economic growth are the symptomatic outcomes.

Finally, East and West differ in their expectations of eternity, thoughts of which must lie in the subconscious of even the non-religious person. In Eastern thinking, heaven or the ultimate condition (Ti'en in Chinese, nirvana in Buddhism and Hinduism) is a state beyond the senses and beyond physical existence. In the West, heaven is far more materialistic, a place where the body will be resurrected – and presumably fed, clothed and catered for, especially in the paradise of Islam. The goals of the two religious philosophies thus lie at the antipodes – the one an utter rejection of physical being, the other a desire to possess physical being for all eternity. The consequences of the latter for modern Western thought are taken up in later chapters.

It would be a mistake to assume that because the number of adherents to conventional religions is declining people no longer need what religion has always supplied: an ultimate meaning or explanation for life. In the West, the experience of inner emptiness has become the norm. People try to fill it by seeking 'success' or 'material consumption' or 'psychic experiences' that all too often prove vacuous and unsatisfying.[37] According to theologian Paul Tillich, the fact that these new, surrogate meanings fail to satisfy us is an 'expression of the presence of the divine in the experience of utter separation from it'.[38] Our very emptiness signals our need for transcendental meaning.

Meantime new belief systems are emerging which attempt to substitute for the cosmological and ethical functions of the old religions. The most pervasive of these are science and political ideology.

Science as Religion?

There is a great tendency to view science as a secular replacement for religion. Science after all 'explains' the universe. The miracles of modern technology, created and understood by the high priests of science, incite feelings of awe and wonder among the uninitiated masses. How convenient to believe with Jacob Bronowski that if only the tolerance and empiricism of science were applied to the rest of human affairs, solutions to our most difficult problems would be in sight.[39] At first glance, science seems to offer both cosmological meaning and an ethic based on empirical rationalism. Neither in fact is true.

In cosmological terms, much has recently been made by 'new-age' thinkers of the physicists' latest dilemmas in trying to describe the universe. It is no longer clear where reality stops and the scientist's imagination takes over. The ultimate universe, once thought to be explainable by science, now seems to be mere mental abstraction, not explainable on empirical grounds at all. Yet even if this were *not* the case, could cause and effect explanations of physical existence ever answer the questions we began with? 'Why are we here?' 'What is life for?' Does knowing that the universe is held together by squiggley threads or that DNA encodes the information for physical existence in any way supply a transcendent meaning to life? Science does not answer questions where *purpose*, rather than cause and effect, is involved. In fact, science specifically abjures meaning and intent and to that extent denies purpose to life. Sacred meaning we must supply for ourselves.

The same applies to the notion that science can tell us what to do. First, there are fundamental limits to its ability to make useful predictions about the complex ecosystems and social systems that most interest us. Says computer modeller Donella Meadows:

> We know very much less than we think we know. I mean this profoundly. After years of constructing little worlds on the computers, worlds in which we know all the rules because we have made them, we have learned the severe limitations of the human mind to understand the behavior of any system with more than three species in it, or more than two interacting economic markets, or more than one renewable resource. In short, our minds are unable to simulate any systems we are really interested in. And the computer, while it can help, can still not encompass the full complexity of the real world.[40]

But even if we *could* make accurate predictions about, say, the effects of rainforest destruction or nuclear war on global climate, we are still faced with the moral decision about what to *do*. On this, science is silent. It provides information only on the probability of outcomes, not which outcome is preferable. Beyond science there must always lie a moral 'given'.

Political Ideology as Religion?

In past times, among the bloodiest wars that were fought were holy wars. Although such struggles exist today, the threat of the most terrible war ever – a nuclear holocaust – comes from two secular powers, the United States and the Soviet Union. Yet they do not compete directly for any crucial resource. Their struggle is over political ideology, which has the same emotional functions as religion in knitting each society together and giving it an ultimate purpose.

Under communism, the proletariat are the chosen people and the class struggle is their holy war. The predicted historical inevitability of their success and the coming utopia of universal justice and concord parallel the Old Testament Messianic promises of God's kingdom on Earth; the 'prophets' of communism are Marx, Engels, Lenin and Mao. The inner paradox of Marxism lies in its *belief* that the outcome of the struggle is inevitable, while the struggle itself must consciously be waged through the practice of Hegelian dialectics – a dialogue between thesis and antithesis, leading to a new synthesis that accommodates both. Yet as philosopher-economist Robert Heilbroner points out:

All that dialectics gives us is an understanding of class struggle and alienation as intrinsic elements in the movement of societies through time. As to any 'finality' to these movements, dialectics say nothing, *even though Marx himself (and innumerable Marxists after him) believed in the impending end to class struggle and perhaps to alienation.*[41] (emphasis added).

Communism's faith in an inevitable utopian future has until recently given Soviet society an outlook as rigid as any religious dogma. All events were interpreted through the 'scripture' of Marx and Lenin. Whoever disagreed with the current interpretations was denounced as 'counter-revolutionary', and banished from society. This equivalent of religious persecution seems to be lifting under the new outlook of *glasnost* introduced by Mikhail Gorbachev.

The capitalist West possesses its own dogma and mysticism which is explored in detail in Part III. America, the 'leader of the free world', has its own prophets, the Founding Fathers, whose inspired words in the Declaration of Independence and the Constitution have been extrapolated and reinterpreted to legitimise those modern economic and social institutions that are widely believed to be the *moral* basis of America's phenomenal technological and material success. There is a tendency to equate democracy and political freedom with private enterprise and a free-market economy. Indeed, if America's recent enthusiasm for the economic changes being cautiously introduced in the People's Republic of China is any indication, her ideology places free-market capitalism on a higher rung of the ladder of values than it does the right to vote. The most important freedom seems to be economic, not political, and any who question America's economic institutions are dismissed with the label 'communists'.

Whether they be based on religious tenets or political ideologies, all societies require a sense of purpose and sacred meaning. This is our human dilemma.

A Return to Religious Fundamentalism?

In America, as her global powers wane, there has been a reversion on the part of conservative thinkers toward religious fundamentalism. The words 'one nation *under God*' have been inserted into the Pledge of Allegiance to the flag that is repeated daily by school children, subtly implying that America's cause is morally just because God is on her side. Over the mass media, the evangelical religious sects promote uncritical patriotism as next to godliness, and recent American presidents have been closely associated with such movements. In the words of theologian Richard Shaull:

> [T]he carpenter of Nazareth has been transformed in the United States into a 'superstar', the object of near idolatrous worship by those who are seeking divine validation for their threatened values and a way out of their frustrated search for personal fulfillment. [42]

But a different set of American religious leaders has turned this claim to religious piety through 180 degrees: If the United States is indeed 'under God' then its political acts must be subjected to the most searching moral scrutiny. The Reverend Robert McAfee Brown would say, in a hungry, troubled world, if God is on the side of the oppressed, where is America in relationship to God? [43] This is a question that many Christians are asking more and more. While the evangelical fundamentalists have held the lime-light, the established Christian churches have become actively concerned with economic, political and social justice. From taking a stand against the arms race, to support for Central American political refugees through the sanctuary movement, to the pastoral letter of the Catholic bishops decrying an economic system that ignores and even denies its poor, [44] America's mainstream churches are forcing critical self-examination. It is they who are paying attention to Jesus's fundamental teachings.

What is happening in North America is small compared to the ground-swell change in the Catholic Church of Central and South America. The Church which began as a haven and promise for the poor became after Constantine the conservator of the political *status quo*, and nowhere were the powerful better served by it than in Latin America after the Spanish conquest. But then came Pope John XXIII who, like Jesus, championed the poor, emphasising the right of all humans to a decent standard of living. Following Vatican II, the Second Ecumenical Council convened by John from 1962 to 1965, a new spirit flowed through the Church. As journalist Penny Lernoux writes:

Vatican II widened the floodgates by establishing two radically new principles: that the Church is of and with this world, not composed of some otherworldly body of celestial advocates, and that it is a community of equals, whether they be laity, priest, or bishop, each with some gift to contribute and responsibility to share.[45]

These revolutionary ideas rapidly took root among the Latin American priests who worked among the desperately poor and the politically oppressed. Searching back to the fundamentals of Judeo-Christian belief, they came up with an entirely new perspective, popularly known as Liberation Theology. The new vision rests on a reinterpretation of the message of the Old Testament and the life of Jesus. Yahweh, through His prophets, constantly denounces corruption, oppression and injustice. The god of the Jews, states one liberation theologian, has been 'the symbol of a single-minded pursuit of an egalitarian tribal system'.[46] Another argues that '[t]he poor are the only legitimate interpreters of the biblical text, since that text belongs to the historical memory of the poor'.[47] Authority lies not in the hierarchy headed by the bishop of Rome, but in the words of holy scripture.

With this vision of Jesus as their model, thousands of Latin American priests and nuns, supported by about half the bishops, have established grassroots base-communities among the poorest and most oppressed peoples. Their creed is self-help through faith. As Shaull says:

> For Jesus the oppressed could only be liberated if they were the agents of their own liberation. The Messiah would not rob them of their responsibility for creating history and working out their own destiny, even if this required a long historical struggle.[48]

Such a struggle is indeed underway. As the poor seek peaceably to organise themselves, to form cooperatives, to become literate, they and their clergy are coming under heavy criticism from Rome and the most terrible repression from their own governments. (Accounts of the grisly, heart-rending details are documented in Penny Lernoux's book, *Cry of the People*.) Liberation Theology is frequently labelled 'Marxist', and blamed for communist guerrilla activities. This the bishops of Santiago, Talco, and Copiapó, Chile, vigorously deny:

> Some people believe the 'liberation of Christ' is synonymous with guerilla activity or armed subversion. Such people have never heard the Gospel of Jesus.[49]

The outcome of this struggle on the continent that is home to more than half the world's Catholics still hangs in the balance. What it is about is nothing less than whether society shall be held together by hierarchical authority or by egalitarian participation of all the people.

During this same period, a different though parallel fundamentalist resurgence has been occurring in the Middle East. The development of the gigantic oil fields around the Persian Gulf following the Second World War brought a sudden and massive infusion of foreign ideas and institutions into Muslim society, as well as a flood of new wealth. Foreign entrepreneurs with capitalist notions of property, profit and interest negotiated with Muslim heads of state. Although consequences have varied from one country to the next, none has escaped the unsettling pressures.

Nations reacted in various ways. Turkey, Iraq and Egypt developed more secular governments pandering to outside support – mainly American and Soviet, often both. In both Iraq and Egypt, however, secularism is increasingly unpopular with the masses. A swing of the pendulum back to more traditional Islamic views was responsible for the assassination of Egypt's Anwar Sadat, whom many considered far too pro-Western. Fear of such fundamentalist resurgence no doubt contributed to the well-armed Iraqi leader's decision to invade Iran in 1980, in the hope of quickly crushing the Ayatollah Khomeini's government before it could export its fundamentalist revolution to its neighbours.[50]

Except for Syria and Iran, the Middle Eastern Muslim countries are governed by wealthy members of the traditional ruling families. Associated with the old, hierarchical 'Establishment Islam', they have done little to ensure that the new wealth is fairly distributed among the increasing numbers of poor. Until recently, a sense of loyalty prevented political unrest, but over the past decade or so, a great wave of populist ferment has been surging across Establishment Islam. According to James Bill, professor of government at the University of Texas:

> The fundamental impulse for resurgent Islam comes from the grass-roots of society ... It is a force generated by the mass citizenry, those referred to as the downtrodden and the deprived. Sweeping upward from the angry, alienated and frustrated, Populist Islam has now penetrated the middle classes ... It directly confronts the various ruling elites in the Muslim world, the Islam of Sadat, of the Al-Saud family of Saudi Arabia, and of Mohammad Zia ul-Haq and Gaafer Nimeiri of Pakistan and Sudan.[51]

Populist Islam comes in two forms, Sunni fundamentalism, based primarily on traditional *Arab* Islam, and Shi'i fundamentalism, with its centre in *Persian* tradition. It is the latter – whose primary distinction is its connectedness with the early Imams, the twelve immediate successors of Muhammad, and especially with his grandson, Imam Hussein, martyred in AD 680 – that the Establishment rulers most fear. For Shi'ite Muslims, with their belief in the sumptuous Islamic paradise, martyrdom in the footsteps of Imam Hussein is an honour. Hence their extraordinary militancy. Their best-known religious leader is Iran's Ayatollah Ruhollah Khomeini, who instigated the 1978–9 revolution that deposed the American-supported Shah. Establishment rulers in the Middle East keep a close watch on the many Shi'ites living outside Iran lest they be inspired to imitate the Iranian revolution. The governments of Kuwait, Bahrain, and the United Arab Emirates regularly employ their security forces against their Shi'ite citizens; they are much more tolerant of their less militant Sunni fundamentalist critics.

From an outsider's point of view, however, both forms of Populist Islam have identical goals: the expulsion of foreign influence and the achievement of social justice through a return to basic Islamic lifestyles. America is unpopular for its uncritical support of Establishment Islam and particularly for its continuing support of an increasingly belligerent Israel. At the same time, however, Russian advisors are intensely disliked, sometimes even assassinated, and communist ideology is abhorred. As Professor Bill points out, it is a grave mistake to interpret the events in the Middle East in terms of the East–West conflict:

> The Iranian revolution, the assassination of Sadat, the Lebanese civil war, and all the problems associated with the Palestinian issue cannot be explained in terms of Soviet or communist involvement.[52]

In both Latin America and the Middle East, religious fundamentalism is deeply entwined with popular concerns for social justice.

SOME CONCLUDING THOUGHTS

The world today is in a state of massive religious and moral flux. To some, religious thought seems anachronistic; to others it is indispensable. By now it should be clear that the two primary functions of religion in human existence cannot be dispensed with. We need to know where we stand in relation to the universe – to know 'meaning'. And we need a set of social rules to live by. Secular societies attempt to disconnect the latter from the former, establishing through politics laws that will create and maintain social cohesion. What often escapes our awareness is that laws must be based on *beliefs* about what a society should be and *these are inevitably derived from the cosmological understanding of our religious past.*

Given the deep importance that religion – or its secular counterpart, political ideology – has for the human psyche it is not surprising that different belief systems arouse our passions. How often has one heard that religion and ideology have done more harm than good in human history and should be dispensed with? But we cannot dispense with these, our deepest beliefs and assumptions, although with great effort we can modify them. Part IV of the book takes up this process of modification.

8 On Acquiring a Worldview

> The task of our generation, I have no doubt, is one of metaphysical reconstruction ... Our task – and the task of all education – is to understand the present world, the world in which we live and make our choices ... Education which fails to clarify our central convictions is mere training or indulgence. For it is our central convictions that are in disorder, and, as long as the present antimetaphysical temper persists, the disorder will grow worse. Education, far from ranking as man's greatest resource, will then be an agent of destruction, in accordance with the principle *corruptio optimi pessima.* *
>
> E. F. Schumacher, 1974[†]

INTRODUCTION

One day I found myself in the library searching back and forth between the QP and R sections for the P books. Finally, in desperation, I asked the two student librarians: 'Where have you put "P"?' 'Try the floor below!' they replied. To their great amusement, I wandered off, repeating the alphabet to myself. Until we become senile (or have a temporary lapse of memory), the alphabet and simple arithmetic are part of our mental furniture that we automatically use to make sense of the world. Everyone shares these mental images, and we communicate on that assumption. It is ridiculous, laughable, not to know them. Yet in Moscow, my problem would not have arisen; the Cyrillic alphabet contains no 'que' to confuse with 'pje'(π). In China, there is no alphabet at all. Characters in Chinese dictionaries are listed under the 200 or so different sound symbols they contain, followed by the number of brush strokes needed to write the character.[1]

The point is that we each live and act on the basis of the mental model of the world that we have in our heads. Not only does this mental model include the sequence of letters in the alphabet; it encompasses everything else that we 'know', including the cosmology of one's culture discussed in Chapter 7. As scientist-philosopher Michael Polanyi reminds us:

> Human thought grows only within language and since language can exist only in a society, all thought is rooted in society.[2]

This statement is extremely important, for it reminds us that *no fact, no idea, no thought can ever be wholly free from cultural bias.* The mental

---

*'Worst of all is the corruption of the best.'

world in which we live comes to us in large part from the society into which we are born. From the day of birth on, a child has transmitted to it an enormous but distinctive body of knowledge, concepts, symbols and rules which provide it with this lifelong framework for dealing with the world. It is the individual's cultural worldview. In this chapter, we inquire into how much such worldviews are acquired and *how they change*. We shall need to examine both the stages of human mental development and the methods of cultural transmission. But before tackling these, we need to philosophise a bit on what thinking is, and agree on our definitions of knowledge and worldview.

THE NATURE OF THINKING

'What causes a thought?' Our common experience tells us that we all have thoughts, but no one can precisely explain what a thought is or where it comes from. Yet the generation, use, and transmission of thoughts is the very essence of human experience. Another universal experience is that everyday events all seem to have a cause. Even when a cause is not obvious, we assume there *is* one – that events do not happen 'by themselves'. On this notion, we have built a complex body of scientific knowledge. So pervasive is the idea of cause and effect that we even extend it into the area of moral causes: 'What have I done to deserve this?' asks the cancer victim; 'What act of mine *caused* my illness?' We do this even when we are capable of distinguishing between rational and irrational causes.

If we extrapolate this idea of causality to its ultimate limits, it encompasses all of space, all of time, and all human thoughts and actions. It is the source of all the inevitable but still exasperating 'But *why*, Mummy?' questions of a small child. If I have an idea, what *caused* it? Can anything ever happen without a prior cause, and if not, is there any sense in which I – or you, or anyone at all – can be said to have free will and the ability to make choices? How we resolve this problem for ourselves makes a big difference in the way we set about educating children and otherwise influencing human behaviour and historical events.

On Causality and Free Will

Religions have always assigned first causes to supernatural beings. But the God of the Judeo–Christian–Islamic tradition has created a particular paradox for believers: He stands alone and is omnipotent. Not only did He create the world, He continues to control all that happens in it. Yet (these religions also insist), humans have free will. But how can we be both controlled and still free? This paradox has ___r been satisfactorily resolved. Here is St Thomas Aquinas's attemp___ ___he thirteenth century:

> Man's turning to God is by free choice; ___ ___us man is bidden to turn himself to God. But free choice can be t___ ___ to God only when God turns it . . . It is the part of man to prepa___ ___soul, since he does this by his free choice. And yet he does not d___ ___s without the help of God moving him . . . And thus even the ___ ___movement of free choice, whereby anyone is prepared for recei___ ___he gift of grace, is an act of free choice moved by God . . . Man's ___ ___ration for grace is from God, as mover, and from free choice, as ___ ___d.[3]

For Muslims, the paradox is equa___ ___ifficult; the Qur'an says in one place that all events are predestine___ ___God and elsewhere that human beings are free to obey God or no___ ___s theologian Richard Cavendish:

> In effect, it seems, men do n___ ___ve free will, but they think they do, and in making decisions which ___ ___ believe are their own they become responsible for their decision___ ___ are justly rewarded or punished by God.[4]

The rise of modern science with its enormous explanatory power did not resolve the paradox, but merely transferred it from an omnipotent deity to undirected accumulations of energy-matter. As mathematician-philosopher Bertrand Russell wrote in 1914:

> Given the state of the whole universe throughout any finite time, however short, every previous and subsequent event can theoretically be determined as a function of the given events during that time.[5]

Once again, human thoughts and actions, given that they arise from assemblages of matter-energy – namely, *us* – are assumed to be predetermined; if only we could measure *everything* at once and make the necessary calculations, we could explain all, even our thoughts. And modern supercomputers are just the tool we need to tackle this.

As noted in Chapter 1, there are two objections to this notion. First, it is inherently impossible to know *precisely* the properties of anything, no matter how simple it is; there is always some inescapable error in our measurement: whatever we measure is always uncertain. While the error of a given measurement may be small enough to allow us to predict the properties of the measured object a short time later, the uncertainty of our prediction grows with time. If we observe a simple system at time, t_0, we can predict its condition fairly accurately at time, t_1, less accurately at t_2, and even less so at t_3 and so on. Gradually, the information we initially possessed is dissipated within the system. This continuous loss of information (and hence of predictability) is inherent in Nature. As the French physicist, Leon Brillouin, puts it:

> Causality must be replaced by statistical probabilities; a scientist may or may not believe in determinism. It is a matter of faith, and belongs to metaphysics. Physical discussions are unable to prove or disprove it.[6]

Yet even knowing precisely the motions of all the atoms of the universe would help us not one bit in understanding the causes of our thoughts and actions. The functioning of our brains, although they are no doubt made up of moving atomic particles, cannot be explained by atomic physics *in a way that our minds can grasp*. The world is organised on a hierarchical basis, and a causal explanation appropriate at *one* level is simply not appropriate at another. As noted earlier, complexity of organisation gives rise to new, 'emergent' properties not possessed by the component parts in isolation.[7] We cannot understand the usefulness of, say, a book by knowing the structure and positions of the molecules of ink and paper which it comprises. Such reductionism destroys the very meaning of the subject under study. Science leads us to conclude only that thoughts and actions arise from special but real properties of a complex organ, the brain. The fact that these are properties which we cannot fully describe (and may never fully understand) does not invalidate the logical necessity of their existence.

Our next, related question is: Can this brain generate thoughts and actions *on its own*? Or does it make responses only to the environment? Can thoughts arise in the absence of external causes? Psychologist-turned-philosopher B.F. Skinner says, No, there is no such autonomous self:

> We can follow the path taken by physics and biology by turning directly to the relation between behavior and the environment and neglecting supposed mediating states of mind. Physics did not advance by looking more closely at the jubilance of a falling body, or biology by looking at the nature of vital spirits, and we do not need to try to discover what personalities, states of mind, feelings, traits of character, plans, purposes, intentions, or the other perquisites of autonomous man really are in order to get on with a scientific analysis of behavior.[8]

For Skinner, who devised the famous 'Skinner box' for rewarding or punishing the behaviour of experimental rats and pigeons (Figure 8.1), *all* behaviour, human as well as animal, is environmentally determined (except for such trivial, genetically programmed reflexes as blinking and coughing): 'An experimental analysis shifts the determination of behaviour from autonomous man to the environment'.[9]

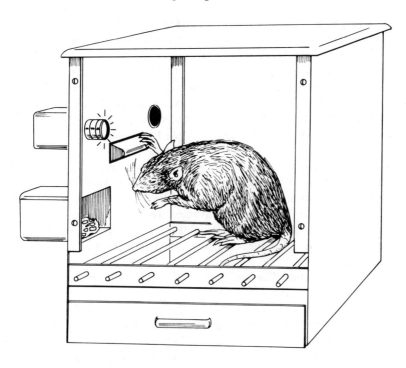

Source: M. E. Clark, *Contemporary Biology*, 2nd edn (Philadelphia: W. B. Saunders, 1979) Figure 14.13, p. 345.

Figure 8.1 Experimental rat in a 'Skinner box'.
The 'Skinner box' was designed to study behaviour modification (learning) through the use of rewards and punishments. Here a rat learns that food can be obtained by pressing the lever only when the light is on.

A whole school of educational philosophy has grown up based on Skinner's theory of 'contingent reinforcement' – the environmental outcome of behaviour reacts back on the doer. Although every schoolchild is aware that brownie points and black marks are frequently used to manipulate its behaviour, Skinnerians argue that such operant conditioning is the *only way* that learning occurs. Furthermore: 'To take advantage of recent advances in the study of learning, mechanical devices must be used'.[10] This theory denies the possibility of spontaneous thought, and effectively exempts moral responsibility in human behaviour. If I am the product only of my genes and my environment, I cannot help what I do; but as part of *your* environment, I affect your actions. I cannot be responsible for how I acted on you, nor can you be responsible for what you did as a result. Certainly, the environment, also incapable of spontaneous thoughts, is not morally responsible. Saying, as Skinner does, that the environment that controls us is largely of our own making is, by his own view of things, a circular argument and solves nothing. (Unfortunately, Skinner tries to have his cake and eat it too, by making us not responsible for our acts, yet capable of controlling them.)

This inconvenient state of affairs is eliminated if we agree that human thoughts and actions are more than inevitable responses to the environment, modified only by past reinforcing experiences. The alternative is to assume that conscious awareness and thought are *internal* properties of the functioning brain. As cognitive psychologist Piaget said:

> [C]onsciousness... makes use of specific notions that are quite outside physical or physiological causality... [F]or example, 2 is not the 'cause' of 4, but its meaning 'implies' that $2 + 2 = 4$, which is not at all the same thing. [Our conscious awareness thus permits perception of meaningful, yet abstract, relationships;] it is the very basis of the formal systems on which our comprehension of matter depends.[11]

And here is how Albert Einstein, one of the most perceptive scientists of all time, described thinking:

> It seems that the human mind has first to construct forms independently before we can find them in things... [K]nowledge cannot spring from experience alone but only from the comparison of the inventions of the intellect with observed fact.[12]

We need to *invent* an interpreted world in our minds in order to understand – and react to – the external world, and we use a semeiotic shorthand of words, symbols, images, and numbers to do this. Furthermore, this inventive ability (otherwise known as *imagination*) permits us to construct worlds *that have never existed and may never exist!* (The once widely held vision that the world is flat is a case in point.) It is these internal imaginings of what the world *might* be like that science historian Thomas Kuhn calls 'scientific paradigms'; they are the working visions that scientists use to guide them as they explore external reality.[13] Regarding free will, our freedom to choose among behavioural options, we now see that *it is precisely this ability to construct multiple mental visions of what a future world might be like that provides us with the opportunity for choice.*[14] In the discussion that follows, it is assumed that both an inner mental world and free will are part of human nature.

The Inner World

We obtain information about the outside world through our five senses, yet as the testimony of trial witnesses repeatedly shows, we do not register all we see and hear. We *select*. We ignore non-significant signals. For instance, if you are really concentrating when reading, you do not 'hear' the noises around you unless someone speaks your name or some other significant word. We internalise only what we pay attention to, and to a far larger extent than most of us realise, what we pay attention to is what our culture tells us is important. Culture selects a world for us!

Incoming information is of two sorts. There is direct observation, where we ourselves hear, see, touch, taste or smell and so 'know'. And although our senses can fool us (as children we learn about optical illusions and magicians' sleight-of-hand tricks), in general they are pretty reliable, and we discover ways to test their reliability. Our other source of information is second hand. It is what *other* people report to us. It is on *their* authority that we accept the validity of what they communicate. In fact, other people are far and away our greatest source of knowledge. Not only do they tell us of things we can never experience; much of the meaning and importance of our primary observations of the world are interpreted to us by others. Our understanding of history, our religious beliefs, our scientific explanations about how the world works are almost entirely conveyed to us by others. This is what it means to be a cultural animal.

Purveyors of knowledge usually try to demonstrate their reliability. Historians provide 'documentation' – letters, laws, proclamations; clergy rely on holy scripture; scientists describe detailed experiments. Our decision to accept such secondary information depends considerably on whether we trust the qualifications of our informants – which in turn may be determined by the stage we have reached in life. In general, knowledge assimilated at an early age is less amenable to subsequent critical alternation than is information gathered later on. *Such questioning of secondary knowledge is a major factor in deliberate cultural change*, which is what this book is all about.

The information and ideas entering the brain and stored in its 'memory' through interactions among networks of neurons do not form photographic images of reality, but rather constitute diffuse 'maps' to which we make reference with each new experience. This inner, generalised map is thus constantly adjusted, not only by external experience but also by internal reflection or 'thinking'. Psychologist Jean Piaget believes such manipulations are logical – that is, they are based on the kind of stepwise thinking that leads to an inevitable conclusion. But he focused mainly on those limited aspects of knowledge which indeed are logically obtained – namely, the basic spatial and temporal relations learned by young children.[15]

More recent studies in cognitive science indicate that much of our 'thinking', our manipulation of facts and ideas to formulate our inner map, is less logical than intuitive. We create categories and generalisations about reality that are linked up on the basis of untested *beliefs and assumptions* – some of which we form for ourselves, others of which are taught us by our peers. Our everyday reasoning relies on the recognition of similarities or analogies between two situations, and we are constantly referring from immediate experience back to the inner map *and its assumptions*.[16] Thinking, pondering, philosophising, are thus not so much logical manipulations of 'facts' as readjustment of beliefs and assumptions to see 'what the world of reality might be like'. This is what scientists, sleuths and inventors have done when they suddenly shout 'Eureka! – I have found it'. It is what generates not only art and music but also bridges, buildings and new industrial processes. It is the thrill of such mental insights that drives creative experience, including education at its best. Finally, it is this kind of thought process that allows us to imagine beforehand what might happen – *to imagine the future*. But creative imagination cannot work in a vacuum; it requires a mind packed with pertinent information to draw upon, to be tallied up and fitted together in new ways. And it requires a willingness to undertake something called critical thinking, the 'critiquing' of our beliefs and assumptions.

In a highly uniform culture, all adults share a nearly identical worldview, but this is not true in complex modern societies, where there are substantial differences not only between ethnic or socioeconomic groups, but also between generations. For example, to be an effective teacher, I find I must constantly struggle to make the concepts in *my* imagination meaningful in the framework of my students' inner world. The job of communication is to locate points of similarity between worldviews, and build on them.

It is the development of this inner world that is the central subject of this chapter. It is a world composed not just of selected facts and beliefs about reality, but also of what is more important and less important, what is right and wrong, good and bad, desirable and undesirable. How do we learn who we are and how to behave? And how do we change and modify that inner world?

COGNITIVE DEVELOPMENT

Some time after birth, the cells of the human brain stop dividing and the nervous system contains all the neurones it will ever have. The further anatomical development of the brain during childhood consists of the formation of synapses – junctions between brain cells – and facilitation of these neuronal pathways. These processes are greatly dependent on a rich supply of incoming stimuli from the sense organs and from the proprioceptors located in the muscles and tendons of the body. Both in animals and in humans, sensory deprivation or motor incapacitation hinders normal mental development.[17] Our minds cannot 'grow up' if isolated from their surroundings.

Although everyone begins as a baby and grows into a perceptive adult, no one, not even the neuroscientists, knows much about the mechanics of thinking and learning. Not even the computers used to model brain functions can handle information in ways every small child can. So far, computers can only obey 'rules' – like those for playing chess or diagnosing certain diseases – that have been borrowed from experts. But computers lack tacit knowledge – the ability we have to recognise in an instant the whole of something – a human face, an accident about to happen, the solution to a murder mystery. These are matters of intuition, of making sense of diffuse patterns, not of following a set of logical rules that can be programmed into a computer.[18] Without a good model of how we think and learn, it is difficult to study how our brains develop or to agree on the best way to teach children new concepts.

Nevertheless, most child psychologists agree that the growing child goes through a sequence of cognitive stages, measured roughly by increasing capabilities in perception, language use, and reasoning power. Development comes about by continuing interactions among three factors. There is the physical child itself: its maturing brain, sense organs, and muscular system, that develop best in an enriched environment. There is the child's *active* exploration of that environment, which helps it, through direct sensory contact, to create an inner image of the world. And there is interaction with others, who constantly challenge the child's inner image of the world, forcing it to reconsider its assumptions. For Piaget, growing up is emancipating oneself from one's immediate perceptions, becoming free to extrapolate from what is *now* observed to past and future images stored or invented in the mind.[19]

Without going into the details of these stages (which in any case are somewhat in dispute) we can summarise a few major findings with regard to child development that may help us in thinking about the processes of cultural transmission. First, young children are not miniature adults; they live in their own world, where reality is quite different than for an adult. For an infant, when a person or object disappears, there is no assurance it will ever return. Such early games as peek-a-boo and hide-and-seek are far more than just entertainment; they teach the concept of the permanency of objects. All things that move are believed by a young child to be alive and to have minds of their own. (Recall from *A Child's Garden of Verses* Robert Louis Stevenson's charming poem: 'I have a little shadow that goes in and out with me, and what can be the use of him is more than I can see.') Children also invent playmates and make-believe worlds by which they actively live out their inner reality. Besides contributing to physical development, a child's play is its way of practising and assimilating experience, of learning how to construct an inner reality. This sort of mental activity is virtually absent when a child is passively entertained, as when it sits staring at a television programme.[20] Active play, both alone and with others, is essential for normal development.

Along with cognitive development goes moral development. Although on occasion quite young children spontaneously express deep compassion, especially when illness or death strike a family, on the whole their dependent physical status is coupled with moral dependency as well. They need constant guidance. Young children fear making mistakes – even when accidental – and need reassuring praise for 'good' actions. By school age, however, this simple world of reward and punishment has progressed to a pragmatic stage in which 'fairness' and 'rules' are needed to create a functioning society. In the schoolyards of England for example, Iona and Peter Opie have observed a rigorous sort of verbal morality:

> The schoolchild ... conducts his business with his fellows by ritual declaration. His affidavits, promissory notes, claims, deeds of conveyance, receipts, and notices of resignation, are verbal, and are sealed by the utterance of ancient words which are recognized and considered binding by the whole community.[21]

'Cross my heart and hope to die', 'Giving's keeping, taking back's stealing', and 'Finders keepers, losers weepers', and time-out signals such as 'Kings!' or 'Exes!' are all part of the code of oral legislation. Such rules are obeyed to obtain social acceptance and avoid censure. The fear of rupturing precious social bonds thus plays an enormous role in developing ethical behaviour.

Many adults reach an even higher level of moral behaviour that is independent of such Skinnerian consequences, what psychologist Lawrence Kohlberg calls principled morality,[22] where one's views of right and wrong are internally determined and may differ from those of society. Draft resisters and conscientious objectors to war are familiar examples of people expressing this level of moral development, but we all exhibit it whenever we are acting 'according to our conscience'. *The potential for principled action is essential for critiquing and changing worldviews.*

The Western tendency to establish developmental 'standards' for children creates unnecessary difficulties. Rates of development differ, both with the child and with the environment. Those growing up in 'relaxed' rural environments may lag behind their 'pressured' urban peers. Furthermore, standards of intellectual ability are laden with deep cultural biases. There is the emotionally charged question of IQ tests and differences in inherited intelligence among races, now largely shown to be non-existent.[23] But ethnocentrism still too often leads us to false conclusions about the mental capacities of peoples with worldviews different from our own. This is a problem we deal with further in Part IV. Here we can conclude that the optimum development of a child, whatever its culture, depends on rich interactions with its surroundings and on supportive and patient adults. Deprived of either, it may remain intellectually or emotionally underdeveloped. Which brings us to the ways that cultures are transmitted to the next generation.

PRINCIPLES OF CULTURAL TRANSMISSION

There are two routes by which information can be transmitted from one generation to the next – via genes, or via culture. Genetic transmission can occur *only* between parents and offspring, and all of the information is transmitted at one time in a single package – the fertilised egg. Whenever favourable genes arise, they spread only slowly through the population, a process requiring many generations. By contrast, cultural transmission can occur between any two members of a population, it continues throughout an individual's entire lifetime, and the rate of transmission of new ideas can be very rapid indeed. New parodies of Walt Disney's 'Ballad of Davy Crockett' swept by word of mouth across the schoolyards of England in a matter of days,[24] and modern electronics make transmission of information globally instantaneous.

Are Genes Involved?

Although people are prone, in their ethnocentric way, to assign genetic differences (usually deficiencies) to members of other cultures, there is no established body of evidence for genetic differences in intelligence among cultures or for the genetic transmission of culture-specific behaviours. Although two sociobiologists at Harvard have theorised that certain cultural traits are indeed inherited – such as what phobias we have, the relative proportions of our several emotions, and how we make decisions – they offer no evidence that such inheritance in fact occurs nor do they suggest feasible means for testing their theory.[25] As critics have pointed out, 'their theory can have little heuristic value because it is open to only weak tests at best'.[26] Here we shall assume that the cultural specifics of human behaviour (as distinct from its total potential range) are *learned* or *invented*, and are not genetically transmitted. While we can agree with the many mothers who will swear that their children were each born with a particular personality, this is not the same thing as saying that an entire society owes its particular cultural character to commonly shared genes.

Routes of Cultural Transmission

Besides our own senses, what we know about the world plainly comes from our parents, our teachers, our friends, the media. But which aspects come from which sources, and how much does each influence the way we think and act? Attempts to analyse these features of cultural transmission are relatively new, and it should not surprise us to find evolutionary geneticists at the forefront in developing theoretical approaches.[27] It is too soon for any useful quantitative data to have accumulated, but some qualitative predictions can be made about cultural stability and rates of cultural change for different routes of transmission.

Three possible routes for transmission of cultural traits are shown schematically in Figure 8.2. *Vertical transmission* occurs directly from parents to children, from one-to-one, or one-to-few. Since an entire generation is required, cultural traits transmitted vertically are slow to change. This is even more true if grandparents as well as parents share in child-rearing. If the society is fairly uniform, then all children will naturally share similar cultural traits, and the society remains stable over long periods. But if the society is pluralistic, there will be a high level of heterogeneity of vertically transmitted traits within the population, which will be slow to disappear. Among a sample population of students at

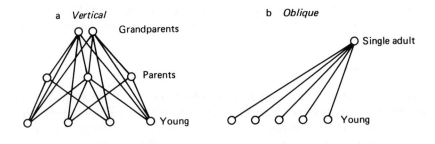

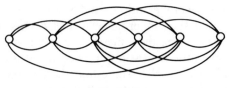

c   *Horizontal*

Source: M. E. Clark.

Figure 8.2 Routes of cultural transmission.
The routes of cultural transmission described by L. L. Cavalli-Sforza and his colleagues (*Science*, 218 (1982) pp. 19–27).

a: *Vertical transmission* – occurs within a family, extended family or small village. In hunting–gathering and other traditional societies this is the primary route of cultural transmission. The presence of strong grandparental influence conserves traditional values. In modern nuclear or single-parent families, vertical transmission loses some of its strength.

b: *Oblique transmission* – one significant adult influences many young. Such adults include teachers, religious leaders, youth leaders (scoutmasters and so forth), politicians, media commentators, and other role models (Jesus, Hitler, Gandhi and Churchill fit this category).

c: *Horizontal transmission* – peer influence can cause both rapid change (fads) and conservation (superstitions, myths) of values. Because acceptance or ostracism by peers is a most powerful influence on behaviour, this form of transmission can effect rapid cultural change.

Stanford University, it was found that mothers had an enormous influence on their children's religious affiliation, particularly among Jews and Catholics, while both parents significantly influenced the political views of their children. Traits reflecting food preferences or daily habits rather than deep-seated beliefs were much less influenced by parents.[28]

In *oblique transmission*, cultural traits are passed to children by non-parental adults. This group includes the extended family, teachers, ministers, political leaders, and today, the mass media. For uniform cultures, such as small hunting–gathering societies, or isolated social classes within a larger social system, many like-minded adults influence each child and once again the cultural trait is stable over time and homogeneous in the subset of the society. But in most larger, pluralistic societies, one important adult, such as a school teacher or film star, influences many children, in which case new traits may spread rapidly throughout the younger generation, regardless of the diversity of their parents' views. This is particularly exemplified by history's greatest teachers, people like Confucius, Jesus, and Gandhi. Their 'pupils', although not all children, looked to them as figures of parental authority.

Social leaders who deliberately decide to make rapid cultural changes introduce their ideas widely among the younger generation. This was what Hitler did when he developed his *Jugend Bewegung* (youth movement). The Boy and Girl Scout movements transmit numerous cultural traits to the young and have been used to facilitate rapid cultural change. During the 1960s civil rights movement in the United States, every effort was made to eliminate anything – such as songs with the word 'pickaninny' – that might be construed as a racial slur, and to develop racially integrated programmes. In the USSR and China, all children are encouraged to join the Young Communists, an organisation that stresses the precepts of their present governments, and most grade school children proudly sport the red ties that signify membership. Since formal education of the young is a powerful tool for cultural change, school curricula quickly become politic-ally heated when subjects relating to deep-seated beliefs and values are introduced, such as sex education or Creationism or school prayer. What is truly amazing, however, is the relatively little public concern paid to the influence of the mass media – especially television – on young minds. We shall return to questions of the schools and the media shortly.

Cultural transmission can also be *horizontal* when ideas are passed between peers or siblings of similar age. We have already noted examples, from the schoolyards of England, where the rate of transmission of new ideas can be almost instantaneous. The spread of new fads and cults among the young seems to occur ever faster in this electronic era, and sometimes they have lasting effects on the whole society. Two powerful political movements of the 1960s were peer-maintained rebellions of the young. After Chairman Mao encouraged China's youth to take over society, the

Cultural Revolution was largely sustained by peer transmission and peer pressure. Less violent but more autonomous was the rebellion of American university students against the general insensitivity of society reflected in the giant 'multiversities'. Both episodes had permanent consequences for their respective society's worldviews.

Rates of Cultural Change

Even before Alvin Toffler's book *Future Shock* appeared in 1970, the average Westerner was aware of the rapid rate of cultural change. But few reflect on it critically, merely accepting change as an inevitable part of living in the bustling modern world. No serious consequences are contemplated, either for the individual or for society as a whole. After all, the human being is practically infinitely plastic: just think of all the diverse cultural conditions under which people have survived. This easy assumption of relatively painless change is possible, I believe, because no one stops to ask how societies (rather than individual people) really work nor how change affects them. There are, after all, few historical examples to support our hasty assumption of infinite social plasticity. Perhaps the person who best understood the meaning of change for human societies was Margaret Mead, who recorded her thinking about this in her book *Culture and Commitment*. Much of what follows comes from that source.

Through almost all of human existence, rates of cultural change have been sufficiently slow that a child growing up knew in advance what its adult life would be like. One's career, marriage, and place in society were predictable. 'The past of the adults is the future of each new generation; their lives provide the groundplan!'[29] In the minds of the young, there was no other possible way to be. Every individual, young or old, embodied the whole worldview of the society. No one questioned the fundamental assumptions about how life should be lived, not because conformity was imposed intentionally – in the manner of modern totalitarian systems – but simply because it never occurred to people to think otherwise: 'That's the way we do'. Even when cultural contacts made alternative worldviews available, they were almost always considered inferior. Ethnocentrism – preferring one's own culture – is not new; what is new is the concept of 'progress'.

Few of the remaining traditional societies are untouched by the encroaching world and its seeming insistence on change. Many have consciously resisted, with varying degrees of success. Only small enclaves of the aborigines of Australia and of native Americans remain intact. But in North America, several highly religious farming communities – the Hutterites, the Amish, the Dukhabors – have intentionally isolated themselves, rejecting modern technology yet still remaining prosperous.[30] The current bloody struggles in the Middle East are about maintaining ancient Islamic traditions in the face of intrusive change. As Mead has emphasised, stable, unchanging cultures have always required close contact among three generations.

But in most contemporary societies, change is the order of the day. Such change and the concomitant loss of cultural stability are hastened whenever people become highly mobile, and nuclear rather than extended families are the norm. The grandparents are no longer the models. Instead, each parental generation has to work out for itself new ways of doing things. The process has been most marked for families who have migrated to a new culture. Their young acquire their worldview not from their parents and grandparents, but from the institutions – the army, the schools, youth groups, and the mass media – of the new culture. Oblique transmission dominates, and parents relinquish the control of their children to others. It is a phenomenon that typifies America's past and present history, but it also occurs when foreign (usually Western) cultures infiltrate and take over traditional societies.

Today, however, there is an additional factor in cultural change at large in the world: 'the expectation that each new generation will experience a technologically different world'.[31] Despite this expectation, parents cling to the belief that their children will miraculously retain the basic cultural values that they themselves had, that only the mechanical details and work skills of their lives will be altered. As Mead noted:

> [C]hildren in our own and many other cultures are being reared to an expectation of *change within changelessness*. The mere admission that the values of the young generation, or of some group within it, may be different in *kind* from those of their elders is treated as a threat to whatever moral, patriotic and religious values their parents uphold with ... unquestioning zeal or with ... defensive loyalty.[32]

We are, of course, doing as human beings have always done, clinging to our old worldview, to tradition, to what is familiar, because in fact it is built into our personalities. It is the basis of all the beliefs, biases and assumptions by which we 'walk across the room'. But it is unrealistic. New technologies are making many old values anachronistic. In a sense, we are trying to fit 'steel axes' into an outdated worldview. We teach our children that the unbounded economic growth necessary to maintain a free-enterprise system is still possible in a world of limited resources; that the West is so strong that it can isolate itself from the economic demands of the billions of poor in the world, although thanks to the electronic media we have created, those poor are fully informed as to how 'the rich' live. We continue to teach our children, through our schools, through sports, and the job market, that human nature is inherently competitive and that, in an age of nuclear weapons (which are not very hard to make), it is still possible to protect one's wealth and culture through the threat of violence. Our worldview is out of touch with the new realities our own technologies have created.

The young today are uneasy. They know that parental values are out of touch. They are also quite aware that they are being moulded to fit a productive slot in the social system. In talking with students, one senses they accept this as inescapable necessity, not with any belief that it is the right way for the world to be. There is deep, unvoiced cynicism; the proferred ethics do not match the reality. By treating life as merely a game with rules that a person can opt in and out of, Western culture trivialises both the individual and society. [33] And as Mead reminds us, even the rules we are forcing the young to play by are socially destructive:

> Resistance among the young [to modern society] is also expressed by an essentially uninvolved and exploitative compliance with rules – rules that are regarded as meaningless. Perhaps those who take this stand are the most frightening... [for they come] to regard all social systems in terms of exploitation. [34]

Is this not indeed our social message to the young? Don't try to question or understand – just compete and succeed! No more 'fair play'; no more social conscience. [35] This is the ultimate in individualism: complete alienation.

We are now in an era where social values are disregarded, rejected, or at best treated with cynicism by the new generation. There is no conscious effort to examine values or even to ask *how* to critically evaluate – to 'critique' – them. Yet these are the very skills that we should be developing in our young. What routes have we for tackling this task?

LEARNING TO 'CRITIQUE' VALUES

In Western industrial societies, the worldviews people hold are being fashioned less and less by parental values and beliefs and more and more by the oblique transmission of ideas from three sources: political leaders, commerical media, and the schools. In a sense, the extraordinarily centralised, special-interest power represented by the first two can be balanced only by what takes place in the third. Whatever politicians and other media users tell us is 'true' and 'good', our educational system must teach us independently to criticise and evaluate.

Before proceeding to this critical role of education in modern society, we need briefly to examine the influence of politicians and the media on how we see and interpret the world.

Politicians: 'Telling It Like It Isn't'

Ideally, political leaders are elected to high office for their qualities of wisdom and diplomacy. Above all, their judgement is to be trusted. Political leaders become accepted figures of authority. Their facts, their representations of reality, are 'true', even if one disagrees with their policies. Ideally, that is the case. In practice, we are far from it, as contemporary voter cynicism and widespread apathy indicate. In an age when a clear grasp of reality is more necessary than ever before, political leaders are failing us in two ways: they distort the truth; and they withhold the truth.

Richard Lamm, the environmentally-minded former Governor of Colorado, has said 'I know some politicians who as captains of the *Titanic* would persuade the passengers they were only stopping for ice'.[36] When the people prefer not to know the magnitude of the problems they face, the politician who paints the rosiest picture gets elected. Unpalatable truths are anathema to leaders in any political system, but they can be fatal in democracies unless leaders of every stripe agree to face them, or the public demands they be addressed. Today, neither is yet the case.

Various techniques are used to avoid facing the hard issues – limits to growth, environmental destruction, economic instability, growing debt, and so on. Most commonly, they are dismissed as 'unimportant', 'exaggerated', as 'nothing to worry about'. They are pushed onto the back burner, while public attention is deflected to such emotionally gripping issues as national security, national morality, and national prestige. Given the power of the spoken word and television's ability to reach an entire nation, modern politicians are more and more substituting rhetoric for substance to maintain their positions of leadership. It is a technique not unknown in the past. Historian Norman Baynes writes:

> Hitler from the first recognized the preponderant significance of the spoken over the written word: the force which set in motion all the great historical avalanches, whether political or religious, was from the beginning of time the spell of the spoken word.[37]

Today's 'teflon' world makes it even easier for politicians to maintain power through appeals to a 'higher' authority (Providence, God), to national pride, and to fear of a common enemy, while ignoring the hard issues at hand.

George Orwell, who died as the television age was beginning, was deeply aware of the medium's potential misuse by politicians. Already distrustful of the way language was being distorted by speech writers, he foresaw, in his magnificent novel *Nineteen Eighty-Four*, how the mass projection of a pre-selected 'reality' could create the desired version of 'truth' – and hence of behaviour – in a whole society.[38] Although Big Brother is not yet in charge and *newspeak* is not yet in common use, there is no doubt that television provides opportunities both for verbal rhetoric and for presenting the truth as selected by the camera's eye as if it were *all* the truth – opportunities that public leaders have not been slow to utilise.

Beyond merely controlling what is presented in the media, politicians can and do create biased images for the public – as when Ronald Reagan recently likened the Nicaraguan rebels (the Contras) to America's Founding Fathers. Even worse is their growing tendency to conduct government in secret. In the mid-1980s, secret illegal acts by both British and American governments were publicly exposed. MI5, the British secret service had tried ten years earlier to subvert its own government by insinuating that Prime Minister Harold Wilson was a Communist sympathiser.[39] And under President Reagan, the American National Security Council, supposedly an advisory body, conducted secret illegal operations in several countries and then deliberately lied to the press, to Congress, and even to the President. Governments that allow their own agents to act in secret deceive not only the supposed enemy, but their own people. As Norman Cousins asks:

> What is meant by 'covert' operations? Does it mean that an agency of the US government should be free to engage in assassinations or the overthrow of other governments? Does it mean that activities contrary to the traditions and institutions of the American people are acceptable because this is supposed to be the only way to cope with potential enemies? The moment we acquiesce to such an approach, we do to ourselves what no enemy in our history has been able to do to us.[40]

The distortion of reality of which the democracies accuse dictatorships, their politicians now readily employ at home. The above are only the most recent and well-publicised examples. The Vietnam War was a case where American generals concealed the true dimensions of US involvement from Congress and the public, as the Pentagon Papers revealed. Politicians today deliberately and repeatedly distort reality to the point where public understanding is paralysed. There no longer is such a thing as 'public policy'.

The Bias of the Media

That politicians in free societies can distort reality to the degree they do is in large part a consequence of who decides what is 'news' and how that news is interpreted. Despite the existence of a 'free press', the vast majority of people living in Western democracies glean their information from a handful of newspapers and television networks, whose description of the world is highly selective. The problem is particularly severe in the

United States; on the whole, European journalism has traditionally been more openly politicised and more frankly diverse, functioning as a true 'fourth estate' overseeing the establishment. Except for the restrictive British Official Secrets Act, European media are generally freer than their American counterparts, whose brief escape from establishment oversight is now virtually ended.[41]

American journalism emerged from the patriotic fervour of the Second World War into the Cold War and Korean conflict abroad and McCarthyism at home. For several years the flamboyant senator's unsubstantiated accusations of communists in high places went unchallenged by reporters who duly reported all of them as 'news'. Only in 1954 did CBS commentator Edward R. Murrow finally adopt an adversarial stance, something no one else had dared to do. It was the first sign of an awakening press, but the process was slow. When in 1957, *New York Times* correspondent Herbert Matthews discovered Fidel Castro still alive and heading his guerrillas in Cuba's jungle, he was widely attacked as a 'meddler' who should have left well enough alone. In Vietnam in the early 1960s, reporters like David Halberstam and Harrison Salisbury filed story after story about what they saw actually taking place, and the media and public at home became deeply split between those who believed their first-hand reports of a solidly based and independent communist movement, and those who preferred the official accounts from Washington of a war America was winning. Finally, after the Tet offensive in 1968, when the Vietcong mounted a surprise attack on the US embassy in Saigon, CBS anchorman Walter Cronkite discarded the 'neutrality' of the networks, urging that the United States should get out. Journalists began giving opinions different from officialdom.

Meantime on the home front, a new phalanx of the press began *independently* to report on the civil-rights movement in the South, taking their cues not from official reports of the white establishment, but from their own on-the-spot observations. Said the *doyen* of American professional journalism, columnist Walter Lippmann:

> This growing professionalism is, I believe, the most radical innovation since the press became free of government control and censorship. For it introduces into the conscience of the working journalist a commitment to seek the truth which is independent of and superior to all his other commitments – his commitment to publish newspapers that will sell, his commitment to his political party, his commitment even to promote the policies of his government.[42]

Through most of the 1970s, a sense of socially sensitive, politically independent journalism dominated, with the investigative reporting of the Watergate scandal the hallmark of the decade. But then the pendulum began to swing. Journalists were said by some to be out of control, not reflecting 'dominant' social values. The establishment claimed they were subversives, society's adversaries, not its guardians. Media management, itself a profit-making part of the same establishment, reasserted its bureaucratic control. Said Michael O'Neill, editor of the New York *Daily News*:

> We should make peace with the government. We should not be its enemy... We are supposed to be the observers, not the participants – the neutral party, not the permanent opposition... We should cure ourselves of our adversarial mindset. The adversarial culture is a disease attacking the nation's vital organs.[43]

And so American journalism has returned to where it had been before. George Kennan has called it the greatest 'militarization of thought and discourse' since the Second World War.[44] Journalism historian James Boylan sees the media has having given in entirely to what he calls 'supernews' – events such as the American invasion of Grenada, the Soviet shooting down of a Korean airliner, and the US bombing of Libya, where the massive outpouring of official 'information' renders all countering voices impotent. Supernews now comes not only directly from official channels, but also circuitously, by way of the CIA (which runs the biggest news service in the world) and the US Information Service, both of which plant news items abroad that are then picked up by the media back home;[45] and by way of White House controlled but secretly and privately funded contractors such as the Latin American Public Diplomacy Office, whose job is to 'educate' the American public according to the way the President wants.[46] As Boylan concludes:

> The standard that provided incentive for coverage of Vietnam and Watergate, that of journalism that could stand the test of history, has been placed in jeopardy by supernews. In Central America, American journalists have exposed serious shortcomings in American policies and clients, but over the years the government has successfully overcome such details and has won its main points, to the extent that by 1986 official premises – that, for example, a government in Central America constitutes a major security threat to the United States – underlay many news stories. Such assumptions were effectively tested by reporting from Vietnam; in the case of Central America, by contrast, supernews has made it possible for official policy to triumph over mere fact.[47]

The success of supernews is furthered by the fact that all of the major US media are now part of far larger corporate structures, and that televised news, in particular, in competing for ever-higher ratings, focuses more on the charisma of its expensive anchorpersons and the images of the world's newsworthy personalities than on the substance of events and the social processes underlying them. Supernews is being coupled with a kind of entertaining non-information.[48] Americans today who wish a clearer picture of the world, including their own country, would do well to search out the small, investigative weeklies and the best of the foreign press.

In any examination of 'truth' we are always faced with the problem of inevitable *selection* and *interpretation* of information discussed at the outset of this chapter, and this unconscious, culturally determined bias in our vision applies equally to the way 'news' is identified, gathered and written about. As David Halberstam has pointed out:

> [D]espite all the fine talk of objectivity, the only thing that mildly approached objectivity was the form in which the reporter wrote the news, a technical style which required the journalist to appear to be much dumber and more innocent than in fact he was. So he wrote in a bland, uncritical way which gave greater credence to the utterances of public officials, no matter how mindless these utterances.[49]

The reporter does not even evaluate for the reader the quality of the news source, nor inquire what forces might lie behind the 'facts' being reported. As journalist Henry Mayer points out:

> [A fiction has arisen] that there is a separation between the editorial and the front page. There is an expectation that 'news' is a commodity that we can obtain without additives, that the reporter's and owner's views can be kept out of the delivery system.[50]

As this is simply an impossibility, it would seem wiser that the media – as much of the European press in fact does – present stories as documented interpretations of events, replete with opinions clearly identified. Surely this would alert readers to the presence of intentional bias, rather than pretending as now that it does not exist, as well as forcing into the open the existence of inescapable unconscious bias as different reporters interpret the same set of events. The current uncritical approach opens the door to the propagation of deliberate falsehoods by whoever captures the attention of the media.

Admittedly, in a changing world, staying as completely informed as possible demands considerable time and effort. It can also be personally unnerving whenever one's old beliefs and assumptions come into question. Finally, it is socially destabilising: when old institutions and the values underlying them are criticised, social cohesiveness is weakened and power struggles ensue as new values and institutions emerge. It is such social readjustment that those who object to 'adversarial journalism' most fear, for they are generally the most favoured under present institutions. Yet if Part I has shown us anything, it is that present institutions are leading us toward an unsustainable future.

The Promise of Education

The political openness and critical journalism needed to facilitate the kinds of deep changes we require in our shared worldview will come about only when enough people demand them. A society that indulges in struthianism – that is addicted to Monday night football, to betting on the soccer pools, or to whatever is the national 'circus' – is sure to decay as surely as did the ancient Romans. But becoming an able participant in substantitive change requires special skills and long years of training – it requires being educated for the task.

In every society, the transmission of knowledge to the young involves two things, the teaching of technical skills and the imparting of shared traditions and values. If the society is a static, egalitarian culture, what is taught is seldom questioned on either count. But in today's changing, hierarchical, technological societies, education is a highly political act on both fronts, and hence raises deep passions.

More than ever before, knowledge means power. Possessing specialised, often secret, knowledge brings social power, prestige and wealth. Consider the secrecy that envelops knowledge about national security, marketing strategies, or production technologies – the most powerful forces in the modern world. Any specialised knowledge confers power on the elite group of 'experts' who possess it, and education offers a passport to such power. This is what the struggle for equal access to education is all about.

The function of passing along shared traditions and values is, if anything, even more political, for it is upon these that the whole structure of a culture is based. The entire choice of curriculum, what subjects are taught, what 'facts' are learned, creates the inner worldview of the child. When educators attempt to question or modify the accepted set of values, they become a force for political change. When they fail to raise questions about accepted norms, but merely pass them along, they are supporting the political *status quo*. *Education can never be apolitical, 'objective' or 'value neutral'*: it is – and ever must be – a political endeavour. It either *moulds* the young to fit in with traditional beliefs, or it *critiques* those beliefs and helps to *create* new ones.

'Mould-to-Fit' versus 'Critique/Create'

Throughout most of human existence, cultural change has been imperceptibly slow and the young have been carefully taught the beliefs and values of their parents. Rapid change was the result of violence – conquest or catastrophe – not of deliberate educational shaping. Given the need for social cohesion, 'mould-to-fit' is the natural form of cultural transmission. In Western history there have been but two major periods of *conscious* social change, when societies deliberately 'critiqued' themselves and created new worldviews (see Figure 8.3).

This first occurred in the Greek city states, particularly Athens, during the Golden Age of Pericles (500–400 BC). In the burgeoning Hellenic world of commerce and industry, the old 'givens' became suspect. Along

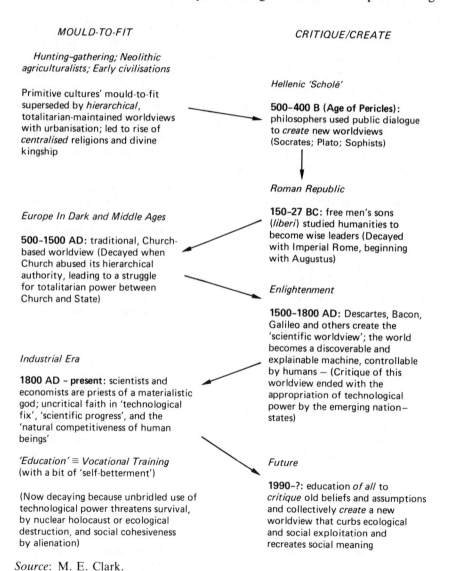

MOULD-TO-FIT

Hunting-gathering; Neolithic agriculturalists; Early civilisations

Primitive cultures' mould-to-fit superseded by *hierarchical*, totalitarian-maintained worldviews with urbanisation; led to rise of *centralised* religions and divine kingship

Europe In Dark and Middle Ages

500–1500 AD: traditional, Church-based worldview (Decayed when Church abused its hierarchical authority, leading to a struggle for totalitarian power between Church and State)

Industrial Era

1800 AD – present: scientists and economists are priests of a materialistic god; uncritical faith in 'technological fix', 'scientific progress', and the 'natural competitiveness of human beings'

'Education' ≡ *Vocational Training* (with a bit of 'self-betterment')

(Now decaying because unbridled use of technological power threatens survival, by nuclear holocaust or ecological destruction, and social cohesiveness by alienation)

CRITIQUE/CREATE

Hellenic 'Scholē'

500–400 B (Age of Pericles): philosophers used public dialogue to *create* new worldviews (Socrates; Plato; Sophists)

Roman Republic

150–27 BC: free men's sons (*liberi*) studied humanities to become wise leaders (Decayed with Imperial Rome, beginning with Augustus)

Enlightenment

1500–1800 AD: Descartes, Bacon, Galileo and others create the 'scientific worldview'; the world becomes a discoverable and explainable machine, controllable by humans – (Critique of this worldview ended with the appropriation of technological power by the emerging nation–states)

Future

1990–?: education *of all* to *critique* old beliefs and assumptions and collectively *create* a new worldview that curbs ecological and social exploitation and recreates social meaning

Source: M. E. Clark.

Figure 8.3 A schematic history of Western education.

with music and drama, there flourished dialogue and argument – and there arose the σχολή, the first schools. There the disciples of Socrates and Plato taught the young to philosophise in order *to create social change*. As educational historian Bernard Mehl writes:

> The early Greek philosophers succeeded in starting a new way of directing their thinking. They asked a different class of questions and were led to an entirely different line of thought.[51]

All too soon, a competing school of Sophists took over, however, teaching their students not how to become philosophers, but how to acquire skill in the uses of logic and rhetoric to gain important positions in public life. The Sophists became promoters of a new *status quo*; an elitist democracy survived for a time, but creative social change slowly waned and then disappeared for 1500 years. Not until the Renaissance and its sequel, the Enlightenment (which we shall take up in Part III), did Western culture again subject itself to critical philosophical thought. Out of that ferment emerged our modern worldview, which has remained largely unchallenged for two centuries.

This does not mean that during those long historical periods when mould-to-fit was the educational mode no social or technological changes were occurring. Quite the opposite. As historian Leften Stavrianos has pointed out, the so-called Dark Ages were in fact a time of considerable social reform and technological inventiveness.[52] And no one would deny that during the past 200 years there have been great social, political, and technical changes in the West, including of course the ideas of Darwin and Marx. Yet in each of these periods, change has evolved out of – and, more importantly, *has been justified by* – the received worldview. While social and political change is possible within a particular worldview (especially in our own, where change is a built-in expectation), that change is strictly limited in scope. It must not offend any of the major beliefs, assumptions, or tenets of the worldview. It is 'change within changelessness'.

But as we have seen, the Western worldview that now dominates so much of the world has grown maladaptive; changes in it *must* take place. Given that revolutionary violence as a means of forcing cultural change is rapidly becoming obsolete, we are left with deliberate change by the process of 'critique/create'. It is a process societies must teach their young how to undertake. Here we briefly examine the ability of the American educational system to accomplish this.

Two centuries of American education

The criticisms of Enlightenment thinkers boldly attacked the *status quo*, forcing a new worldview. But as historian Henry Steele Commager reminds us, only America could provide virgin soil for such thoughts:

Europe could not realize the Enlightenment without first sweeping away centuries of privilege, toppling Monarchies, overthrowing Churches, shaking society itself to its foundations and even shattering it. There the Revolution was, in Jefferson's magnanimous phrase, a spectacle of 'infuriated man, seeking through blood and slaughter his long-lost liberties'.[53]

The Enlightenment's great levelling idea – that all individuals possess equal rights and *become* the state through their active participation in it – meant that for the first time in history, universal public education was essential. The 1789 French Declaration of the Rights of Man claimed education at public expense a universal right, and in America, Thomas Jefferson became the champion of universal public schooling:

> I know no safe depository of the ultimate powers of the society but the people themselves; and if we think them not enlightened enough to exercise their control with a wholesome discretion, the remedy is not to take it from them, but to inform their discretion by education.[54]

In principle, the democracy of the Enlightenment demands universal education to ensure that every citizen may fully participate in critiquing and guiding the evolving social order. The primary purpose of education is to prepare men and women for political thought and action. In practice, this goal has fallen far short. American education critic Sheldon Wolin bluntly summarises the situation:

> Throughout its history America has negotiated the difficulties of trying to acknowledge the democratic principle in politics and education without accepting it. Those who have ruled America have shaped a political system so that while it has incorporated certain formal elements of democracy and fostered the growth of public education, it has prevented both the democratization of education and the education of democracy from being realized.[55]

Despite Jefferson's oft-quoted words, he and the other Founding Fathers were essentially aristocratic elitists.[56] The nation's governance was to be entrusted only to educate men of property, and the educational system proposed for the masses was, at best, limited in scope. There was a brief flurry of egalitarianism during the Jacksonian era with regard to access to the higher orbits of the American educational system, followed by a gradual expansion, completed about 1890, of free public grade schools for all children. But to this day, a two-tiered system of a limited education for the masses and higher education for the few has persisted in America. A parallel situation has existed throughout most of Europe, alleviated only recently by modest reforms in England and France. In America, *de facto* segregation of ethnic groups in ghetto neighbourhoods and in poorer school districts has exacerbated this disparity in educational quality. Finally, the higher education received by the few is paid for by all the taxpayers, with private universities receiving an even higher proportion of their support from federal coffers than state institutions.[57]

Not only has the educational system been two-tiered, enabling only a minority to qualify for positions of power and prestige; it has also, and more importantly, failed to teach critical thinking at any level, leaving both the elite and the masses philosophically bereft, unable to evaluate the basic premises of their own society. From top to bottom, the thrust of American education has been one of 'mould-to-fit' – mould the young, the minorities, the immigrants to fit unquestioningly into a socioeconomic hierarchy they neither understand nor control.[58]

A few educators of the twentieth century reacted against this 'mould-to-fit' pattern. In the United States, John Dewey began to stress education to meet the needs of the individual child rather than those of an amorphous society. It was an idea supported by other progressivist educators, such as Count Leo Tolstoy in Tsarist Russia, Jean Piaget in France, and Arnold Gesell and more recently Carl Rogers in America. For a time, the goal of educating the individual was broadly adopted in America since it did not challenge the commonly held worldview. On the contrary, by promoting each individual to 'become all she or he can be', it democratised and legitimised the entire existing socioeconomic hierarchy. Education was facilitating the individual's opportunity to climb to the top, but the necessity of the hierarchy was not questioned. This appealing vision has been rescinded in recent years because it was poorly applied. The 'needs' of the child were misinterpreted to mean unguided, autonomous development, which of course failed and resulted in a generation of self-centred underachievers.

But Dewey's vision was quite different. He cared not only about the individual child, but about that child's ability actively to participate in a democratic milieu. He viewed schooling as a miniature staging ground for adult life and encouraged the direct practice of democracy and social decision-making in the classroom. With the encouragement of teachers, children were to set up their own rules and govern themselves. School was to be the learning centre for a critique/create society. So far as I know, this practice, which began to be widely adopted in the 1930s, is now virtually abandoned. Teachers today, as Michael Parenti says, dare not raise questions of social criticism in the classroom:

> Teachers in primary and secondary schools who wish to introduce radical critiques of American politico-economic institutions do so at the risk of jeopardizing their careers. High school students who attempt to

sponsor unpopular speakers and explore progressive views in student newspapers have frequently been overruled by administrators and threatened with disciplinary action.[59]

American history and politics are taught as sterile facts, not as ideas that real people have pondered over and debated. As Frances FitzGerald, in describing American history textbooks, notes:

> [A reader] would have to conclude that American political life was completely mindless. For instance, the texts report that Thomas Paine's *Common Sense* was an influential pamphlet without ever discussing what it says. The fact that Paine was an internationalist who believed that the Revolution would spread back to Europe is another well-kept secret . . . But it is not only radical or currently unfashionable ideas that the texts leave out – it is all ideas, including those of their heroes . . . Even Thomas Jefferson and Alexander Hamilton are insubstantial, their ideas on government reduced to little more than a difference on the merits of a national bank. As for the Puritans, the texts manage to describe that most ideological of all communities without ever saying what they believed in.[60]

Americans grow up with only the vaguest notions of what the ideas *are* that they are supposed to believe in. Schools simply skirt around the ideas on which the worldview is based, and hence Americans have no tools with which to critique or evaluate it. Furthermore, children are taught that theirs is a country without deep problems. Everything is resolvable within present institutions and beliefs. Hence they are not in a position to be political participants in any real sense: they are imbued with a form of mindless 'citizenship', to become rubber stamps of the *status quo*. Whether deliberately or not, schools are moulding children to fit the accepted worldview without a murmur. Critical thinking is not in the curriculum.

Higher education also manages to sidestep direct examination of the beliefs and assumptions on which the American worldview is based, and avoids any serious consideration of where America and the world as a whole are headed, and why. On these matters the academic curriculum required of most graduates is virtually silent. The reasons are twofold.

Students do not see universities as places of higher education, but as a ticket to a 'piece of the pie'. More and more, they present themselves as clients to be shaped and turned out as a suitable, saleable product in the competitive job market. They prefer business, but medicine or engineering are popular; teaching is not. So intent are they on their goal that their time and effort spent enquiring into other things – history, philosophy, science – is resented. This entire educational thrust is being supported by private corporations who have greatly increased their funding of higher education. As the 1980 Sloan Commission report noted: 'Higher education is now a big business'.[61] With the growing vested interest of students, faculty, and the private sector of the economy in maintaining the socioeconomic *status quo*, universities are abandoning their deeper function – to reflect on and evaluate society itself, mirroring back to it what it most needs to know: namely, how adapted is it to survive in the future? Where shall the philosophers come from, if not academe?

Internal forces have further weakened this function of the universities, as demonstrated by Allan Bloom in his book, *The Closing of the American Mind*.[62] Through increasing fragmentation of disciplines and hyperspecial-isation of faculty, the world presented to students is a meaningless kaleido-scope of disconnected images – a bit of science, a sampling of history, a glimpse of art, a taste of literature, a smattering of economics. Instead of constructing a coherent inner world that students can evaluate and analyse and *think about*, today's so-called General Education programmes merely fill up a prescribed volume of the curriculum required to graduate, but they make little sense.

Yet even now, beginnings are being made among small groups of college faculty on many campuses to recapture the concept of collegiality and to put the disciplines back together into a meaningful whole. In their growing recognition that knowledge that liberates must be alive to the here-and-now, they are constructing alternative, integrated programmes that are not afraid to raise questions of beliefs and values – indeed, to introduce their students to critical thinking. Most heartening of all is the response of the students. Most of them welcome, with a sense of relief, a meaningful inquiry into the world and its future. For the first time, someone is talking to them about things that really do matter.

SOME CONCLUDING THOUGHTS

As in every previous age when fundamental beliefs and worldviews were being challenged, the years ahead will bring tension in academe and passionate rhetoric outside. It will be harder and harder for the 'average person' to escape political awareness, and the struggle for the 'average mind' will be titanic. Resistance will come for two quite different reasons.

First, as this chapter has shown, it is psychologically hard to change one's worldview. Our deepest beliefs and assumptions, built into us since earliest childhood, are 'us'; they define the very meaning of our lives. To deny them is to deny ourselves, or at very least our parents with whom we are often so powerfully bonded. For this reason, the deliberate changes that we make in our worldview will necessarily be *selective* and *evolutionary* rather than *cataclysmic* and *revolutionary*. By selecting what is viable and good from our past (and there is a great deal), we will retain a sufficient part of ourselves on which to build a new, better adapted worldview. The 'cup' of our culture is not yet broken, it merely needs reshaping to hold water better.

The second source of resistance will of course be those who gain most from the present worldview, the *status quo*. The creation of a new worldview that denies the legitimacy of their special privilege is something they will surely resist, and they will use the beliefs and assumptions of the old worldview to support their case. They will tell us that attempts to regulate pollution or wasteful consumption are denials of 'freedom', that forming cooperative institutions is a form of 'communism', and so on. With regard to the educational changes required to establish a critique/create mode of worldview transmission, we can expect them to raise the conventional myths about what education ought to be in a democracy. It must be 'apolitical' and 'value-free', presenting only 'the facts'. The state, after all, may not impose any belief system, especially through state-supported schools. When controversial questions arise, they must be treated in a 'balanced' fashion.

Yet as this chapter has shown, 'facts' are never free from bias. Whoever has control of the curriculum picks and chooses what information students will learn. Value judgements are implicit in every decision, and what is omitted is automatically made to seem less important, less valuable, than what is included. And refusing to examine the moral basis of political matters *is in itself a political act*; it puts an implicit stamp of approval on the *status quo*. It is impossible for education to be apolitical, and attempts to make it so are attempts to maintain the present social system.

In thinking about changes in the bias of our educational curriculum we might well begin with three 'sacred cows' that psychiatrist Judd Marmor believes threaten our survival: (1) the free-enterprise/profit system that demands aggressiveness and competition; (2) nationalism that views one's own country in self-righteous terms; and (3) war that is deemed a glorious and patriotic institution, while peace is subversive. We conclude with Marmor's prescription:

> If the organized killing and exploitation of men by other men is to be rendered obsolete ... every element in the acculturation process that shapes our perceptions and our goals, beginning in early childhood and continuing throughout life, should reinforce the value systems of cooperation, social concern, and non-violence. Not only the toys and games of childhood, but our textbooks, our history books, our encyclopedias, and our mass media need to be oriented toward the ennoblement of man's peaceful and cooperative accomplishments rather than the glamorization of his battles or of his individualistic acquisitions of power ...

> It may be argued that what I am advocating is a kind of massive indoctrination of people and a rewriting of history that is incompatible

with the ideals of a free society. Such an argument fails to take into account the fact that no society is without a set of mores that it imposes, either implicitly or explicitly, on its members. The glorification of free enterprise, of ethnocentric nationalism, and of war all represent value systems that have evolved out of the historical necessities of earlier eras. We have now entered upon an era in which these are no longer adaptive. In view of this development, it becomes imperative for man to consciously attempt to modify such outmoded value systems and to replace them with others that have greater adaptive usefulness. To endeavor to plan and change our educational, child-training and mass communication systems, as well as our social institutions, along principles that enhance the dignity, security, and potential of human life is not only an adaptive necessity – it is also compatible with the most cherished values of a free society.[63]

Before we can begin to eliminate these 'sacred cows' (and a few others, as well), we need to know how we came by them. It is the origins of our present worldview that we now turn to in Part III of this book.

Part III
Possessive Individualism: 'Whence Comes Our Western Worldview?'

The values of the modern Western worldview, based on notions of 'science', 'freedom' and 'progress', now dominate the global dialogue, in parallel with the political and economic power which that worldview has unleashed. Yet it is this same worldview that is largely responsible for many of our global problems, from environmental degradation to human exploitation, which in turn triggers population growth. It also causes a pervasive sense of powerlessness and alienation in those very people who are its supposed beneficiaries, yet they – we – seem unable to discover the source of this malaise.

Part III explores the source of our modern values as they developed historically over the past few centuries and analyses their social consequences. Chapter 9, 'From God to Man: Origins of the Western Worldview', traces the relationship between the individual and society from Medieval times through the Reformation and the Enlightenment, which introduced the 'Age of Reason', to the present era of possessive individualism. The contributions of the Protestant reformers, and the social philosophers from Hobbes to Jefferson, create a logical sequence of ideas for our modern vision of ourselves. Chapter 10, 'The Cult of Efficiency', shows how that original vision of the liberated human being was further moulded into 'rational economic man', a new kind of person who would naturally desire technological and organisational efficiency, thus paving the way for modern corporate capitalism. It examines how putting efficiency at the top of our list of values turns people into servants of, not masters over, the technological system they have invented.

Finally, Chapter 11, 'Alienation – The Loss of the Sacred', explores the heavy social costs of our competitive and materialistic value system that sets individuals apart in 'splendid isolation', free to create themselves. The harvest is rising ennui, disillusionment and violence, as people vainly seek a substitute for lost social meaning.

9 From God to Man: Origins of the Western Worldview

Man, like the gen'rous vine, supported lives;
The strength he gains is from th' embrace he gives.
On their own axis as the planets run,
Yet make at once their circle round the sun;
So two consistent motions act the soul;
And one regards itself, and one the whole.
Thus God and nature link'd the gen'ral frame,
And bade self-love and social be the same.

Alexander Pope (1688–1744)[†]

INTRODUCTION

In his 'Essay on Man' the English poet Alexander Pope, a pious Catholic, was eloquently pleading for a retention of the social order ordained by the Church. It is man's God-given nature to be social, said Pope, and therefore self-fulfilment is possible only in relation to others. He reacted in horror at the notion emerging from the Age of Reason that humans are basically selfish, greedy and lonely and the purpose of social institutions is merely to regulate these propensities. The true satisfactions in life that Pope believed came only from meeting the obligations of reciprocal affection were being replaced in peoples' hearts and minds by the false satisfactions of personal wealth and power.

As British historian R.H. Tawney relates, in just two centuries, from 1500 to 1700, '[t]he theory of a hierarchy of values, embracing all human interests and activities in a system of which the apex is religion, is replaced by the conception of separate and parallel [religious and secular] compartments, between which a due balance should be maintained, but which should have no vital connection with each other'.[1] How was a system of social values that had stood for a thousand years turned upside down so rapidly, such that the greed of usury became the logic of finance; the sin of avarice, the virtues of production and consumption; and the worthiness of charity, the vice of rewarding slothfulness? Where, indeed, did today's ideas of 'freedom', 'individual rights', 'rationality' and 'efficiency' come from, and how were they justified?

So natural is it for us to assign such ideas top priority that we seldom question their origins or their validity. We accept *our* worldview just as automatically as did the Havasupai Indian, who when asked to explain the 'whys' of his culture, replied with great patience and firmness 'That's–the–way–we–do!'. But the worldview of traditional societies did not need to be explainable. Established over centuries, changing only imperceptibly to adjust to gradual changes in the surrounding environment, such world-views 'worked' empirically. They produced a stable relationship both within society and between society and Nature. But today both our social and physical surroundings change almost daily, mostly in unforeseen ways, and almost entirely in response to our own activities. As Margaret Mead warned, we no longer can afford the luxury of uncritically accepting our worldview, assuming that it will somehow continue to 'work' in the face of such stupendous change. Yet this is exactly what we strive to do by focusing on technological fixes rather than assessing the adequacy of our basic beliefs and assumptions. The primary task of this generation is to discover *not* how to build gardens in space or to mine magnesium nodules from the seabed, but *how to re-order our hierarchy of social values to ensure a continuing sensitivity to both Nature and the psychic needs of human beings.*

In our schools and universities we virtually ignore any examination of either the psychic or the environmental impact of our worldview. Although textbooks are regularly updated to reflect currently accepted social and polit-ical opinion, they seldom examine where either the old or the new opinions came from, and they never evaluate them in relation to modern global problems. Yet if we are ever to gain conscious control over the arms race, pollution and resource depletion, we must first comprehend the origins of our present worldview. This is the first step toward introducing the 'critique/create' mode into our educational system, and hence into our society.

THE INDIVIDUAL AND SOCIETY

Modern Western democracies are based on the notion that the indi-vidual's right to freedom is the most important of all social values and the function of the State is to preserve that right. As shown below, this is not as simple an idea as it seems, and putting it into practice is even more difficult. Our immediate concern is with the pinnacle on which it places the individual, leaving society as a whole in second place. In the history of worldviews, this is a most unusual and recent development.

The notion of 'individual rights' would be inconceivable to most people of the past. Hunting–gathering societies were changeless and non-hierarchical. 'Rights' were never a question. People had roles and duties within society that gave their lives meaning. Their religious beliefs and assumptions and the rules that followed from them, far from being constraining, provided a framework for living much as the rules of grammar provide a framework for speech. As anthropologist Dorothy Lee puts it:

The intricate set of [cultural] regulations is like a map which affords freedom to proceed to a man lost in the jungle.[2]

Within the boundaries of the worldview, the individual creatively fulfils his or her role. People do not try to dominate others, even small children. Yet it does not occur to parents that they are 'permitting' their children to be free; a child's autonomy is respected from birth. Lee describes the situation that once existed in many American Indian societies:

The individual, shown absolute respect from birth and valued as sheer being for his own uniqueness, apparently learns with every experience to have this same respect and value for others.[3]

In preliterate societies, then, there was remarkable respect for the individual. Someone who failed to follow the rules was criticised, but seldom punished or coerced. There was little tension between the individual and society. If personal clashes could not be resolved within a group, there was fission and reformation into new groups, but the basic rules were not something to be challenged or escaped. The question of being more free or less free simply did not arise. Neither *freedom from* nor *coercion by* one's society was a comprehensible concept.

With the advent of large, agriculturally-based hierarchical societies in which the rulers were often living gods, there were multitudes of people who were conspicuously unfree, indeed, 'non-human'. Nor did these slaves meekly accept their lot of working to exhaustion, if we believe the staves and whips depicted in Assyrian and Egyptian artwork.[4] As already suggested, the major religions very probably arose to mitigate the worst excesses of these earliest 'civilisations', introducing for the first time into human consciousness notions of *social ethics* which had been present but unarticulated in primitive worldviews. The grand prevalence of the Golden Rule attests to this ancient need. But in these new megasocieties, religion did not eliminate the tension between the individual and the social hierarchy; it made it manageable. Where the few manipulate the many, sheer force is seldom sufficient, but a divine authority who mitigates the worst excesses of power yet whom mere humans cannot question serves the purpose.

The thought that flowered in ancient Greece raised questions about the validity and wisdom of divine authority. Philosophers began envisioning societies based on human rules. During this first period of 'critique/create' in pre-Periclean Athens, a new school of thought arose that attempted to use rational argument in arriving at its new views of society.

The first step was to counteract the oppression of hierarchies by turning the control of society over to ordinary citizens. Thus was born δημοκρατία – democracy, or rule of the people. Of course 'the people' were limited to the well-to-do nobles, but this was a remarkable advance over the preceding notion of a divinely ordained hierarchy. In the sense that all citizens in a society have equal importance in making the social rules, Greek democracy resembled the egalitarianism of the primitive societies that had existed for tens of thousands of years. But it had the quite new charge of challenging accepted beliefs and of establishing by general agreement a new set of social rules – indeed, a new world-view.

Two philosophical schools emerged, vying for the education of the sons of Athen's elite, the citizens of the new society, but neither led to universal participatory democracy. The pragmatic Sophists, for whom the preservation of the elitist position of their charges became the goal, focused on rhetoric rather than careful reasoning. The other school, focused on the theoretical thinking of Socrates, Plato and Aristotle, abandoned democracy in favour of an ideal hierarchy free from corruption. In his *Republic*, Plato imagined a community composed of classes of people with different social functions working together as a giant organism. Rulers were to live separately from others in a garrison without family ties or private property, free from temptation to corruption or personal power. As lifelong philosophers, they would be ideally suited to make rules for the benefit of the whole. While Aristotle criticised the details of this ideal republic as an impossible (and in some respects unwise) Utopia, he strongly advanced the idea that a society is a single living organism which supersedes the individuals that compose it:

> The State is by nature clearly prior to the family and to the individual, since the whole is of necessity prior to the parts.[5]

Aristotle's social organism required the internal unity of a single worldview, and in this respect resembled preliterate organic societies, but his society retained hierarchy and inequality – indeed, these were essential. It was the role of social philosophers to act as physicians to the State, objectively prescribing better rules to improve its health. These law-givers however were outside the law.

> Legislation is necessarily concerned only with those who are equal in birth and power; for men of pre-eminent virtue there is no law – they are themselves a law. Any one would be ridiculous who attempted to make laws for them.[6]

The relative merits of these two conceptions of an ideal society – one run by widespread democratic participation, the other by an elitist group of 'wise men' – that originated in Athens over 2300 years ago are still argued by modern political scientists. They are divided mainly on the source of authority in society: all the people, or wise men? The differences are critically important during periods of rapid social change, such as our own, when rules, and even whole belief systems, are being reshaped. In the one case, everyone has a more or less equal say in the shaping process, and moral emphasis is placed on the rights of the citizen *vis-à-vis* society as a whole. Such societies are variously known as 'individualistic', 'atomistic', or 'democratic' societies. Nevertheless, individuals agree to surrender some degree of autonomy to a select group of law-givers and, as is the case in modern industrial democracies, such societies may still possess extremely powerful hierarchies which, although they claim to be more economic than political, play powerful but often unrecognised roles in generating new beliefs and worldviews. This point is taken up in later chapters.

In the other case, often termed an 'organic' society, new rules are made very much along Aristotelian lines. An elite group of 'wise men' makes decisions for the good of society as a whole, and not to preserve the well-being or rights of any particular individual or group. For Aristotle, the pragmatist, whatever worked was 'good'. But later thinkers, from the Catholic Church in the Middle Ages to Thomas Hobbes and Karl Marx, introduced moral authority to legitimise the new rules they created. If the masses are taught that the law-givers are wiser than themselves, they not only submit to the laws but often strongly support the society as a whole. While organic societies are necessarily hierarchical politically, in terms of the distribution of wealth, welfare and social status, they can, at least in theory, be quite egalitarian.

Once a reasonably satisfactory worldview has been worked out, large organic societies, like their preliterate predecessors, can be enormously stable through long periods of time. Since they utilise the mould-to-fit mode of cultural transmission, there is no perceived need for social change. It is not surprising then that most human societies have been so constructed. Everyone accepts, as given, her or his place in the social organism and the authority that decrees it. One has a 'station in life'.

Just this sort of organic society persisted for over a thousand years in Europe. Although threads of the democratic concept that once flourished in some of the Greek City States surfaced from time to time during the Macedonian and Roman Empires, it was finally buried after the fall of Rome in AD 476. With the demise of the *Pax Romana*, a feudal polity descended over Europe. The Catholic Church, drawing heavily on the works of Aristotle, established a holistic organic society of unquestioned authority.

The story of how that Medieval society, where all of whatever station saw themselves as part of a divinely ordered (if not always divinely functioning) interdependent community, was turned 180 degrees, to become one of individuals each acting specifically in his or her own best interests, is the subject of this chapter.

THE MEDIEVAL WORLDVIEW

It would be irresponsible to paint the thousand or so years from the fall of Rome to the first major stirrings of religious unrest in Europe as uniform and uneventful. A thriving trade through the Levant to the Orient stimulated the growth of commercial centres around the Mediterranean and even in central Europe. Italian cities became financial centres and craft industries flourished across Europe. But England remained a bucolic backwater, largely untouched by the major events of the times. Yet it was here, following the 'discovery' of the New World, that the complete conversion from the old organic to the new individualistic worldview happened first and most rapidly, and much of our attention focuses on it.

Life in Medieval times was precarious; half the babies born died before reaching adulthood and even adults were victims of plagues and strife. Under such constant threat of death, life might well have been insupportable without faith in some sort of afterlife, and this the Catholic Church offered. But afterlife came in two forms, salvation and damnation. Those who died without the sacraments and forgiveness of the Church were condemned to eternal suffering. So powerful was the fear of hell that sinners almost always recanted on their deathbeds and, if wealthy, bequeathed much of their estate to the Church or the poor. Nor did the Church hesitate to use its powers of excommunication (which meant automatic damnation) as well as the Holy Office for the suppression of heresy to maintain both spiritual and worldly supremacy. Throughout most of the Middle Ages the Church of Rome was far wealthier and more powerful than any of the feudal princes or budding monarchs of the secular world of Europe.

Yet it was neither mere fear nor sheer power that brought about universal acceptance of the Medieval worldview and its rigid social hierarchy. The whole of society saw itself as a single entity resembling a human body, where each social class was like one of the several bodily organs with its own job to perform; it was an idea reflected in the social structure of the times.

Feudal communities were relatively small, fashioned around a fortified manor inhabited by the lord and his retainers. Nearby lived artisans and craftsmen, and in scattered villages were the serfs who worked the land. A cadre of priests ministered to the community and in the neighbourhood there might well be a monastery with its own vast lands. The whole community was welded together by reciprocal duties and obligations. The peasants tilled the land for the manor folk and in return received protection from marauders, the right to graze animals on the lord's pasture land, petty cash for their produce with which to purchase household items they could not make for themselves, and above all, succour and sustenance in times of hardship. Coopers, weavers, tanners and smiths belonging to the manor received similar security in return for their loyal labours. Such arrangements were neither arbitrary nor spontaneous, being defined and ordered by the moral authority of the Church which permeated all life's activities. As Tawney writes:

> [S]ocial institutions assume a character which may almost be called sacramental, for they are the outward and imperfect expression of a supreme spiritual reality.[7]

Town life also fell under the moral authority of the Church, which peered into commercial activities, heard the confessions of those who took usury or otherwise engaged in sharp business practices, and insisted that every parish collect alms for its poor. Each person had to contribute according to his station and received no more than his due; disputes and injustices were resolved by the ecclesiastical courts. The Church *was* society. John Wyclif, a fourteenth century English theologian, envisioned the organic society in which he lived this way:

> The Church is divided in these three parts, preachers, and defenders, and ... laborers ... As she is our mother, so she is a body, and health of this body stands in this, that one part of her answer to another, after the same measure that Jesus Christ has ordained it ... Kindly man's hand helps his head, and his eye helps his foot, and his foot his body ... and thus should it be in parts of the Church ... As divers parts of man served unkindly to man if one took the service of another and left his own proper work, so divers parts of the Church shall never be whole before proportions of her parts be brought again by this heavenly leech* and [by] medicine of men.[8]

This was the worldview of Medieval Europe. Each social stratum had its own divinely ordained obligation to the social organism, the Church. The hierarchy of priests, with the Pope at its head, was the interpreter of God's will, the ultimate authority. The feudal lords ensured that God's will was enforced. Merchants were necessary for the exchange of goods, and peasants to till the soil. Economic inequality was naturally necessary if each class was to perform its rightful function. Since the mere accumulation of money had no demonstrable social function, usury and other forms of greed were stigmatised. (Trading centres imported Jews as money-lenders to handle transactions forbidden to Christians, providing Shakespeare with the evil character of Shylock in his play *The Merchant of Venice*.) Master craftsmen, although belonging to monopolistic guilds,

---

*In Medieval times leeches were used medically to suck out 'bad blood'.

believed in just salaries and prices – on pain of eternal damnation. As Tawney states:

> [If] peasants must not encroach on those above them, [then neither may] lords... despoil peasants. Craftsmen and merchants must receive what will maintain them in their calling, and no more.[9]

If great inequality exists, it is for the purpose of social stability, and both oppression and greed must be reined in. It must have been impossible for the average artisan or labourer, who until the end of the Middle Ages did not even possess a surname,[10] to think of himself as anything but God's servant, here to do His will as explained by his betters.

By the fourteenth century, however, seeds of discontent were sprouting. There were bloody peasant uprisings in France and England. Too much oppression, especially in the face of blatant opulence in the Church and the royal courts of Europe, fuelled the growing rumblings of equality:

> Matters cannot go well in England until all things shall be held in common; when there shall be neither vassals nor lords, when lords shall be no more masters than ourselves... Are we not all descended from the same parents, Adam and Eve?[11]

These were the sentiments of John Ball, a travelling priest who with Wat Tyler led the English Peasants' Revolt in 1381. Serious cracks were developing in the social organism known as Holy Mother Church.

THE METAMORPHOSIS (1500–1700)

In little more than two centuries, the all-pervasive authority of the Catholic Church had been replaced by a secular authority favouring the rights and powers of an entirely new class, the commercial bourgeoisie, whose worldview is still very much with us. Long before the Protestant Reformation, forces were already at work weakening the Church. Its own internal squabbles over power and doctrine had led to schisms and doubts. In the face of the growing economic power of princes and kings, the Church needed ever more wealth to maintain its own status, so sectors within it began engaging in the once proscribed practices of usury and even oppression of the poor. In doing so, they lost for the entire Church much of its moral authority. Its own priests, particularly the mendicant Franciscans, began to question papal morality. No wonder that the fourteenth century outlaw, Robin Hood, who successfully defied both the ecclesiastics and the secular Sheriff of Nottingham, became a hero among the poor.[12] To aid in keeping track of events, a chronological list of important events is given in Table 9.1.

Table 9.1 Major events in England (1500–1700)

Major events		Reigns		Important people	
1381	Peasants' Revolt	1399–1509	Early Tudor kings	1473–1543	Copernicus
1536	Dissolution of the monasteries and establishment of Anglican church	1509–1547	HENRY VIII	1483–1546	Martin Luther
		1547–1553	EDWARD VI		
		1553–1558	MARY I (Bloody Mary)	1509–1564	John Calvin
		1558–1603	ELIZABETH I	1561–1626	Francis Bacon
		1603–1625	JAMES I	1564–1642	Galileo
1650?	Quakers founded by George Fox	1625–1649	CHARLES I (executed)	1573–1645	William Laud, Archbishop of Canterbury (executed)
1651	*Leviathan* published		CIVIL WAR	1596–1650	René Descartes
		1649–1660	Commonwealth	1599–1658	Oliver Cromwell
			RESTORATION	1588–1679	Thomas Hobbes
1666	Great Fire destroys London and ends last big plague	1660–1685	CHARLES II	1632–1677	Benedict Spinoza
		1685–1688	JAMES II (deposed)	1632–1704	John Locke
1687	*Principia* published		GLORIOUS REVOLUTION	1642–1727	Isaac Newton
		1689–1702	WILLIAM III and MARY		
		1702–1714	ANNE		
		1714–1727	GEORGE I	1688–1744	Alexander Pope
1776	*The Wealth of Nations* published and American Declaration of Independence	1727–1837	Later Hanoverian kings	1723–1790	Adam Smith
				1724–1804	Immanuel Kant
				1743–1826	Thomas Jefferson
				1748–1832	Jeremy Bentham
				1808–1873	John Stuart Mill
				1818–1883	Karl Marx

254   *Possessive Individualism*

Economic and Political Currents

Despite the terrible scourges of the Black Death, the population of
Europe continued to climb through the Middle Ages so that by 1500 the
former wastelands – the potentially arable marshes, forests and moors –
had been cleared and were inhabited. There was no spare land left. The
Renaissance, the vigorous 'rediscovery' of Greek and Roman culture
during the preceding two centuries, had stimulated both the arts and trade.
The latter, in turn, created new wealth. The beginnings of science – the
notions that the world was round and perhaps not the centre of the
universe – were slowly permeating thought. And a large, heterogeneous
'middle class' was arising, comprising not only merchants, but the wealthier
artisans and guildsmen in the cities and the yeomen or untitled gentry in the
countryside. For some time a small fraction of serfs had been buying their
release from bondage and acquiring modest free holdings of farmland; as
their numbers slowly grew, some became quite wealthy and formed a significant political and economic faction. The new middle classes did not fit neatly
into the old Medieval social organism with its reciprocal duties and obligations: without realising it, they were becoming independent of the old order.

International banking was developing on the continent, especially at the
Bourse in Antwerp, which stimulated a growing commercialism. Capital
began to be consolidated in the hands of a relatively few wealthy
merchants, who used it to undertake larger business ventures than ever
before. A consequence of growing trade in England was a sky-rocketing
demand for woollen cloth for export. To obtain more land for raising
sheep, the larger landowners began enclosing the common lands that had
been used for centuries by peasants for farming and grazing. On other
lands high rents were suddenly imposed. What once was the peasants'
'right' to free use of land was removed, and the dispossessed were reduced
to begging or moving to towns in search of a living.[13] The social and
economic hardships of this 'depopulation' drive horrified the Church, and
even at times the State. In 1553, sixteen-year-old King Edward VI, with
the aid of his tutor, issued 'A Prayer for Landlords':

> We heartily pray thee to send thy holy spirite into the hearts of them
> that possess the grounds, pastures, and dwelling-places of the earth, that
> they, remembering themselves to be thy tenants, may not rack and stretch
> out the rents of their houses and lands, nor yet take unreasonable fines and
> incomes, after the manner of covetous worldlings . . . but so behave
> themselves in letting out their tenements, lands and pastures, that after this
> life they may be received into everlasting dwelling places.[14]

But despite the efforts of subsequent kings and the newly formed Anglican
Church to maintain a semblance of social order, all ended abruptly with the
Civil War and the execution of Charles I by the Puritans.

But two other contemporary threads are also important. The first was the increasing strength of nation-states and of the monarchs that ruled them, the second the growing dissatisfaction among the clergy in Europe with the ever more dissolute Princes of the Church in Rome, the Pope and his cardinals.

The economics of nation-state wars

By the sixteenth century, when the coalescence of tiny fiefdoms into sizeable nations was almost complete, war had progressed from Medieval raids and skirmishes to large battles waged by powerful monarchs. Unlike the feudal lords who raised unpaid armies from among their vassals, the princes and kings of the new nations needed cash to go to war. Like their continental peers, the Tudor kings of England first exhausted the resources directly under their control: they sold crown lands, they collected higher land rents, they let political offices for a fee, and they levied taxes. When those sources proved insufficient they became indebted to the continental money-lenders, who made a profit out of war by loaning money to both sides. England's King Henry VIII, in his religious dispute with the Pope over his divorce from Catherine of Aragon, hit upon a solution to both his marital and financial problems. Not only did he break religious ties with Rome and establish an independent Anglican Church which agreed to nullify his marriage; he also abolished the wealthy Catholic monasteries, appropriated their extensive lands and sold many of them to pay his debts. At one stroke he achieved the right to marry again and the wherewithal to finance his wars.

The new Anglican Church was not as revolutionary as one might suppose. It clung for a long time to the moral tenets of organic Catholicism. But in the century between Henry's death and the Civil War which brought Oliver Cromwell and the Puritans to power there arose a complex power struggle among Church, King, nobility, wealthy merchants, untitled landlords, and the poorer but politically free yeomen and artisans. While battles were reshaping the political face of Europe, their financing was reshaping the distribution of economic power. Political power, religious ethics, traditional morality and economic conventions were all in flux. The old organic society stood in disarray but there was nothing to take its place. At that moment there emerged on the European continent a new social vision – that of the Reformation or what later was called Puritanism.

The continental reformers

Although the Anglican Church made some changes in the old Catholic doctrine, these were nothing compared with those unleashed by Martin Luther in Germany and John Calvin in Geneva. In their efforts to reform the Catholic Church, neither man wished to weaken the hold of religious authority on society as a whole. Indeed, it was the recent moral laxness of the Church that they so heartily condemned. Their target was greed – 'the avarice which was thought to be peculiarly the sin of Rome'.[15]

In Germany, Luther decried with equal vigour the Papal exploitation of the Church and the capitalist merchants' exploitation of craftsman and peasant. He argued *not* for individual freedom but for a strict adherence to the rules of the early Christian Church. His appeals to conscience and brotherly love among all were incompatible with the exploitation he saw going on around him. But – and here is a point of significance – everyone had a *duty* to work; idleness and covetousness were equally unforgiveable sins:

> Pilgrimages, saints' days and monasteries are an excuse for idleness and must be suppressed.[16]

Luther was the father of the Puritan work-ethic.

But Luther's eyes were on the past. For him, the class-structured society was impossible to change. After all, St Paul himself had implied that master and slave must remain such in this world, even if in Heaven they were equals. By returning to the ideals of the early Church, Luther threw out the entire Catholic hierarchy that had grown up. God did not speak to individuals through the intermediary of priests and a pope, but directly. Kings and states were to be stripped of religious authority. Since, according to Luther, God's authority could be known only through one's own conscience, this left the temporal world of business and politics without any moral guidelines. As Tawney puts it, in one fell swoop, Luther 'emptied religion of its social content and society of its soul'.[17]

John Calvin, on the other hand, attempted to retain religious authority over secular matters. Calvin, too, was anxious to reform life on Earth, but unlike the Church, the nobility or Martin Luther, he came to terms with the 'new economics' of mercantilism. For him, the burgeoning profits accruing to the middle class from trade and finance were morally legitimate:

> Whence do the merchant's profits come, except from his own diligence and industry?[18]

Wealth, though, must not be accumulated at others' expense, or for ostentation, but only for *services acceptable to God*. Usury was acceptable, but *only* at a fair rate of interest on profits earned by the money loaned; it must never be taken at the expense of the poor. Like Luther, Calvin saw hard work as Godlike and idleness a sin. If self-discipline, thrift, sobriety and frugality should perchance result in material gain, then that is clearly part of God's plan. Although only a few – God's chosen – would be touched by His grace and saved, material success through diligence was more and more taken as a likely sign of such grace. A stamp of approval was thus put upon commercial enterprise – and, simultaneously, poverty, which had once been next to Godliness, or at least was held to be pitiable, was now condemned as the just reward of the shiftless and idle – those of 'bad character'.

Calvinism was primarily an urban sect, especially attractive to members of the rising merchant class. It began as a curious mixture of intense individualism and fervent moral socialism. In its strongholds, in Geneva, Scotland, and the New England colonies, Puritanism saw all social life as a gigantic monastery where every detail – from frivolous dress, to family quarrels, to commerce – was under constant supervision of the spiritual dictatorship. As Tawney puts it: 'In the struggle between liberty and authority, Calvinism sacrificed liberty, not with reluctance, but with enthusiasm'.[19] Perhaps the pinnacle of Puritan moral fanaticism was reached during the witch trials in colonial Massachusetts, which Arthur Miller so movingly dramatised in his play *The Crucible* (a parable of the communist witch-hunts being conducted in the 1950s by Senator Joe McCarthy). In early puritan America the old sins of covetousness and avarice were thus severely suppressed. But in Anglican England, where the Puritans were long a political minority living under a hostile State and Church, the social aspects of Puritanism (except for a brief time under Cromwell) never became ascendant, and it was the belief in individualism that took hold and flourished.

The Shaping of the Protestant Ethic

The British offshoots of Calvinism, the Presbyterians in Scotland and the Congregationalists (and later the Quakers) in England, eschewing any ecclesiastical hierarchy, set up independent churches which *elected* their own elders and ministers. Said the Puritan leader Robert Browne in 1581:

Any group of Christians should have the right to organize itself for worship, formulate its own creed on the basis of Scripture, choose its own leaders, and live its religious life free from outside interference, acknowledging no rule but the Bible, no authority but Christ.[20]

Instantly recognising the implied danger to her absolute sovereignty, Queen Elizabeth I moved to suppress such republican tendencies, sending to the Tower of London any members of Parliament who proposed reforming or replacing the Anglican Church from which came her divine right to rule. Thus were the political battle lines drawn – between freedom of religious conscience on the one hand and, on the other, the power of an authoritarian monarch to adjudicate both moral and secular law. Out of this struggle was born our contemporary worldview.

Puritanism, like the Catholicism it replaced, was a way of thinking that pervaded all life – family, business and politics alike. It was by no means a monolithic movement, as it encompassed many diverse non-conformist groups all struggling for religious freedom. Among its ranks were those who wished to impose on society a rigid collectivism enforced by the iron discipline of a truly reformed Church, as existed at Geneva. At the other end of the spectrum were the Levellers who, 'abhoring the whole mechanism of ecclesiastical discipline and compulsory conformity ... endeavored to achieve the same social and ethical ends by political action'.[21] Drawn mainly from the ranks of independent yeomen, these democrats formed a large part of Cromwell's rebel army during the English Civil War, and for a time it seemed that their goal of extending the vote to every propertied man in the Commonwealth, no matter how small his holding, might prevail. But times were not yet ripe for so much political equality, which seemed like anarchy even to most Puritans. Only gradually was power to be wrested from the authoritarian head of a religious State and vested in the hands of the people.

After the brief and unhappy decade of the Commonwealth, the Restoration of the Stuart kings to the throne of England brought renewed persecution of the Puritans. One of them, the Quaker William Penn, pleaded for religious tolerance and was eventually given lands in America for his followers to settle. Another, the Puritan preacher Richard Baxter, urged a return to the strict morality of early Calvinism, preaching hell-fire and damnation for unjust usury and other sharp business practices. But his words fell largely on deaf ears. The Puritan commercial classes pursued business in their own way, frequently overstepping the restrictive trade laws of the Tory-controlled Parliament:*

> The gaols were crowded with the most substantial tradesmen and inhabitants, the clothiers were forced from their houses, and thousands of workmen and women whom they employed set to starving.[22]

Thus reflected an anonymous historian on the period of Charles II. Persecution of the business class threatened economic collapse, and religious tolerance began to seem like economic wisdom.

The next step toward democracy came with the bloodless 'Glorious Revolution' of 1689 when Catholic James II, the last absolute monarch of England, was deposed and a new constitution gave England a Protestant 'King-in-Parliament'. Supremacy of an elected Parliament over the Crown

---

*Tory politicians favoured the monarchy and the landed aristocracy and supported the old Catholic moral concepts still held by the Anglican Church. The Whig opposition favoured the new middle class commercialism, a more democratic Parliament, and religious freedom.

was finally secured. The engineers of this coup, however, were not the Puritans but the landed Anglican aristocracy, who had their own quarrels with the King. In one of history's ironies, the imported new King whom they installed as secular head of the Anglican Church, Prince William of Orange, was himself a Dutch Calvinist. It was he who finally brought religious tolerance to England's government, unwittingly paving the way for a new social morality. As Tawney tells us, the life of trade, once considered reprehensible, would soon be seen as 'service to God and the training-ground of the soul'.[23]

The nub of the new Puritan ethic was the belief that each person has a particular 'Calling', ordained by God. Fulfilment of one's Calling is a Christian duty, the conscientious discharge of which 'is among the loftiest of religious and moral virtues'. Thus was Luther's original view that a Calling is one's station in life against which it is impious to rebel turned on its head. Says Tawney:

> On the lips of Puritan divines, it is not an invitation to resignation, but the bugle-call which summons the elect to the long battle which will end only with their death.[24]

The more successful one is at one's Calling, the more moral virtue accrues to one's soul. Success is a sign of being among God's chosen, and if one's Calling is business, the accumulation of wealth is the most certain mark of success. Such logic would label poverty as due to moral weakness, best treated by severity. As Tawney puts it:

> A society which reverences the attainment of riches as the supreme felicity will naturally be disposed to regard the poor as damned in the next world, if only to justify itself for making their life a hell in this.[25]

The 'disgracefulness of poverty' accounts for the debtors' prisons and brutal Victorian workhouses so castigated by Charles Dickens, and for the continuing inhuman treatment of vagrants in this century, vividly described in George Orwell's *Down and Out in London and Paris*. Nor are the poor even today regarded as blameless for their lot by conservative politicians wherever they are in power: poverty is still widely considered a self-inflicted debility.

THE RATIONALISATION OF CAPITALISM

If the Puritans managed to give religious justifications for the rise of capitalism and the right of the bourgeoisie to dominate society, it was the pundits of the Age of Reason who provided a rational explication. The

Renaissance had already triggered renewed interest in philosophy, especially in the techniques of rational argument, and the tender new discipline of science was sprouting its wings. Copernicus, Kepler and Galileo had cast grave doubt upon the Biblical notion that man and the Earth are at the centre of God's universe, and inspired by Galileo's experiments on bodies in motion, scientist-philosopher René Descartes developed a mechanistic view of life which he felicitously subjected to divine wisdom, thus uniting reason and faith.[26] Newton's laws of gravitation and other physical forces, elucidated in his *'Principia'*, finalised the idea that observable events, whether in the heavens or here on Earth, had explainable causes beyond the whims of divine will. This did not of course mean that God could not have ordained the physical laws in the first place; it merely implied that thunder and lightning, plague, and crop failures had explainable direct causes other than the wrath of God.

As historian C. S. Lewis tells us, the new worldview discarded forever the belief that Earth began in a perfect state of order and harmony – the Garden of Eden – and since the Fall has been in a continuous state of decay: 'All perfect things precede all imperfect things'. Long before Darwin, the reverse image of an 'upward' *progression* in cosmic matters had formed in peoples' minds. The simplest leads *to* the complex; the lower *to* the higher; the lesser *to* the greater: 'The starting point is always lower than what is developed'.[27] As philosopher-historian Edgar Zilsel points out, *this complete revolution in the way people perceived the changes going on in their world was essential for our present worldview*:

> Today this idea or ideal of progress, both in the evolution of the world and in our understanding of it seems almost self-evident. Yet no Brahmanic, Buddhistic, Moslem, or Catholic scholastic, no Confucian scholar or Renaissance humanist, no philosopher or rhetore of classical antiquity ever achieved it. It is a specific characteristic of the scientific spirit and of modern Western civilization.[28]

It was against this emerging backdrop of a linear and mechanistic universe that the seventeenth century political philosophers developed a new vision of society. Following the methods of science and logic urged on them by that herald of The Age of Reason, Francis Bacon, they searched for an initial premise on which to base their arguments, and came up with an assumed *presocial* state of humankind – the so-called 'State of Nature'. Today we know that human beings from their earliest beginnings were strongly bonded, highly social organisms, so the imagined primitive condition of selfish, isolated individuals is a seventeenth-century fiction. But it was not an unreasonable image for writers to hold who lived in a troubled era and who had no knowledge of evolution or of the cooperative lives led by their hunting–gathering ancestors. The views they developed of human nature and the rights of the individual are still very much with us!

Thomas Hobbes

It was in the writings of Thomas Hobbes, particularly his *Leviathan* published in 1651, that the notion of possessive individualism as the moral basis of society was first made explicit. Hobbes asked two questions: What is human nature? and How shall society be organised to accommodate this nature? To arrive at human nature, Hobbes could not explore humankind in all its cultural diversity or make inferences from a fossil record. His imaginative solution was to strip civilised humankind of all its social arrangements, thus creating a fictitious species that never has existed nor ever could exist. He then endowed his isolated human being with an overriding independence of will and propensity to compete. Strongly affected by the rise of materialistic science and its mechanical explanations, Hobbes viewed precivilised humans as self-moving systems of matter in motion – as rational, calculating, self-correcting machines – *directly opposing the motions of all other individuals.* There was war of all against all. (Here we have a preview of the unremitting competitiveness that characterises social Darwinism and Richard Dawkins' 'Selfish Gene'. Hobbes was but one of many Egoists of varying shades, including Descartes, Spinoza, Freud and Camus, all of whom believed that self-interest was the fundamental human trait and prime mover of human action.)

Hobbes then argued that each person owns himself and has no obligations to society except those he voluntarily enters into. In a State of Nature, there were no contracts and every person was a potential threat to every other: 'the life of man [was] solitary, poore, nasty, brutish, and short'.[29] To create social order, Hobbes imagined a system of formal contracts – the marketplace. By mutually agreed voluntary transactions in the market, each individual could freely sell – or refuse to sell – both his property and his labour to another. Both are things each person owns and has the moral right to dispose of as he pleases. (Neither Hobbes nor his successors were much concerned about the political rights of women.) Political arrangements are solely for the protection of property, both goods and labour. It is only to secure their property rights, their persons and their belongings, that people willingly subject themselves to the sovereign power of the State, which has no other justifiable function, and certainly not that of imposing moral obligations between one person and another. The paternalistic arrangements of a Church-run organic society were out. By starting with a *materialist model* of human nature Hobbes deduced the necessity of a *market model* of society. He sidesteps the whole question of human bondedness: love is mere appetite, and shame and guilt simply grief from loss of reputation. He never explains why an entirely selfish person would feel concern about his reputation in the first place.

Hobbes gave moral force to his social model by employing the notion that if something *is* a fact of life, then it *ought* morally to exist. This was revolutionary for political theory, for it derived morality from *scientific facts* rather than religious dogma. Few modern thinkers any longer try to derive what *ought* to be from what *is* (although it is still a popular approach among the public at large). As British political scientist C. B. MacPherson notes, among moral philosophers '[i]t has become axiomatic in recent years that no moral principle can logically be deduced from any statements of fact'.[30] But in the seventeenth century, Reason was replacing Faith as the source of moral authority, and so Hobbes made the attempt: 'No traditional concepts of justice, natural law, or divine purpose were needed. Obligation of the individual to the state was deduced from the supposed facts, as set out in the materialist model of man and the market model of society'.[31]

Thus, according to MacPherson, did Hobbes morally justify unlimited accumulation of wealth by a few individuals in a market society as simply the natural outcome of voluntary transactions among free men. By investing the possessive market society with a natural moral authority, Hobbes defined the social value or worth – the 'goodness' – of each person solely in terms of the market-place:

> The *Value* or WORTH of a man, is as of all other things, his price... For a man's labour also, is a commodity exchangeable for benefit, as well as any other thing.[32]

The valuing of others as ends in themselves, quite apart from their market value, who by their very existence create meaning for myself, lies quite outside Hobbes's calculus.

John Locke

Hobbes's blatantly materialistic view of human society did not sit well with the leading figures of his time, who were still grappling with the social morality imposed by the old organic Catholic doctrine. For this and other reasons, his ideas were never well received in his lifetime. It was, rather, to the more agreeable arguments of John Locke that people turned.

The son of a Puritan who expounded the ideas of popular sovereignty and representative government, Locke was a liberal in his era, and his writings, particularly his *Second Treatise on Government*, were intended to justify the 'Glorious Revolution' with its wresting of power from an absolute monarch and transfer to the electorate in Parliament. With enormous skill he managed to uphold the concept of moral equality among men while still justifying the accumulation of wealth and power by the bourgeoisie.

Locke begins by asserting that land is a common gift to all men from

God, to be held in stewardship. But he gives to each individual the sole 'ownership' of his body and his labour, and whatever each person wrests from the Earth belongs to him. If a man ploughs fallow land and makes it productive, the land in its changed condition belongs to him, as well as its produce. Like Hobbes, Locke had no place in his moral order for the sharing so common among preliterate peoples or the reciprocal moral commitments of Medieval society. If you picked up an acorn or raised food, they were automatically yours by right:

> But the *chief matter of Property* being now not the Fruits of the Earth, and the Beasts that subsist on it, but *the Earth it self* . . . I think it is plain, that *Property* in that too is acquired as the former. *As much Land* as a Man Tills, Plants, Improves, Cultivates, and can use the Product of, so much is his *Property*. He by his Labour does, as it were, inclose it from the Common.[33]

Locke did put restrictions on how much one person might own:

> The same Law of Nature, that does by this means give us Property, does also *bound* that *Property* too . . . As much as anyone can make use of to any advantage of life before it spoils; so much he may by his labour fix a Property in. Whatever is beyond this, is more than his share, and belongs to others. Nothing was made by God for Man to spoil or destroy.[34]

Locke further assumed that land was available in unlimited amount; one had only to seek it.[35] In case England were becoming too crowded, he advised moving to the New World, ignoring any inconvenience to either emigrants or Indians.[36] Locke's belief that in his day there was enough land in the world to go around was perhaps correct, given the strict limitations he placed on taking only what you personally could use and assuming all hunter–gatherers turned to agriculture. But his theory scarcely justified the existence of the wealthy English landowners and merchants of his time.

Locke proposed to justify their possession of wealth in two ways. First, the accumulation of wealth is essential for the defence and financial stability of the nation in its competition with others:

> The chief end of trade is Riches & Power which beget each other. Riches consists in plenty of mooveables, that will yield a price to a foraigner, & are not like to be consumed at home, but espetially in plenty of gold & silver. Power consists in numbers of men & ability to maintaine them. Trade conduces to both these by increasing y^r stock & y^r people, & they each other.[37]

This was the old mercantilist theory that said that money is wealth and wealthy nations are powerful because they can maintain armies. Locke was stating what every sovereign knew: war is costly and demands a strong economy.

Second, Locke justified the enclosure of excessive amounts of land and other economic inequalities by an early form of 'trickle-down' economic thinking. To make his point he contrasts the life of 'savages' in North America with that of the English labouring class:

> There cannot be a clearer demonstration of any thing, than several Nations of the *Americans* are of this, who are rich in Land, and poor in all the Comforts of Life . . . And a King of a large and fruitful Territory there feeds, lodges, and is clad worse than a day Labourer in *England*.[38]

The poorest Englishman was better off than the chiefs of uncivilised tribes, and the reason lay in England's superior economic organisation. Her economic system, although it produced great inequality, also produced so much wealth that even the poorest had a better life as a result. Moreover, wealthy landlords and merchants were simply supplying wage-earners the opportunity to exercise their 'natural right' – namely to sell the only property they possessed, their labour. Locke thus deftly converted the absolute *necessity* for half the families in England to have jobs into the free exercise of their *natural right* to obtain the rewards of their labours. The wealth of the rich beneficently creates a livelihood for the poor. This basic presumption is still very much a part of the Western worldview.

Like other Puritans, Locke placed the cause of class inequalities squarely on the wage-earners and paupers; they had failed to make the best use of their God-given abilities. Locke argued that there is a natural equality of men, and that the lot of wage-earners or unemployed is the just result of a wilful failure to exercise their equal capacities. Furthermore, having fallen to the bottom of the heap through their own inaction, those without property had no political rights. Wage-earners were also clearly unfit, through want of leisure and motivation, for education:

> 'Tis well if men of that rank (to say nothing of the other sex) can comprehend plain propositions, and a short reasoning about things familiar to their minds, and nearly allied to their daily experience. Go beyond this, and you amaze the greatest part of mankind.[39]

Having once fallen through their own wickedness, the poor were never to be given the opportunity to rise again. With Professor MacPherson, we can conclude:

> With the Restoration [of the British monarchy] the idea of political right for the poor dropped out of sight again, and the idea of their moral

inadequacy was raised to the status of economic orthodoxy . . . The view that human beings of the labouring class were a commodity out of which [national] riches and dominion might be derived, a raw material to be worked up and disposed of by the political authority, was typical of Locke's period. So was the political corollary, that the labouring class was rightly subject to but without full membership in the state, and so was the moral foundation, that the labouring class does not and cannot live a rational life [of independent thought]. Locke did not have to argue these points.[40]

Thus did Locke's theories give the seal of moral approval to the rights of private property owners to amass wealth, this being generally beneficial to the welfare of the State and to its citizens of every rank. Yet the State retained considerable control over commercial transactions, it being deemed necessary under mercantilist theory for a nation to subsidise the production of exports and to place a duty on competitive imports in order to maintain the home economy and ensure a favourable balance of trade; the accumulation of gold and silver was a nation's primary goal. It remained for the Scots economist Adam Smith, writing almost a century later, to plead for a completely unregulated, *laissez faire* economy whereby the entrepreneurial class would finally attain complete autonomy.

Adam Smith

In his lengthy treatise *An Inquiry into the Nature and Causes of the Wealth of Nations*, published in 1776, Smith castigates the protectionist policies instituted by governments under mercantilist theory, policies that were intended to foster home industries. According to Smith, such policies are neither just:

It is the industry that is carried on for the benefit of the rich and the powerful that is principally encouraged by our mercantile system. That which is carried on for the benefit of the poor and the indigent is too often either neglected or oppressed.[41]

nor are they effective:

It is thus that every system which endeavours, either by extraordinary encouragements to draw towards a particular species of industry a greater share of the capital of the society than what would naturally go to it, or, by extraordinary restraints, force from a particular species of industry some share of the capital which would otherwise be employed in it, is in reality subversive of the great purpose which it means to promote. It retards, instead of accelerating, the progress of the society towards real wealth and greatness; and diminishes, instead of increasing, the real value of the annual produce of its land and labour... [The sovereign state is thus totally unsuited for] superintending the industry of private people, and of directing it towards the employments most suitable to the interest of the society.[42]

Smith argued that the three sole duties of the sovereign state are to protect the society from external violence, to promote justice, and to establish beneficial public works that the private individual has no incentive to undertake; it is *not* to interfere in the economy. This should be left to the individual who naturally invests his capital where it will bring the greatest return. The wealth of society, being composed of the summed wealth of individuals, will naturally increase as a consequence:

As every individual, therefore, endeavours as much as he can both to employ his capital in the support of domestic industry, and so to direct that industry that its produce may be of the greatest value; every individual necessarily labours to render the annual revenue of the society as great as he can. He generally, indeed, neither intends to promote the public interest, nor knows how much he is promoting it. By preferring the support of domestic to that of foreign industry, he intends only his own security; and by directing that industry in such a manner as its produce may be of the greatest value, he intends only his own gain, and he is in this, as in many other cases, led by an *invisible hand* to promote an end which was no part of his intention.[43] (emphasis added)

Adam Smith has long been recognised as the greatest exponent of the concept of *laissez faire* – an unregulated, private enterprise economy which, through the natural workings of a competitive and unfettered marketplace, will automatically accomplish the just distribution of a maximum amount of wealth within a nation. Implicit in Smith's reasoning are two fundamental ideas that are still strongly held today: the state is an entity quite distinct from the individuals that it comprises, and all wealth is of equal social value (that is, it is the monetary value that counts, not the use or function of what is produced). The correctness of these assumptions are examined in later chapters.

The Utilitarians

The final break with the old religion-based moral philosophy occurred independently but almost simultaneously in Germany and England. While Immanuel Kant in East Prussia was developing his categorical imperative (do only that which you would wish to become the general law – a modification of the Golden Rule) without recourse to divine authority,

Jeremy Bentham in Great Britain was formulating the concept of Utilitarianism. As Hobbes had earlier, Bentham turned to human nature as a starting point. For him, the central goal of human action is to seek pleasure and avoid pain, and this 'fact' formed the basis of his morality. But Kant and Bentham, unlike earlier philosophers, related morality to its consequences in *this* life, not to some future Heaven or Hell.

For the Utilitarians, the 'good' was whatever brought about pleasure or happiness and relieved pain. In its social context it represented a revolt against the earlier Christian ethic of self-sacrifice, which had more and more come to benefit, at least *in this life*, the few at the expense of the many. For Bentham, social justice was defined by maximising the total social good – the summed happiness of all of the individual members of society.[44]

Although in theory Utilitarianism has much to commend it, in practice it has two serious flaws. It demands a means not only of *defining* happiness, but also of *quantifying* happiness. To define happiness in a moral sense requires a definition of 'desire'. John Stuart Mill, a latter-day disciple of Bentham, argued in his essay on Utilitarianism that whatever *was* desired was *ipso facto* 'good' and therefore one had a moral right to attain it. But as Bertrand Russell points out, this is confusing the issue: if we have no choice but to follow our desires, then morality does not exist. All behaviour is predetermined, there is no such thing as moral choice, and everything is equally good and bad.[45] But everything that we desire cannot be morally correct, only what we *ought* to desire. These are quite distinct. If free will and choice exist, then some forms of happiness (fulfilled desires) are morally illegitimate.

Quantification of happiness is an even more formidable problem. To maximise something, you have to be able to *measure* it. Casting about for a method of quantifying happiness, the Utilitarians naturally hit upon the most quantifiable item in sight – the monetary value of one's possessions. And so, if it wasn't already, material wealth became equated with that which all persons, by their nature, most desire. Since according to Utilitarian thinking, whatever is desired is, by definition, 'good', happiness and moral worth could at last be objectively measured, finally confirming the Puritans' earlier hunch that these were equated with material success.

The Liberal Democrats

The early Utilitarians were unconcerned about liberty. Jeremy Bentham denounced the articles in the French revolutionaries' *'Déclaration des droits de l'homme'* as either unintelligible or false, and sometimes both.[46] It was to be the American revolutionary movement that formulated a completely new *political* model of the emerging Western worldview.

Filled with Locke's theories on the natural rights of man and disgusted with the persecution practised by the Anglican Church in his native Virginia and with the arbitrary use of political authority by the British, Thomas Jefferson took pen in hand and became the inspired spokesman for a new politics. In the American Declaration of Independence he deftly weaves together the Puritan notion of religious individualism, Locke's 'natural rights', and the budding Utilitarian concept of happiness, with the new idea that political authority comes *from the people themselves*, not from God, the Church, the King, or any foreign Parliament:

> We hold these truths to be self-evident, that all men are created equal, that they are endowed by their Creator with certain unalienable Rights, that among these are Life, Liberty, and the pursuit of Happiness. That to secure these rights, Governments are instituted among Men, deriving their just powers from the consent of the governed.[47]

Eleven years later the Founding Fathers were to lay down the procedural details of such a self-run democracy in the *Constitution of the United States* and its first ten amendments, *The Bill of Rights*.

Although generally believing in the rule of the majority, Jefferson realised that it could unjustly tyrannise minorities,[48] and more than that, it could be morally in error. This concern, I believe, underlay his many exhortations to educate all the people:

> Enlighten the people generally, and tyranny and oppressions of body and mind will vanish like evil spirits at the dawn of day. Although I do not, with some enthusiasts, believe that the human condition will ever advance to such a state of perfection as that there shall no longer be pain or vice in the world, yet I believe it susceptible of much improvement, and most of all, in matters of government and religion; and that the diffusion of knowledge among the people is to be the instrument by which it is to be effected.[49]

And so American representative democracy, based on the notion of individual freedom and rights with all their implications for social and economic life, became the living model of the new worldview. Moral decisions were removed from the word of God and invested in the majority of Men. Nowhere in this highly individualistic worldview, moreover, is any moral responsibility implied over and above the legal responsibility to obey the law. So ended organic society.

POSSESSIVE INDIVIDUALISM: THE REALITY

By 1700, after two centuries of social revolutions, both physical and intellectual, the new notion of possessive individualism was well and truly launched. As far as the poor went, it seems to have worsened rather than bettered their lot. In the England of 1688, 30 per cent of families, one-quarter of the population, were utter paupers, without access to land or a job, who depended on alms or parish assistance for survival. Almost all who earned wages as servants or labourers could not adequately support their families and depended on charity. All told, at least half the population required some degree of public support.[50] The shift from an economics based on mutual social obligations to one based purely on monetary exchange created not only a new wealthy class of bourgeoisie, but also a growing proletariat of labourers unable to accumulate the necessary capital to free themselves from abject dependence on others. (This is occurring today in Third World countries, as we shall see in Chapter 13.)

The new worldview, despite the moral arguments used to justify it, thus did little to create social and economic justice. Whatever progress was to be made in this direction occurred with the rise of Socialism in the nineteenth century, which led to the trade union movement and other institutions for providing political clout and economic security to the labouring classes.

What, then, did the new worldview accomplish? To begin with, it totally changed people's image of their place in the universe. From being servants of God (whatever their worldly station) each contributing to the life of His Church on Earth, they became isolated, competing agents. Human nature, once made in the image of God and innately capable of brotherly love, became a self-centred consumer of pleasure, the sole purpose of whose Reason was to further its own existence. If an image of God still lurked in human nature, it was one of omnipotence rather than compassion. Purpose in life shifted from serving God and society, and thereby oneself, to serving oneself first, and thereby – almost by accident – society. Fulfilling one's own desires became the most legitimate goal of existence. Where the God-given right of an individual to exist once put the other members of society under an *obligation* to care for her or him, in the new worldview no such social obligations are assumed. Each person has a *natural right* to preserve his life; hence his labour is his own, to sell if he chooses, and so is any property that he improves by his labour. But no moral obligations toward others exist beyond the contracts of the marketplace.

Finally, if God exists, the new worldview decrees that He speaks only through each individual's conscience, never through religiously ordained social laws. Nevertheless, worldly success, particularly material success, probably indicates divine approval and is a sign of great moral virtue. Entrepreneurs and corporate heads who create a means of livelihood for others are natural pillars of society. And those at the bottom, since they lack the necessary moral virtue, deserve their fate. The materially successful, who have come more and more to control the political system that legitimises economic and social arrangements, have thus become the beneficiaries of a moral authority that ensures their continued success. Tawney makes this point rather sharply:

> Few tricks of the unsophisticated intellect are more curious than the naive psychology of the business man, who ascribes his achievements to his own unaided efforts, in bland unconsciousness of a social order without whose continuous support and vigilant protection he would be as a lamb bleating in the desert.[51]

Consciously or unconsciously, despite the 'natural rights of all men', the practical rules which govern the daily activities of a capitalist society are largely made by and for those with material wealth. This is one of the inevitable outcomes of accepting the premises upon which our current worldview is based.[52]

It will be said that this picture of the modern Western worldview is too rigid, too 'black and white'. And so it is. It ignores many areas of altruism: it ignores thousands of institutions – the Churches, the Scouts, the Salvation Army; it ignores organisations like Oxfam; and it ignores government-sponsored social welfare programmes. Yet the tenor of Western society, especially as it has grown and evolved in the twentieth century, is indeed as Tawney describes it. Otherwise it is hard to explain our almost universal concerns with money, competition and material consumption; we are bombarded with them daily, on all sides, from birth onward. We *believe* this is how people normally are, that these are the major concerns of 'human nature'. And when the average person thinks about the problems that exist in the world, both in terms of environmental and human exploitation, the tendency is to throw up one's hands in despair: 'You can't change human nature!'

Yet, as Part II has shown, this pessimistic view is unfounded. In fact, if it resembles anything at all, the 'State of Nature' described by Hobbes, Locke and the other seventeenth century writers resembles the pitiable African Ik tribe, a collection of human beings reduced to a subhuman existence by a world which has closed its doors to them. Neither Hobbes's or Locke's primitive man nor the Ik are accurate models of the first human beings; rather, both are non-viable forms of human existence, whether imagined or real, created for the convenience of others. Individualism, materialism, and competitiveness are nurtured in us, woven implicitly into the very fabric of our worldview. And once accepted, this view of course tends to create the very kind of human behaviour that it assumes is inevitable. Just as with the Dobu, the Hopi, or the Yanomamö, our worldview becomes its own self-fulfilling prophecy.

SOME CONCLUDING THOUGHTS

Over the centuries, the attributes of possessive individualism – the notions of property rights, individual conscience, materialism, independence, competition – that at first were designed to free people (or at least some people) from oppressive social constraints, have evolved and changed in ways not foreseen. Perhaps none has been carried to such an extreme as the concept of freedom. The notion of individual freedom, which began at the end of the Middle Ages as *freedom from* arbitrary control over one's right to own property, has gradually been converted into the *right to* do whatever one pleases. Thus does the corporate owner of a factory justify his 'right' to discharge noxious wastes in order to make a living in the competitive marketplace, the Off-Road-Vehicle owner, his 'right' to run his dune-buggy over the fragile desert, and the teenage rock fan, her 'right' to play her transistor radio at top volume in public places. As technology continues to increase our ability to interfere destructively with the lives of others, our options are to adapt to chaos, or to impose on ourselves an ever-burgeoning list of regulations, or – which is surely preferable – voluntarily to adjust our concept of individual rights.[53] To do so, however, will require some major changes in our worldview.

Finally, the idea of 'absolute freedom', freedom carried to its ultimate and absurd conclusion, attempts to free the individual from all culture and tradition, leaving her or him in a social limbo. Philosopher Mary Midgley explains it well:

> *Freedom*, that most general of negative words, looms over us. It no longer means just the absence of a few specified evils – slavery, oppression, terror – or the entrance ticket to certain specified goods – attractive political and personal choices. It no longer means freedom *from* or *to do* anything particular. It has spread itself to cover the isolation of the individual from all connection with others, therefore from most of what gives life meaning: tradition, influence, affection, personal and local ties, natural roots and sympathies.[54]

Instead of seeking fulfilment through participation in society, people are taught to look *inward* for satisfaction: 'Be yourself!' 'Do your own thing'. For a time, it was fashionable for American schools to abandon any sort of compulsory programme and invite students to invent their own curriculum and participate in it only when they felt like it – with disastrous educational results. Although in recent years the educational pendulum has somewhat reversed itself, the purpose has not been to create social bonds nor to generate a meaningful worldview, but to mould students – like products – for a specialised niche in the national economy.

Thus does the worldview of possessive individualism, as it has evolved in the age of technology, create for us a curious paradox. Instead of being economically free, we have in fact become completely dependent for our very existence and livelihood on enormous productive institutions over which we exert no personal and little political control. In that sense, we are

Possessive Individualism

socially *less* free than most hunting–gathering people, who had some say in the running of their society.[55] At the same time, in order to demonstrate our freedom, we dispense with the very social bonds and commitments that give life its meaning: we undertake a form of voluntary alienation, that we then label 'freedom'. These two aspects of our modern paradox are pursued in Chapters 10 and 11.

10 The Cult of Efficiency

Representative Tilson: You found, as I understand it, that at 38 pounds
to the shovel that was not an economical load?

Mr. Frederick Taylor: Not an economical one if it was too heavy a shovel
load and prevented the man from doing a proper day's work.

Mr. Tilson: That is, your dirt pile grew as the size of your shovel went
down?

Mr. Taylor: The pile of dirt shoveled in a day grew larger and larger as
the shovel load starting with 38 pounds per shovel went down until we
reached a 21½-pounds shovel load, at which load the men did their
largest day's work, and then again the dirt pile grew smaller and
smaller as the shovel load became lighter and lighter than 21½
pounds.

> Testimony of Frederick Winslow Taylor
> before a Committee of Congress,
> 30 January 1912[†]

INTRODUCTION

Efficiency! It pervades our thinking. How can we get the most *output* of
goods and services for the least *input* of energy, labour and materials?
What is the most 'efficient', the most 'economical' way of doing things?
When 'wealth' is the supreme goal, the more efficiently wealth is produced,
the better. For the sake of efficiency, we subordinate human values that in
other times, in other societies, in other worldviews, had far higher priority.
The origin, development and logical consequences of our modern devotion
to wealth and efficiency are the subjects of this chapter.

The original notion of economic efficiency must have depended on an
awareness of *time* – not in terms of years or days, but in terms of hours and
minutes. Curiously enough, it was not workers, but monks in the
monasteries and cathedrals of Medieval Europe who first kept track of the
hours – the canonical hours for the daily prayers that marked their lives.
The earliest mechanical clocks were made for this purpose, and ancient
survivors still keep time in the cathedrals of Salisbury (1386), Rouen (1389)
and Wells (1392) (Figure 10.1).[1] Living by the clock, which is second-
nature today, was thus quite recently acquired. Modern people *tell* time,
save time, *spend* time, *keep* time, *pass* time, *mark* time, *waste* time. They
work from eight to five. They spend 50 minute hours in classes. This
all-pervasive consciousness of the passing of time is an absolute prerequisite

Source: Drawing by the author.

Figure 10.1 The clock at Wells cathedral, Somerset, England, built in 1392 AD.

The outer circle of 24 hours is monitored by a moving sun that reaches its apogee at 12 noon (straight-up) and is in the antipodes at midnight. A star-pointer tells the minutes of the hour on the Arabic numerals of the middle circle. The inner circle rotates daily, while the inner disc rotates in the same direction but slightly slower. Beside an opening in the disc showing the waxing and waning of the moon, is a trident which points to the day of the lunar month on the inner circle of numbers. The second opening contains the image of Phoebe, the moon, which remains always upright as the disc rotates; the words *Sic peragrat Phoebe* mean 'So progresses the moon'. In each corner are the four cardinal winds, supervised by angels, and the whole background is a dark blue sky with golden stars.

A Latin inscription around the moon reads 'This circular dial represents the actual universe in miniature.' In other words, the clock embodies the medieval belief that Earth was created at the centre of the universe, and is orbited by moon, sun and stars.

for the building of an 'efficient' society. It is a fundamental part of the Western worldview.

A special characteristic of efficiency is that it deals only with what can be *counted*. It asks: How much? How big? How many? How fast? How long? When efficiency takes over, 'goodness' is defined in numerical terms. Whatever is to have value must somehow be converted into a measurable *quantity*. What cannot be counted is 'of no account'; it is outside the system. (Today's yearning for more emphasis on the *quality* of life – an outcome of our frustration at having to exclude the non-quantifiable from our mental values scale – is discussed in Chapter 11.)

Since science deals in measurable entities, it seemed only sensible to apply scientific measurements to the analysis of the production of goods. At the turn of the century, the American, Frederick Winslow Taylor, equipped with stopwatch and sliderule, began measuring the efficiency of factory labourers. The old practices that employers had used to stimulate productivity – the sticks and carrots of threats and incentives – were replaced with analytical time and motion studies that better fitted the labourer's movements to his task. 'Taylorism', formally known as Scientific Management,[2] was so successful in increasing output that it spread from factories into every other aspect of organised life. Today no institution is exempt from the principles of efficient management. Schools, hospitals, investment firms, supermarket chains, political campaigns, even the televised evangelical churches, are 'managed' by professionals trained to organise the most efficient operation possible. We all are taught to 'manage our lives efficiently'. Our main complaint about government is that its particular form of bureaucracy seems *inefficient*, not that it is remote or undemocratic. We live in a systematised, managed world.

So interwoven into the fabric of society is this notion of efficiency that it seems 'only common sense'. The fact that this belief in the supreme rightness of efficiency is just that, a *belief* based on certain arbitrary assumptions, escapes us. Some of these assumptions were discussed in Chapter 9: the inevitable competitiveness and selfishness of 'human nature' (Hobbes); the 'natural rights' of individuals to possessive ownership of their labour and their property (Locke); the notion that economic self-interest automatically benefits the whole of society (Adam Smith); the tendency for the world to 'progress' (Darwin[3]). Never mind that these assumptions (as shown in Part II) are either arbitrary (Locke), wrong (Hobbes), misleading (Darwin, in the sense that 'higher' is 'better', a notion Darwin himself never advanced), or require enormous qualification (as Smith himself recognised); they are held by most Westerners. It is from these assumptions, and others now to be explored, that our worship of efficiency derives – a concept which totally shapes Western society and drives it in its present direction. It is part of our notion of 'rational behaviour'.

'RATIONAL ECONOMIC MAN'

At the turn of this century, a group of 'neoclassical' economists was reshaping economic theory along quantifiable, 'scientific' lines. For this purpose they needed a predictable human being, one that would behave in an economically consistent or 'rational' pattern. Now the word, rational, strictly speaking, applies to whatever seems sensible or reasonable. In order to create a *reasonable* economic agent, the new economists enlarged upon Jeremy Bentham's notion of utility. On entering the marketplace, a 'rational' person spends her or his limited resources on whatever mix of items provides maximal utility or pleasure. The 'value' of anything now becomes a measurable quantity, the amount of purchasing power given up for it:

> To satisfy our wants to the utmost with the least effort – to procure the greatest amount of what is desirable at the expense of the least that is undesirable – in other words, to *maximize* pleasure, is the problem of economics.[4]

So spoke the English neoclassical economist Stanley Jevons. All one needed was a 'hedonimeter' in order to have a perfect measure of economic value.

Naturally no such instrument exists; there is no *scientific* way to test whether Rational Economic Man exists or not. (As related in Chapter 11, common experience shows it is *not* an accurate image of human nature but, like the Second Law of Thermodynamics, there is no way to prove it from logical principles.) Nevertheless, almost all mainstream modern economic theory is based on this assumption. The economists' imagined human is an isolated being, acting purely out of personal self-interest, like Robinson Crusoe on his island. He is unaffected by those about him, and he is a lightning calculator of all the costs (in money) and all the benefits (in pleasure) available to him in the marketplace. Finally, if purchasing power expended is the measure of 'pleasurable value' or 'utility', then presumably things of equal price are substitutable in terms of social value. This subsuming of diverse social benefits under the single yardstick of monetary value leads to serious discrepancies between the perceived and the actual well-being of societies. Another implication is that what is 'free' is of no value. The 'free' services of Mother Nature and the 'free' services of mothers working at home are undervalued or even ignored in a monetarised economic calculus of goods and services. These obvious deficiencies in modern economic thinking are discussed further in later chapters.

Despite all this, the traits assigned to Rational Economic Man have become the basis not only of conventional economic theory, but also of the development of efficiency-oriented economies. While each consumer is maximising personal pleasure by 'efficiently' allocating personal income, so each producer is maximising income by the most efficient production process possible. Thus does a 'rational' society in theory maximise its aggregate well-being and so ensure Bentham's goal of 'the greatest good for the greatest number'.

As the German sociologist Max Weber observed,[5] a perfectly functioning 'rational' society would automatically be just: over a lifetime, one would take out exactly what one put in. But ideal conditions seldom exist, and concerns about social justice are still evident in the writings of such early economists as Smith, Mill and Bentham. Not so today. Moral value, not being amenable to quantification, leaves the focus of modern economic argument solely on the *efficiency* with which goods and services are produced. And theory further ordains that efficiency requires a *laissez faire*, free-market economy, regulated only by competition; the most efficient producer wins out in the competition for buyers. To act otherwise is to be 'irrational', a term neoclassical economists apply to societies that are less concerned with efficiency than with other social goals – that produce goods and services according to the rules of custom (traditional societies like the Hopi), or to meet religious requirements (the Islamic states), or to fulfil the goals of a centrally planned economy (as in the Soviet Union).

What are the properties of this 'rational' economics which dominates our thinking, moulds our lives and defines our social purpose?

CHARACTERISTICS OF A 'RATIONAL' SOCIETY

If a society is composed of 'Rational Economic Men' (no one has yet described 'Rational Economic Persons'), for whom maximising pleasures purchasable in the marketplace is the primary 'good' and work a disutility, a 'cost' to be minimised – in other words, a society whose primary social goal is *efficiency* in the production and consumption of wealth – what will it look like? What are its structural characteristics? How is it ordered? What is its morality? What behavioural principles does efficiency demand?

The Structure of a 'Rational' Society

There is perhaps no more clear example of productive efficiency than was given two centuries ago by Adam Smith in his classic description of a pin factory. Although today pins are turned out in automated factories, untouched by human hands, the principle he describes is crystal clear; his pin factory is a microcosm of industrial society.

The effects of the division of labour, in the general business of society, will be more easily understood by considering in what manner it operates in some particular manufactures... To take an example, therefore, from a very trifling manufacture ... the trade of pin-maker; a workman not educated to this business ... nor acquainted with the use of the machinery employed in it ... could scarce, perhaps, with his utmost industry, make one pin in a day, and certainly could not make twenty. But in the way in which this business is now carried on ... the whole work ... is divided into a number of branches. One man draws out the wire, another straights it, a third cuts it, a fourth points it, a fifth grinds it at the top for receiving the head; to make the head requires two or three distinct operations; to put it on is a peculiar business, to whiten the pins is another; it is even a trade by itself to put them into the paper; and the important business of making a pin is, in this manner, divided into about eighteen distinct operations... I have seen a small manufactory... where ten men only were employed ... [who] could make among them upwards of forty-eight thousand pins in a day.[6]

By *specialising*, men increased their collective productivity hundreds, perhaps thousands, of times. No wonder 'pin money', once an important item in a housewife's budget, became a trivial sum allotted his wife by a stingy husband!

How does specialisation develop?

The technology of production

To begin with, there must be *Technical Know-how*: How is something produced? Are there more efficient manufacturing processes and ways of utilising labour? Modern inventors, scientists, and engineers all contribute to 'R & D' – the research and development of ever more efficient means of production. This was Frederick Taylor's genius in the management of the human component of production; he was an engineer of human action, the father of scientific management.

Next is the need for *Organisation*. Each person has a single, often minute, job within the total enterprise. If specialisation is to result in efficiency, these diverse specialists must be integrated by yet another class of specialists, the managers. There is hardly any aspect of Western life – certainly no institution – in which the idea of *Management* is not all-pervasive. Besides organisations *per se*, we now manage highways, rivers, national parks, even wilderness areas, and are working on how to manage the weather, the seas, the atmosphere and space, in order that Nature, too, shall be made to fit into a controlled, orderly, 'rational' pattern.

Finally, any productive enterprise, whether 'rationally' organised or not, requires a certain amount of the physical wealth and labour of a society for its initial construction. Part of the economic capacity of society is channelled from the immediate consumption of wealth into the production and maintenance of *Capital*, the physical means for producing future wealth: factories, office buildings, railroads and so on. All industrial societies – capitalist, socialist or communist – have a requirement for capital.

In command economies, as in Eastern Europe, the state fixes wages and prices, thus controlling how much wealth goes to capital investment. In the 'rational' – or capitalist – societies, capital comes from individual decisions to forego immediate rewards and invest money, either directly or through savings, in new ventures. An expectation of later rewards – in interest or profits – provides the incentive. In a command economy, national saving is compulsory (the money is never given to people in the first place), and ideally the benefits of investment accrue equally to all as better products at lower costs. In a capitalist economy, saving is voluntary, and only part of the benefit is distributed to consumers as a whole; the rest goes to the *owners* of the capital as their particular reward. In modern America, most capital is supplied by corporations themselves, who reinvest much of their profit rather than distributing it to the 'owners' – the stockholders.[7] (The consequences of this for the autonomous growth and power of corporations is considered below.)

Private property

If personal incentives – based on the premise of a Rational Economic Man concerned only with maximising his personal pleasure – are the most efficient means of producing wealth, then private property is essential to efficiency. Many theorists blame the lower *per capita* productivity of command economies on an absence of the incentive arising from private ownership of wealth; by getting higher personal rewards, people contribute more effort and all society benefits: Adam Smith's 'invisible hand' is at work. But wherever growth of material wealth is the primary social goal (including both capitalistic and communistic societies), there is a constant struggle between increasing the *efficiency in the production of wealth* through the incentive of private property and increasing the *fairness in its distribution*. This tension between efficient production and just distribution contributes to almost every problem facing the world today.

Contracts

Modern industrial efficiency demands a level of flexibility in social arrangements utterly unknown in earlier traditional societies. In the potlatch system of the Northwest Indians or in the feudal society of Medieval Europe, the distribution of wealth was based on *status*, and one's status was fixed for life; new institutional arrangements were unheard of. But the Enlightenment introduced the idea of 'social contracts': social arrangements among individuals that are *mutually* agreed upon and may be *mutually* broken. Economic arrangements suddenly became much more flexible, and money, the connecting bridge between two independent sets of contracts – one's contributions to society and one's rewards from it – took on a new importance. The flexibility available in a contract-structured society is essential for efficiency. Indeed, it may be this as much as the lack of material incentives that holds back the productivity of command economies, where the freedom of individuals to make and break contracts is considerably limited.

The Morality of a 'Rational' Society

The Western capitalist view that material rewards are proper incentives to productive efficiency is a direct outgrowth of the utilitarians' pursuit of pleasure. *If* one's highest moral duty is self-gratification and *if* possession and consumption of material goods are gratifying, then ever-increasing wealth becomes the central goal in society, the means of acquiring status.[8] What then are the moral components of a rational society?

If the most efficient way to produce goods is to have highly trained specialists, carefully organised into a productive machine, bound by legal contracts, and motivated by the desire for material gain, then two attitudes about human relations must prevail: (1) egoism is not only tolerable but desirable, and (2) individual people are best treated as impersonal objects, designed to perform a specific function within the giant economic machine as efficiently as possible. As noted, the mass educational system in the United States has always emphasised moulding the young to fit economic niches and now, as critic Sheldon Wolin observes, American higher education is perfecting this art:

> After World War II, the growth of the nation's economic and political power, and the determination of its ruling groups to compete for global supremacy, were reflected back upon universities – in the form of pressures and incentives to concentrate upon developing scientific research, technical skills and methods, and the forms of professional knowledge that aid in social control (law, medicine, public health, social welfare, and the management sciences). The result was the radical alteration of the purpose of the university and college, from education [for critical thought] to the pursuit and imparting of knowledge [for personal and national power]. It was at this point, when humanistic education was being replaced by technical knowledge, that the masses went to classes.[9]

Europe lagged about a decade behind, but it too changed emphasis from a broad humanistic education for a few leaders to specialised practical education for ever-growing numbers. Thus are the young honed to fit into the machine.

This concept of objectivity in human relations requires a few words. To obtain its highest level of efficiency, a materialistic system must treat people not in terms of their personalities, their relationships to others, or their needs, but solely in terms of their ability to function in a specified niche in the productive machine. When an economically 'rational' system is working ideally, it eliminates both favouritism and prejudice. Each person has an equal opportunity to strive for whatever niche he or she may choose, *and* to succeed – or fail. A strictly 'rational' society has no safety nets for failures: the unlucky, the untrained, the disabled, the unwilling. It is up to each person to discover what she or he can do best, without false encouragement and without any expectation of social security if that specialty becomes obsolete or is superseded by a machine. This is the logic of a truly efficient society. It has no room for bondedness or compassion.

A humane society, on the other hand, has pockets for failures, finds work for its less competent members, pays attention to its disabled, retrains redundant workers. It cushions the inability of individuals to fit smoothly into the productive machine. To the extent that a society does any of these things, it is behaving 'irrationally', not 'rationally'. After all, some individuals are giving up personal utility for the sake of others, something 'Rational Economic Man' is not programmed to do.

In modern capitalistic societies, during periods of economic expansion when the demands of greed which drive the system are being reasonably met, there is something left over for the misfits and failures. Only then does the morality of economic 'rationality' extend to humane concerns. But when the economy falters, 'efficiency' again takes over as the moral basis of social decisions. We see this occurring in the uncertain economies of the 1980s. Support for public welfare declines and capital is diverted into the private, profit-making sector ('supply-side economics'), and college students swing away from careers in teaching and social work towards more lucrative jobs in business, engineering and computer programming.[10]

What are the logical outcomes of this pursuit of efficiency?

LOGICAL CONSEQUENCES OF ECONOMIC 'RATIONALITY'

When the central social goal becomes the efficient production of an ever-increasing supply of material wealth, certain developments automatically follow that affect the organisation and character of society. These consequences also affect people's personal lives, an aspect taken up in Chapter 11; here we concentrate on the broader aspects of the drive for efficiency.

Standardisation

The American physician and poet, Oliver Wendell Holmes, wrote a delightful verse, 'The Deacon's Masterpiece', about 'the wonderful one-hoss shay, that was built in such a logical way it ran a hundred years to a day'. By taking pains to build all parts equally well, the deacon ensured that his custom-made carriage could never break down. Nor did it. It simply wore out, all at once, on the way to church, dumping the current parson on the ground:

– What do you think the parson found
When he got up and stared around?
The poor old chaise in a heap or mound,
As if it had been to the mill and ground!
You see, of course, if you're not a dunce,
How it went to pieces all at once, –
All at once, and nothing first, –
Just as bubbles do when they burst.[11]

Unlike the deacon's carriage, modern automobiles are produced by the thousands and all too often have parts that give out. But because large numbers of cars are identical, it is possible to produce and stock spare parts. It is cheaper to mass produce cars and the spare parts to keep them running than produce a 'perfect car' initially.

Precision-produced interchangeable parts were first introduced by the Springfield and Harpers Ferry armouries who made rifles for the Americans during the War of 1812. It may have turned the tide of history, for the British, who were anxious to regain their lost colonies, were plagued throughout the war with jammed guns. Being custom made, they were not easily repaired. But it was not until the 1880s and 1890s that standardised parts spread throughout Western manufacturing, greatly facilitating the mass production of items like sewing machines and bicycles.

Finally, Henry Ford, after the turn of the century, introduced the assembly line into the organisation of modern industry.[12] Not only were the parts standardised, so were the jobs, and it took its toll on workers. In 1913, the annual employee turnover at Ford's plant was 380 per cent; labourers intensely disliked the inhuman demands of the assembly line. Only when wages were doubled did they stay. But people soon became replaceable, like the spare parts they manufactured. In modern industry, individual workers, managers and even presidents are expendable. Nothing serious occurs when someone dies, quits, retires or is fired. The system goes on, readily able to fill the empty niche.[13]

The Technological Imperative

In a world where competition and efficiency are the watchwords, old manufacturing processes are constantly being supplanted by new ones producing 'better' products. The switch from vacuum tubes to silicon chips and printed circuits in the electronics industry is an obvious case in point. Often the 'new, improved' product is only marginally better than the old, but the constant jockeying for markets creates an unremitting pressure for product development which has several important consequences.

In certain fields like automotive and aerospace engineering, new technologies are becoming ever more complex and costly to develop. In most industrial countries, much of the R & D is government-subsidised. Furthermore, as technologies become more complex the lead time required to gear up for commercial production also lengthens. At the turn of the century, Ford could alter the design of his Model A or Model T in a few days. Not now. A new automobile is designed several *years* before the first one rolls off the assembly line. If anything goes wrong, millions of dollars and years of time are forfeited. For giant undertakings requiring government backing, such as new nuclear reactors, supersonic transports (SSTs), or giant military missiles, once public funds are committed, it is almost impossible to stop a project – even an obvious mistake. Britain's cancellation of its Blue Streak missile in the 1960s caused a national scandal, and it took eight years before America's SST was finally scrapped, at a cost to the public of over $1 billion.[14] 'Star Wars' is America's current such project.

Long lead times and high development costs mean that even 'successful' small companies have difficulty surviving. Either they grow rapidly or they are 'taken over' by one of the giants. In the United States, for example, the hundreds of thousands of small business firms either make highly specialised products for an elite market or are in the trade and servicing sectors of the economy. Nearly a quarter of small companies fail in their first year and 80 per cent fail within five years.[15]

A second consequence of the technological imperative in a competitive economy is that 'knowledge is power'. Francis Bacon three centuries ago meant power over an untrustworthy environment, but today it means power over competitors. Trade secrets are jealously guarded and certain areas of science are rapidly going underground. In 'hot fields', there is often a race to patent a new discovery; knowledge is private property, to be owned, bought and sold. Even universities rush to hire scientists who might create patentable genes or establish profitable liaisons with pharmaceutical and chemical firms, and many have knowledgeable patent officers on the lookout for lucrative research. Whoever wins the patent gains the profits!

In the field of food production, scientists, traditional plant breeders, and corporations such as Bayer and Monsanto are rushing to patent genetically engineered varieties of plants.[16] Europeans have traditionally held that plants should not be patented nor should private companies obtain monopolies over important foodstuffs. But now a struggle is underway between chemical giants in Germany and Switzerland who invested in recombinant

DNA research and wish to patent new genes useful in agriculture and food production, and traditional plant breeders who see their standard plant varieties, infected with a single new gene, suddenly becoming the property of chemical companies. The notion that knowledge is private property, already prevalent in America, is growing in Europe, where the World Intellectual Property Organisation is located in Geneva.

The technological imperative also means that as products and techniques become more powerful, their potential impacts on society and the environment grow accordingly. Not only do R & D costs soar (sometimes making a project, once under way, virtually unstoppable); so do the costs of regulating technology, of monitoring its effects, and of repairing the damage caused by unexpected failures or side-effects. Thalidomide, the tranquilliser unwittingly taken by pregnant European women who then bore limbless babies in the late 1950s, resulted in millions of dollars being expended on special services for thousands of bright but physically disabled people (Figure 10.2). Among environmental side-effects are acid rain, the pollution of drinking water, the destruction of forests and lakes, and growing numbers of toxic waste dumps, all unpaid overheads of modern technology. Compared to the several hundred billion dollars needed to clean up the

Source: Redrawn by R. Carruthers from P. Grant, *Biology of Developing Systems* (New York: Holt, Rinehart & Winston, 1978) Figure 20.1, p. 661.

Figure 10.2 Newborn female infant with phocomelia ('seal's limbs') resulting from exposure to the tranquilliser thalidomide during foetal life.

major toxic waste sites already in the United States, the government's allotted 'Superfund' will barely scratch the surface. In the coming decades it will cost California alone $40 billion for the consequences of its toxic wastes.[17] Just *monitoring* America's air pollution comes to hundreds of millions of dollars annually;[18] this includes none of the costs of *reversing* it or *repairing* the damage to buildings, people and forests.

Major industrial accidents now rival natural disasters. In April 1947 the town of Texas City was destroyed in a single day when fertilizer in the hold of a ship exploded, setting off a chain of explosions in nearby ships, in oil refineries and chemical plants; the fires continued for a week. Energy equivalent to an atomic bomb had been released, killing or injuring one-quarter of the residents, and doing $60 million worth of damage.[19] In December 1984, more than 2600 people died and 200,000 were injured by deadly fumes that escaped from Union Carbide's pesticide producing plant in Bhopal, India. Settlement costs may approach $1 billion, although $10 billion would be more just compensation; it would easily bankrupt the company.[20] Legal limits are being placed on corporate financial responsibility as the dangers increase. When in 1957 American utility companies were unable to buy private insurance against damages from nuclear power accidents, the Federal government passed the Price–Anderson Act guaranteeing a half-billion dollars to victims of each accident. But if the courts awarded higher damages, neither power company nor government could be sued, even if the accident were caused by sheer negligence.[21]

The technological imperative ineluctably creates new and more costly social overheads which are acknowledged only *after* the technologies are in place. As French social philosopher Jacques Ellul says:

> History shows that every technical application from its beginnings presents certain unforeseeable secondary effects ... These effects exist alongside those effects which were foreseen and expected and which represent something valuable and positive.[22]

The business world naturally baulks at assuming responsibility for these 'external' costs, arguing that to do so would 'slow progress' and 'endanger the economy'. Accountability is thus 'inefficient', and today it is the public who pays.

A final consequence of the drive to technological change is that returns on investments are decreasing as technologies become more rapidly outmoded. Indeed, this may be a primary cause of today's massive corporate borrowing.[23] There is a growing 'competitive overhead' as companies leapfrog over each other in the technological stampede. This together with growing environmental and social costs may soon reveal the technological imperative as increasingly *inefficient*.

Corporate Autonomy

It was Harvard economist John Kenneth Galbraith who first showed in his book, *The New Industrial State*,[24] how the technological imperative and the cult of efficiency gradually forced a complete change in organisational structure of the top corporations that have largely determined the direction of American society in the last half century. And where America has led, other industrial societies have not been slow to follow.

The driving forces behind this change have been the increasing complexity of technology and the longer lead time and higher costs of research and development. Too much capital was at stake to leave decisions in the hands of a single entrepreneur, or to leave the acquisition of raw materials or the sales of the finished products to the whims of the marketplace. Corporations had to become 'rational'. Wise planning was the first step, and increasing control over the market was the second.

The corporate metamorphosis

In the 1930s, Henry Ford brought his company to the brink of financial collapse by refusing to listen to advisors, taking total power in his own hands and making one disastrous decision after another.[25] After his death in 1947 power shifted from one man to many, but it did not go to the Board of Directors elected by the company's owners, the stockholders. Instead, an entirely new network of executives, engineers, marketing specialists, designers and accountants took over. Decisions were now made in small groups, each set of specialists contributing its expertise in multiple committee meetings, providing pieces of information to be fitted into the overall picture. An organised system replaced the old entrepreneur, and its combined expertise was capable of far better decisions. The Ford Motor Company quickly recovered.

This shift in corporate structure, depicted schematically in Figure 10.3 has occurred in almost every major corporation in the world. In the old entrepreneurial company with relatively few white-collar workers, the blue-collar labourers often were in a confrontational relationship with their employer. At the end of the nineteenth century, American workers, following their European counterparts, began to organise in a struggle against the monopoly power of employers and 'the horrors of industrial capitalism'.[26] Unions were formed and there was an upsurge of industrial violence. The Socialist Party and other reformist groups began to grow. But by the late 1930s and early 1940s, the larger American corporations, sometimes with the help of government, had generally come to terms with the giant unions.

Automation, although increasing industrial productivity, has greatly diminished the need for unskilled labour, while the need for skilled

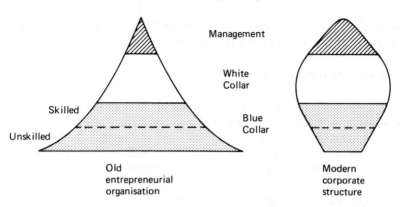

Source: M.E. Clark.

Figure 10.3 Contrast between workforces in old entrepreneurial organisations and modern corporations.

Automation has removed many blue collar jobs, especially of the unskilled sort, while the proportion of white collar jobs – office and engineering staff, marketing personnel and so forth – has greatly increased. The old entrepreneur at the top of 'his' company has been replaced by an entire network of decision-making specialists, the modern management.

specialists has sharply increased. The impact on higher education has been profound. In the United States, for example, since the turn of the century, the number of college students has increased 36 times, and the faculty, 38 times,[27] while the population as a whole has little more than doubled. Explosive growth in 'practical' disciplines followed on the heels of the Second World War. The marginally educated in today's industrial societies are becoming 'hardcore' unemployed.

Corporate power has also shifted. It once lay with the entrepreneur and the voting shareholders; those who made the corporate decisions stood to gain the most financially. In today's corporate giants, however, the decision-makers at the top, like everyone else, are salaried (albeit generously). Unlike the old entrepreneur, they do not directly benefit from corporate profits. Some go to shareholders, but most are reinvested in the company. Corporations are not run, as is widely supposed, by their shareholders, whose power is negligible. The Board of Directors is usually nominated by top management who in turn are hired by the Board: a reciprocal arrangement. By reinvesting profits, corporations become autonomous, free from external controls. Says Galbraith: 'It is hardly surprising that retained earnings of corporations have become such an overwhelmingly important source of capital.'[28]

The primary goal of the decision-maker in the big corporations subtly shifted from maximising profits to maximising stability and control over the environment in which the corporation functioned, thus ensuring corporate immortality. Infinite security now supplants mere wealth as the goal of the elite cadre of decision-making specialists.

The history of corporate growth

For a corporation in a competitive economy, the best way to ensure immortality is to grow. The first step is to *take over competitors*. In most major industries only a handful of companies dominate the market of any given country.[29] In the United States, a vast country of 240 million people, there are but three major automobile companies and seven major oil companies. Over the years, smaller companies have been taken over by the larger. Oldsmobile was absorbed by General Motors, Dodge Brothers by Chrysler. Duesenbergs, Grahams, Hudsons, Willys-Knights, and Studebakers have all been swallowed up. An even more startling contraction occurred among the dozens of American oil companies. The result of this *horizontal integration* is oligopoly, the capture of a vast market by a few giant firms.

The next step in corporate security was to gain control over suppliers and distributors by merging with them. Giant food chains for example now own their own fields, their own processing plants, and their own trucks, as well as their own outlet stores.[30] This kind of *vertical integration* occurs in many industries.

The third step is to *diversify*: buy up unrelated companies that are, or can be turned into, profitable enterprises. Thus do corporate giants cushion themselves; when the sales of one product flag, those of another will compensate; the conglomerate as a whole will always have a steady flow of earnings and management's immortality is assured. This particular strategy is being practised today by many industries. The big oil companies, having bought up all their smaller competitors, are now turning to alternative energy supplies. In the United States, they own 55 per cent of coal and 35 per cent of uranium, and are entering into the areas of geothermal and solar power.[31] One strategy of the Reagan administration's supply-side economics was to encourage corporations to reinvest their earnings and become more competitively productive, but many used the newly tax-exempt earnings to take over other profitable enterprises instead. While such a practice increases short-term corporate profits, it does nothing to improve the productive health of the national economy.[32]

The final step in achieving corporate security is to become a *multinational corporation* (MNC), building production plants throughout the world and attempting to control not just a national but the global market. At first, most MNCs originated in the United States, but today many are European and Japanese.

Historically, then, corporate management has increasingly controlled both production and the market through greater and greater economic *centralisation*. On a global scale, economic power is being concentrated in the hands of fewer and fewer people.

The myth of the market

Courses in economics, and almost all textbooks, still insist that the consumer is sovereign: by the exercise of free choice this product is bought and that one is not. In the free marketplace, prices are regulated by a balance between supply and demand, and kept low by 'healthy competition'. That is the theory.

As Galbraith points out, however, market theory works only for that half of an industrial economy run by the myriad small businesses that really do compete with each other to supply services and specialised products on a small scale. But it is not at all true of the giant corporations that dominate economic life. They leave as little as possible to chance and the vagaries of the consuming public. Corporations in the same industry seldom engage in suicidal price wars; they discreetly settle upon similar prices for their products, that will bring a respectable level of earnings over and above costs. The illegal collusion of 'price fixing' is not at all necessary.[33] They then set out to *control* the market, to ensure that the public buys their product. In short, they advertise massively: on television, on neon billboards, in throwaway mailings, on the radio, in magazines and newspapers – even in the sky over football stadiums and crowded beaches. And the public responds – by the millions. The same story is enacted in every free-market industrial economy around the world.

'Advertising exists only to purvey what people don't need. Whatever people do need they will find without advertising if it is available.' So says Jerry Mander in his book, *Four Arguments for the Elimination of Television*.[34] Advertising *creates* 'needs' and then sells us products to fill them. It labels body odours as disgusting and sells us deodorants; it points to rings around collars and toilet bowls as signs of housewifely incompetence and sells products to remove them; it tells us we cannot cope with our daily lives and sells us tranquillisers; it sells us junk foods that make us obese and follows these up with fad diets and appetite-restricting pills to cure the consequences. Says Mander:

> By entering the human being's inner sanctum, our inner wilderness, advertising effectively pulls our feelings up out of ourselves, displays them and sells them back to us like iron from the ground. Our inner feelings are transmogrified into a new form – commodities. We desperately seek to get them back, and pay high prices for the privilege.[35]

The prime consequence of this constant bombardment by advertising is chronic dissatisfaction. People with discretionary income – money left over after their basic survival needs are met – are often easy prey to such artificially created 'needs'. They will even take extra jobs and go into debt to satisfy them. People hooked on advertising thus serve a double function for the corporate giants: they are a source of willing labour and a steady, secure market. Without advertising, the propensity to consume would never have expanded as it has. Growth, the force that drives the giant corporations, 'requires that people will work without any limiting horizon to procure more goods'.[36] The process is enhanced if the goods are soon made to seem obsolescent – last year's cars, clothes, kitchens – or are so shoddy they fall apart. 'The ideal world for advertisers would be one in which whatever is bought is used only once and then tossed aside. Many new products have been designed to fit such a world'.[37] In the wake of the modern consumer there is a flotsam of Kleenex, empty containers, paper plates, plastic spoons – a whole melange of single-use garbage.

The 'planned economy'   The West today has a 'planned economy'. It is not planned by the state, as in Eastern Europe, but by the major corporations, which in turn are driven by technological change and the incongruous demand for efficiency in production and obsolescence in use. Just when needed, technology provided the mass advertising media to bring it all about:

> The [industrial] planning system is profoundly dependent on commercial television and could not exist in its present form without it. Economists who eschew discussion of its economic significance or dismiss it as a wicked waste are protecting their reputation and that of their subject for Calvinist austerity. But they are not adding to their reputation for relevance.[38]

Thus does Galbraith flatly deny the claims of economists such as Paul Samuelson that '[t]he consumer is, so to speak, the king . . . each is a voter who uses his votes in the marketplace to get things done that he wants done'.[39] So long as this notion, however patently false, is uncritically believed and widely accepted, then any regulation of corporate marketing is shunned as undue restriction of freedom. If the economy is 'shaped by the votes cast by the free consumer', then commercials, environmental abuse, and pollution are all the *consumer's* fault. The corporation is morally absolved for it is merely serving society's demands.

'The economy' has become the central – one might even say the sacred – focus of capitalist societies, as political theorist Sheldon Wolin shows in writing about the United States in the 1980s:

> 'The economy' has emerged in the public consciousness as a sharply outlined, autonomous entity, the theater in which the destiny and

meaning of the society will be worked out ... The state of the nation becomes meaningful only when we are able to talk about it as 'rates' of various kinds – rates of 'inflation', 'interest', 'productivity', 'money supply', 'capital formation', and, last but least, 'unemployment' ... What can hardly be doubted is that economics now dominates public discourse ... It prescribes the form that 'problems' have to be given before they can be acted upon, the kinds of 'choices' that exist, and the meaning of 'rationality'.[40]

The 'health' of the economy is a major concern of the state, which manipulates money supply and interest rates, taxes and government spending, and regulations on business; it bales out sinking giants such as Chrysler and Lockheed; it maintains a 'credible' military force to protect economic interests abroad. The economy is planned by the corporate giants but the state ensures that their plans do not fail.

The Government–Industry Complex

This close relationship between government and business grew out of the inability of industries to cope with fluctuating demands in an unregulated market. During the first half of this century there were increasingly severe cycles of boom and bust. The ever-more 'efficient', heavily mechanised industries were inflexible and hence fragile in the face of a fickle market. Manipulating the market by advertising worked well only in times of growth. Repeated business cycles culminating in the stockmarket crash of 1929 and the long years of depression required more drastic action. The British economist, John Maynard Keynes, proposed that the state should act as a long-term buffer of economic cycles, preventing both too rapid growth and catastrophic collapse.[41]

Keynes believed that depressions were caused by an unwillingness of entrepreneurs to invest capital in the absence of obvious consumer demand. With so many unemployed, where was the market for even *more* goods to come from? But if demand were stimulated by the state, new investment would automatically follow. Keynes argued that lowering taxes and printing money to increase purchasing power and lowering interest rates to increase borrowing were not enough. The state itself should become a direct investor in corporations, thus generating new jobs.

Although Western governments have willingly manipulated money, interest rates and taxes in order to regulate business cycles (Figure 10.4a), none has ever undertaken the 'comprehensive socialisation of investment' recommended by Keynes. In the 1930s' depression, governments 'primed the pump' through direct welfare payments or by creating jobs on public projects, such as Franklin Delano Roosevelt's Works' Progress Administration. After the Second World War, socialist governments in Europe

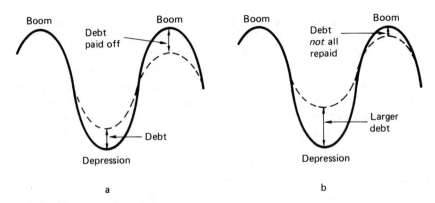

Source: M.E. Clark.

Figure 10.4 Keynesian theory for government moderation of business cycles.

    a: Theory states that a depression should be eased by government stimulation of
       the economy, through lowering taxes and interest rates and priming consumer
       pockets with welfare payments or government jobs. Debt incurred during this
       period is recouped during the next 'boom' period; the boom is tempered by
       raising taxes and interest rates to slow investment.
    b: In practice, restraining good times is politically more difficult than cushioning
       the bad ones, so national debts tend to grow after each cycle.

directly nationalised unstable private industries important to public welfare: health, utilities, transportation and others. And in the 1980s,
governments have attempted through 'supply-side economics' to stimulate
corporate investment indirectly through tax breaks and deregulation,
which (as shown later) has largely failed.

    Limited government intervention in the economy is now widely accepted
and the swings between boom and bust have been somewhat damped. Yet
the tendency has been to go into debt during slack periods without fully
repaying such debts during peak years (Figure 10.4b), with the chronic
results of a growing national debt and inevitable inflation. The consequences of this – an increasing debt which demands *economic growth* at a
time when resources are becoming *more and more scarce* – are considered
in Part IV.

    For any government to wield the impact demanded by Keynesian
theory, it must be a major participant in the economy. In the United
States, the federal government disposes of 24 per cent of the GNP; when
state and local expenditures are added, the total approaches 34 per cent.[42]
Moreover, government bureaucracy interlocks with that of the giant
corporations.[43] In particular, the state has become, as President Eisenhower
warned it would, a major and reliable market for industry, especially in the
areas of space and defence. Furthermore, governments pay for most of the
costly 'basic' research carried out in universities and national laboratories

that industry later makes use of. (The American computer industry has complained bitterly that their Japanese rivals are outdistancing them in supercomputer technology owing to the massive technological cooperation they are getting from the Japanese government.[44]) Another subsidy to industry is the taxpayers' support of education, particularly the costly higher education that turns out the necessary white-collar specialists. The military also produces skilled professionals for the private sector: airline pilots are mostly ex-air force pilots, and many nuclear power-plant operators in the United States were trained on the navy's nuclear-powered ships.

Finally, most industries depend absolutely on the provision of public services. A striking example is the automobile industry, whose astounding growth could not have occurred without government funds for the construction of highways. The airlines have benefited from the billions spent on the development of military aircraft and the installation by governments of navigational facilities.[45] And the nuclear power industry has been heavily government-supported in every country where it exists. With few exceptions, governments have become not the regulators of the corporate system, but its servants and handmaidens. In undertaking responsibility for the economy, the state has committed itself to supporting whatever course of action most favours the economic health of the industrial giants upon whom capitalist economies now primarily depend. As Galbraith says: 'The goals of the industrial planning system . . . become the goals of . . . society itself'.[46] Corporate needs *become* our needs. What is good for a nation's industry is good for the nation's people.

CONSEQUENCES OF 'EFFICIENCY'

It would be hard for anyone to deny that material well-being has become the central concern of most industrialised societies. This chapter and the preceding one have reviewed the step-by-step progression from the organic society of the Middle Ages to our present dedication to individualism, material consumption and economic efficiency. Yet it is a peculiar sort of efficiency we worship, since we are encouraged to be wantonly wasteful of the very things we insist be produced efficiently. As is shown in Part IV, modern economies are actually 'uneconomic', for they encourage ever-increasing *throughput* of goods with an ever-diminishing useful *lifetime*, precisely the opposite of rational economic behaviour! We are unwittingly acting out a logical paradox because we have unconsciously distorted the meaning of what we say we value.

The 'Quantoid Worldview'

In life, some things can be counted and others cannot. Those things which matter most – beauty, faith, love, joy, friendship, self-expression –

are immeasurable. There is no way to *count* them. They are not market-
able. As soon as we put a price on them, they are prostituted and debased.
Yet as a society, this is exactly what we do. Only what can be bought and
sold is given value. Indeed, it is this counting syndrome, which arises out of
the cult of efficiency, that makes material goods so desirable: 'happiness'
becomes a *number* – the number of bathrooms in one's house, of cars in
the garage, of Valentines received at school, of promotions up the
corporate ladder of success, of awards, achievements, accolades. What are
your school test scores? What is your income? How much? How many?
This pervasive obsession with number is a 'quantoid worldview' (*quantum*
plus *-oid*, meaning 'like'). Its opposite is a 'humanoid worldview', about
which more later. Here is philosopher Henryk Skolimowski's assessment
of our predicament:

> The quantitative civilization that the West has developed is at once a
> great achievement of the human mind... and a great aberration, for
> we have attempted to reduce all qualities to easily quantifiable physical
> quantities.[47]

Even the social sciences that supposedly deal with 'human' behaviour have
become 'objectified' and 'scientific', as the following examples suggest.

Almost everyone has heard of 'goals and objectives'. It is the catchphrase
of modern management. Setting goals and objectives and then estimating
how effectively they have been met is how we judge the goodness of our
actions, how we demonstrate *accountability*. It is the modern extension of
Taylor's 'scientific management'; only by *counting* can we determine our
success. We tally-up the results: the achievement scores of high school
students; the percentage of cancer patients cured; the increase in average
per capita income. Any goal whose outcome cannot readily be counted
tends to be excluded from consideration as a possible social good; at best,
it receives a low priority among competing goals.

Quantoid thinking also dominates government decisions. Consider the
'risk–benefit' analyses performed by law-makers when legislating the
development or regulation of particular technologies. In the United States
Congress, the question of whether to ban the artificial sweetener saccharin
from soft drinks was argued in purely quantitative terms: how many
prematurely dead, overweight and diabetic people would there be if it were
banned versus how many excess fatalities from bladder cancer if it were
not. All discussion revolved around numbers: body counts, dose levels,
thresholds of carcinogenicity, and probabilities of morbidity. No one
suggested imaginative, non-numerical solutions, such as encouraging
changes in eating habits or offering sweetener-free drinks, as in Canada.[48]
On other occasions legislative cost–benefit analyses are conducted, as
English technical critic Arnold Pacey points out, 'by imposing economic

reference points on everything, perhaps by assigning a money value to a botanically unique habitat or an ancient church'. As Wolin puts it, there is a universal 'faith that practically any public concern can be reduced to economic categories'.[49] If it doesn't have a 'price' it cannot be important.

Finally, quantoid thinking assumes that a technological fix exists for every human problem. If there are too many people, all we need are better contraceptives to stop the population explosion, and more 'efficient' crop plants to feed the present overflow. If technology causes pollution, all we need is a new technology to reverse it. Science and technology are the basis for progress and prosperity. This is one of the sacred shibboleths of modern Western thinking, and it deeply pervades our entire worldview. Skolimowski remarks on the matter thus:

> For this reason not only technocrats and simple-minded technicians with vested interests in technology, but also Nobel Prize winners and people of superior knowledge and exquisite refinement . . . unblinkingly insist that what has been spoiled by technology *will* be cured by technology and can *only* be cured by technology . . . The power of the myth of technology is so great and so dangerous precisely because it has pervaded the recesses of our Western mentality. Technology has become our physical and mental crutch to such a pervasive and perverse degree that even if we realize it devastates our natural and human habitat, our immediate reaction is to think about another technology which will mend it all. Technology is a state of Western consciousness.[50]

So deeply is society hooked on technology that when things go wrong the first cry is to upgrade, modernise and reinforce the scientific/technological establishment. That was the response to Sputnik; that is the current response to a flagging economy. Yet the lessons of too rapidly applied past technologies are ignored: clumsily organised cities; energy-extravagant skyscrapers; destructive agriculture – all are far from ideal technologies upon which, however, we have become utterly dependent. The past cannot be easily dismantled. To paraphrase that most thoughtful of American historians, Lewis Mumford[51]: Because we have used knowledge – the technology that science provides us – as a *substitute* for philosophy, we are in trouble on two counts. We have no guide as to what technologies we *should* attempt to develop. And we run the risk of becoming trapped – even more deeply than today – by technologies of our own making. Without moral guidelines, we are passive agents at the mercy of the technological imperative, which in turn is driven by the cult of efficiency.

The Erosion of 'Freedom'

Another seldom-recognised problem with technology is the way it limits our freedom. 'Freedom' is a big word that is used rather casually to mean many things, all of them supposedly 'good'. When we talk about the 'free market' of a *laissez faire* economy we assume without reflection that a regulated market must be 'bad' because it is 'unfree'. It seldom crosses our mind that absolute freedom for all to do as they please results not in universal freedom but its opposite – general chaos. Stopping at traffic lights, paying our taxes, advertising truthfully, and not smoking in elevators are all limits on our personal freedom *to do* things. These limits are obvious and we accept them because they make social life possible. But there are other limits to freedom imposed by large, centralised technologies that restrict us in ways we scarcely realise. Some restrictions affect us as individuals, moulding us to fit the demands of a technologically-oriented market economy, about which more in Chapter 11. Others limit the 'freedom' of industrial society as a whole.

Efficiency and the technological imperative have created centralised, heavily energy-dependent social organisations, the giant conurbations whose very existence is locked into the continued functioning of a narrow selection of high-tech systems. Instead of being *independent* through technological diversity and redundancy, modern industrial societies are rigidly *dependent* on a few complex technologies. Being overspecialised, they are extremely vulnerable, as recent incidents in the United States attest.

The brief oil crises of 1973 and 1979, when oil imports were temporarily cut off, created widespread chaos, and a number of cases of violence occurred. And in the New York City blackouts of 1965 and 1977, the whole of greater New York was plunged into utter darkness. The 1965 blackout stranded millions of people in pitchblack subways and skyscrapers for up to 13 hours, and during the 1977 episode, arson, vandalism and looting were severe over large areas of the city.

But it is not just Murphy's Law – that whatever can go wrong, will go wrong – that industrial societies face. Highly centralised technologies are at the mercy of small handfuls of terrorists. Less than 20 determined people could cut off power to the entire northeastern United States for weeks at a time.[52] As terrorism rises in parallel with worsening global injustice, industrialised societies are likely to be increasing targets for such disruption.

Despite these gentle warnings, society seems not to notice that extreme dependence on a relatively few 'high' technologies is neither a form of freedom nor insurance for long-term survival. Instead, its dependence limits society's ability to adapt by limiting its choices. 'Hi-tech' societies live on the edge of crises, yet to dissolve and disband their fragile, centralised systems in which so much has been invested seems impossible.

On asking 'How did we get into such a predicament?' it is clear that it did not come about in any *democratic* fashion. Citizens did not vote on how cities should be developed, or on how centralised the technological and economic powers should be on which they now so completely depend. The pretence made by politicians that this utter dependence of ordinary citizens on social forces over which they have no control is 'democratic' because it emerged from a supposed 'free market' is a major cause of widespread political cynicism and voter apathy. The connections between big government and big business become increasingly apparent as the corporate funds shunted into political action committees (PACs) and lobbying efforts continue to grow.

In the United States, a single giant management system permeates the entire society. Although still loosely knit, it comprises the major corporations and government bureaucracies, between which top people move with ease. As Galbraith succinctly states: 'Men will look back in amusement at the pretense that once led [people] to refer to General Dynamics, Lockheed and A.T. & T. as *private* business'.[53] This super complex of management exists not for the general welfare, but for the sake of *its own continued existence*, its own immortality. It is a dilemma that faces *every* industrial society – including the USSR and its Eastern European satellites. As Galbraith points out, if the human aspects of industrial society are to regain their rightful place, then 'the monopoly of the industrial system on social purpose must be broken'.[54] The *freedom* to choose the future must be returned to ordinary people.

This leads to a look at some ethical consequences of the cult of efficiency.

'O Tempora! O Mores!'

In 1947, the American playwright Arthur Miller wrote 'All My Sons', a drama about the aircraft industry during the Second World War. The protagonist, a mercenary entrepreneur, maintains his profit-level by selling defective engines to the air force. In flight, the planes fail and 21 pilots die. One of the industrialist's own sons, a fighter pilot and the intended beneficiary of his father's greed, kills himself when he discovers what his father has done.[55]

That was fiction. In 1967, employees of the B.F. Goodrich Company, a leading American manufacturer of aircraft wheels and brakes, knowingly designed and built inadequate brakes for a new air force plane, and deliberately falsified the factory test records that showed the brakes to be dangerously inadequate. Only after they failed when tested on aircraft, with almost fatal results for the pilots, were the brakes redesigned. At a United States Senate investigation, in the face of damning evidence, Goodrich officials denied any intent to defraud. The company was never charged with wrong-doing and still builds aircraft brakes. The two conscience-striken engineers who 'blew the whistle' on the company lost their jobs, while the others involved were rewarded with promotions.[56]

Such stories are not uncommon and they also occur in government agencies.

A recent example in the United States of unscrupulous mismanagement by government occurred in the field of nuclear power, reported by Daniel Ford in *The Cult of the Atom.*[57] For decades the Atomic Energy Commission and its successor, the Nuclear Regulatory Commission (NRC), suppressed the findings of their own Advisory Committee on Reactor Safety, which repeatedly warned of poor reactor design features, of potential failures in cooling systems, and of faulty fuel rod structures, any one of which could cause a serious core meltdown – a 'China Syndrome'. In 1975, the NRC's Rasmussen Report was finally released with an up-front summary describing reactor safety in glowing terms. Only after independent scientists spent several years carefully analysing the detailed data in the back pages of the report and published scathing criticisms, citing the summary as 'outrageously misleading', did the NRC repudiate its own report.[58] Two months later, the incident at Three Mile Island power station occurred, which almost became a China Syndrome. It was caused by precisely the kinds of failures the safety committee had warned about which had been ignored, suppressed or downplayed by one agency chairperson after another.

How could the public have remained so ignorant? The answer is *secrecy*. To obtain access to government documents, Ford and his colleagues at the Union of Concerned Scientists filed hundreds of Freedom of Information Act requests and undertook two major law suits. As with the Goodrich incident, potential whistle-blowers on the safety committee had to keep silent or end their careers. Those in charge desperately wanted nuclear energy to be a success. They had urged Congress to commit the nation to a nuclear future. Millions of taxpayers' dollars had already been spent. The utilities were sold on the idea and anxious to purchase plants. Corporate giants such as General Electric and Westinghouse had tooled up and were producing and selling reactors. The total momentum was enormous – it *had* to work!

The technological mindset, acting through a consortium of state and corporate bureaucracies, drove an entire society into an enormous, costly and unsafe technology. The public was not free to choose; even Congress was long kept ignorant of the truth on matters of safety. Instead of open, democratic planning there was planning in secret by 'experts' who were both ardent promoters and the supposed regulators of a powerful and potentially dangerous technology. But who in a complex management system is to blame when things go wrong? Everyone points a finger at everyone else when secrecy is lifted and faults are exposed. Critics inside the organisation lose their jobs – often their careers – if they fail to accede to group dishonesty, or even to honest majority decisions that they believe are in error.

One can well ask: Are the autonomous goals of organisations such that they lie outside of or beyond the usual code of responsibility that ordinary individuals live by? Philosopher John Ladd thinks so.[59] Corporations, being legal entities but not real people, are not bound by personal codes of behaviour. But philosopher Kenneth Goodpaster disagrees, arguing that corporations (and presumably all other managed entities) can and should exercise moral awareness.[60] Given that the locus of power and planning functions in industrial societies lies not with the public but with the autonomous management systems of corporations and government bureaucracies, it is hard to discover how the future can have any moral guidance whatsoever unless Goodpaster's view prevails. Whoever makes the decisions *must be held morally responsible*. Otherwise the future of any society will be determined only by what is best for the unidentified decision-makers. Whatever few checks and balances still exist will totally cease to function.

The public generally seems to agree with Goodpaster. In recent decades there has been a blossoming of grassroots organisations whose major goal is to force corporate accountability. In the US, in addition to the Sierra Club, Friends of the Earth, the Environmental Defense Fund and others, which have played major roles in insisting on corporate and state accommodation toward the environment, there have evolved groups who focus directly on corporate ethics and the relations between industry and government; Common Cause and Ralph Nader's Center for the Study of Responsive Law and his numerous Public Interest Research Groups are notable examples. The government has had to create a whole new battery of regulatory and monitoring agencies. Besides the older Food and Drug Administration, there are the Environmental Protection Agency and the Occupational Safety and Health Agency, to mention but two of the more important ones.

Litigation against corporations for criminal negligence has seldom been successful. Corporations, although frequently required to pay damages for employees or consumers killed or injured through careless design and production processes, have scarcely ever been held *criminally* liable. Legally, the corporation *is* an individual, but precisely *which* individual is to be punished for criminal wrong-doing? In 1985, in a landmark case an American judge found three executives guilty not of accidental death but of murder in the death by cyanide poisoning of one of their employees. The man, a Polish immigrant, was not given proper safety gear, and although the corporate executives were aware of the dangers, they did nothing. In his concluding statement, the trial judge Ronald Banks added the following unprecedented words:

> I also find that to state that a corporation cannot be convicted of a crime because it has no mind, and it cannot therefore have a mental

state ... is totally erroneous. It is my belief that the mind and mental state of a corporation is the mind and mental state of the directors, officers and high managerial personnel, because they act on behalf of the corporation, for both the benefit of the corporation and for themselves.[61]

For the first time corporations which long enjoyed the property rights of individuals were made to bear the same level of moral responsibility. The conviction, if not overturned on appeal, carries 25 year prison terms for the defendants.

Industry reacted to public criticism, particularly in the 1970s and early 1980s, by initiating enormous propaganda campaigns to assure the public of its social responsibility and contributions to the general welfare. The following corporate image ads from the glossy pages of *Scientific American* magazine speak for themselves. Large amounts of tax-deductible advertising dollars were spent to tell the reader that:

Amoco urges the public to make 'smart choices', namely to deregulate private enterprise and allow it to develop energy resources on federal lands (October 1981).

McDonnell Douglas's 'Evolutionary technology ... helps keep down the costs of America's defense' (October 1982).

Exxon has developed a 'safer, more economical way to get millions of barrels of [offshore] oil from ship to shore to you' (August 1974).

Polaroid films are proving invaluable in the 'conservation of art treasures' (August 1974).

Westinghouse announces 'plans for a multimillion dollar Environmental Center ... on Chesapeake Bay' to solve environmental problems with 'industry know-how' (January 1974).

Alcoa reminds you that 'the aluminum can is America's favorite recyclable beverage container' (January 1982).

Union Carbide assures us that 'Today, something we do will touch your life' – and reminds us that most of the 21 billion hot dogs Americans ate in 1974 were wrapped in casings they had manufactured (November 1975).

Not only do corporations want to 'look good'; they also campaign for public support for the unleashing of their technological urges; they encourage 'progress' and 'deregulation' and constantly eulogise science and technology. They are selling the image of the high-tech corporations as the foundation pillars of an industrialised society. As science journalists

David Dickson and David Noble further observe, corporations distrust democratic tendencies that might get in their way:

> In a corporate-dominated political system such as that of the United States, demands that the direction and utilization of science and technology be determined within the institutions of popular democratic politics represent a direct challenge to the power of private corporations.[62]

They cite case after case of leaders in American science and technology – Dr John Kemeny, Dr Lewis Branscomb, Dr Edward David, Dr Simon Ramo, Dr William O. Baker, Dr Jerome Weisner – who regard broad-based democracy as inappropriate to a high-tech society. For Weisner, a former president of the Massachusetts Institute of Technology and presidential science advisor, technology cannot be moulded to social needs; on the contrary, it is society that must be moulded to fit the demands of the technological imperative. For him, democracy 'leads to a paralysis of decision-making'.

In order to maintain as much control as possible over the democratic process, corporations and other special interest groups in the United States spend millions of dollars to help elect candidates favourable to them. For the 1986 Congressional elections, candidates spent some $450 million, of which $132 million came from Political Action Committees.[63] Over half of these were associated with the business sector. Furthermore, much of the money is now given *after* an election, thus avoiding pre-election disclosure of donations by candidates and ensuring that the legislators continue to cooperate with corporate goals once they are safely in office.

SOME CONCLUDING THOUGHTS

A titanic struggle is developing between an unorganised, often manipulated and poorly informed but massive majority and a small, rich and powerful management clique over the control of the planning and decision-making processes that will shape the economic, political and cultural future of industrialised nations. Through massive advertising corporations shape the public's worldview, and by extravagant lobbying and election support they control the makeup of the government. As long as the public of any nation remains passive, then most certainly their worldview will reflect and serve the values of science and technology, and of the military–industrial complex and of the government bureaucracies that serve both. These unelected 'leaders' are the management elite – the high priests of the 'cult of efficiency'. Theirs is a highly quantoid worldview with scarcely any non-quantifiable human characteristics, for there is nowhere for these to enter their social dialogue.

Meantime, for an increasing number the promised materialist dream seems empty, a psychic illusion. The supposedly contented masses are restless, unsatisfied, alienated and questioning. In Chapter 11, we turn to the plight of the ordinary citizen living in the modern, efficient, high-tech world, a consumer without a purpose. In the very bleakness of the dream's tarnished reality lies the hope that the technological imperative is in fact escapable after all.

11 Alienation: the Loss of the Sacred

> One of the fundamental needs of men, as basic as those for food, shelter, procreation, security and communication, is to belong to identifiable communal groups, each possessing its own unique language, traditions, historical memories, style and outlook. Only if a man truly *belongs* to such a community, naturally and unselfconsciously, can he enter into the living stream and lead a full, creative, spontaneous life, at home in the world and at one with himself and his fellow men.
>
> A theme of Isaiah Berlin, summarised by Roger Hausheer, 1980[†]

INTRODUCTION

Ours is an age of scientifically based technology. Most Westerners consume 20, 30 or even 40 times more energy and resources than any earlier human society. We live in luxury once reserved for kings. We live long lives, mostly free from physical pain. Most of us are never hungry except when on a self-inflicted diet. We are seldom too cold or too hot; we do not have to do much hard physical work. We have amusement at the touch of our finger-tips. We are, quite literally, immersed in the wonders of technology – yet we are increasingly less satisfied.

In the midst of all this material splendour, the human soul wanders as in a desert, parched, empty, undernourished, asking: 'Is this all that life is about? Are men and women simply consuming machines, to be worked *for* and catered *to* by other machines – robots in factories, computers in banks, CAT-scanners in hospitals?' Like chickens in an automated henhouse, our every physical need is mechanically accommodated. Yet although we feel concern for the captive hen which can no longer range freely in the barnyard, we fail to see that our own malaise results from a similar kind of captive material security that we neither create nor control. Having largely freed ourselves from the vicissitudes of Nature that humans once experienced, we find ourselves enveloped in an artificial environment that seems to give us not more and more, but less and less satisfaction. Surely there is more to life than achieving the perfection of a safe cocoon in which to travel from birth to death.

'The Meaning of Life': That is the great insoluble mystery that our earliest ancestors had to struggle with from the moment they first asked: 'Why birth?' 'Why death?' 'Why life of any sort?' Peoples of every kind have engaged in generating meaning out of the fact of existence. Nor has the mystery declined in any way today. Neither science nor the modest control over Nature that technology provides has lessened our need to understand: 'Why am I alive?'

Primitive people, with their animistic religions, found meaning in a world of living spirits. Today, the world's great religions offer meaning at the end of life – with promises of a blissful oblivion, of complete understanding, or of an eternal Heaven – and in *this* life, they establish rules for living that, by preparing one for the ultimate, give a point to existence. The prescribed rules vary from culture to culture but the goal is the same: *To know why we are.* Culture explains how to live and die. Communally shared beliefs provide a sense of security and personal strength. Through our social bonds we find meaning.

That was the case everywhere until the past few centuries, when the trend to 'rationalism' in the West denied the role of mystery and wonder in human existence. All became logical, quantitative and material. As long as pleasure is fulfilled, why worry? We are born to enjoy life and it is our duty to do so. Suffering and death are inconveniences that science is working to abolish – and surely one day death will be postponed indefinitely.

Not only does Western culture sidestep the meaning of life and death; it also denies a purpose to society other than promoting the welfare of the individual. Each 'Rational Economic Man' is put on Earth to make the most of his personal 'opportunity'. We are isolated, independent, autonomous, self-created beings. 'Human nature' in Western eyes has undergone a complete metamorphosis from a culture-centred being to a totally self-centred being. In our pursuit of the twin goals of freedom and efficiency (which are really incompatible) we have thrown away the sacred meaning by which societies are defined. The cultural norm of Western society is *to do away with culture.* Clinical psychologist Paul Wachtel writes:

> [W]e are continually undermining the basis for our shared cultural life and attempting to substitute the authority of our own urges for all previous authorities, whether parental, political, or moral . . .
>
> The continual questioning of all cultural givens and all cultural demands has . . . its own programmatic ethic, but an ethic of a rather different sort. What is being created is essentially an anticultural culture, a culture in which good and evil are viewed simply as deceptions. We are to live 'with a minimum of pretense to anything more grand than sweetening the time'. In the new culture, 'men will have ceased to seek any salvation other than amplitude in living itself'.[1]

Disconnected from others, people must independently invent their own life meaning while remaining helplessly trapped by gigantic social institutions devoid of moral direction. Our lonely state is captured well in these words of Allen Wheelis, from his book *On Not Knowing How to Live*:

> I sit inside my skull and look out as a frightened man from a moated castle. Me in here and the world out there. We negotiate, we make deals, exchanges, but we are not one. I am an entity, complete. Never do I lose sight of where I stop and the world begins. With sleepless vigilance I patrol the edges of selfhood, warn visitors away. I am independent within this domain, but am dying. It is my wholeness that destroys me.[2]

In seeking absolute freedom, we leave ourselves alienated from everything – from Nature and from one another. With its central focus on the consumption of material wealth, Western society is inexorably destroying both the *environment* and the *human psyche*. The former has already been discussed. It is the destruction of the human psyche that occurs when 'individualism' and 'freedom' are carried to present extremes that is examined in this chapter.

THE SYMPTOMS

Dispassionately to judge one's own society while living in the midst of it is difficult at best. For the easily biased or prejudiced it is impossible. In their opera *The Mikado*, Gilbert and Sullivan included on the Lord High Executioner's list of people 'who never would be missed':

The idiot who praises, with enthusiastic tone,
All centuries but this, and every country but his own.[3]

They might equally have included the idiot who praises only his own country at the present moment. To avoid the extremes of idiocy, however, it is not necessary to praise the well-advertised material benefits of industrial society in order to examine the psychic costs, seldom mentioned publicly, which are paid for these benefits. The goal of the critic is not to scold but to heal.

The symptoms exhibited by those industrial societies that prize possessive individualism and efficiency have reached their highest expression in the United States: most of what follows is therefore based on data from there.

An Epidemic of Violence

Violence is nothing new in human societies, yet one would expect that an 'advanced' society would have *reduced* violence if it really were advanced.

The opposite seems to be true; modern societies are becoming not less but *more* violent, with the United States far in the lead. One in a thousand Americans will be murdered in the next ten years; one in four will be beaten, robbed or raped. Until 1984, murder was the fastest-growing cause of death in the United States. The murder rate doubled in 20 years and other violent crimes – rape, assault and armed robbery – quadrupled. Reported cases of child abuse increased 200 per cent in eight years.[4] Since 1965, violent crime in the United States has increased by an average ten per cent a year,[5] although it now appears to have reached a steady, high plateau. Wealthy, 'free' America is the most criminally violent country in the world. According to Handgun Control, Inc.:

> In 1980, handguns killed 77 people in Japan, 8 in Great Britain, 24 in Switzerland, 4 in Australia, and 11,522 in the United States.[6]

In addition to violence directed against others, self-inflicted violence is also on the increase. Suicides typically occur among the elderly, those who have lost all their loved ones or are suffering terminal illness; among this group, suicide is high everywhere. In the United States, however, there has been more than a doubling since 1960 of suicides among those 15 to 24, for whom, after car accidents and murder, it is the major cause of death (see Figure 11.1).[7]

The young also account for a larger proportion of violence toward others. In the decade after 1965, the average age of murderers in the United States fell from about 30 to nearer 20.[8] Violence is also continuing to rise in the schools, where over 100,000 teachers a year are assaulted and nearly 10,000 are raped; where $600 million in vandalism occurs, 20 million thefts, and 400,000 destructive acts, much of them arson.[9] The most rapidly increasing targets of violence are children aged one to four, a fact that reflects rising violence in the home. Psychologist Arnold P. Goldstein reports:

> Not only are America's streets, schools, and mass media witness to an increasingly violent and dangerous population – so too, are its homes. Spouse abuse, child abuse, and an especially nasty behavior unamusingly called 'granny bashing' are more and more with all of us today.[10]

Besides physical violence, there is a rise in litigiousness – a legalised form of social violence. Instead of resolving disputes amicably, or at least maturely, by discussion and arbitration, more and more Americans are hauling one another off to court. In 1981, there were more than 13 million civil suits in the United States (compared to a little over one quarter million in Japan, which has one-half the population). According to Mark Cannon, administrative assistant to the Chief Justice of the United States:

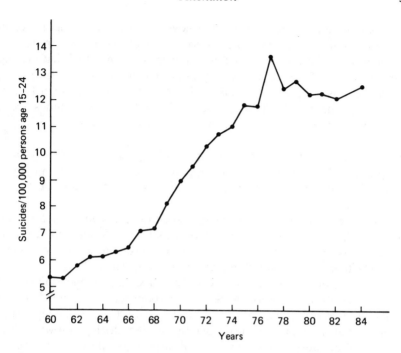

Source: United States Department of Health and Human Services, National Center for Health Statistics, *Vital Statistics of the United States* (published annually) Vol. II, Mortality, Part A, Table 1.9; data converted to a graph by M.E. Clark.
Figure 11.1 Suicide rates among youth in the United States (1960–84). 1984 is the last date of published figures.

Litigiousness appears the symptom of more basic changes in American attitudes toward community. The breakdown in community consensus, shown in our churches, schools, and neighborhood organizations, has brought a greater emphasis on adversarial procedure. Likewise, there is less trust in the nonlegal dispute resolution forums that were once at the heart of our sense of community.[11]

There are other statistics about this continuing increase in violence and people's fear of and responses to it: the rise in privately owned handguns, the new neighbourhood watch committees, rising vandalism in public places, overcrowding of gaols, and court calendars that are full up for months ahead. But the point is clear enough: growing violence signals, in this richest of all nations, a spreading discontent.

Disillusion

Daniel Yankelovich heads a company that studies social trends, and one of its recent findings is a widespread disillusion with the promise of the

American Dream. There is mistrust of government, lack of confidence in the future, and gloom about the economy. Fiscal uncertainty, rocketing housing costs and other economic pressures make people feel that playing by the rules is a fool's game; you only get cheated.[12] Not surprisingly, apprehension is widespread. Yankelovich reports that more than half of America's young people in 1979 'held fretful and anxious attitudes about their lives'. Compared to youth in the late 1950s, there was

> a steep rise in all the behavioral symptoms of increased anxiety – greater frequency of headaches, loss of appetite, trouble in sleeping, upset stomachs and higher levels of feeling nervous, fidgety and tense.[13]

In the author's experience with college students, the symptoms are little changed today. Other surveys have shown that more Americans reported themselves as 'very happy' in 1957, when living standards were considerably lower than today, than they have at any time since.[14]

Curiously enough, side by side with their feelings of uncertainty and inability to control the future, Americans harbour feelings of hope and an unwillingness to forego any of their material goals. There is a deep desire to believe that the United States economy and the ideology on which it is based will ultimately 'work', and that all aspects of life can be simultaneously improved: material consumption, environmental quality, social justice, human health, economic security, and personal freedom and fulfilment. Yankelovich calls it the 'We-expect-more-of-everything' outlook. He further adds that people deem it their 'moral right' as human beings to have all their wants fulfilled. Two opposite attitudes co-exist: one anxious and fearful, the other brazen and demanding. 'Maintaining these two states of mind simultaneously creates an almost intolerable level of tension and confusion', reports Yankelovich.[15]

For some, the tension becomes intolerable and the hurdles of life insurmountable. Those who do not take to violence against others or themselves become social dropouts. The number of homeless men and women sleeping in the streets of urban slums has been increasing steadily. The city of Los Angeles officially estimates 38,000 homeless; across America there are as many as 2.5 million. London is full of them. Most are poor and unemployed; some are alcholics; many suffer from mental illness. The rise in their numbers is due to a combination of growing social alienation together with increased unemployment and a decrease in the availability of low-cost housing and of welfare benefits. The well-intentioned closing of many state mental institutions in order to 'rehabilitate' the mentally ill into society failed; funds never emerged to do the job. As with the other unfortunates that society ignores – the destitute aged, the chronically ill, the underemployed – the psychological misfits and disillusioned dropouts remain an alienated and forgotten class.[16]

The Inward Quest

In America, the past decade or so has been dubbed the 'me-first' generation, in England, the 'I'm alright, Jack' era. It marks the end of the post Second World War trend toward social justice throughout the West and a swing to more personal concerns. It is characterised by both political conservatism and moral liberalism. As college students drift away from teaching careers and seek lucrative jobs in business and engineering,[17] their personal commitments to others – in marriage, as parents, at work – are declining. People are becoming more and more alienated from others, more and more focused on themselves.

Yankelovich describes people as empty ice-cube trays, full of 'needs' that require filling: 'emotional needs', 'sexual needs', 'material needs', 'the need to be challenged intellectually', 'the need to assert oneself', 'the need to keep "growing"'.[18] Furthermore, they are afraid to make permanent commitments for fear of losing their freedom. As sociologist Robert Bellah and his colleagues observe, 'friendliness' is replacing true friendship, and in marriage '[t]he present ideology of American individualism has difficulty ... justifying why men and women should be giving to each other at all'.[19]

This inward quest of radical individualism has grown into an 'I-owe-it-to-myself' ethic; Americans have come to regard 'the self' as a sacred object. This explains both the fascination with pop psychology and the desperate need for psychotherapy, as lonely individuals struggle to overcome their emptiness through personal reconstruction. The 'human potentials movement' begun by A. H. Maslow has spawned a vast popular literature: *Looking Out for #1*, *Self Creation*, *Pulling Your Own Strings*, and *How to Be Your Own Best Friend*. In a penetrating analysis, Yankelovich points out the necessity for 'stepping off Maslow's escalator' of hierarchical needs – our dream of a self-actualised pinnacle of successful 'beinghood' – and recognising instead that we are *not* culture-free beings acting autonomously, and therefore that our search should be directed outwards, not inwards.[20]

THE DIAGNOSIS

If the symptoms of our age are violence, disillusion, and feelings of meaningless and loneliness, what causes them? What is wrong with our Western worldview, dedicated as it is to notions of individual rights and freedoms; to a free marketplace; to the natural necessity for competition to attain maximum efficiency and general wealth; and to the inevitable drive toward technological change? Where have we gone wrong?

The locus of our problem, I firmly believe, is our absolute and single-minded pursuit of freedom. The word 'free' is used arbitrarily and without reflection whenever one wants to justify anything: the 'free' world, the 'free' market, the 'free' spirit. Anything that suggests 'unfreedom' is bad. Constraint, commitment, bonds, responsibility, tradition – to the extent these interfere with freedom, they are suspect. To diagnose our present age, then, it is necessary to examine its most valued concept, 'freedom'.

The Shibboleth of 'Freedom'

During the 200 years since the American Declaration of Independence and the French Declaration of the Rights of Man, the Western notion of freedom has taken on a far different meaning than it originally had. Indeed, the eighteenth century revolutionaries spoke not of 'freedom' but of 'liberty': 'Life, Liberty, and the Pursuit of Happiness'; 'Give Me Liberty or Give Me Death'; 'Liberty, Equality, Fraternity'. Liberty meant freedom *from* arbitrary or oppressive rule and political injustice. It gave people the freedom *to* make their own society, decide on its rules and administer its justice. But the visions of the Massachusetts Puritan John Winthrop, and of Thomas Jefferson 150 years later, far from picturing a free-for-all of competing individualists, envisioned a consensual society of responsible political participants seeking equality and justice for everyone. America's political revolution succeeded because the colonies had already eliminated the gross social inequities of Europe, at least among whites. Theirs was a *society* of equals, working *together* to build a new kind of political system.[21]

In addition to freedom from arbitrary rule which led to the establishment of various forms of democratic government, the peoples of Europe eventually followed the Americans' lead in demanding freedom from constraints in matters of belief and faith. Almost imperceptibly however, this notion of freedom from *religious* constraints has been expanded to freedom from cultural constraints of almost any kind except those precisely specified in law.

In frontier America the notion of rugged individualism took hold. Not only did the Puritan work ethic prescribe strict individual responsibility; it further implied that material success was a sign of heavenly approval. Freedom was now expanded from political and religious freedom to the economic freedom to pursue private, individual goals without social hindrance. After all, Adam Smith's 'invisible hand' put a stamp of social virtue on economic self-seeking, and the concept of a 'free' marketplace was soon made synonymous with a politically free democracy. As economic wealth grew out of the Industrial Revolution, the idea that owners of private property should be free from any constraint on their pursuit and accumulation of wealth permitted the nineteenth century entrepreneurs and the twentieth century corporations to amass enormous power. And as shown below, that power has been used by way of the media to reinforce the supposed identity between 'free' enterprise and a 'free' democracy.

But the expansion of the concept of 'freedom' has led even further, to a sense of complete freedom *from* any and all social constraints – from responsibilities toward others, from cultural norms, even from the conventions of politeness and courtesy – that has led to the 'anticultural culture' described at the outset of this chapter. More and more are Western societies – and American society in particular – becoming assemblages of unhappy, alienated individuals proud of their freedom from dependence on others, unaware that it is the rejection of their mutual relatedness that creates their unhappiness. Our increasing goal of absolute freedom is creating in us the very 'human nature' that Hobbes so incorrectly assigned to primitive humans: self-centred, greedy, brutish, one-against-all. No wonder there is such a tendency to believe he was correct!

Absolute 'freedom' means alienation

We generally think of freedom as the opposite of involuntary subjugation, and this indeed *is* its political meaning: 'liberty'. But freedom can equally be the opposite of promise, vow, contract, commitment, obligation – all those understandings that bind us to others and create a society. What young man wants to hear from the lips of his beloved that he is 'free'? Who wants to be 'let go' from a job, excommunicated from church, or exiled to another country? Such 'freedoms' are signals of *rejection*, the breaking of bonds that are the very substance and meaning of life. It is these precious conditions of 'unfreedom' that our growing notion of 'individual freedom' is inexorably eroding.

As both Yankelovich and Bellah's studies show, it is our reluctance to undertake long-term commitments for fear of losing our personal freedom that causes much of today's widespread ennui and alienation. People first lose their willingness to commit themselves; then they become anxious, sceptical, mistrusting, paranoid. Bellah and his colleagues describe in depth the widespread demand for psychotherapy, which however fails because it, too, denies the necessity of permanent community:

> Only occasionally do we find therapists who recognize that 'community' is not a collection of self-seeking individuals, not a temporary remedy, like Parents Without Partners, that can be abandoned as soon as a partner has been found, but a [permanent] context within which personal identity is formed.[22]

There is an uncanny resemblance here to the Ik tribe described earlier, where a whole society disintegrated when its interpersonal bonds collapsed. Indeed, both Webster's and the Oxford English dictionary give the second meaning of alienation as 'mental derangement; insanity'.

In small hunting–gathering societies and early neolithic settlements, personal identity came from membership in a group with a single, deeply shared worldview; one's place in society and in the cosmos was never in doubt. If personalities clashed, groups fissioned and reformed, but one still *belonged to* and was *immersed in* the same culture. To invent a new set of social arrangements or independently live a different lifestyle was beyond imagining.

Yet the deep social commitments entailed by traditional societies did not at all constrain important personal freedoms. In her illuminating book *Freedom and Culture,*[23] anthropologist Dorothy Lee shows how having important and sacred duties to perform – having *extremely significant social commitments* – is precisely what unlocks the greatest imagination and stimulates the highest individual achievements. The whole emphasis in such societies is on the freedom necessary optimally to fulfil one's sacred function.

By contrast, in giant modern societies where personal ties are weak and impermanent, a sense of belonging must be achieved in quite different ways.[24] In place of real bonds we use *symbols*: awards, honours, income, consumption level, prestige, titles. These serve as surrogate bonds, reassuring us that we *do* belong and *are* an accepted member of society. Our exaggerated concerns for recognition, approval, status, and 'success' are substitutes for truly belonging.

This artificial form of social identity has two great psychological failings. First, since belongingness is never based on *real* bonds it is always insecure. Underneath lies a constant fear of alienation; we are never sure that we are still valued. As psychologist Paul Wachtel observes, we are forced into utter self-sufficiency:

> Our view of the self is that it is 'portable', it can be carried around from place to place, fully intact, and then plugged in whenever necessary.[25]

Second, since all these surrogates for real social bonding are measures of one's *relative position* in an impersonal society, they generate unending competition. At a parade, when one person stands on tiptoe to see better, his purpose is promptly defeated when others do likewise. Similarly, a person seeking identity by achieving a higher level of material consumption or social recognition or power is defeated when too many others attain the same level. An unceasing struggle of 'one-upmanship' ensues that further alienates us from one another, threatening even the few *real* bonds we do have. One who is competing for 'success' can hardly fail to be jealous when a colleague receives an award or a sibling gets unusual parental recognition. Old 'friends' are readily discarded when promotion to higher 'status' leaves them behind. Thus do these surrogates for real social bonds frustrate us twice over: they are never 'sufficient' and they often force us to compete with and alienate the few people to whom we are most closely bonded.

Our lives today are thus less and less defined by *permanent* social commitments. Social relations are being reduced to the level of game-theory strategies, an unending sequence of cost–benefit calculations in the ongoing search for social status and personal pleasure. Our individually defined notion of 'success' becomes a substitute for true membership in a society based on sacred meaning. And since self-defined 'success' can be only a pale reflection of our *real* need for a meaningful society, it repeatedly fails to satisfy us.

An important consequence of our need for social standing, whether due to real bonds or surrogate status symbols, is the deep anger we feel when that standing is challenged. Perceived threats to our 'social security' are a major cause of aggression. The role these surrogate symbols of belonging play in the development of militant nationalism is taken up at the end of the chapter.

Two ways to lose 'freedom'

In the course of seeking the social relatedness every human psyche needs we each make a commitment; we give up some of our personal autonomy, our 'freedom', in order to *belong*. What is it we *receive* in exchange for this loss of freedom? When does commitment result in something even more valuable than freedom, namely strong, reciprocal social bonds, and when does it result in something considerably less – a commitment with more loss of freedom than expected and less satisfaction than hoped for?

To approach this question is to examine perhaps the most critical social issue of our time: What are societies *for*? Are they for fulfilling the general human need to belong to a social group that has sacred meaning? Or are they simply pragmatic arrangements for getting collective jobs done and, as Hobbes thought, for organising and controlling people – and perhaps Nature as well? In the case of social commitment through bonding, the individual is rewarded with a sense of purpose; in the other case, the reward is physical security through membership in a powerful, efficient social machine. But efficiency, when carried to an extreme, denies freedom, precludes social bonds and leads to alienation.

'Efficiency' versus 'freedom'   Chapter 10 examined how the pursuit of efficiency in modern capitalist societies has led to a pervasive competitiveness, the dehumanisation of work, and an extreme centralisation of decision-making power. According to historian Lewis Mumford, the once democratic United States is well along the path to becoming another of history's failed 'megamachines' – a superorganised, anti-humanistic civilisation that is destined to collapse through its inability to satisfy the deepest needs of its citizens. Today's level of material consumption has been achieved by a degree of industrial and managerial organisation unknown in human history. Nor does the authority for what happens reside in a popular

consensus but in the centres of bureaucratic management. As Mumford predicted, decisions are made with the primary purpose of extending the power and control of the decision-makers, those at the apex of the social hierarchy, the 'immortal management'.[26]

Instead of *actively participating* in major social decisions – 'How shall cities be organised?', 'What should be the purpose of education?', 'Do we really want massive data banks?' – most citizens of industrialised countries are anxiously preparing themselves *to fit into* an even more competitive, high-tech future. Society is not democratically moulded; rather, people mould themselves to fit a social order created by the decisions of an unknown few. In addition, the more effort and money invested in moulding oneself to fit into a specialised social niche, the more difficult it becomes to change occupations. Says American psychologist Seymour Sarason: '[O]ur society has made it more difficult to change careers than to change marriage partners'. As he further points out, since work is the major means of social identity, the effects on people of being stuck forever in an unsatisfactory profession can be grim:

> [W]e define ourselves, and others define us, largely in terms of our work, and perceived dissatisfaction with our work can be very upsetting and have adverse consequences in all areas of living. Some people do not like to think about it . . ., for others it is a chronically gnawing irritant associated with feelings of impotence or being 'locked in'; and for others a kind of live volcano that literally forces them drastically to alter the shape and direction of their lives . . .

> [W]hen large segments of the population find their work uninteresting or feel unrelated to the products of their work . . ., one has to examine the role of changes in the social–economic–ideological fabric.[27]

Employment in giant, impersonal corporations, the need to become a cog-to-fit-the-machine, the weakening of family ties that used to make boring or stressful work psychologically acceptable, have all taken their toll. Yankelovich puts it clearly:

> [I]n industrial society human relationships grow impersonal, commercialized. Giant institutions (big government, big business, the communications media) destroy what Edmund Burke called 'the inns and resting places of the human spirit' – the smaller, more human-sized institutions of the society: the local church, the old neighborhoods, the small schools, the local shops and family relationships. In modern industrial society we often purchase our material well-being at a high human cost, the chief symptom of which is the destruction of *community*.[28]

Commitment to an efficient 'machine' society decreases the strength and stability of human bonds. We have paid not just an environmental price, but also a human price, for our 'high standard of living'. Indeed, efficiency – valuing only what can be counted – even alienates us from ourselves. We are not sure of our own selfworth unless we can define it in numerical terms: How many acquaintances, how much income, how far up the ladder of success? We no longer know how to value that which cannot be counted, namely the *human* commitment we so deeply desire and need.

Thus does the efficiency of the megamachine at once diminish our human ties and our freedom to change our way of life. Indeed, the two central goals of modern industrial society – 'efficiency' and 'freedom' – are mutually incompatible. This is a major paradox of the system. Moreover, our commitment to efficiency, far from giving us back a sense of social bonding and belonging in exchange for our lost freedom, serves only to alienate us further. We are left with nowhere to go but inwards, seeking to fill our social longing from within. The 'I-have-a-duty-to-myself' attitude is an understandable, if neurotic, response.

If 'efficiency' is so destructive, why do we accept it? This is the enigma that social psychologist Erich Fromm tackles in his book, *The Sane Society*:

> That human nature and society can have conflicting demands, and hence that a whole society can be sick, is an assumption which was made very explicitly by Freud, most extensively in his *Civilization and Its Discontents*.
>
> He starts out with the premise of a human nature common to the human race, throughout all cultures and ages, and of certain ascertainable needs and strivings inherent in that nature. He believes that culture and civilization develop in an ever-increasing contrast to the needs of man, and thus he arrives at the concept of the 'social neurosis'.[29]

Fromm goes on to argue that simply because the majority share certain ideas, feelings and beliefs, this alone does not validate them as sound ones for a society to hold. A sane society is one that meets the basic needs of humankind – which are not necessarily the needs people *think* they have. As Fromm notes 'even the most pathological aims can be felt subjectively as that which the person wants most'[30]. How an entire society may be so completely misled is addressed shortly. First, a closer look is needed at what we do *not* have, namely social commitment at a personal level.

'Personal commitment' versus 'freedom' Although humans, as noted earlier, have a universal psychological need for commitment to a shared worldview, the Western worldview with its central focus on 'individual freedom' is logically committed to *no commitment*. It is anticultural. For the sake of efficiency in material consumption, of belonging to a gigantic nation, and of trying to retain a semblance of individual freedom (whatever

is left over after having efficiently performed our impersonal tasks for the megamachine), we give up lasting commitment to other people.

As pointed out by Staffan Burenstam-Linder in *The Harried Leisure Class*, Westerners expect ever-rising levels of consumption. To consume more, they must earn more – that much is obvious. But consuming also takes time. Being wealthy thus means not only working harder to produce more, but 'playing' harder to consume more; our affluent lives become more and more frenetic.[31]

Americans are the most harried of all. Visitors from abroad complain they are always in a rush, too busy to stop and talk, unable to relax and contemplate life. Just across the border in Canada, casual conversations at check-out stands and on street-corners are far more common. In less affluent countries, leisure is spent in almost continuous conversation. In Greece, men discuss politics in coffee shops while women gossip across the narrow streets. On summer nights in China, people sit outdoors on stools, chatting in the evening shadows. In central Africa, they gather in shops to visit and share a ritual glass of frothy strong, sweet tea. The poorer a society, the more time its people seem to have for social intercourse.

The average American, meantime, when not racing about from one place to another, is sitting at home five hours a day 'consuming' television; even when leisure time allows for social interaction, it is shunned.[32] Families scarcely talk together; even public bars are antisocial places. In Britain for example, a lively community life exists in the brightly lit corner pubs, where neighbours meet to drink beer and converse for two or three hours. In dimly lit American bars, the patrons huddle in silence, staring at football on jumbo television screens. Even when opportunity exists, there seems to be a taboo on social interactions.

This extreme isolation of the individual even in the physical presence of others that characterises modern affluence stands in stark contrast to the strong social bonds fostered throughout most of human history. Anthropologist Thomas Gladwin describes the social bonding of precolonial peoples who lived in what is today called the Third World.

> The extended family was . . . far more than a means of linking people into a cooperative work force and a source of mutual support in time of need. It was an educational institution, often placing children in the care of their grandparents rich in the knowledge and wisdom of maturity. It was a judicial system in which elders, who knew everyone intimately, resolved grievances and disputes in preference to meting out harsh punishments. It was a forerunner of psychiatry, giving comfort, advice, sympathy and tolerance to people who were troubled. Through ceremony and tradition it linked people to the forces of nature upon which their existence depended. The extended family thus fulfilled the basic needs – material, psychological, and spiritual – of every person within it.[33]

The individual in such a society was never in doubt that he or she *belonged*. One might experience friction and unhappiness, but one could not be ignored; it was impossible to be a non-person, a mere cipher, a social security number without a personality. Although one could not escape from such a society, neither could it throw anyone out. Strong bonding and reciprocal commitments have characterised most societies since humankind emerged. It is the alienating social systems of massive, hierarchically ordered 'civilisations' like our own that are 'abnormal' in human existence.

While the West thus fears any loss of 'personal freedom' through commitment and so creates a psychic vacuum, it willingly abandons aspects of 'personal freedom' in the pursuit of efficient production and consumption. What social institutions reinforce the beliefs that create this odd state of affairs?

The Reinforcement of Alienation

Our Western worldview has been premised on an image of isolated individuals endowed with a basically nasty human nature that civilisation has gradually moulded into a still selfish but rational entity. Now although basic human nature is not at all as Hobbes described it, this fundamental image of isolation, competitiveness, aggression and greed is so embedded in our thinking that almost everyone naturally assumes it is a 'fact' about which little if anything can be done. *This presumed image of man is the underlying basis of most of what is wrong with Western society – both in terms of our behaviour toward our environment and toward one another. And our worldview reinforces that image, making it a self-fulfilling prophecy that only 'proves' how correct the assumption was.* This point can be demonstrated in several ways.

The psychologists' 'man'

So ingrained is the Hobbesian image of human nature that it has been accepted uncritically by almost the entire profession that proclaims itself the experts on the subject – namely, the psychologists. Like other social scientists, they *assume* a polarity, a dichotomy, between the individual and society. The good of the group is opposite to that of the individual. The group constrains and the individual struggles to be 'free'. The fact that everyone is bent on joining groups, on belonging to groups, on not being left out of groups, is treated almost as a sign of neurosis, the result of negative things like 'immaturity' and 'peer pressure' and curable by group therapy. By accepting this artificial dichotomy, the discipline of psychology has chosen to focus solely on the individual, as though a person's actions could be understood in isolation from those of his or her particular culture. As Seymour Sarason implies in his book, *Psychology Misdirected*, Western psychologists, having grown up immersed in the Western worldview, unthinkingly accept and reinforce its image of individualistic man, selfish, competitive, and inner directed.[34]

When the first experimental psychologists put not a family of rats into a maze, but a *single* rat, and then claimed to say something about how rats learn their way about in their natural surroundings, they revealed their unconscious assumption that all animals behave like Hobbesian man – isolated, separable, discrete. Of course, analysing what *one* rat or *one* person does is far simpler and appears more 'scientific' than sorting out the complex behaviour of a group of animals or people, but it is doubtful whether such studies tell us much about real-world behaviour.[35] Only in the past few decades have European ethologists and a smaller group of American colleagues begun to observe the spontaneous social behaviour of both animals and humans and provide new insights into the processes of socialisation.

Sarason concludes that psychology, by atomising its study of human nature, has repeatedly failed to offer useful answers to such culture-dependent questions as What is intelligence? and How can we measure learning? Furthermore, by studying Western man* out of the context of his particular culture, as though he represented a generalised model of all human nature, psychologists seem to have 'scientifically proved' what everyone already 'knew to be correct', namely, that humans are naturally aggressive, greedy and self-centred, and that such tendencies can be overcome only with great effort and often great sacrifice as well. In fact, it was to provide guidelines for resolving this supposedly natural conflict between the inner-directed self and the external demands of society that Maslow, Rogers and other modern psychologists developed their hierarchy of psychological 'needs'. In doing so, they greatly advanced the psychologists' basic belief in 'individualism'. And what the experts scientifically tell us we are, we of course must be!

'Value-free' education

Another institution reinforcing our modern tendency toward alienation is the educational system. In addition to institutionalising individual competition in the classroom by proscribing cooperative group learning, Western education, taking its cue from the 'freedom of the individual', makes two assumptions: the goal of education is to 'unfold' the individual, and something vaguely called 'citizenship' is to be imparted without explicitly examining social values. A few words first about the 'flowering' child.

The 1960s' counter-cultural movement with its increasing stress on the rights of individuals led, especially in America, to the idea of 'free' or

---

*'Man' is used here intentionally. Until recently, little attention has been paid to the nature of 'woman'.

'open' schools. Children, placed in 'enriched' environments untrammelled by imposed guidelines or values, would invent their own curriculum and emerge as liberated adults. Here are some attitudes of its practitioners:

> Let the child be himself. Don't push him around. Don't teach him . . . Don't force him to do anything.[36]

> The Godlike role of teachers in setting goals . . . is no longer morally tenable . . . 'Who are we to say what this child should learn?'[37]

As Lawrence Kohlberg makes clear, given these attitudes a child can only conclude that 'liberty' means that moral values are *arbitrary and everyone's values have equal standing*. There is among free-schoolers, he writes, 'the belief that what children *do* want, when left to themselves, can be equated with what they *should* want from an ethical standpoint . . . [W]hatever children do is right.'[38] This extraordinary extrapolation into the classroom of specific political rights guaranteed to adult citizens devloped partly in reaction to the traditional notion, actively promoted by B. F. Skinner, that the school's job is to inculcate rigid cultural norms. As Kohlberg further notes, this latter route assumes that a society's values are all for the best and leaves no room for social improvement; it uncritically accepts that *'what is'* is *'what ought to be'*. Indeed, either extreme fails; both 'Skinner and [the free-schoolers] agree it is better for the child to be a happy pig than an unhappy Socrates'.[39] Although belief in the 'spontaneously developing child' has waned in recent years, the notion that the explicit examination of values is not appropriate to the school curriculum persists.

As shown earlier, however, education can be neither value-free nor value-neutral. Since not everything can be taught, *curricular selection* inevitably takes place. What is taught and what is omitted, even when we are only considering 'facts', thus establishes a particular bias, a particular worldview in a child's mind. Moreover, the very act of ignoring values in the school curriculum is *in itself* a value-laden statement that says 'values are valueless', 'life can be lived without values', or 'any old values will do'. Such a message can lead only to confusion and lifetime uncertainty for children so taught, and must bear considerable responsibility for the outbreak of violence, despair and anxiety among young adults described earlier in this chapter.

Furthermore, failure to discuss values openly leads to instilling them covertly, what Kohlberg calls 'the hidden curriculum'.[40] A so-called 'value-free' educational system *implies* uncritical acceptance of the current worldview: individualism, competition, consumption, self-centredness, conformity to the megamachine. These are automatically accepted as 'good', along with whatever basic values are implied in textbooks on history, science or social studies. Both approaches, then – the Skinnerians' indoctrination with the *status quo* and the free-schoolers' supposedly 'value-free' education – tend to converge on the same result: an un-examined, automatic acceptance of contemporary Western values.

So powerful is the feeling that values education does not belong in American schools, for example, that the Supreme Court, in its Schempp decision, seemed to imply that teaching values is unconstitutional. According to Kohlberg, one could interpret this decision as 'in effect prohibiting the public school from engaging in moral education' because the Court defines 'religion as embracing *any* articulated credos or value systems, including Ethical Culture or Secular Humanism, credos that essentially consist of the moral principles of Western culture'.[41] Here is an extreme example of the logic of a culture committed to no commitment, of an anticultural society! In finding this decision, the honourable judges quite ignore that the Constitution itself and the law that stems from it imply a set of moral values that is *not* synonymous with any established religion (although its tenets overlap those of many religions). Kohlberg concludes with the irrefutable argument: 'The school is no more committed to value neutrality than is the government or the law'.[42]

Until value concerns are consciously introduced into the school curriculum (as they are in numerous other countries, including the Soviet Union[43]) the American educational system will continue reinforcing the alienating tendencies of the Western worldview. It is a trend that various Western educational systems are further abetting by de-emphasising the 'impractical' liberal arts while stressing 'practical' skills training and career preparation – all intended to fit the 'products' of our schools into a useful niche in the economic machine.

The impact of the media

A third institution having an important alienating role is the media. Convention has it that a free press, including not only newspapers and magazines but also radio and television, is the 'watchdog' of society, the unfettered critic who keeps an eye on government, uncovers corporate misbehaviour and exposes social injustice. Its members do investigative reporting; they inform and alert the public and generally keep things honest. While this is to some extent true – such exposés as the Pentagon Papers and the Watergate investigations were indeed the result of bold actions by the public press – on the whole, there is little diversity of opinion among the major Western media, particularly in the United States. Multiple television channels, magazines, radio stations, daily papers and book publishers seem to imply diversity, but most are owned by or affiliated with one of a few corporate conglomerates who themselves are of a single mind. As American communications analyst Herbert Schiller observes:

> Though no single program, performer, commentator, or informational bit is necessarily identical to its competitors, *there is no significant*

qualitative difference ... [All] are selected from the same informational universe by 'gatekeepers' motivated by essentially inescapable commercial imperatives. Style and metaphor may vary, but not the essence.[44]

The tacit goal of all this uniformity is to maintain the *status quo*. Being in thrall to their corporate advertisers, commercial networks cannot stray far in their programming from the values that promote the private enterprise system. Moreover, each event – each riot, or stockmarket plunge, or government scandal, or educational issue – is treated in isolation, unrelated to anything else. The public are treated like passive observers to be entertained by the external world rather than as concerned participants in their own society.

Beyond even that, those who oversee the media reflect their own preferred image of the world. As Jerry Mander, author of *Four Arguments for the Elimination of Television*, reminds us, they even create 'news' events that never happened, such as the Gulf of Tonkin incident, 'seen' on television and believed by most Americans. The North Vietnamese were reported to have fired on American ships; this 'news' was carried by every media outlet in the nation and convinced both Congress and the public to let President Johnson escalate the war. Says Mander:

This event was later exposed as only one of the many non-events pushed through the media to sell us that war. It occurred to me that the very fact that this could be done at all – fictional news about fictional military events expanding faraway wars that no one watching the images could observe firsthand – was cause for serious alarm about the power of the media to pursue fictitious realities.[45]

How could such a thing happen? To what extent is what we see invented or altered purposely to deceive? Mander documents numerous cases in his text and other examples continue to be uncovered.[46] By providing a fictitious or deliberately biased 'truth', the media, even in a 'free' society, can cordon the viewer off from reality.

In addition to deliberate misinformation, the very nature of television as a medium distorts the programming. To watch television – something the average American spends as much time at as at his or her job – one must sit motionless in a dimly lit room with eyes locked on the set and focused at a constant distance whether the picture is a close-up of mascara-treated eyelashes or a panoramic view of the alps. The viewer is passive; information travels in one direction only, *from* the 'box' *to* you. Since the human body is built to *respond* to stimuli, television is inherently boring. To retain your attention, certain tricks are employed.

Mander calls such tricks 'technical events' – changes in camera angle, zooms, overlays of shots and so on. They constantly jerk the viewer's attention back to the tube. Non-commercial television has an average of two or three technical events per minute while commercial programming has eight to ten. During advertisements, which are least engaging of all, there are 20 to 30, one every couple of seconds.[47]

The need for constant change on the screen means that thoughtful, deliberate or subtle programmes have a low audience rating. High action and overt emotions are what draw viewers. News events where the protagonists are low key and reasonable thus tend to be neglected while those involving confrontation and violence receive undue attention. The medium *selects* the message: it reinforces the worst of what exists in society. It is obvious why those with a grievance commit acts of terrorism – they draw global attention to their cause. Violence, flamboyance and rhetoric attract viewers; reasoned arguments do not.

An extraordinary amount of violence, both real and fictitious, is shown on television, particularly American television. There is no doubt that specific cinema and TV films – *Psycho, Boulevard Nights, The Warriors, The Deerhunter* – were directly responsible for outbreaks of homicide and suicide, so exactly did the real violence mimic that on the screen. There is less evidence of a direct connection between the general rise in real violence and that shown generally on television. Yet considering that between ages five and fifteen the average American child watches the killing of 13,000 people; that 97 per cent of children's cartoons include acts of violence; that in a typical year on prime time TV a person witnesses 5000 murders, rapes, beatings and stabbings, 1300 acts of adultery and 2700 sexually aggressive comments – considering these things, can one seriously claim that rising violence is not at all due to this medium? As Michael Nagler observes, to the TV viewer other people become little more than life-size dolls to be bashed about and multilated at will.[48] Is there any greater form of social alienation?

In addition to violence there are other forms of alienation that television imposes on society. As already mentioned, TV advertising is enormously successful at appropriating our emotions, converting them into commodities and then selling them back to us. Even more seriously, TV is creating new role models for the young. They are coming to admire Alexis, the predatory heroine of 'Dynasty' and the unscrupulous J. R. Ewing of 'Dallas'. Herbert London of New York University says about students' attitudes:

> There appeared to be consensus that if you can get away with certain actions – as television heroes regularly do – that translate into wealth, status, and power, then the stigma of immorality is mitigated by rewards. Do the ends justify the means? Yes, say writers of television programs, and yes, say students who watch television regularly ... The bounds of appropriate behavior have been stretched to incorporate deceit, adultery, blackmail – behavior frequently condoned on television.[49]

The hours spent watching television are hours forfeited; they could have been spent in *social interactions* with others. 'Watching TV together' may be better than not being together at all, but it is no substitute for direct, interactive social intercourse. During the hours hyperactive children watch television they are suppressing their already exaggerated need to engage in physical activity. Far from being a 'cure' for hyperactivity, television is one of its prime causes.[50]

If Mander and Schiller are right, television is a powerful way of manipulating people, of implanting uninvited images, chosen, processed and presented by strangers, into the minds of viewers. Given the likelihood they are right, the public in general is extraordinarily inconsistent in its attitudes toward 'freedom' and 'values'. People worry prodigiously over whether their children are being 'brainwashed' with someone else's values in schools, in America to the point where even the Supreme Court intervenes. At the same time, there is barely a murmur of concern about TV programming and its effects on children's beliefs and values, let alone their physical and mental health. Presumably, children are 'coerced' at school, but 'free' when watching TV. Yet until age 12, the average American child spends more hours per week freely watching TV than it is required by law to spend in school. The peculiar assumption is that schools – an accessible, decentralised public institution – are dangerous and must be monitored in case they intrude on individual freedom, whereas the media – distant, highly centralised, privately controlled institutions that are freely let into the living room – must not be regulated since that would infringe upon freedom of speech and free enterprise. This curious inversion of where trust should be placed in a democracy appears again in later chapters. Suffice it to say here that such anti-democratic, non-participatory logic is a major symptom of the alienation that besets our time. What is big, distant and abstract is valuable and trustworthy; what is small, local and tangible is suspect and a suitable target for our discontents.

The Rise of Militant Nationalism

It is possible to depict the three social themes touched upon in this chapter as the three corners of a triangle (Figure 11.2a): individual *freedom*, economic *efficiency*, and social *commitment*. In an ideal society, these three are in mutual balance, represented here by their equal distances from the centre of the triangle. Excessive emphasis on one apex of the triangle inevitably means de-emphasis of the other two. As we have seen, however, of the three, human nature requires social commitment *above all else*. Indeed, throughout most of human existence it has been the social factor that ensured the other two. Through their bonds with others, people not only were able to form an effective economic group that ensured survival; by virtue of the shared values and shared worldview that

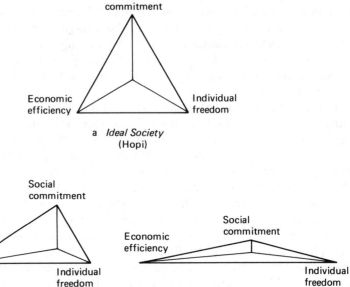

a   *Ideal Society*
(Hopi)

b   *Organised Megamachine*
(China)

c   *Capitalist Megamachine*
(United States)

Source: M.E. Clark.

Figure 11.2 Balance among three social themes: economic efficiency, social commitment and individual freedom:

  a: *The ideal society* – the three themes are in optimum balance. (The Hopi Indians, when free from external stresses, have long been such a society.)
  b: *The organised megamachine* – economic efficiency dominates, mainly at the expense of individual freedom; social commitment is still important. (The People's Republic of China is a typical example of such a society.)
  c: *The capitalist megamachine* – individual freedom is over-emphasised along with economic efficiency, causing social commitment to shrink drastically. (The United States is a typical example of such a society.)

gave life its meaning, they also found creative freedom – a creative *purpose* – they would not otherwise have.

It is thus not surprising that when our bonds with others are threatened, we become angry, hurt, resentful and aggressive. This is an important cause not only of personal aggression – a false-hearted lover, a competitor for our job, a dishonest friend – but also of intergroup aggression. When outsiders threaten the bonds *within a group* they elicit an aggressively defensive response that often brings the group closer together. *It is not so much that strangers may harm us physically but that they may harm our relationships.* This explains why aggressive acts between groups of human beings frequently occur over non-material things. Not only do threats to territory and resources stir anger; so do threats to the beliefs, the ideology, the very worldview that bonds people together. Even the most 'democratic' of societies finds it difficult to tolerate internal debate on such matters, and certainly not the intervention of outsiders.

All giant industrial societies suffer from a high degree of impersonality and alienation. Every such megamachine puts management efficiency at the head of its list of priorities, weakening both personal commitments and individual freedom (Figure 11.2b). In capitalist societies where 'individual freedom' is the watchword, personal commitments suffer even more; people are further alienated (Figure 11.2c). As their real bonds fall away however, human beings still require an illusion of social fellowship and belonging. In addition to the symbolic social recognition people accrue from awards, wealth and status, they also seek symbolic surrogate groups to attach themselves to, *in lieu* of a true community of relatives, friends and neighbours. Herein lies the origin of the passionate loyalty to giant nation-states.* As Yankelovich says: 'The nation-state [fills] the vacuum created by the dislocations of industrialism'.[51]

'Patriotism' has become a ready abstract substitute for non-existent personal bonds; the ideology of nationalism substitutes for a genuinely shared and understood worldview. As Mumford observed, when meaning is lacking in one's personal life it can be symbolically found in the grandiose schemes of one's country. It is now widely recognised that the massive demoralisation of Germany as a *nation* following the punitive Treaty of Versailles in 1919 paved the way for Hitler to gain power and create the society he did. The symbolic accomplishments of the mega-machines of the past – the pyramids of ancient Egypt, the roads and triumphal arches of imperial Rome – find parallels in modern space programmes and arms build-ups. 'Status' among the world's nations is conferred by such projects. Why else did America, for example, send men to the moon when an unmanned spaceship could have collected moonrocks at a fraction of the cost? Because the meaning for millions of affluent yet discontent Americans would not have been the same. 'Our' man on the moon, 'our' enormous military presence in the world, 'our' extravagant consumption made possible by 'our' high-tech megamachine – all swell the pride of national identity in each American bosom and lend surrogate meaning to a loosely shared worldview. The commitment becomes not a

---

*Nationalism globally takes many forms in an almost continuous spectrum from a sense of tribal bondedness where personally committed members share language, cultural history and religious values, to the giant, continent-sized megamachines characterised by personal alienation and a paucity of shared values, where the sense of belonging is abstract and symbolic. Both react to external threats, but only the latter *require* such threats to maintain social cohesion. Iran, Costa Rica and many other developing countries exemplify the former extreme; the United States and the Union of Soviet Socialist Republics, the latter.

real one to each other, but a symbolic one to the megamachine – 'our' country.[52]

Thus does one's nation come to symbolise all one cares about in life; its flag, its emblems, its anthem become emotionally connected to home, family and friends. Why else would 200 million Americans care passionately about the fate of 50 hostages held for over a year in Iran, while remaining utterly unconcerned for the hundreds killed or maimed each day at home owing to violence in their own country? Why else, during my decade living abroad, did I repeatedly discover uncontrollable tears welling into my eyes whenever I heard the *Star Spangled Banner* or saw the American flag displayed?

Unlike emergent tribal nationalism in the Third World that is based on true cultural bonds that naturally draw people together, the symbolic nationalism characterising large, amorphous and highly diverse societies is fragile. To maintain internal cohesion political leaders often exaggerate external threats, locating somewhere in the world an 'evil empire' against which the nation must pull together in order to defend itself. This explains how the United States and the Soviet Union, which share no common borders and do not materially confront or conflict with one another in any critical way, have become so mutually suspicious that they are locked in the most physically deadly and economically destructive arms race the world has ever known. Each megamachine in its own way has created an alienated society which is all too ready to turn its inner frustrations onto an external enemy. And of course, this externally directed anger becomes its own self-fulfilling prophecy, constantly magnifying the danger of war. This thread is pursued further in Part IV.

SOME CONCLUDING THOUGHTS FOR PART III

While it is undeniably true that Western capitalism has produced untold material wealth and a remarkable degree of political freedom, it has done so at very high costs. By concentrating its attention on the central regions of the entire means–ends spectrum of human concerns (Figure 11.3), the Western worldview endangers both the environment, the *ultimate means* that supports and nurtures us, and our own spiritual need for meaning, the *ultimate end* of conscious existence. Being singlemindedly concerned with production and consumption, our pragmatic, quantoid vision of life is devoid of both 'head' and 'feet'. It assumes that human beings will soon have totally mastered Nature and have no higher needs in life than can be attained by such mastery – namely the unending consumption of an infinitude of material wealth.

Our belief that we can master all Nature however is an illusion which, as shown in Part I, leads us to destroy Earth's resources without concern for

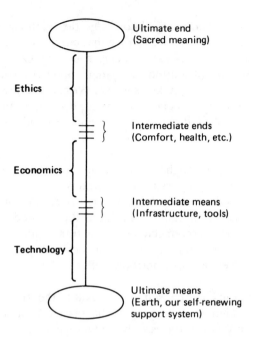

Source: Adapted from H. Daly, *Steady State Economics* (San Francisco: W.H. Freeman, 1977) Figure 1, p. 19.

Figure 11.3 The means–ends spectrum of human concerns.

Modern industrial society has concentrated on technology and economics, largely ignoring the Earth, the ultimate means of our existence, and the spiritual meaning of that existence, which derives from ethics, philosophy and religion.

the future. We unwittingly charge up these costs to our 'Environmental Credit Card' Account and then award ourselves first prize for cleverness. Philosopher Linda Holler, borrowing a quote from Santayana, makes clear the weakness of our misplaced faith in our power to command Nature:

> 'When our will commands and seems, we know not how, to be obeyed by our bodies and by the world, we are like Joshua seeing the sun stand still at his bidding; when we command and nothing happens, we are like King Canute surprised that the rising tide should not obey him; and when we say we have executed a great work and re-directed the course of history, we are like Chanticleer attributing the sunrise to his crowing.' (Santayana)

> So long as the illusion [that humans are the centre of the universe] could be maintained, we could afford to act like Joshuas. But our relationship to the world is in reality more like King Canute's. And, like Chanticleer, we look increasingly ludicrous attributing the sun's rising to our crowing.

When [finally we learn] to acknowledge the relational nature of our existence (as we now must), we can see that our mastery is an illusion. In fact, the modern world is experiencing a new realization of absolute dependence; we are absolutely dependent upon the connections that support our existence and make us whole. We can interpret that absolute dependence in religious, social, economic, physical, or biological terms, but we must interpret it, attend to it, and respond to it in a fitting manner.[53]

We must recognise, through a change in our worldview, our utter dependence on and relatedness to the natural world.

At the other end of the means–ends spectrum, our anticultural 'commitment to no commitment' has placed the isolated individual on a pedestal above all else, disconnected not only from Nature but from fellow beings as well. The individual is his or her own 'end' without relation to others. Again, Holler puts our position well:

[T]he world outside the self is irrelevant, the small world which remains is intended as something to be mastered, and other people appear as objects with services to be rendered.[54]

We do not *belong* to each other; we are alienated, isolated, alone. And without recognising it we have been systematically struggling to make our human psyches fit into this unsatisfactory, antisocial arrangement. Mander sums up what we have done to ourselves:

We have had to re-create ourselves to fit. We have had to re-shape our very personalities to be competitive, aggressive, mentally fast, charming and manipulative. These qualities succeed in today's world and offer survival and some measure of satisfaction within the cycle of work–consume, work–consume, work–consume. As for any dormant anxieties or unreconstructed internal wilderness, these may be smoothed over by compulsive working, compulsive eating, compulsive buying, compulsive sex, and then our brands of soma: alcohol, Librium, Valium, Thorazine, marijuana and television.[55]

Our prescription must be this. Life is *not* to be lived in isolation, either from Nature or from one another. We are all part of a whole we did not create, but on which we entirely depend. Today's Western worldview which denies all relatedness is maladaptive. Instead of continuing the attempt to mould ourselves and even Nature herself to fit a flawed cosmic vision, it is time to reconstruct that vision, to rediscover sacred meaning.

So far this book has identified the errors of our thinking, seeking out their historical origins. Part IV begins the search for change. How can the awareness of our misperceptions be applied to future decisions? What guidelines do we need? What comprises true 'meaning'? Surely it requires something beyond mere hedonism or sheer power; it requires participating in a sacred *social* vision.

In some societies, it is filled by religion. For others, it is filled by a deeply shared cultural identity. The sense of participating in a grand social experiment, even at great personal cost, meets one of the deepest needs of the human soul, one of personal transcendence that neither absolute 'freedom' nor untrammelled 'consumption' can fulfil. There is a need to belong to and live for a great social idea, an idea so majestic that it lends meaning to the most ordinary life. The pervasive alienation among affluent Western peoples is an indication of how little individualism and material-ism satisfy this most important of all human needs.

The goal is to foster among all human societies those visions that permit sacred meaning to prevail in people's lives while at the same time sidestepping the violence that too often erupts between differing world-views; our growing technological cleverness will soon *forever preclude* the dedicated use of force among all peoples as it now does the superpowers. The last part of this book develops one possible way in which such a goal might be met.

Part IV
New Modes of Thinking: 'Where Do We Begin?'

If aspects of the Western worldview are a central cause of many of our global problems, how can we begin to change the way we think? What assumptions about the world and how it works need to be critically examined and modified? What are the valuable parts of the Western vision that need to be preserved and built on? Most difficult of all, how do we rediscover sacred meaning?

Chapter 12, 'Rethinking Economics', exposes the physical, logical and moral fallacies of conventional economic thinking and shows how democratising and decentralising economic systems can begin to heal our errors. In Chapter 13, 'Defusing the Global Powder Keg', the history of politico-economic relations between the First and Third Worlds is recounted, which explains how the current economic dependency, exploitation and crippling debt of the developing countries came about. It suggests that economic autonomy offers a politically and socially more stable path for the excolonial nations to pursue than does Westernised 'development'.

Chapter 14, 'Politics: Worldviews in Action', underlines the role politics plays in the process of changing social values, and shows how super-large societies, with their centralised hierarchies that require rigid ideologies and external enemies to cohere, are maladaptive. Again, the solution seems to be decentralisation and democratisation along the lines of the Green movements. Chapter 15, 'Nuclear "Defence" – or Conflict Resolution?' is a continuation of the argument in Chapter 14. By our own cleverness, we have 'outlawed' war as a means for resolving ideological and other differences, so we must begin to take negotiation and non-violent defence seriously.

12 Rethinking Economics

> I shall argue that it is the capital stocks from which we derive satisfactions, not from the additions to it (production) or the subtractions from it (consumption): that consumption, far from being a desideratum, is a deplorable property of capital stock which necessitates the equally deplorable activities of production: and that the objective of economic policy should not be to maximize consumption or production, but rather to minimize it, i.e., to enable us to maintain our capital stock with as little consumption or production as possible.
>
> Kenneth Boulding, 1949[†]

INTRODUCTION

Though couched in the esoteric language of economics, this perceptive statement simply means that it is not consumption which provides economic well-being, but the *use* of things like cars, shoes and houses that give long-lasting service. Yet most economists encourage precisely the opposite: 'economic growth' – an ever-expanding rate of production and consumption, with all the negative environmental outcomes they entail. Why are they unable to reconcile their understanding of economics with the realities of a finite Earth – our 'spaceship Earth' as Kenneth Boulding first called it? Without reconstructing in detail the tenets of conventional economic thinking, which in any case is in considerable disarray, this chapter examines some of its major fallacies, so that economic ideas about the world can be re-created on a sounder base.

This is by no means an easy task. As the focus of politics, following in the footsteps of 'scientific efficiency' and rapid technological change, has shifted from moral to material concerns, nations have come to define themselves in *economic* terms. The words of Sheldon Wolin are worth repeating:

> [E]conomics now dominates public discourse . . . [There is a universal] faith that practically any public concern can be reduced to economic categories . . . Economics thus becomes the paradigm of what public reason should be. It prescribes the form that 'problems' have to be given before they can be acted upon, the kinds of 'choices' that exist, and the meaning of 'rationality'.[1]

Since 1969, when the first Nobel Memorial Prize in Economics was given,[*] the aura of scientific objectivity surrounding economics has expanded. This

*This prize did not originate from the estate of Alfred Nobel but from the Central Bank of Sweden that persuaded the Nobel Committee to call the award the 'Nobel Memorial Prize in Economics'.

333

plus its jargon gives the subject a mystique, placing it outside public under-standing, the special province of 'experts'. These experts have made certain assumptions that they expound in the form of 'economic laws', which are then taught as 'facts' and accepted without question. Whether intended or not, their 'objective truths' are actually values in disguise, as humanistic economists Mark Lutz and Kenneth Lux have pointed out:

> The most effective way to transmit values in a society, say, from one generation to the next, is to do so without letting on that it is *values* being transmitted; they must be disguised as objective truths. In recent times science has been the medium through which this was done, and economics has shared heavily in the process. The claim made that economics is a value-free or positive science has been the perfect cover for the set of values which does in fact exist in economic theory.[2]

This chapter attempts to define 'economic well-being', and to identify where it comes from and how to obtain it non-destructively – without injury to the environment or fellow humans. It demonstrates the need for an about-face in our economic thinking, away from 'consumption' and 'growth' and toward 'economising'. New attitudes will entail new institu-tions as well, particularly a more democratic control over those economic decisions that affect society as a whole. But first it is necessary to explore the errors underlying modern economic thought: the *physical errors* about the nature of wealth; the *logical errors* regarding our measurement of well-being; and the *moral errors* over the distribution of wealth. (Chapter 11 has already treated the psychic alienation stemming from a modern 'economic' worldview that regards work as a disutility to be avoided and material consumption as a satisfactory substitute for social and spiritual meaning.)

PHYSICAL ERRORS OF MODERN ECONOMICS

Economics deals with the exchange of goods and services between members of a community, with the production and distribution of material wealth in a society. It defines certain kinds of values and certain kinds of scarcities. Yet what is generally neglected – or at least not emphasised – is that *all* economic activity ultimately depends upon a throughput of matter and energy. But a glance back at Figure 3.7 shows that a modern industrial economy – human beings and their artifacts – depends upon the *flow* of matter (natural resources) and energy (mainly fossil fuels). Energy is used but once while resources ideally are recycled, although too often they pile up as environmental pollutants.

As shown in Part I, economic growth has occurred by an exponential increase in the throughput of matter-energy. *Per capita* consumption of resources has been doubling every generation or less. Now it would seem obvious that the amounts of matter and energy available put absolute constraints on our ability to produce and consume. Yet such a well-known and respected economist as Lester Thurow flatly denies this:

> Often a fallacious 'impossibility' argument is advanced to imply that we have to limit economic growth. The argument usually starts with a question. How many tons of these or those non-renewable resources would be needed if everyone in the world now had the consumption standards enjoyed by those in the United States? The answer is designed to be a very large mind-boggling number which convinces you that something has to be done to limit American consumption and that others can never achieve our standard of living.
>
> What the question ignores is the fact that the rest of the world cannot have a U.S. standard of living until it has a U.S. standard of productivity. While consumption would go up by a large amount if this were true so would production. The world can only consume what it produces. When the rest of the world has our standard of living, they will be producing the extra resources necessary to have it. The relative prices of different products would undoubtedly change if this were true. We undoubtedly would be forced to shift away from an oil economy faster and do more recycling of materials if the rest of the world were growing more rapidly; but economic advances in the rest of the world do not depend upon cuts in our consumption.[3]

When read aloud to ecologists this incredible statement regularly elicits a sharp gasp of disbelief. Thurow seems to argue that the bottleneck to increasing global consumption is simply a lack of know-how in backward nations and once they have that they can readily attain Western levels of production. The problems of global resource and energy scarcity discussed in earlier chapters are apparently to be got over by 'shifting away from an oil economy faster' – as though some suitable alternative were already in reach – and doing 'more recycling of materials' – as though this were not an energy-requiring process. When leading economists are capable of such confusion, is it any wonder that politicians and the public are misled?

Since there in fact *is* a real limit on the amount of matter/energy available to an economy on a continuing basis, is there any way to increase economic well-being while maintaining a 'steady-state' throughput? Can an economy improve while living safely within its environmental income of renewable resources? The answer to this is a qualified 'yes'. To comprehend this, one must examine the sectors of the economy from a matter/energy perspective.

Sectors of the Economy

It is popular to talk about the economies of the future as 'post-industrial', 'service-oriented' or 'information-based', as though the need for matter/energy will markedly subside while we all swap backrubs, accounting services and musical entertainment. The implied assumption is a decline in industrial production, although it is vigorously denied that there will be any decrease in physical living standards. Economist Herman Daly calls this 'angelizing GNP' – imagining that people no longer require real matter/energy to maintain themselves.[4] A glance at the kinds of economic goods required by people exposes the fallacy.

The things people consume fall into three classes of economic utility. First are the 'perishable' items, things whose usefulness consists in their consumption. Among these are food and beverages; the energy for cooking and warmth, and running machines; medicines, soap, matches; chemicals like bleach, fertilisers and lye. Other 'perishables' – plastic cups, paper goods, non-returnable containers – are mere luxuries of our throwaway economy, but every society has a minimum requirement for those in the former category to maintain the well-being of its people.

Second are the capital stocks of society, what Boulding reminds us provide most of our material satisfaction. Capital stocks are things that are valuable because they can be used over and over again: they last. Capital stocks in the home consist of the house itself; furniture; consumer durables such as cars and refrigerators; instruments and tools such as lawnmowers and vacuum cleaners; and personal items like books, sports equipment and clothes.

There are also capital stocks that play a role in production and distribution of wealth: the plants and machinery used in manufacturing; the buildings and hardware of the business world; and the roads, railways, reservoirs and utilities that provide the economic infrastructure. Both kinds of capital stocks are valuable *in their use*, not in their consumption or decay. We are economically better off building the sturdiest capital stocks possible and maintaining and repairing them carefully to prevent their breakdown – their 'consumption'. Where capital stocks are concerned, we should be decreasing economic throughput, not increasing it. Yet it is by production *and consumption* of cars, clothes and television sets that we now judge our well-being. The reasons for this pressure to *consume* rather than *preserve* is explained below.

The third class of economic goods is 'services'. There are the direct services we obtain from educators, doctors and hospitals, entertainers, government bureaux, lawyers, garbage collectors, legislators, ministers, firemen, and the military; and there are indirect services we obtain from the producers and distributors of consumables, and from the producers and maintainers of stocks such as engineers and road crews. It is often assumed that services place minimal demands on the economic throughput of matter/energy, yet this is far from the case. Every service requires certain physical amenities where it is performed. Surgeons require elaborate operating rooms and expensive equipment; even a psychiatrist must have an office, furniture, lights, a secretary, a telephone. Services consume matter/energy, and in certain instances, as in medicine, new technologies are increasing the amounts consumed per service rendered.

Steady-state 'growth'

Returning to our question – Can an economy improve while staying safely within its environmental income of renewable resources? – imagine an economy supplied only with a steady income of 'renewable' energy: solar, wind, geothermal. To maintain itself indefinitely, some of that energy must be reserved for recycling waste matter – garbage, old cars, 'junk' metal – whatever nature does not recycle 'for free'. Clearly the less garbage to be recycled the more energy is left for useful services such as running machinery or transporting people. Perishable goods except those essential to survival should be minimised or eliminated. Throwaway products consume energy twice: in their production and again in recycling. The goal is to *minimise* throughput of matter/energy.

Similarly, there should be as little throughput as possible of capital goods. Machinery and transportation systems that require little energy to produce and run and that last for a very long time are preferable to capital goods that are in constant need of repair and tend to decay rapidly. It is the production energy *times* the length of trouble-free, efficient service that counts, not merely the production cost itself. Even though Oliver Wendell Holmes's 'wonderful one-hoss shay' might have cost twice as much to build as another carriage, its ability to give repair-free service for a hundred years made it more 'economic' than five carriages each lasting only 20 years. Economising is not just a matter of purchase price; it is, as Boulding says, a matter of long, reliable service.

Finally one can ask: When is a new, 'more efficient' technology really economic? When should a factory be stripped and its plant replaced? The answer in ecological terms has to be: When the same service over the lifetime of the new factory and its products can be obtained for less total throughput than the old factory and its products could have provided.

To summarise, throughput is an evil, not a good, and rapid obsolescence is a cost, not a benefit. *Time* is thus an important economic factor: the longer something lasts, the better. Yet economists talk only about the present, or at best the short term; five years is a long time and 25 an eternity. It is production efficiency rather than performance efficiency that interests them. If a house can be built more cheaply, it doesn't matter if it lasts 20 years or 200 years; sale price is what counts. This habit of discounting the future is taken up again later. First it is necessary to understand money in order to see why so much emphasis is placed on increasing economic throughput.

'Wealth, Virtual Wealth and Debt'

People with giant bank accounts are considered 'wealthy'. Paul McCartney is reputedly 'worth' $500 million, Bob Hope, $200 million, and Michael Jackson is approaching $1 billion.[5] But money itself is not wealth. Money is merely the ability to *demand wealth* – real matter/energy that has utility. Of course, the people named above *do* tend to own real possessions, lots of them. If they chose instead to live very simply, they would not be wealthy, but they *could* be if they chose to exercise their demand rights. Money is thus a *claim* on wealth.

This distinction between wealth – *real* things which can be used – and money, which Nobel Prize-winning physicist Frederick Soddy clearly made in his 1926 book, *Wealth, Virtual Wealth, and Debt*, is one that few people appreciate. Money is only a means for exchanging wealth; in itself it is pretty useless stuff. In the old days, people could make jewellery out of gold and silver coins, but paper money has no more intrinsic 'use' than a scrap of wastepaper.

Money is really a *national debt* – a promise of a fixed amount of wealth that a government makes to its people. Soddy calls it 'virtual wealth', an amount of wealth people can demand, but which does not really exist. Money, however, is a debt that does not bear interest. If you keep 1000 dollar bills under your mattress for 10 years, they will not have acquired more value simply because you did not spend them.

Credit, or 'There is no such thing as "minus two pigs"'

You can either make a purchase with cash or you can buy 'on credit', incurring a debt. But if you pay cash you pay less; with a loan, one has to pay back principal plus interest. During the loan period, the lender gives up no real wealth at all, merely the claim on it, and there are two 'owners' of the car or house for which you borrowed the money. You physically possess the wealth but the lender legally owns it. On paper, the lender appears wealthy, yet what he 'owns' is a minus quantity. If I loan you two pigs, perhaps to get a hog farm started, then you owe me two pigs. But I am not richer beause of this debt; I am poorer. Here is how Soddy explains it:

> Wealth is a positive physical quantity, but debt is a negative quantity. It has no concrete existence, and is to the physicist an imaginary quantity. If we deal with numbers, then we may with great appropriateness give them either sign; but in physics, which deals with real quantities, we can only do so with caution. The positive physical quantity, two pigs,

is something anyone may see with their own eyes. It is impossible to see minus two pigs. The least number of pigs that can be physically dealt with is zero. Plus two pigs at least must be taken for granted before we can, even for reckoning, make use of the imaginary quantity, minus two pigs. Though we may, with the utmost mathematical purism, deduct two from one and have left minus one, we cannot deduct two pigs from one pig and have left minus one pig . . .

Debts are subject to the laws of mathematics rather than physics. Unlike wealth, which is subject to the laws of thermodynamics, debts do not rot with old age and are not consumed in the process of living. On the contrary, they grow at so much per cent per annum, by the well-known mathematical laws of simple and compound interest.[6]

Economists, however, have tended to persist not only in calling money 'wealth' but credit also. They appear to have created wealth out of nothing, in defiance of the laws of physics, since credit, like money, is a legal claim on real physical wealth.

Consequences of growing debt

In all societies, the creation of capital goods – houses, factories, highways – demands that a certain fraction of material resources, of energy, and of labour be temporarily diverted from other economic enterprises. To this purpose, 'capital' is accumulated in the form of money or credit, so that the needed components can be purchased. In command economies, such as the USSR, capital is raised by taxation or other government action, while in capitalist economies, it is usually borrowed, with interest, from private investors. If the 'loan' is a direct one, the investor receives ownership stocks in the enterprise that entitle her to a share in all profits henceforth. If the loan is made via a bank, the investor receives compound interest on her deposit. Provided the enterprise survives or the bank does not fail, the investor is assured of an income – either as profit or interest – for which she does nothing at all.

To illustrate this accumulation process, Soddy tells the story of a man who loans the government a nine inch ball of gold at five per cent interest per annum. If the earnings are paid in gold and are in turn reloaned to the government, in just 600 years (24 generations) his heirs would have legal claim 'to a golden ball equal in size to the earth'.[7] Here is yet another example of exponential growth!

Banks today have extraordinary latitude in converting money (virtual wealth) deposited with them into loans (debts), for they only have to keep a small fraction of each deposit on hand to meet sudden withdrawals. For a US checking account, it is 12 per cent. If a depositor banks $1000 the banker can loan $880 to another person, who perhaps deposits it for a few days. Now the banker can loan $775 of this to a third borrower, and so on. By paper transactions entirely within the bank, $1000 soon becomes $7000, on which the banker earns interest while paying the original depositor interest only on the original $1000. As Soddy says:

> Purely fictitious money, which the nation has not authorised the issue of, is fictitiously lent without anyone giving it up, and then creates perfectly genuine deposits and legal claims upon the community's market for the supply of wealth.[8]

Conventional economists are fond of pointing out that money, and particularly financial markets, create a climate in which wealth and productivity are enabled to 'grow'. They seldom mention the other consequences of turning debt into a commodity: (1) economic growth has to keep up with the exponential growth of interest; (2) if such economic growth fails to occur, inflation ensues; (3) it results in the accumulation of wealth in the hands of fewer and fewer people. In Soddy's words, the clear goal of the capitalist is to get 'the community into his debt and prevent repayment, so that the ... community must share the produce of their efforts with their creditor'.[9] The capitalist permanently rents his capital which, unlike capital goods or land, never wears out, but continues to accumulate 'wealth' forever.

Owing to its exponential nature, what seems a small rate of growth in debt suddenly begins to sky-rocket. Soddy warned of this, too:

> It is characteristic of the dizzy virtues of compound interest that they are not at all dizzy at the start. It is only after they have been in operation a certain time that they show any disposition to become marvellous and to transcend the bounds of the physically possible ... This means that a system might show no signs of breakdown for a century and yet become absolutely impossible during the course of the next.[10]

Soddy's impossible moment appears close at hand. The global economy, after several decades of borrowing fictitious wealth at high interest rates from the financial markets, has failed to grow at anything like the necessary rate to pay back the interest, let alone the principal. Almost every nation is entangled in the international net of mutual indebtedness and there is no obvious way to stop its exponential growth. As their collateral erodes, prospects for the owners of debt look bleaker.

Most are familiar with the heavy indebtedness of developing nations. Third World countries, encouraged to borrow by Western bankers during the massive influx of petrodollars in the 1970s, find themselves unable to meet their interest payments, let alone the principal. In all, over $1 trillion is outstanding, and more than half the debtor countries spend more than 20 per cent of their export earnings to service their debts. For several – Argentina, the Philippines, Uruguay, Israel, Mexico – interest owed each year exceeds total export earnings by 150 to 200 per cent! Furthermore, for most developing countries, export earnings are declining, not growing.[11] The bankers' solution so far has been to 'reschedule' debts, to loan debtors more and postpone payments further into the future – the capitalist's desired situation of permanent indebtedness. But the possibility of debt repudiation always looms in the background.

Yet the Third World is no longer alone as the world's greatest debtor. That dubious honour is now challenged by the wealthiest nation of all. In 1985, for the first time this century, the United States became a debtor nation and is rapidly running up its foreign debt toward $1 trillion. If reached as expected in 1989, annual interest payments will be $100 billion. Although US investments and loans abroad, nearly $800 billion, should offset much of this, the latter are mainly in nearly bankrupt developing countries while debts are owed to West Germany, Japan and oil-producing countries.

But a trade imbalance is not the only US financial problem. The nation as a whole has public and private debts exceeding $7 trillion, or $35,000 per person. Its predicament serves as a model for numerous other capitalist nations, particularly Great Britain, and so bears examining. Nearly half the debt – $3 trillion – is owed by business, which owing to tax laws finds it cheaper to raise capital by borrowing than through investment shares. Although most is used to upgrade productive capacity, more and more goes into risky takeover bids. Consumer debt, in home mortgages and other long-term instalment loans, totals $2.6 trillion, but with declining incomes, defaults are growing. And the Federal debt now exceeds $2 trillion, with the government borrowing up to 50 per cent of national savings. One-tenth of the national income goes to service that debt, almost half of which is owned by foreigners. The instability of so much uncertain debt is marked by increasing bank failures – up from 48 in 1983 to over 150 in 1986 – and increasing volatility in the stockmarket. As Alfred Malabre, economics writer for the *Wall Street Journal* observes, the outlook is terrifying:

> Taken as a whole, public and private debt in the U.S. now exceeds $7 trillion and, through the magic of compounding, that figure could easily double by the end of this decade if present trends persist. Looking further ahead, total debt could well reach some $57 trillion by 1995 and, again through compounding, $448 trillion by 2015. It's unreasonable to suppose that the Fed or any other governmental or private institution could contain the trouble that would inevitably arise in such a debt-burdened future.[12]

The Fallacy of Misplaced Concreteness

The mistaken notion that money *is* wealth is an example of what Alfred North Whitehead called the 'fallacy of misplaced concreteness'.[13] Not only does it invite us to believe that the growth of debt is the growth of wealth; it also causes us to assume that because money circulates indefinitely in one direction, that the goods and services for which it is exchanged therefore flow indefinitely in the opposite direction. Even current texts give the impression that wealth flows in a circle and is self-regenerating. Here are economist Herman Daly's comments on two sentences from an introductory textbook:

> The student is told, with the emphasis of bold print, that, 'The flow of output is circular, self-renewing, self-feeding.' Note that the flow of *output*, not [merely] purchasing power, is considered circular. The first study question at the end of the chapter is, 'Explain how the circularity of the economic process means that the outputs of the system are returned as fresh inputs.' One might as well ask an engineering student to explain how a car can run on its own exhaust, or ask a biology student to explain how an organism can metabolize its own excreta![14]

All that *circulates* is purchasing power. Real wealth flows in a unidirectional fashion. It enters the economy from the environment through expenditure of energy on matter, and it leaves as useless remnants, often widely dispersed. The *only* purpose of money is to facilitate the processes of production, distribution and maintenance of wealth in a society of specialists. The fact that money circulates merely indicates that different people contribute to the various stages between the initial extraction of wealth from the surroundings and its final decay.

LOGICAL ERRORS OF MODERN ECONOMICS

Beyond the purely physical errors in our economic thinking that emerge out of imagining that money is wealth, there are also logical errors. In particular, we confuse ourselves about such basic economic matters as 'costs' and 'benefits', about 'efficiency', and about 'freedom' and 'competition'. Nowhere do we so mesmerise ourselves than in calculating our material well-being in terms of that hallowed quantity, the GNP. By carefully considering what is and is not summed up in this sacred number we begin to see exactly how misleading a quantity it is.

Gross National Product

The first thing to notice about the Gross National Product is that it is a number – something *countable*. It fits in with our 'quantoid worldview'.

GNP clearly belongs to Western, 'scientific' thinking. It adds up the *monetary values* of three things: personal consumption of goods and services; government expenditure on goods and services; and investments in productive capital assets. The rules for calculating these components are complex and full of arguable assumptions, but those need not concern us. Our concern is with more obvious and important matters.

First, GNP is a summation across a whole nation – in the case of the United States, 240 million people. In comparing nations, *per capital* GNP is the standard of measure, but it is a single number, an average. It tells us nothing about the size of the disparity between rich and poor.

Secondly, GNP focuses on the *consumption* of resources. It is counting the rate of flow of wealth *through* the economy as the measure of well-being. What we already own and use but have not yet destroyed is not part of GNP. We thus measure economic well-being in terms of consuming, *not* conserving. The more things we use up, destroy and replace, the better off we are! This view seems to border on insanity. But first, what does it mean to add up all consumption in terms of a single yardstick: dollars, or rubles, or francs, or pesos?

Apples = Oranges = Pacemakers ≠ Mother's Love

At some time most of us learn that you cannot add up $2a + 3b$ – you cannot meaningfully add up apples and oranges. But GNP does exactly this, and by adding up the monetary values of everything, quickly becomes almost meaningless. A football hero or a rock star contributes far, far more to the GNP than the head of state. One more bomber has the same 'value' as a new hospital. And the $50.00 a student spends on textbooks counts just as much as the $50.00 worth of pornographic comics sold downtown. GNP is blind to anything but monetary value. If somebody wants something and is fool enough to buy it, it has economic 'value'. There is no distinction between life-saving, life-enhancing, and life-destroying consumption.

Yet if GNP is meaningless in what it adds together, it is even *more* meaningless in all the goods and services it neglects. In the past, many transactions that today are included as part of GNP were not 'paid' for and could not be assigned monetary value. Even in modern economies, there are innumerable services that people perform for themselves and others that are not counted in the GNP. Mothers who watch over their own children, clean their own homes, and prepare their family's meals are performing economic services. So is every man who mows the lawn, repairs his own plumbing, and renovates the kitchen. There are also great armies of volunteers who provide dozens of community services without pay. All these economic benefits never show up in GNP.

In fact, a considerable part of the recent GNP 'growth' in Western

economies has come about through the 'commoditisation' of these previously 'free' services. Mothers who used to stay home now go out to work.
Not only are their *paid* services now counted as part of the GNP; so too are
the salaries they pay baby-sitters and housekeepers. Likewise, the home
maintenance work formerly done by the unpaid handy husbands is now
suddenly added to the GNP as 'earnings' by gardeners and plumbers. Nevertheless, GNP still excludes a large quantity of freely given nurturing services;
in the United States, it was estimated to be worth $300 billion in 1969.[15]

Another economic asset that is never counted because no one pays for it
is the multiple functions performed by Nature, particularly the purification
of air and water and the recycling of wastes. Only when we exceed Nature's
capacity to clean up after us and have to pay the consequences do we begin
to value what we have ignored. Only then do we see the fallacy in assuming
that what is not owned has no value.

Costs are 'benefits'

A closer look at the goods and services that make up GNP reveals that
many things we pay for are actually 'costs', not 'benefits'. Take advertising.
It 'benefits' only the producers of goods; it is a wasteful overhead of
competition, amounting in the American economy to about $100 billion a
year. Moreover, most advertising encourages the very thing we should be
discouraging: throughput. Clothes, cars and furnishings are discarded
because they are 'out-of-fashion'; everything must be 'up-to-date'.

Many other GNP 'services' are really avoidable costs: hospital bills due
to avoidable industrial accidents and environmental pollution; repairs to
things that prematurely break down. As someone once colourfully put it,
everytime there's an automobile accident, the GNP goes up!

This wonderful calculus means that GNP – our supposed well-being –
increases not only with every automobile accident, but with every case of
arson, divorce, murder, or occupational illness. As long as someone
'makes money' out of the consequences of social misery, the GNP goes up.
We could get by, in a saner society, with a great deal less of certain
'services': psychiatrists, juvenile detention centres, police and prison
officers, lawyers and courts, and the gigantic welfare agencies with their
layers of bureaucratic overhead and dehumanising computers.

Besides social costs, today's GNP includes a growing number of environmental costs, incurred in monitoring and cleaning up the damage created
by our massive industrial throughput. In the United States, the Environmental Protection Agency, the Food and Drug Administration and dozens
of other federal, state and local agencies are charged with monitoring food,
air and water, and regulating the sources of fumes, toxic chemicals, sewage
and other effluvia. The *costs* of meausuring pollutants, of putting antismog
devices on cars, of installing chimneystack-scrubbers, and of litigation
when there are disputes are all *added to GNP*!

The list keeps growing. More and more of what society pays for is to keep things from getting worse. Althogh the future costs of cleaning up toxic waste dumps is staggering, in terms of GNP it will count as a 'benefit'. No wonder people have had to double their energy consumption in less than two decades without feeling they have anything to show for it. [16]

Costs are 'externalities'

Besides the costs we pay people to take care of and therefore add to GNP, there are the costs that no one pays for – yet. These we are charging to the future: the increasing social injustice, both within industrial societies and globally, and the continuing deterioration of the environment that we keep adding to our Environmental 'Credit Card' Account. Economists call these uncounted costs 'externalities' – they are *external* to their balance sheets. Acid rain, the pollution of rivers and groundwater, the destruction of forests and topsoil, never show up in the profit/loss columns of businesses nor in tabulations of the GNP. Anything without a dollar sign has no 'economic' value and is ignored. Externalities rate but a few paragraphs in 1000-page introductory economics texts.

Clearly we need a new kind of economic 'balance-sheet' drawn up in terms of human well-being, now and in the future. It will include all those costs and benefits that no one now pays for, and it will give proper attention to *plus* and *minus* signs! GNP as now calculated does not measure human well-being; it tells us only how fast we are using things up.

Misplaced 'Efficiency'

The logical steps in the pursuit of 'efficiency' – the division and specialisation of labour; the need for an organising management; the standardisation and mass production of goods; and finally, the need for centralised control over the supply of raw materials and over the market – have led inexorably to the giant, depersonalised, immortal management systems, the huge national and multinational corporations that now control most of the global economy. We uncritically assume this is the price of industrial affluence. Yet careful examination shows the present system is far from efficient in terms of human well-being. Besides the numerous costs, either ignored as 'externalities' or added in as 'benefits' in the GNP calculus, there are other overheads not yet considered.

One cost has been the growth of big government. In order to monitor and control the worst excesses of big business – to enact child labour laws, a minimum wage law, anti-monopoly legislation and literally thousands of other laws; to maintain tax and welfare systems in order to redistribute the wealth that otherwise accumulates in the hands of a few; to establish and administer rules for trade, commerce, banking, and even for government itself – governments have grown enormously, multiplying their bureaucracies many times. In every industrialised nation there are agencies to regulate airports and airwaves, to monitor advertising, allocate water resources, and keep track of patents. Every new bureau, every new regulation becomes a *cost* that must be paid both to service and to control the huge power of centralised industrial systems. Often falsely painted as excessive government interference with the free-enterprise system, it is in fact an *overhead* of that system. Such functions of government are *costs of production*, and must be recognised as such. When the developing countries produce goods more 'cheaply', it is not necessarily because they are more 'efficient', but because these costs are paid instead in terms of human misery and environmental deterioration, as 'externalities' not added to the price. 'Price' is a poor measure of true cost. [17]

Even the giant corporations are losing out as they attempt to grow ever bigger. Their early gains in efficiency from specialisation and organisation are now being dissipated in the giant overheads of their top-heavy bureaucracies. The spiralling costs of management are becoming a net liability. Futurist economist Hazel Henderson explains their predicament:

> Belgian information theorist Jean Voge... verifies that the logic of efficiencies of scale in production are meeting diminishing returns and bogging down in the even larger information and coordination costs they incur, resulting in increasing bureaucratic sectors... Voge demonstrates what E. F. Schumacher and I had asserted: that when industrial economies reach a certain limit of centralized, capital-intensive production, they will have to shift direction to more-decentralized production technologies and decentralize economic activities and political configurations, using more laterally linked information networks, if they are to overcome the severe information bottlenecks in excessively hierarchical, bureaucratized institutions. [18]

Another factor in our declining socioeconomic 'efficiency' is the dehumanisation of the work force. If people are to *be* productive, they must *feel* productive; work must be a utility, not a disutility. This fact, long ignored by American management, has been taken into account by management in a number of European companies. In such plants as Phillips Electronic Company in England and Volvo in Sweden, workers on the shop floor work as teams, setting their own pace and cooperating together during production. The result is fewer absences, higher productivity, and greater worker satisfaction. Everyone gains by a more humanistic arrangement.

One industrial country that has been remarkably free of worker alienation is Japan. It is not that workers there have more creativity in their jobs, but that they have a strong sense of 'belonging' to a family in the workplace. Company identity results in extraordinary worker cooperation and productivity. Far from decreasing workers' incentive as Western thought would have it, lifelong employment security and promotion with seniority rather than via competition increase it.[19]

In other words, centralisation, specialisation and competition when carried to present extremes are, even in the strictest economic terms, cost *inefficient*. Only centralised power is being promoted.

The Overheads of 'Competition'

Economists are fond of telling us that 'competition' in a free-market system is the most efficient way to improve social well-being. Competition is what keeps prices down while ensuring that supply meets demand, and consumers get the best products possible. This theoretical model may be approximately true when the number of players is the statistically huge number envisioned by Adam Smith. But today, almost every major sector of the economy – except personal services, and certain specialist producers and retail outlets – is controlled by a few giant corporations. Under years of competition, single entrepreneurs evolved into faceless management; management took over control of resources and markets; and finally there emerged the giant conglomerates. What began as competition for relatively more efficient production ended up as competition for absolute economic power. This new kind of competition has somewhat different consequences, depending on whether it occurs within a national economy or across national boundaries in the global arena.

Within a single economy, it became apparent that cut-throat competition among a small group of firms in the same business was mutually suicidal. In a price war, the consumer might 'win' while it lasted, but the end result was the likely bankruptcy of all the competitors and an abrupt end to the consumers' heyday. Far better for all concerned that the big companies tacitly agree on prices and their 'fair shares' of the market. In many important areas of production where oligopolies reign, competitive pricing is the exception. Market shares are maintained by advertising and product 'improvement'. As price competition has faded, technological competition has taken over.

In a capitalist economy the maintenance of *throughput* is essential for the steady flow of profits and interest that drive the system. Obsolescence is vital 'for a healthy economy', and what better way to create obsolescence than with 'new', 'improved' products? The research and development of new products and the capital stocks needed to produce them all require investment that almost inevitably means the new product is more costly than the one it replaces, even if it is 'better'.

The consequences of this constant competitive innovation are twofold. One is a new kind of inflation, labelled 'cost-push'. The expectation is that when an old product is replaced, the new one will naturally cost more. One's standard of living *ought* constantly to get better. Salaries and wages *ought* to rise to accommodate this. Not only are rising costs of these constant 'improvements' devastating for those living on fixed incomes: welfare recipients, retired veterans, and the elderly; they also put pressure on the economy to 'grow', to constantly speed up *throughput* and so generate the profits to pay off the last investment and begin the next cycle of 'innovative improvement'.

Whereas oligopolies within a country can often suppress the consequences of excessive competition in technological innovation by buying up new patents or taking over new, innovative companies,[20] this does not work in the international economy. Here there is fierce and growing competition, both in terms of technological change and prices. Indeed, to cut back on their labour costs, many multinational corporations are relocating from rich countries to poor countries, leaving widening pockets of unemployment behind. The Western industrial economies are battling each other – the North American economy, the Japanese economy, the European Common Market – for shares in one another's markets and for control over new Third World markets. In each country, economists are urging increased investments in scientific and technological research and development 'to keep *our* economy competitive'.

Supposedly consumers everywhere will benefit from all this competition. Yet a close look at who benefits from this highly centralised, enormously powerful economic system uncovers a rather shaky moral base. Instead of a future world of five billion or so happily consuming people all living in a technological paradise, the system as now structured is creating a few wealthy centres of economic activity fiercely vying for each other's markets, and the remainder of the globe, stripped of its resources, an impoverished home for hundreds of millions.

MORAL ERRORS OF MODERN ECONOMICS

So far we have considered economic systems from a macroeconomic perspective, looking at the total flows of production and consumption. But macroeconomics largely ignores how goods and services are distributed – whether people share more or less equally or whether some gain at the expense of others. To perceive the degree of moral justice in a capitalist economy one must examine the categories of economic actors.

The Economic Actors

Capitalism begins with 'investors', persons who contribute some of their purchasing power to the expansion or upgrading of productive capacity. At some future time they expect to have *more* real purchasing power, from interest or profits, than they originally invested. At seven per cent per year, their wealth doubles every ten years. Such expectations can be met only if the economy actually *does* produce more; 'successful' capitalism thus demands constant economic growth. Few people in capitalist societies are actual capitalists, however. In the United States, the wealthiest seven per cent of people own nearly 90 per cent of all stocks and bonds held by individuals.[21] Other large investors are banks and financial institutions. These groups continuously accumulate, as profits and interest, a constant fraction of economic throughput, regardless of the nature of what is being produced and consumed and by whom.

Another set of actors is corporate management, those who make the economic decisions about what is produced, how, and where. The heads of giant corporations not only control most of the corporate wealth but, being the highest paid category of persons, they own most of the private wealth as well. In the United States, constant centralisation of economic power has resulted in 200 corporations controlling half the nation's business assets, employing one-third of its labour force, with sales amounting to three-quarters of the total. Political economist Thomas Dye estimates that around 3500 corporate leaders in business, banking and insurance, many of whom sit on several corporate boards, control the bulk of the economy.[22]

Both investors and management benefit from economic expansion, from decreased governmental safety and environmental regulations, and from decreased taxation. To minimise production costs, they relocate abroad where labour is cheap. To minimise uncertainties, they 'hedge their bets' by merging into diversified conglomerates, through both friendly and hostile takeovers. And they automate, less to economise than to eliminate labour uncertainties. As historian David Noble writes:

> The ideal automated factory resembles an artificial world with little or no human intervention. It consists of an army of tireless robots, uninterested in coffee breaks or strikes, who can work 24 hours a day, monitored by a handful of individuals.[23]

The great mass of workers – the bread-winners, wage-earners, and salaried professionals – make up the vast bulk of actors in the economy. Within this group are also the independent professionals, small business-men, family farmers, and others whose 'profits' are the personal incomes they pay themselves from the sale of goods and services. These people depend upon continuous employment for their purchasing power, which in turn creates the economic throughput that supports investors and corporate management. But while workers and professionals are heavily dependent on the decisions made by the latter two groups for their jobs, they have negligible input into those decisions. A few own stocks; some have a small amount of savings or equity in a home; most however are net debtors. Their only economic power lies in choosing what products to buy and in electing mayors and legislators who will vie with other mayors and legislators in bringing about local economic benefits. But that is *all* they can do.

A further group is the retired persons, living for the most part on fixed incomes of pensions and savings. In our aging societies, their numbers are increasing. At the bottom is the most economically disenfranchised group of all, the unemployed, who survive on welfare or charity. The growing plight of workers, the aged, and the unemployed is dealt with below.

The single most powerful economic actor is government, particularly the central government. Slowly but surely governments have come to play an increasing economic role, to the point where today they are held responsible for the smooth running of the national economy. Their earliest functions were simply to set rules by which free markets operated and to run public services such as the post office and national defence. Over the years, such services have grown, and in many countries, utilities, communications, medical services and coalmines are owned and operated by government, giving it enormous control over those sectors of the economy. As major employers and consumers of goods and services, governments play a large economic role.

But besides being economic participants, governments have increasingly become manipulators of the economy. This point, already raised in Chapter 10, requires only a brief comment here. Modern capitalism has built into it structural faults that generate recurrent bouts of depression, inflation and social injustice which governments try to ameliorate by speeding up or slowing down economic growth and redistributing wealth from the rich to the poor. These manipulations, however, appear to be less and less successful. As Hazel Henderson notes:

> The facts that unemployment remains high during inflation and that prices remain high in recessionary times of tight money and expensive credit indicate clearly that these problems are now built into the structure of mature industrial economies.[24]

The Central Problem of Capitalism

It is not generally appreciated that the revolutionary British economist, John Maynard Keynes, whose theories of economic manipulation of taxes, interest rates and spending by governments have been widely put into practice, also advocated a future economy where capital no longer commanded interest. When productive capacity was sufficient to meet society's needs, only small amounts of capital would be needed to maintain or modify it. Wrote Keynes:

> If I am right in supposing it to be comparatively easy to make capital-goods so abundant that the marginal efficiency of capital is zero,

this may be the most sensible way of gradually getting rid of many of the objectionable features of capitalism. For a little reflection will show what enormous social changes would result from a gradual *disappearance of a rate of return on accumulated wealth*. A man would still be free to accumulate his earned income with a view to spending it at a later date. *But his accumulation would not grow.*[25] (emphasis added)

Furthermore, Keynes saw socialised capitalisation as the best solution to the ongoing need for small amounts of capital:

[I]t will . . . be possible for communal saving through the agency of the State to be maintained at a level which will allow the growth of capital up to the point where it ceases to be scarce.

I see . . . the rentier aspect of capitalism as a transitional phase which will disappear when it has done its work.[26]

In Keynes's view, ideal capitalism would lack capitalists. Wage-earners, having their basic needs fulfilled, would save enough earnings with the State for their retirement and such savings would provide the interest-free loans necessary for new investment. Interest-bearing debts would disappear.

Not only has Keynes's vision not come true, but the 'objectionable features of capitalism' have now spread around the globe and grown to unmanageable proportions, leading to our impending crises. There has been a centralisation of economic power in the hands of investors and management, who are closely allied with elected government and manipulate its decisions. Several interlocking factors are responsible for this state of affairs.

In our modern Western tradition, the main function of government has been the protection of private property. Indeed, the United States Constitution was designed with this purpose explicitly in mind, as James Madison so clearly stated in the Federalist Paper No. 10. Because the wealth of the few 'by rights' ought to be protected from the greed of the many, simple majorities must not be allowed to prevail. As political economist Creel Froman points out,[27] this creed has survived unscathed for 200 years, leading to the massive centralisation of wealth that now exists in the United States (as well, of course, as in other capitalist nations). He describes in detail the tremendous overlap between corporate management, financiers and government officials. In America – and probably in most capitalist countries – no political party can achieve power without the tacit approval of the corporate sector which effectively controls the economy.

Through the exponential growth of interest, capitalism accumulates wealth in the hands of fewer and fewer persons. In the absence of enough economic growth to provide significant 'trickle-down' to the poorer classes, there is an inevitable *transfer* of purchasing power from the poor to the rich. Sooner or later, government intervention is required to redress the inequality, either by progressive taxation and redistribution as 'welfare', or by permitting inflation. In the latter case, the poor, who are always in debt to the wealthy, can pay back what they owe in cheaper currency. In the 1980s, inflation has been checked by tight government controls on money, and welfare spending has been sharply cut in both Britain and America, with a consequent widening of the gap between the rich minority and the poor majority. Moreover, the newly created jobs that replace those lost to automation or exported overseas by multinational corporations are low-paying service jobs that often leave full-time workers with incomes well below the poverty level. Since the 1970s, the 'average' income has progressively lost in purchasing power, and the gap between an ever smaller group of very wealthy and a growing group of very poor is getting wider in almost every capitalist country.

In America the problem has grown so bad that the Catholic bishops denounced the situation as intolerable in a recent pastoral letter on the United States economy:

> [U]nequal distribution [of income and wealth] should be evaluated in terms of several moral principles we have enunciated . . . In view of these norms we find the disparities of income and wealth in the United States to be unacceptable. Justice requires that all members of our society work for economic, political and social reforms that will decrease these inequities. [28]

This indictment of inequality within the world's wealthiest nation becomes even more severe when one considers the disparities between rich and poor nations, a subject taken up in Chapter 13.

Western governments have further compounded the central problem of capitalism by making their own exorbitant demands on capital. In particular has been the maintenance of a 'warfare' economy since the onset of the Cold War in the late 1940s. During the past decade, military expenditures have soared and are still rising. Instead of paying its way through taxes, the United States government has chosen to borrow to pay its military bills, thus maintaining the scarcity of capital for productive investment and ensuring that the wealthy continue to have a high unearned income on the interest they collect. Wage-earners' taxes thus line wealthy pockets twice over: first as interest and eventually as principal on government debts.

Finally, the ability of Western economies to generate the excess wages and profits needed to pay back the interest owing on all their multiple public and private debts is rapidly waning. There is already an overabundance of productive capacity. Those who can afford them are satiated with cars, TV sets, and refrigerators. Those who want them are not earning enough to buy them. The system, having transferred all its discretionary wealth to the glutted affluent, cannot keep itself going. Either there will be massive inflation, destroying the wealth of the rich and the subsistence of the poor as it erases the debts of the masses, or massive deflation that brings all economic activity to a halt. Or, maybe, governments will finally 'take charge' and mandate highly planned and regulated economies.[29]

What is abundantly clear is that modern economic argument, deduced from elaborate mathematical theories untested by reality, is largely useless. In the words of Nobel prize-winning economist Wassily Leontief:

> Year after year economic theorists continue to produce scores of mathematical models and to explore in great detail their formal properties; and the econometricians fit algebraic functions of all possible shapes to essentially the same sets of data without being able to advance, in any perceptible way, a systematic understanding of the structure and the operations of a real economic system.[30]

TOWARD THE FUTURE

If we are to arrive safely at some future state of the world where human beings live within their global income and economic justice prevails, we shall need to change in two major ways. First we shall have to give up the notion that wealth is an end in itself, something to be admired and sought after. And second, we shall have to restructure our economic thinking to incorporate questions of justice in both the present *and* the future. Destruction of Earth's resources today is a burden being placed on future generations. Indeed, the future has become a 'new frontier'. Having outgrown the resources of Europe, of North America, and now of the Third World, industrialised societies are turning to the future as a new 'colony' for continuing their exploitative growth. If the future no longer counts, then you can do things today that you otherwise could not do.

Discounting the Future

Modern industrial societies, accustomed to rapid economic growth, find it hard to imagine an alternative future. The tendency is simply to say the future does not count. Take, for example, the notion of credit discussed at the beginning of this chapter. If I borrow $1000 today and promise to pay it back, at seven per cent annual interest, at the end of 10 years, I will in fact have to pay back $2000. What I am *really* saying is that a dollar ten years from now will be worth to me half what it is now. In 20 years, it will be worth one-quarter as much, and so on. (This is why debt, in the absence of economic growth, creates inflation; when the future actually arrives, no one wants to make the promised sacrifice.)

But real wealth, unlike money, does not inflate. What I 'borrow' from the future in terms of eroded topsoil or polluted groundwater can only be paid back by putting *at least as much* real matter and energy back into the system as I took out of it earlier. I cannot put half as much back and get a repaired Earth. The supposition, however, is that some new technological fix will come along to patch things up. It is not up to *me* today to worry about *them* tomorrow. If they are not clever enough to succeed and the human race comes to an end, so what?

Economist Robert Heilbroner quotes an unnamed British colleague who put this modern 'rationalist' position quite plainly:

> Suppose that, as a result of using up all the world's resources, human life did come to an end. So what? What is so desirable about an indefinite continuation of the human species, religious convictions apart? It may well be that nearly everybody who is here on earth would be reluctant to die, and that everybody has an instinctive fear of death. But one must not confuse this with the notion that, in any meaningful sense, generations who are yet unborn can be said to be better off if they are born than if they are not.[31]

This convenient argument – that it is the responsibility of 'the future' to look after itself despite what we do today – says bluntly that morality is merely a matter of personal convenience and expected reciprocity; it is just one more economic cost–benefit analysis. Embodied in this attitude is the final dehumanisation of a people dedicated utterly to self-interest. In its very emptiness, it is the beginning of the hopeless ethic of the Ik. It arrogantly supposes that we owe nothing to the past, that we have in fact created ourselves from scratch. It forgets that *we are our ancestors' future*! This truism was long ago understood by Native Americans, and is embodied in The Great Law of the Six Nations of the Iroquois: 'In our every deliberation, we must consider the impact of our decisions on the next seven generations . . . on those faces that are yet beneath the ground'. The industrial age is unique in human experience in its gross disdain for the future.

Moonrocks, Space Dust, and Other Nonsense

A great many who *do* care about the future still pin their hopes on some undiscovered technological frontier, and among the young in particular, that frontier is located in space. The moon will provide scarce resources. Feather-light orbiting factories, powered by solar energy, will manufacture goods from spacedust, and food will be grown in encapsulated space gardens where the sun never sets. These flights of fancy, published in serious technical journals as well as popular science glossies, view the 'colonising' of space as purely an engineering problem, with a smattering of space psychology and space medicine thrown in. Completely ignored are such matters as the military vulnerability of space stations. Will we have solved the problem of war by then? What happens when crew members violently disagree? Won't such highly-interdependent groups require either super-human restraint or the level of authoritarian control of a military combat team? And what about the absolute necessity for fail-safe systems in a 100 per cent technological environment? The cost of providing the relatively simply safety systems needed for our Earthbound nuclear power plants has proved so prohibitively expensive that we make do with less than adequate safety – as recent accidents have demonstrated.[32]

As far as permanent space colonies are concerned, recent evidence from extended space flights indicates that prolonged periods in space are hazardous to health. Two major dangers so far uncovered are frequent but erratic proton showers emitted by solar flares that would be disabling or even lethal to workers outside their space vehicles, and, even more serious, the massive loss of strength in bone and muscle that results during prolonged weightlessness. In just one week, adolescent rats lost 45 per cent of their bone strength and 40 per cent of their muscle tissue. Older rats showed only slightly less damage in this brief period.[33] Although the rates of loss are slower in adult humans, serious problems have been experienced by astronauts who have spent several weeks in space. The human body is adapted for life under a particular set of physical conditions, namely those of planet Earth.

The space-tech literature glosses over these and other tough questions. It never asks how economic, social and political injustice, alienation and meaninglessness in human life are to be resolved through space technology – or by any other technology for that matter. To colonise space successfully we shall first have to have mastered our non-technical, human problems. And if we solve them, we will not need 'space' – we will be living in harmony with each other and with Earth already.

Democratising the Economy

Once we admit that indefinite economic growth is not feasible and that constantly adjusting welfare payments is not a satisfactory solution to economic injustice, we begin to look about for radically different alternatives. We discard the flawed idea that the ultimate goal of life is the consumption of wealth, and begin to restructure our economic institutions to serve the *new* goals of economic and social justice, of human bondedness and shared experience. We shall have escaped from our destructive belief in the power of economic growth and technological fixes to 'solve' our problems.

How shall it be done? Why by democratic planning, of course. By communities of people agreeing on what they want and devising institutions to facilitate their goals. The word 'planning', however, usually sends chills down the spines of 'free-enterprise' Westerners. It smacks of state control and loss of freedom – or so those who stand to lose most from any restructuring of economic institutions have repeatedly told us.

But how much freedom to participate in the nation's economic decisions does today's average citizen really have? The answer is virtually none. As political scientist Creel Froman demonstrates in his book, *The Two American Political Systems*, the open political system in which citizens democratically elect representatives to do their will is superseded by the corporate political system that secretively controls public policy-making, both domestic and foreign. By making 'the economy' the highest good and putting the rights of private property above all else, capitalist nations tacitly place decision-making power in private hands.[34] No one is more 'unfree' than the individual who feels powerless to control the major economic and political decisions that affect his or her life.

The fear of planning nurtured in us is a smokescreen. We already have planning. It is carried out by large corporations and the financiers who invest in them. Advertising is planning; it 'plans' that a certain market will exist. If advertising did not work as 'planned,' no one would pay for it. Indeed, as shown in Part III, planning is a hallmark of all centralised industrial economies. The real questions are, *who* does the planning and what are *their* economic and social goals?

The problems of centralised planning

We are encouraged to believe that the only alternative to the centralised capitalist system is a centrally-planned command economy such as that of the Soviet Union. The seldom-mentioned advantage of such an economy is that it avoids the profit-driven necessity for growth and the propensity to sink into debt and runaway inflation. But it shares two weaknesses with Western capitalism that make it equally prone to failure in the long run.

First, although Marx recognised the importance of natural resources in the development of wealth, he paid far more attention to the contribution of labour, so that environmental issues tend to be forgotten in Marxist thinking and planning. Indeed, Marx's dialectical materialism forces a society to focus on a continuing accumulation of material goods. Hence the Soviet Union has problems of environmental destruction, pollution and resource depletion parallel to those of any highly industrialised society.[35]

Second, in its search for a rapid increase in material wealth, the Soviet economy has introduced the same sorts of giant technologies, with the same alienating effects on workers and the same built-in management overheads as exist in capitalist systems. The Soviets merely centralised their decision-making even more than did Western economies, and hence have suffered more from the consequences of managerial gigantism; it is an error they are now beginning to correct.

That centralisation is economically wasteful and socially destructive, far from being a dismal thought, offers great reason for hope. By *eliminating* costly organisational overheads, we might well produce sufficient worthwhile material goods while living well within our environmental income and enjoying our work as well. We begin to see the advantages of smaller, decentralised economies, organised according to ecologically meaningful regions, economies which have alternative goals to those of today's gigantic, centralised economic systems. E. F. Schumacher called it 'economics with a human face'.

An alternative economics

Future economies will function within their environmental incomes. Today's multinational corporations that move freely about the globe searching for ever-greater profits have no incentive to do so, but *local* people do. Future economies will concentrate on maintenance of capital stocks, not the ever-increasing throughputs demanded today to pay unending interest to private investors. Future economies will provide fairly-paid jobs for everyone. Economic decision-makers today have no such obligation. When the economic decisions that affect people are not made by the people themselves, both they and their environment are likely to suffer. To correct all this, future economic decisions must be made locally, by the people concerned. This translates into decentralised, participatory democracy. As Yale University sociologist Charles Lindblom points out, the anti-democratic behaviour of large private corporations has been a remarkable anomaly in Western democracies:

> It has been a curious feature of democratic thought that it has not faced up to the private corporation as a peculiar organization in an ostensible democracy. Enormously large, rich in resources, the big corporations... command more resources than do most government units. They can also, over a broad range, insist that government meet their demands, even if these demands run counter to those of citizens... Moreover, they do not disqualify themselves from playing the partisan role of a citizen – for the corporation is legally a person. And they exercise unusual veto powers. They are on all these counts disproportionately powerful... The large private corporation fits oddly into democratic theory and vision. Indeed, it does not fit.[36]

While the precise structure of future democratically organised local economies cannot be specified, since each local region and each local community must evolve according to its own environmental constraints and social goals, some broad principles can be identified.

Public ownership of capital The first step is to eliminate the constant drain of wealth to the private owners of capital which necessitates the unceasing growth of throughput. When capital is owned by society as a whole, both the rate of accumulation of capital and the kinds of new enterprises to be undertaken become social decisions. New investments are no longer made to generate profits for one sector of society, but to fulfil such social goals as maintaining the environment, providing medical care and schools, renovating urban centres, researching new energy technologies, and so on.

Sixty years ago, Frederick Soddy argued that all productive capital should become public property, which enterprises then rent at a fixed rate. Profits over and above those for plant maintenance, salaries, and materials costs are subject to taxation according to the needs of society, as is any other income. The original private owners of capital stocks are bought out by government, their taxes being paid in shares of their capital holdings. Eventually, private ownership of productive capital is thus eliminated.[37]

The rents paid on public capital provide the means for funding new enterprises that are compatible with both social goals and environmental constraints. Since optimisation of profits is no longer the prime criterion for investment, public funding would generate the socially needed activities that today's profit-motivated investors are loathe to support. Although government owns the capital stocks, it does not run businesses. These are left in the hands of private management, which is free to develop each enterprise in its own way. Such a system does not parallel the command economies of Eastern Europe, which rigidly specify all details of an enterprise, but rather resembles the way government-owned enterprises have been independently managed in several West European countries. Applying for public funds to begin a new enterprise could be done in a manner similar to the way scientists apply for funds to support their research. Each project would meet selected criteria of social worth and environmental sensitivity, and the proposers would naturally need to demonstrate skills commensurate with the task undertaken, but there would be enormous latitude for initiative and creativity.

Public participation in decisions Competition, rapid change and technological efficiency – as shown in Part III – demand order, organisation and action. The most efficient way to run an army is to have a few clever strategists in charge and a mass of unquestioning, obedient followers. If every decision is put to a vote, the battle is lost before it is engaged. More than ever, the global economy resembles a battlefield – Hobbes's world of 'all against all'. Only today it is giants who are competing: Chrysler against Ford, GM, Nissan, Hyundai, Renault and BMW; America against Japan, Europe and Korea. Everyone is in an economic war with everyone else. Under the global influence of unrelenting competition, the whole world is becoming an economic armed camp, run on the same autocratic premises that have lent their logic to ever more efficient war-making.

If societies insist on engaging each other in all-out economic competition, then the rules of battle are appropriate and the consequent destruction and suffering commensurate. But, as with modern warfare, it is increasingly doubtful whether economic competition has any ultimate 'winners' or whether the social and environmental sacrifices made in the name of economic efficiency make sense at all any more.

The alternative to 'efficient' top-down decision making by a select few is bottom-up decison making by a slow, argumentative, participatory process where everyone has a voice. Public dialogue is not the language of the battlefield, but of the Greek agora, the Roman forum, the New England town meeting. Popular participation and dialogue work slowly – as do the mills of the gods – but like them, they grind exceedingly fine; they produce long-lasting decisions with a moral content that embodies the best will and wisdom of the time. It is out of this sort of dialogue that new social visions emerge: new shared values, new worldviews that provide the meaning in life which is missing today.

Economic decisions, being based on a set of values, are in fact *political* acts. And if new values are to be brought into the process, it will require *active* political participation by people generally. A small group of 'experts' cannot restructure an entire society. Broad-based participation requires *decentralisation* in order to happen: local communities making local economic decisions, based on the twin aims of social well-being and environmental stewardship. Only through participatory democratic politics can the exploitation of both people and the environment be eliminated. As alternative economists Gar Alperovitz and Jeff Faux argue: 'The rule is that if a public function can be performed at a more local level, it should be'.[38]

A natural consequence of local autonomy in economic planning is the emergence of local economic self-sufficiency. This is an important result not only for economically depressed communities within giant industrial nations; it is also critical to the healthy, independent development of Third World countries, as shown in Chapter 13.

Worker self-management   Even the smallest economic production unit of all, the firm or company, will have the hallmarks of democracy rather than today's employer–employee hierarchy. Ideally, workers will democratically elect a management team to make company policy and carry it out with the help of a hired director. As workers gain a say in the organisation of their own labour, the traditional friction between the goals of employers and those of employees will disappear.

Some beginnings

When first met, such ideas as public ownership of capital and worker self-management sound so revolutionary they tend to be dismissed out of hand. Yet workable examples already exist. Once something *has* happened, one can imagine it happening again in a wider context.

State financing of enterprise is not new; it had a respectable 20 year history in the United States, in the form of the Reconstruction Finance Corporation (RFC). Begun by Herbert Hoover to rescue floundering corporations in the Great Depression, the RFC expanded under Roosevelt, who eventually used it as an instrument of economic planning during the Second World War. By 1945, the government owned iron and steel plants, half the aluminum capacity, and almost all synthetic rubber production. Only under Eisenhower was the agency abolished. Proposals for an equivalent federal lending institution have recently been urged by Felix Rohatyn, the financier credited with rescuing New York City from bankruptcy, by Senator Edward Kennedy, by the big labour unions, and by economists Lester Thurow and Robert Reich.[39] If the federal government of the world's largest democracy can act as a source of investment capital, there is surely no logical barrier to local governments undertaking the same service.

An example at the local level of a grassroots cooperative economy is the small Basque community of Mondragon. It began when a Catholic priest founded a technical school there. In 1956, its first graduates formed an industrial cooperative; 20 years later, there were nearly 60 worker-managed cooperatives, producing a variety of high-technology goods and employing 14,000 people. The capital for these enterprises originates from the workers, and is deposited in the local bank, a savings and investment institution controlled by the cooperatives themselves. Since all these institutions – the bank, the school and the manufacturing firms – are part of the same network, the whole behaves as a single democratic, highly egalitarian community. Wage differentials between the highest and lowest paid workers never exceed a ratio of 3:1 – in sharp contrast to industrialised democracies where it is 10:1 or even higher.[40]

Among single companies, successful experiments in worker ownership–management have existed for decades. Best known is the Scott Bader Commonwealth Ltd, a chemical manufacturing company in England. Its founder always had an aversion to one man hiring another for wages, so in 1951, he and his workers together formed a commonwealth, with a binding constitution and agreed-upon rules. Among these were a narrow range of wages from top to bottom, a limitation on company size, and a limited profit level. At least 60 per cent of earnings go for taxes and reinvestment, and 20 per cent are given to local charitable institutions; only 20 per cent are paid to worker-owners as bonuses or dividends. No member can be arbitrarily fired, and the Board of Directors is responsible to the worker-owners as a whole – the Commonwealth. It is a mini-democracy that has proved remarkably successful for over a quarter century, during which time it has spawned three similar but quite independent companies.[41]

Regarding economic decentralisation, the Swedish parliament in the 1980s has pursued a policy aimed at developing local democratic control of regional industries. Instead of take-home wage increases, workers' increased earnings are paid into a public fund in each of Sweden's 24 counties. Administered by a citizen-elected board, the funds are being used to buy shares in local industries, which eventually will be publicly controlled. Local economic decisions will then be put in the hands of a democratically elected public board. Despite resistance from Sweden's corporate heads, the idea was accepted; 90 per cent of Swedish labour is unionised and most has been solidly behind this new idea. Since Sweden is a small country, the plan will indeed make for meaningful local economic control.[42]

In the United States, numerous communities, impoverished by the closing of their major factories and ignored by the federal government, have been forced into economic self-help programmes. Repeatedly the key has been local self-sufficiency and self-reliance. By utilising local resources, including empty plants and unemployed labour, they are able to substitute costly imports of energy, raw materials, and manufactured goods with locally produced equivalents. The city of St Paul, which imports 85 per cent of its fresh produce, is stimulating local growers by sponsoring farmers' markets. It also is developing technologies to convert local wastes into fuel for industries and homes, cutting down on its fuel imports while creating thousands of jobs. The city of Chicago will rely on local businesses as much as possible for the goods and services needed to run the city.[43] So far unnoticed by the mainstream press, let alone university departments of economics, management, and urban planning, dozens of American communities are quietly undergoing conversion to local economic independence.[44]

A different impetus for community economic reorganisation has come from the peace movement. Because of the arms race, the economies of many cities now rely on the continued funding of local military bases or on military-related contracts to local industries. As Figure 12.1 suggests, it is almost impossible to cut defence spending since it means loss of jobs. To counteract this justifiable fear, numerous American cities that rely heavily on defence contracts – Seattle, San Diego, Boston, the 'Silicon Valley' area south of San Francisco – are developing 'economic conversion projects'. Their goal is to retool factories and retrain workers in order to convert from military-related production to civilian production. By bringing together concerned citizens, workers and management of the affected firms, engineers, local politicians, city planners, and financial advisors, the projects hope to work out the steps by which such economic conversion can occur. The whole community, not just those directly affected, assumes the burdens of economic change. Seventy American cities have already passed Jobs with Peace initiatives and are seeking ways to begin the conversion process.[45]

"AT LAST! A WEAPONS SYSTEM ABSOLUTELY IMPERVIOUS TO ATTACK:
IT HAS COMPONENTS MANUFACTURED IN ALL 435 CONGRESSIONAL DISTRICTS!"

Source: John Trever, Albuquerque Journal, 20 January 1985.

Figure 12.1 The political economy of 'national defence'.

Both economic self-help and conversion projects are local community efforts that could readily evolve into more permanent forms of democratic economic planning and development. A major problem for all communities seeking economic independence is their vulnerability to outside economic forces, particularly through their need to acquire and retain money to facilitate exchange of goods and services within the community. During periods of depression when money is scarce, barter economies tend to spring up locally. People who have no money simply do things for each other. A grapevine develops by which skills, tools and surpluses are put in touch with those who need them. It often means the difference between severe hardship and simply being without much money.

Barter economies, however, are necessarily limited in size. A more flexible alternative is community membership economies known as Local Exchange Trading Systems (LETS). A local form of 'money', recorded in the personal accounts of members, is exchanged by telephone calls to a book keeper; at each transaction, one account is debited and another credited. Along with one's monthly statement come classified lists of goods and services needed and of those available. LETS thus makes it easy for buyers and sellers to get together, helping to alleviate unemployment and scarcity. New members are given an amount of interest-free start-up 'credit' to use before they begin 'earning'. And with modern electronic book keeping, overheads of the system can be kept quite low. LETS is already in use in parts of British Columbia.[46]

Both barter and LETS are economies that are created by people themselves and exist as independent alternatives to the giant economic systems over which individuals have no control. As Schumacher once said:

> The really helpful things will not be done from the centre; they cannot be done by big organisations; but they can be done by the people themselves.[47]

SOME CONCLUDING THOUGHTS

Rethinking economics goes far beyond juggling concepts like 'productivity', 'efficiency' and 'consumption', or worrying about how to manipulate rates of inflation, unemployment and interest. It means changing our whole understanding of what economic activity is *for*. Economic decisions, being value-laden, are *political* decisions. No longer can we pretend economics is an independent science, disconnected from human values on the one hand and physical reality on the other.

Modern competitive economic behaviour, carried on purely for the sake of profits, inevitably exploits both people and their environments and is ultimately destructive of both. Increased economic throughput is no longer an appropriate solution to economic injustice. What is required is the democratic restructuring of economic institutions such that socially agreed upon economic needs are met within the environmental income of each community. This is best accomplished through decentralised economic arrangements that permit local people to match their perceived needs to the natural resources that support them. Local self-sufficiency thus emerges naturally.

Such an economy demands cooperation rather than competition among individuals, and a renewed sense of strong bonds and shared goals. It also demands a renewed sense of respect for our planet. Earth is more than a source of wealth for human exploitation; it is a living entity and we are but members of its total life community. Finding the will and the means to seek a new vision is the historical challenge of our time. Surely we are living in what Hannah Arendt called an 'odd in-between period which sometimes inserts itself into historical time . . . an interval in time which is altogether determined by things that are no longer and by things that are not yet. In history these intervals have shown more than once that they may contain the moment of truth'.[48]

There is an air of expectancy that something is about to happen – and of uncertainty whether we shall survive whatever it is. In Chapter 13 we follow this thread of uncertainty further, from the Western industrialised world, living today beyond its environmental income, to the Third World where the economic and social problems are different, yet closely related.

13 Defusing the Global Powder Keg

In Mexico, you must be either numb or very rich if you fail to notice that 'development' stinks. The damage to persons, the corruption of politics, and the degradation of nature which until recently were only implicit in 'development,' can now be seen, touched, and smelled. The causal connection between the loss of healthy environment and the loss of peasant solidarity... has now been documented by a new, expert establishment. The so-called crisis in Mexico has now provided the peasants and others with the opportunity to dismantle the goal of 'development.'

Gustavo Esteva, 1987[†]

INTRODUCTION

If we look at a map of the world and ask – Where is armed conflict going on? Where are populations growing fastest? Where is there the greatest poverty? – we discover that all maps coincide. Of the 120 armed conflicts taking place in 1987, all but four were in the so-called developing countries of the Third World.[1] Likewise, the highest rates of population growth are located in those same countries, and there also are found the lowest levels of literacy, of *per capita* income and of nutrition, and the highest levels of infant mortality and epidemic disease.

Who *are* these 'Third World' countries? At first, the term encompassed only those emerging states who refused political alignment with either the capitalist West or the communist East, the First and Second Worlds. But eventually 'Third World' came to include all those former colonies, still not industrialised, that were entering the global family of nations in the years following the Second World War. Although Westerners still tend to divide the globe along the East–West, communist–capitalist axis, emphasis in the future will far more probably focus on the North–South, rich–poor axis. The industrialised countries of the North are home to about one billion people, whereas four billion live in the poorer, underdeveloped countries of the South; moreover, during the next 30 years or so, the population in the South will double. 'World' history – as taught in the West – studiously ignores these people, treating them as 'backward', 'uncivilised', and of little importance. Yet the future history of humankind will largely depend on how these people develop and the quality of life they experience. In a world of electronic media, they are no longer isolated and ignorant.

Furthermore, despite their diverse cultures, almost all Third World peoples share one primary and overriding historical experience: *generations of colonial exploitation*. Nor has political independence, mostly achieved in the three decades following the Second World War, brought a significant change in their economic exploitation, today's 'neocolonialism'. The triple threat of exploding population, continuing poverty, and access to modern weapons makes these peoples – by far the great majority of the human race – a powder keg, ready for ignition at any moment.

To perceive the origins of this present state of affairs requires a review of the colonial process and its modern legacy of distorted social structures, economic dependency, and political instability. Such a perspective is prerequisite to understanding the options available to these developing countries as they struggle to resolve their formidable problems.

THE COLONIAL STORY

Since the days of Vasco da Gama and Christopher Columbus, Europeans have colonised, one after another, almost every habitable corner of the world. Of today's major Third World countries, only China and parts of the Middle East were never actually part of a European empire. But they might as well have been, for they became semicolonies, controlled politically and economically by imperial foreign powers.

As pointed out by global historian L. S. Stavrianos, only the capitalist West evolved the sort of worldview that made such overseas colonisation possible. The Chinese, who sailed to distant lands long before the Europeans, were never interested in profits from foreign trade, contenting themselves with acquiring a few curiosities. It was, rather, the emerging European mindset, discussed in Part III, that created overseas imperialism. The mercantilist desire to accumulate wealth fuelled the conquests by Europeans that were to encircle the globe.[2] This period of colonialism, lasting over 400 years, was marked by several forms of imperialism, some of which are still widely prevalent today. It is therefore necessary to identify the various forms in which imperialism occurs.

The Guises of Imperialism

Imperialism generally conjures up geopolitics, a global map in the days before the Second World War when one-quarter of the land mass was coloured pink for the British Empire, and other large areas were green for the French, and so on. By 1900, as Stavrianos puts it:

The greatest land grab in human history ended with the extraordinary spectacle of an Eurasian peninsula dominating the rest of the world.[3]

Table 13.1 Overseas colonial empires in 1914

Countries having colonies	No. of colonies	Area (square miles)		Population	
		Metropolitan country	Colonial countries	Metropolitan country	Colonial countries
United Kingdom	55	120,953	12,043,806	46,052,741	391,582,528
France	29	207,076	4,110,409	39,602,258	62,350,000
Germany	10	208,830	1,230,989	64,925,993	13,074,950
Belgium	1	11,373	910,000	7,571,387	15,000,000
Portugal	8	35,500	804,440	5,960,056	9,680,000
Netherlands	8	12,761	762,863	6,102,399	37,410,000
Italy	4	110,623	591,250	35,238,997	1,396,176
Total	115	707,116	20,453,757	205,453,831	530,493,654

From L.S. Stavrianos, *Global Rift*, (New York: Morrow, 1981), p. 264, originally published in L.S. Stavrianos, *The World Since 1500* (Englewood Cliffs, NJ: Prentice–Hall, 1966) p. 236.

Table 13.1 shows the major colonial empires in 1914. In almost every case, both the colonial area and the subject population were far greater than those of the metropolitan 'mother' country. Colonial governments were usually headed by European nationals, backed by a well-armed military.

This period of *politico-military imperialism* coincided with the gigantic growth of European industrial capitalism, and resulted in simultaneous *economic imperialism* where the resources of the subject country were exploited for the benefit of the imperial power. While living standards in the West soared during the colonial era, nothing of the kind happened in the colonies. But although political imperialism makes economic imperialism simpler, it is by no means necessary. Former colonies still serve the economic enrichment of the Western capitalist nations through the process of neocolonialism.

Finally there is *cultural imperialism*, the most subtle and permanent of all forms of imperialism, for it destroys the cultural integrity of a people, making them subservient to the values and beliefs of an externally imposed worldview. Ethnocentrism, a characteristic of every society, was carried to an extreme by Europeans, with their highly developed belief in their own racial, cultural and moral superiority. This attitude was finally fanned into the searing flame of 'scientific certainty' by such social Darwinists as Herbert Spencer. The 1899 Portuguese 'Code of Work' expresses bluntly the relationship Europeans felt toward their colonial subjects:

> The state, not only as a sovereign of semi-barbarian populations but also as a depository of social authority, should have no scruple in *obliging* and if necessary *forcing* these rude Negroes in Africa, these ignorant Pariahs in Asia, these half-witted savages from Oceania, to work.[4]

Colonists often imposed their own religion as the initial step in cultural imperialism. In Latin America, Spanish soldiers murdered natives who refused to adopt Catholicism.[5] What schooling there was, was carried on in the European tongue, in which all official business was conducted, and the lessons were specifically designed to fit pupils for subordinate positions in the colonial system.[6] Young natives being groomed for higher bureaucratic positions were sent to European schools. This Westernised elite became a subclass of the white colonists, holding minor official positions while acquiring Western tastes and a capitalist worldview. Their goal became to mimic and eventually be accepted by the white population, but few succeeded: the colour-barrier was constantly raised the more 'successful' the non-whites became.

As social psychologist Thomas Gladwin observes, in the face of the white man's self-assured belief in his own superiority, many subject peoples finally came to believe in their own inadequacy. They struggled, and largely failed, to comprehend capitalist values, falling instead into a cultural limbo of discontented submission. As Gladwin notes:

Almost all white men, even those who oppose the capitalist system and its exploitation of the poor by the rich, have taken for granted the universality of capitalist values: that all human beings are at heart individualistic, and all are motivated utimately by their own self-interest. Indeed, it is commonplace by now to accept these qualities as at the core of 'human nature,' and therefore inescapable.

Yet before the white men came, most of the people of the nonwhite world did not live by these values. Their lives and transactions were guided by the moral obligations of kinship and mutual interdependence.[7]

The end of political imperialism did not bring an end to cultural imperialism (which of course has strong links to economic imperialism). Philosopher Sandra Wawrytko summarises the situation in the 1980s:

[T]he most widespread and visible form of cultural imperialism in the world today is without doubt the 'Americanization' of global culture. Begun in earnest in the aftermath of World War II, this movement has been fostered by the spread of American goods and services, technological expertise and social values. The symbolic role of Coca Cola in this invasion is slowly being usurped by the ubiquitous Golden Arches of McDonald's, but the list of influences seems almost endless – from rock-and-roll to home computers, blue jeans to jogging.[8]

Cultural imperialism – the imperialism of beliefs and attitudes – is still in progress. Yet the same electronic media that advertise the material wares of capitalism are also teaching the world's people about the nature of imperialism and the meaning of nationalism and self-determination. Cultural imperialism is becoming more and more widely resented globally.

The Stages of Colonial Dependency

Drawing together all the complex factors that interacted during the colonial era is a formidable – perhaps impossible – task. Nevertheless, certain stages in colonial dependency occurred repeatedly, and were experienced by almost all colonised peoples. They were the development of a money economy, of economic dependency, and eventually of a dual economy.

The money economy

Although Pizarro, Cortes and the others sailed in search of gold and silver, the real value of the lands they conquered proved to be a harvest of cheap labour for extracting abundant raw materials – the basic components needed to generate material wealth. Coopting such resources meant imposing not only military force but also the use of money on societies accustomed to quite different economic arrangements, ones based on kinship relations and tribal traditions. The money, of course, was the species of the conquering nation, and the need for it began with the imposition of taxes. Land that once belonged to the local people was now 'owned' by a distant European monarch who required tribute. This was duly collected by local agents of the crown.

At first, land taxes were paid 'in kind', in produce or handicrafts, which the local chief turned over to the white agent. For acting as tax collector, he received a few coins, and whatever he did not convert into jewellery for adornment he exchanged with white traders for baubles, for cotton cloth, for steel knives. The rest of the people received nothing in return for their tribute except the right to stay on land they already regarded as their own. The cost to Europeans of obtaining colonial wealth was negligible.

Eventually a taste grew among a few native traders for European goods, and they encouraged others to sell more of their produce, in exchange for copper or silver coins. Again, most of the coin became jewelry – such as the lovely filigree pieces common throughout southeast Asia. As 'money', it still had no meaning. The jewelry might later be sold to white traders for food when times were poor, but it was not yet an article of trade, something to be 'saved' or 'invested'.

To augment the flow of wealth back to Europe, colonies were turned into markets for cheap manufactures from the metropolitan country. Propaganda and advertising created a 'need' for Western goods, first for colourful textiles, beads and metal tools; later, for seeds and fertilisers, for firearms, and recently for radios and TV sets, soft drinks and blue jeans. To stimulate their consumption, local people were encouraged to go into debt. In return for money in advance, they promised a sizeable proportion of their harvest later on. The capitalist system first created a 'need', and then a debt. And so a compulsory money economy came to prevail. For most natives, the debt was to become permanent. More importantly, *they had no control whatever over the prices received for their produce or handicrafts.* They became an economically dependent people.

As their debts mounted, people spent more and more time producing cash crops and handicrafts for sale instead of food and goods for themselves. When they ran short of food, they sank even further into debt merely to survive. When they could not pay taxes, the crown sold their lands to white settlers or wealthy natives, who raised cash crops for export. The displaced peasants were forced to work for barely subsistence wages on giant plantations. Whoever failed to find a job there migrated to the growing urban centres to become a servant of the rich, a factory labourer, or if these failed, a beggar.

Economic dependency

Whether based on an economy of small villages and peasant agriculture or an elaborate, hierarchical civilisation, all precolonial societies were economically diverse and internally self-sufficient. The colonial experience – the imposition of an *external* money economy and the switch to dependence on one or two minerals or cash crops – created an extreme degree of economic dependency. Instead of developing a diverse trade with many different countries in the global marketplace, each colony was restricted to one or two commodities and trade with a single metropolitan nation, which set prices and imposed its own justice. The net effect was a massive transfer of wealth from the colonies to Europe.

Attempts by the colonies to diversify their economic base or develop their own manufactures were repeatedly blocked. European nations even forbade their own skilled workmen to emigrate to America or other 'underdeveloped' regions. As early as the 1500s, Portugal refused Ethiopia's request for the means to start up manufacture of swords and muskets and of textiles and books. In the 1700s, England forbade her craftsmen to venture to the independent West African nation of Dahomey to establish their trades.[9] It was the British economist David Ricardo, arguing for the supposed 'comparative advantage' of free trade among nations, whereby each nation specialised in what it could do best, who put a stamp of 'scientific' approval on the wide economic disparity that had developed around the globe by the 1800s. By then, Britain's industrial advantage was so great that countries like Portugal, who were just starting to manufacture, could not compete without protective tariffs. Using Ricardo's justification, Britain insisted on free trade, and Portugal even today remains largely an agrarian nation.[10] Thus does a so-called 'free market' give an extraordinary economic advantage to whoever is technologically ahead, which is why today it is so widely touted by Western business interests.

Yet those who are ahead readily impose tariffs when it seems to their advantage. During the early twentieth century, Britain manipulated tariffs to protect her textile mills from Indian calicoes and Irish linens, helping to destroy the chance for either country to gain economic independence. Mohandas Gandhi, as part of his Indian independence campaign, instituted a programme of home spinning and weaving so that local people could once again provide their own cloth instead of being forced to buy British imports.[11] It was one of many strategies he employed to win back for his people a sense of autonomy and self-sufficiency.

Thus did the Western imperial nations carry out a deliberate policy of coerced underdevelopment. Their goal was to sell *manufactures* in the colonial markets, *not* to promote self-sufficiency. To a great extent that goal, although more subtly applied, remains unchanged today.

Dual economies

The final consequence of colonialism was the emergence of 'dual economies'. Whether the original society had been a hierarchical caste system as in India, an egalitarian agrarian society as in Java, or something in between such as the tribal empires of West Africa, imperialism repeatedly led to the same end result.

In the cities lived the metropolitan nationals who ran the colony and the well-to-do native functionaries who served the colonial system. Together with rich plantation owners, these people formed an economic, political and social elite whose values were those of upper-class Europeans and who often attained a commensurate lifestyle. By the twentieth century, these urban colonial centres were oases of Western technology. Outposts of the Western world, they absorbed whatever wealth remained in the colony, paralleling the economic growth that was occurring in industrialised countries.

In stark contrast was the great majority of people, rural peasants and urban poor, isolated from any access to wealth. With their prior culture disrupted and without access to new skills, they became the 'other' economy, people struggling to survive in conditions of degradation and apathy.

As Stavrianos summarises:

> [T]he new agricultural and industrial technology of the West, together with its banks, joint stock companies, aggressive national monarchs and driving capitalist spirit, gave that hitherto retarded Eurasian peninsula superior economic and military power in its dealings with the rest of the world. The interests of the weak inevitably were subordinated to those of the strong – by mercantilist regulations in the earlier centuries, by free trade after the Industrial Revolution, and by neocolonialism today.[12]

Colonialism destroyed local social and economic patterns by imposing a money economy. By concentrating on a few cash export commodities, it created an economic dependency that was maintained by active suppression of local technological development. And it fostered the emergence of dual economies, of a minority of European-oriented urban elites and wealthy landowners, and a dispossessed majority of illiterate poor. Every recently independent ex-colonial country carries these legacies of colonialism to some degree or other. By examining the histories of three quite different colonial experiences, one can see how all converged toward the same final pattern.

The Americas: 'Temporary' Colonies

The first peoples to be significantly 'colonised' were those in the Americas. The English, Dutch and French settlers in North America – later joined by massive waves of other immigrants – slowly but methodically overwhelmed and destroyed the native tribes. Today's remnants live either in the rugged reaches of northern Canada or on circumscribed 'reservations'. In neither case do they constitute a political unit – a 'nation' – with a recognised voice in global affairs. In North America, colonialism did not exploit the Indians; it simply disposed of them.

The Latin American experience was different. The original Spanish and Portuguese invaders had come to plunder, not to settle. Through arms, disease and psychological intimidation, they quickly destroyed the sophisticated Aztec and Incan civilisations. The Indian population in Spanish America fell in less than a century from 50 million to a mere 4 million, mainly from the ravages of smallpox, typhus and measles.[13]

Yet some conquistadors did stay, and were rewarded by the crown with vast land grants – the haciendas. Since few women accompanied the colonists, a mixed society arose with a hierarchy based on colour. White-skinned Europeans were the landed aristocracy; light-skinned *mestizos*, the middle-class servants, traders and artisans; and the dark-skinned Indians and imported Africans, the serfs who worked the land. This new feudal hierarchy was solidly buttressed by the Catholic Church. As British economist Joan Robinson observed:

> [L]arge land holdings of this type are not propitious to technical development. The owner of a large estate can get a handsome income from a low level of production per acre, simply because he has a large number of acres. This system gives little incentive to landlords to increase productivity through investment and modernisation. Indeed, it is hostile to improvements that would raise the standard of life of the cultivators and make them less helpless.[14]

Thus did Latin America stagnate under this new feudal system. Nor did the political revolutions throughout South and Central America that occurred in the early 1800s change the situation. Although they began with peasants rising up against their local overseers, the revolutions were quickly coopted by the aristocracy under the banner of nationalism. Political independence enabled landowners to escape the European conscience that was being raised against exploitative colonialism.[15] As Stavrianos notes, feudalism quickly returned:

> After independence was won, suppression of the social consequences of this mobilization became the first order of business for the new republics.[16]

The United States, recently independent itself, was relieved to have the European presence expelled from the New World, and proclaimed as much in the Monroe Doctrine of 1823. The descendants of the original Latin American colonists, their numbers enhanced by further immigrants from Europe, became an elite land-owning and urban class that remains in power to this day. With few exceptions, dual economies still prevail in this 'neocolonial' era.

India: The Manipulation of Hierarchy

When the British East India Company began its trade in Indian textiles in the early 1600s, India was ruled by powerful Moguls, but their increasing abrasiveness lost them power, and most of the provinces came under the control of the Hindu majority. The ruling Hindu castes soon grew wealthy on an increasing export trade of tea, indigo, cotton and jute that supplied the growing British manufacturing industries. The local British traders established fortified warehouses, manned by Indian guards, throughout the politically unstable country, and Hindu and British merchants joined together to command total economic power. The East India Company was even given the right of local taxation and control over domestic as well as foreign trade. Gradually the Company was acquiring political power as well as economic hegemony.

So oppressive did the East India Company become that even some of its British employees were shocked. As Richard Beecher wrote in 1769:

> It must give pain to an Englishman to have reason to think that since the accession of the Company to the Diwane [right to collect taxes] the condition of the people of this country has been worse than it was before ... This fine country, which flourished under the most despotic and arbitrary government, is verging towards ruin.[17]

The distant British government complained vociferously, but was powerless to interfere. By the mid-1800s, the Company was the dominant political power. By making loyal followers of the militant Gurkhas and Sikhs, it turned what had begun as economic hegemony into total political control.

Not surprisingly there were numerous uprisings, including the famous 'mutiny' of 1857–8, when native troops and exploited peasants turned against their British masters. This led to the India Act of 1858, by which India formally became part of the British Empire, the 'Star in the Crown'. All these uprisings were successfully suppressed by a handful of British because India lacked central organisation. The Hindu caste system provided a ready-made mechanism for exploiting great masses of people, and the elite class remained divided, separated by a multiplicity of languages and by provincial rivalries which the British actively encouraged. Even during the final struggle for independence, the British made a last attempt to 'divide and rule' by fomenting fear and mistrust between Muslims and Hindus, which culminated in the bloody massacres and the ultimate partitioning of the country after independence.[18]

Yet independence scarcely changed the social and economic structure in India. British-trained elites headed the Congress Party that rode into power on a grand wave of nationalism. It left British investments intact and carried on the same system of government. And a dual economy remained: a wealthy minority of urban professionals and landowning elites, and an impoverished majority dependent on an agrarian-based, commodity-exporting economy.

Java: The Destruction of Agrarian Village Life

Unlike the ancient civilisation of the Indian subcontinent, the peoples of the East Indies islands were organised into loose federations of farming villages under the nominal head of a local chief. The Dutch economist J. J. Boeke captures the essence of precolonial Javanese village life:

> The village is a large household. Not only the sale of land, cattle and houses, butchering and harvest, but also marriage, divorce, education of orphans and the supervision of morals are matters of official concern. The village is not primarily a center of work, of production, but an abode, and only when abiding does a human being truly live.[19]

> Every person in need – from whatever cause – can count on help; such is the power of the communal tie . . . [E]very prosperous person has to share his wealth right and left; every little windfall must be distributed without delay. The village community tolerates no economic differences but acts as a leveler in this respect, regarding the individual as part of the community.[20]

Into this egalitarian world came the Dutch East India Company, even more powerful than its British rival. In the usual pattern, it usurped the land for plantations, employed native workers and established a money economy. Even so, Boeke reports, the Javanese were slow to adopt the capitalists' outlook:

> [There] is absolutely no spirit of rivalry, no competitive struggle. No one grudges another in the least a better price, a higher wage, and no one bothers to ask himself why such advantages fall to that particular man's lot or whether they would be attainable for others.[21]

Relations between the European traders and the local peoples remained purely economic. The Dutch, whether permanent settlers or transient *trekkers*, retained their urbanised Western lifestyle, while the Javanese clung to their own beliefs and cultural patterns. As Boeke says:

> Each lives in a world almost entirely closed to the other – and by closed we mean not only 'unknown' but more: incomprehensible and unattainable.[22]

Only slowly did the capitalist forces at work in Java create a dual economy. Those villagers unable to pay the taxes on their land were forced to seek wages working on plantations or in the urban centres, becoming dependent on the capitalist system. Meantime, better-off natives began to find places as petty government officials, while Chinese immigrants filled a vacuum as merchants and middle-men in the emerging commercial economy. The growing export of handicrafts brought wealth to a few but destroyed traditional village economies. The result once again was a highly polarised dual economy dependent on Western capitalism.

> [T]here are two groups in the Oriental community sharply distinguished in respect to their contact with the West. First, the masses, for whom that contact is an emergency measure, who do not expect to profit from it but who can no longer find adequate means of subsistence in their own environment; and secondly, individuals working their way out of the masses, to whom Western society has become a land of promise, who feel they have outgrown the village.[23]

These three examples – Latin America, India and Java – widely different in their precolonial cultures, all converged under European colonialism toward a final condition of economic dependency within the global commercial system. Each ended up with a dual local economy comprising a small percentage of well-to-do urban elite and wealthy landowners linked to the external capitalist world, and a great mass of poor wage-earning labourers and landless peasants.

Similar stories could be recounted for other colonial societies. Particularly unfortunate was West Africa where, once the sources of gold and ivory dried up, the major export 'commodity' became slaves. Although indigenous 'slavery' existed, it was a form of serfdom where slaves had certain rights and status – totally unlike the conditions imposed later on by white slave-owners. During the period of the slave trade the main exchange merchandise was firearms, which facilitated the capture of more slaves. Any tribes who failed to participate, and so earn weapons to protect themselves, were at the mercy of their predatory neighbours. The whole economy of West Africa became distorted by the slave trade, and on its abolition unsold slaves were pressed into local commodity agriculture, such as palm oil and indigo, in an effort to offset lost earnings.[24]

Another, less widely appreciated, legacy of colonialism is that it created artificial states. Today, there are only 168 'countries' in the world, but there are more than 3000 *nations* – peoples with distinct languages, traditions and worldviews.[25] Colonialism imposed on these disparate peoples an artificial political 'unity' that ignores cultural diversity. The straight lines on a modern map of Africa tell the story. Among the handful of states in sub-Saharan West Africa over 200 languages are spoken, 60 in Nigeria alone.[26]

The multiple problems facing Third World countries as their common colonial legacy – economic dependency, dual internal economies, disruption of social traditions and coercion of diverse cultures into a single political unit – are either ignored or belittled in the West. Ricardo's theory of 'comparative advantage' has been widely praised for making the colonies better off than they would otherwise have been, yet the facts speak otherwise. As Stavrianos observes:

> Whereas the discrepancy in average per capita income between the First and Third Worlds was roughly 3 to 1 in 1500, it had increased . . . to 14 to 1 by 1970. Far from benefiting all parties concerned, the global market economy is widening the gap between poor and rich countries, and at a constantly accelerating pace.[27]

He summarises the 'development' status of the colonies at the end of the colonial era thus:

> All these global economic trends combined to produce the present division of the world into the developed West as against the under-developed Third World. But underdevelopment under these circumstances did not mean nondevelopment; rather it meant distorted development – development designed to produce only one or two commodities needed by the Western markets rather than overall development to meet local needs.[28]

DEVELOPMENT

The West, whose industrial economies still dominate the world, continues to envision an unrealistic future of never-ending economic growth in an increasingly competitive global 'free-market'. The job of Third World countries is to 'catch up', while adhering to the rules laid down by the capitalist system. Development 'specialists' are wont to shake their finger at every developing country and identify *the* reason why it has not leapt ahead as projected in the early days after independence. China 'stagnated' because Mao emphasised regional agriculture and rural health and education instead of heavy industry,[29] while Africa is now 'in crisis' because governments built huge dams and steel mills rather than supporting small farmers.[30]

The truth is few specialists comprehend the complexity of the problems facing Third World countries, or the time frame in which they are developing. So far they have had only one (Mozambique) to four decades (China, India) to accomplish what took the industrial nations more than 200 years: to adopt their social institutions, indeed, their very worldview, to industrial technology, and to complete the demographic transition from high rates of births and deaths to low ones. Furthermore, the global climate in which Third World countries are trying to develop is quite different from that when the industrial nations were developing. In fact, First World nations are poor development models. As shown in Part I, modern industrial economies are 'overdeveloped', functioning well beyond their sustainable income of renewable resources. Not only is their continued economic throughput at present levels unlikely; any presumption that the remaining four-fifths of the world's people will permanently achieve such levels is a cruel delusion, as Joan Robinson has clearly stated:

> [C]omparing countries, there is a clear correlation between GNP *per capita* and the consumption of energy *per capita*. Public opinion in the wealthy countries has begun to recognise that their life style and their technology are excessively wasteful, though no one has much idea what to do about it. For Third World countries to overcome poverty by industrialising on the Western pattern is out of the question. Certainly they need accumulation but they need to direct it into forms suitable to their own situation.[31]

To contrast the present development situation with that of the past we need to inquire into the factors that made Western capitalist development possible.

Factors in Western Capitalist Development

As noted in Chapter 12, capital accumulation is necessary to build the physical stocks that provide the means for achieving material satisfaction. During industrialisation in the West, this accumulation occurred as the result of three interacting conditions, none of which exists in the same way in most Third World countries today.

1. Initial capital accumulation took place because the wealth produced by human labour was not consumed but invested. Entrepreneurs did not squander their profits, but turned them into new productive capital. As Keynes described it:

[T]he capitalist classes were allowed to call the best part of the cake theirs and were theoretically free to consume it, on the tacit underlying condition that they consumed very little of it in practice. The duty of 'saving' became nine-tenths of virtue and the growth of the cake the object of true religion.[32]

The accumulation of Western capital depended on initial abstention from 'conspicuous consumption', yet in many developing countries today, the capitalist classes, imitating their northern contemporaries, are busy consuming the cake rather than reinvesting it.[33]

Even so, economic growth was painfully slow and the working classes of the Industrial Revolution suffered dreadfully. Some emigrated to America and elsewhere, partially relieving the suffering in Europe. But today emigration on a similar scale is not open to the growing destitute in the Third World.

Ultimately the sufferings of the workers led to uprisings, to Marx and Engels's impassioned writings, and to the formation of labour unions. The conscience of governments was stirred by the writings of Charles Dickens and others, leading to laws against the most brutal conditions. Gradually enough benefits of industrialisation were shared with the workers to forestall the revolutions Marx had so confidently predicted.

2. Another factor in Western development was the sudden availability of seemingly boundless energy: coal and oil. Rapid exploitation of fossil fuels has closely paralleled the growth of economic throughput. Cheap energy is simply not going to be available to today's developing countries.

3. A third, often ignored, factor in Western development was the exploitation of the natural resources and cheap labour of the colonies. (The United States had vast resources 'at home', appropriated from the Indians.) This capital, specifically *prevented* from being used to develop the colonies, was instead transferred to the metropolitan countries. Indeed, this form of exploitation continues today, so that capital accumulation in Third World countries, who will never have colonies of their own, is made even *more* difficult.

Factors in Third World Development

Despite their early expectations of 'modernising', most Third World countries today remain virtually where they were before independence – economic 'colonies' of the developed world. If anything, their problems are worse as their populations swell and their national debts mount. This state of 'neocolonialism' has three underlying causes: (1) The existence of an institutionalised dual economy in almost all Third World countries.

(2) The determination of the West to retain unimpeded access to the scarce resources located in those countries. (Continued control over such 'strategic' resources is considered essential, a point taken up in Chapter 14.) (3) The determination of the West to retain economic hegemony over its former colonies. In 1976, the United Nations Conference on Trade and Development summed up the relationship between the industrialised countries and the developing countries thus:

> The fact that the developing countries did not share adequately in the prosperity of the developed countries when the latter were experiencing remarkably rapid expansion indicates the existence of basic weaknesses in the mechanism which links the economies of the two groups of countries.[34]

Any developing country faces several related questions. One is: Whence shall come the accumulation of capital stocks needed for development? A second is: How shall development be defined? Does it mean the raising of living standards of the people generally through improved health care, food production, and widespread education? Or does it mean rapid acquisition of manufacturing technology to generate a competitive economy in the global marketplace, even if only a few benefit? If it is to be a combination of these, what should the relative emphasis be?

As shown in Figure 13.1, land is the fundamental resource for capital accumulation, capable, given the proper technology, of producing more wealth – in food, fibre or timber – than those working it need for themselves. Under proper husbanding such resources should be renewable indefinitely. In some areas, the land may also yield valuable non-renewable resources such as oil or minerals which as long as they last can be exchanged for productive capital goods.

The excess production needed to accumulate capital stocks comes from the country's inhabitants. As the boxes at the top of Figure 13.1 indicate, there are two basic types of socioeconomic systems in Third World countries. The most prevalent, shown on the left, is the direct colonial legacy of a dual economy. An urban enclave of middle-class and wealthy elites and a small number of large landholders form a capitalist clique, often living in extraordinary luxury even by Western standards. They control the government and the economy, and have strong financial connections abroad, including those with multinational corporations. Labour is carried out by a mass of wage-earners, peons and sharecroppers, whose remuneration is often scarcely above subsistence level. Unemployment among the latter may be 30 per cent or more, and little provision is made for their welfare.

At the other end of the spectrum, shown on the right, are egalitarian socioeconomic systems, particularly regarding land ownership. While it is often stated that small landholdings are 'inefficient' from lack of special-

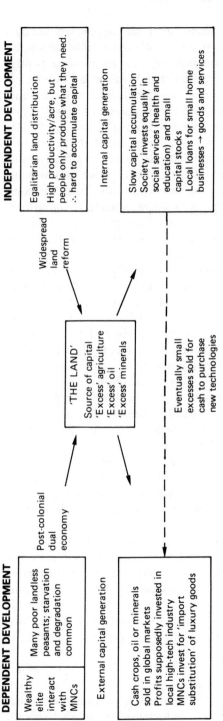

DEPENDENT DEVELOPMENT

| Wealthy elite interact with MNCs | Many poor landless peasants; starvation and degradation common | Post-colonial dual economy |

External capital generation

Cash crops, oil or minerals sold in global markets
Profits supposedly invested in local high-tech industry
MNCs invest for 'import substitution' of luxury goods

'THE LAND'
Source of capital
'Excess' agriculture
'Excess' oil
'Excess' minerals

Eventually small excesses sold for cash to purchase new technologies

Advantage: Theoretically, generates profits for generation of internal capital

Problems:

1. Cash income uncertain owing to fickle global markets; must borrow to feed nation
2. Profits expatriated by MNCs and elites, or consumed by elites, so development fails to occur at expected rate
3. Social unrest requires import of costly armaments to control and repress poor
4. Borrowed capital fails to yield profits, leading to 'debt trap'

INDEPENDENT DEVELOPMENT

Egalitarian land distribution
High productivity/acre, but people only produce what they need.
∴ hard to accumulate capital

Internal capital generation

Slow capital accumulation
Society invests equally in social services (health and education) and small capital stocks
Local loans for small home businesses → goods and services

Widespread land reform

Advantage: Maintenance of internal stability and cultural integrity

Problems:

1. Development is very slow
2. Lack at outset of basic manufactures, such as steel and chemicals

Figure 13.1 Alternative routes of Third World development.
A summary comparison of factors in dependent (neocolonial) and independent development of Third World countries. For detailed discussion, see J. Robinson, *Aspects of Development and Underdevelopment* (Cambridge: Cambridge University Press, 1979).

isation, this is not the case. The productivity of an area of land worked by a given number of people goes up the more equally the land is distributed among them. People work harder for themselves than they do on a plantation for a fixed subsistence wage.[35] Furthermore, mixed agriculture on smallholdings, especially in the tropics, requires far less inputs of chemicals for the same yields than does 'modern' one-crop agriculture.

Although the potential productivity of lands held equally among a peasant population may be high, it is difficult without rewards to obtain production over and beyond the producers' own needs – that is, to accumulate a national excess. Yet if a local community is enabled to improve its *own* capital status by selling its excess, capital accumulation does occur. In China, the communes are encouraged to sell their excess production, partly to the authorities and partly on the open market. The proceeds they can then use to finance whatever the commune chooses: new tractors, a school building, a health clinic, housing facilities, or perhaps a small factory.

A number of Third World countries do not readily fit into the socio-economic spectrum outlined above. Among these are the Muslim states of the Middle East, which never were true colonies and never lost their cultural integrity. Based on Islamic scripture, Muslim societies seem rigid to us, yet their traditions embody a remarkable level of social justice. 'Modern' concepts of economic development do not fit well into their vision of a good society.[36] In a different category are those nations who have attempted – or are attempting – to blend elitist capitalism and egalitarian socialism, including Jamaica under Michael Manley, Chile under Salvadore Allende (both destabilised by external Western capitalist forces), Nicaragua under the Sandinistas (which the United States is trying to destabilise), and Tanzania under Julius Nyerere.

By what routes shall the accumulated wealth be turned into productive capital? Once again, there are two major patterns. On the left in Figure 13.1, excess wealth is exported to developed nations in exchange for capital stocks to build up local industry. On the right, excess wealth is retained *within* the country. Some provides capital resources for local artisans, craftsmen, and small factories. Some is invested (usually by government) in major developments (large factories and infrastructures), and in people (through education, health care and the like). Excess agricultural or manufactured produce that is sold abroad, if any, is exchanged for further capital stocks and expertise.

Both development patterns face enormous internal difficulties, which are hugely magnified by external pressures from the powerful countries to the North. They require examination.

Neocolonialism: the path of dependence

The left-hand side of Figure 13.1 represents a continuation of the economic relationships that existed under colonialism. Typically today the

First World is referred to as the economic 'centre', and the developing countries as the 'periphery', retaining the old sense of inequality and dependency. This neocolonialist path was initially followed by most Third World countries, particularly those with sharply polarised dual economies. In general, they have tried to accumulate capital from the sale of one or two primary commodities.

Problems of single commodity dependency   A global 'free market' for most commodities simply does not exist. Instead, there are many small, economically weak sellers and a few, economically powerful buyers, often oligopolies who have tacitly agreed on the prices they will pay for such items as bauxite, cocoa and bananas. Nor is the demand for primary commodities affected by their price; a fall in price seldom increases demand. Competition among Third World suppliers merely depresses the prices that all receive.[37] Nor have cartels been particularly successful. The rich buyers, who can afford to store the commodities, buy only when prices are low, but the impoverished producers who must sell to survive are bound to sell regardless of price. Although attempts have been made to form cocoa, coffee, tin, copper and bauxite cartels, none has so far met with major success in obtaining higher commodity prices.[38]

The major exception has been OPEC, the Organisation of Petroleum Exporting Countries. The cartel was born when, during a global glut of oil in the 1950s, the seven giant oil-refining companies agreed among themselves to cut the already low prices they paid for crude oil, not once, but twice. This so angered the oil-producing countries that they assembled in Baghdad in 1960 and formed OPEC.

Not much happened for almost a decade after OPEC's birth, except that crude oil prices stopped falling. But in 1969, Colonel Qadaffi of Libya refused to sell oil at such a low price to an independent German company that supplied one-quarter of Europe's oil from Libyan fields. When the big oil giants foolishly refused to supply the 'upstart' German independent, it had no option but to pay Qadaffi's price. Seeing this, OPEC, with control of 80 per cent of world oil production, immediately raised its prices; the worldwide cost of oil soared during the 1970s, with reverberating effects on the global economy. It exacerbated inflation in developed countries and created even greater debts among oil-consuming Third World countries. The oil giants ultimately gained, however. Not only did they pass along to their hapless customers the extra costs of 'crude', but they added a profit margin for themselves as well.

The future of OPEC is uncertain. At the same time that the global economy declined in the early 1980s, lowering the demand for oil, non-OPEC producers such as Britain and Mexico, in need of income to meet their debts, stepped up production. The ensuing glut caused prices to plummet. Agreements among OPEC members to cut back production in order to force prices upward have been only partially successful. And direct military intervention by oil consumers in the Persian Gulf and elsewhere in the event of a perceived 'crisis' is a constant threat.

Not only are Third World producers at the mercy of world commodity prices. They may not even benefit from production in their own country. In Mauritania, for example, foreign capital was used to open up an iron mine, which required building a railroad and a town as well. The nomadic herdsmen who briefly earned wages during construction, when laid off, could not return to their old life and became unemployed. The mine was largely staffed by high-paid foreigners. The economy of Mauritania gained only 20 per cent of the gross income from the mine; the rest of the profits were exported, along with the iron.[39]

Even the land in Third World countries is often under foreign control. Great tracts of land in Mexico are under contract to Del Monte, General Foods, United Brands (the new name of United Fruit), and Safeway, to produce asparagus, cucumbers, strawberries and cantaloupes for the American market. Mexico's vegetable exports now supply 60 per cent of United States' consumption, mainly of foods that landless Mexicans cannot afford.[40] Africa serves the same function for Europe. Coffee and pineapples destined for the breakfast tables of Germany and England are grown on the rich farmlands of Kenya. Says an American observer:

> The real problem in Kenya is that land is used for cash crops instead of for foodstuffs. Kenya has enough productive farmland to feed all of its people without destroying the habitat of the wild animals.* There is no necessity for choosing between the life of a human and that of a gazelle. If a more balanced economic and agricultural system can be created, both humans and beasts will survive on the African plains.[41]

In summary, commodities exported to earn 'cash' for importing foreign goods are often not only externally controlled: in the case of agricultural produce, they also pre-empt land that could be used to feed – as well as employ – local people.

'Investment' of earnings   For development to occur, a country's earnings must be invested in something that will increase its future productivity. With a long-term perspective, such investment would emphasise better

---

*American ecologist Raymond Dasmann, in his book *African Game Ranching*, shows how, by eliminating environmentally destructive cattle and judiciously cropping the migratory wild herbs, East African countries could 'have their game and eat it too'.

health and education of the people as a whole. With a short-term perspective, however, it would emphasise infrastructure, such as roads, dams and power plants, and production capital, such as machine tools and other equipment.

In country after country, particularly those with dual economies, the tendency has been to concentrate earnings in a few 'showcase' projects that promise giant dividends. Dams are a common example. In Ghana, local funds, matched by a World Bank loan, went to build a hydroelectric dam on the Volta River for use by an American aluminum processing company. The company, Valco, was promised energy 'at cost' – a poor bargain for Ghana, since the company imports bauxite and converts it to exportable aluminum, using the cheapest power possible without leaving any significant cash earnings in the country. Although there is some spare electricity for sale, Ghana's earnings from the dam barely pay the interest on its World Bank loan. The major beneficiaries of Ghana's dam are a foreign company and the World Bank, not her own people.

Unfortunately many Third World countries, controlled by a well-to-do, Westernised minority, squander their foreign income on such luxury goods as cars, washing machines and refrigerators. The 'cake' is being eaten rather than reinvested.[42] The wealthy also tend to 'invest' abroad, in Swiss bank accounts or Florida real estate. While seven of the world's largest debtor nations recently accumulated a further $102 billion debt over a three year period, their elite invested $71 billion in foreign securities with guaranteed high earnings.[43]

'Consumerism' also eats up foreign earnings that might otherwise go for investment in the future. Money is spent on artificial 'needs' created by Western advertising: Coke and Pepsi; Western rock-and-roll tapes; Western-style fast foods; Rolex watches and blue jeans. Their costs are huge in terms of average Third World incomes. Says Thomas Gladwin:

> White superiority, by fostering anxiety about a brown man's native competence, thus remains the principal psychological tool with which white men shape brown men in their own image . . . [Thus], Coca Cola has displaced coffee and other local drinks; advertising for Coke subtly exploits the white elitist theme, commonly using, for example, light-skinned models in stylish Western clothes.[44]

One of the most tragic artifically created 'needs' has been for infant powdered milk formula. Abbott Laboratories, Bristol Myers, and Nestlé vie with each other for profits, advertising to new mothers throughout the Third World that bottle-feeding is superior to breast feeding. It is not. Human milk is made for babies. It does not cause colic; it contains antibodies from the mother that protect the infant in its early months against disease; it never causes allergies; it is sterile; and breast feeding establishes important human contact.[45] In poor countries, water which is mixed with the powdered formula is often contaminated, and mothers may have neither the facilities nor the training to sterilise bottles or water. Because formula is expensive, they also over-dilute it. The result is malnutrition and diarrhoea among Third World bottle-fed babies, who have nearly double the mortality rate of breast-fed infants,[46] – all at a tidy profit for the manufacturers. One company even gives away 'free samples' to each new mother, thus ensuring that her own milk dries up so she must bottle-feed her baby.

External 'investment': the MNCs   Desperate for technological know-how, most Third World governments have encouraged investment by multinational corporations. Supposedly the host countries will share in the benefits of capital investment by these giant corporations. The major benefit is jobs. To any country whose economy is so deeply in trouble that it has thousands of unemployed swelling its cities, new factories seem heaven-sent. And indeed, the earnings – such as they are – for factory labour add marginally to the national income. But the products themselves are of no local benefit; they are re-exported for sale in the rich North. Whenever labour costs rise in one place, MNCs simply move their operations elsewhere. Stavrianos documents their behaviour:

> The largest number of American plants abroad are located in Mexico, where the original attraction was a minimum hourly wage of 50 cents for adult males. But when union pressure raised the wage to $1.13, the companies began moving to Haiti, where the minimum wage is $1.30 for a full day's work.[47]

Poor countries have encouraged MNCs to 'colonise' them. In theory, they both create local employment and save on foreign currency exchange by making locally what used to be imported. But most of the items produced are high-profit luxury goods only the rich can afford; neither tariffs nor profit incentives encourage production of machinery or of goods with low profit margins intended for the poor, such as hand tools and cookware. First World capitalists have never been interested in investing in self-sufficiency for the Third World.

This production of luxury goods, moreover, is capital-intensive, hiring only a few well-paid workers, who become a middle class allied politically to the wealthy elite.[48] The dual economy is merely exacerbated without expanding the basic industrial capacity of the country. The spectacular 'growth' in GNP during the 1960s in such countries as Mexico and Brazil went almost entirely to the richest five per cent, while the poor became increasingly worse off.

By definition, MNCs are *foreign* profit-making institutions; they export capital generated from the labour and resources of the host countries.

Annual profits on investments for all United States-based MNCs during the late 1970s averaged 25 to 30 per cent, but only one-quarter was reinvested locally, the rest being 'repatriated' to the United States. Profits on petroleum investments were well over 100 per cent annually.[49] Foreign investments have been a means for extracting wealth, not developing the local economy.

Certain MNCs have also managed to expropriate peasants' lands. When governments, including Brazil and the Philippines, passed laws requiring landowners to register titles to their lands, illiterate peasants often failed to file claims in the required time period. Their ancestral lands suddenly belonged to others, often an MNC. 'Soldiers come, and the peasants are evicted'.[50] Governments of countries with post-colonial dual economies create the conditions from which MNCs can then profit.

Blatant borrowing and the debt crisis When sources of development capital proved insufficient – or were squandered on luxury goods or armaments – Third World leaders turned to borrowing. (Few financial 'aid' transactions to poor countries are outright gifts.) Loans come from three sources: commercial banks; multilateral organisations such as the International Monetary Fund (IMF) and the World Bank; and direct bilateral loans between creditor and borrower governments.

The multilateral lending agencies resulted from the 1944 agreement at Bretton Woods, New Hampshire, when the allies met to decide how to stabilise global currencies after the war. The World Bank, whose funds came from the major Western governments, was intended to make loans to member countries for specific projects – a dam, a school system or a factory. The IMF judged the soundness of loans made by the World Bank, and by private banks as well.[51]

The initial beneficiaries of this system were war-torn Japan and the nations of Western Europe, but by the 1960s the First World had recovered, and financial demands were coming from emerging Third World nations. By the mid-1970s, OPEC's hike in oil prices created a new flood of petrodollars into commercial banks just when it was needed. Loans supposedly would benefit everyone. By stimulating economic growth in developing countries, their standard of living would rise and there would be enough excess to pay back the loans with interest. Expectations for global economic growth were high.

Needless to say, these expectations were not met and the global economy today is riddled with international debts of gigantic proportions. The reasons for failure of growth are numerous and complex. Like foreign earnings, borrowed funds were often squandered by those in power on luxuries and armaments, rather than invested in productive capital.[52] More commonly, the borrowed capital simply did not generate enough wealth. In the 1960s, India, for example, invested most of her loan monies in heavy industry designed to make her a 'world power', and the rest went mostly to buy food; her agriculture simply could not feed her people. Widespread unemployment and poverty remained, draining her internally generated wealth and retarding true economic development.[53] (India's current food self-sufficiency has slightly alleviated this problem.) Other developing countries, too, spent borrowed funds on similar desperate social needs. As Robinson explains, Third World leaders viewed their borrowed funds in a different light from the lenders:

> Governments and economists in the Third World did not take indebtedness very seriously. Many of the newly independent states seem to have felt that their ex-imperialist exploiters owed them some kind of reparations so that aid was only their due. They believed that in the end this would be recognised and the debts would never have to be repaid. Once the promised growth had been realised and the country reached the goal of economic independence, the debts would be written off, either by consent or by repudiation.[54]

Meantime the United States, an overproducer of grain, began using food as a weapon to ensure political allegiance from developing countries. Said Senator Hubert Humphrey:

> [I]f you are looking for a way to get people to lean on you and to be dependent on you, in terms of their cooperation with you, it seems to me that food dependence would be terrific.[55]

The result was a contrived overstimulation of American farms (which today suffer the consequences) and suppression of development of food self-sufficiency in Third World countries. Their output of cash crops for export increased, but food produced for local consumption declined, undercut by subsidised American imports.[56] Meanwhile their populations were doubling. No wonder that spending money on food imports became a way of life for many developing countries. The recent industrialisation of food production in some Asian and Latin American countries has temporarily cured hunger, but not the social dislocation of peasants from their lands.

The outcome of three decades of 'development' is a soaring Third World indebtedness exceeding $1 trillion. These nations have sunk into a 'debt trap'. Today they borrow simply to pay the interest on their earlier debts. Even in 1967, 40 per cent of new loans went for debt servicing; by 1974, it was 70 per cent.[57] Furthermore, the IMF permits new loans, even for debt servicing, only under the most stringent conditions of 'economic reform'. In essence, this means total fiscal 'belt-tightening', cutting back severely on social welfare and rural development programs. As Cheryl Payer observes in her book, *The Debt Trap*:

The International Monetary Fund is the most powerful supranational government in the world today. The resources it controls and its power to interfere in the internal affairs of borrowing nations give it the authority of which United Nations advocates can only dream.[58]

Or, as the magazine *Dollars and Sense* states:

Armed with carrots and sticks, the IMF is both lender of last resort and the world's policeman. The carrots are loans, and the sticks are economic policies the IMF forces on nations in return for help.

These policies are designed to induce austerity. In order to get control of enough money to pay back their loans, Third World governments are required to raise prices and lower wages.[59]

The financial situation of Third World countries is becoming desperate. Not only does austerity lead to social unrest and the need for authoritarian governments. It is also inimical to any future hope for economic independence by Third World countries – and hence any realistic opportunity to escape permanent bondage to the more powerful countries of the North. The IMF's restrictions inhibit local investment; they suppress welfare relief to the desperately poor; they outlaw protective tariffs whereby budding local enterprises might 'take-off'; and they insist on devaluation of local currencies, which means that export earnings are less useful than before to the country's economy, while inflation and unrest at home may soar beyond control.[60] Governments are thus torn between undergoing austerity to obtain more loans and defaulting on their debts in order to prevent strife at home. As global economic pressures grow, so does the possibility of a concerted default by a majority of debtor Third World nations, leading to global financial chaos.

In summary, today's 'neocolonialism' results from a close tie among Third World elites, the financially powerful multinational corporations, and First World governments more interested in economic gain and political alliances than in helping developing countries attain true economic and political independence. Noam Chomsky, an outspoken critic of American foreign policy, states that its logic is quite simple: 'to guarantee the freedom to rob and to exploit'.[61] More charitably, Robinson summarises the situation thus:

[For a Third World country] to confine imports to what is most useful for development would require a strenuous social and political policy to prevent luxury consumption, a high degree of control over the economy and a well-conceived programme of investment to make sure that imported commodities and services are used to the best advantage. A country which tried to embark on such a policy is unlikely to receive aid from the West. In the philosophy of the aid givers, laisser faire and free trade were the road to prosperity; conditions attached to grants of aid were designed to head off any attempts to deviate from it. Moreover, it was convenient for the donor countries that demand generated by aid should be for commodities that they could profitably supply.[62]

Self-sufficient development is simply not considered a profitable undertaking for First World investors.

Independent development

The global economy can develop in two quite different directions. One is toward further economic interdependence, which, however, will require several sweeping changes if collapse into chaos is to be avoided. These include: (1) international regulation of the currently privatised global financial markets to prevent rampant speculation; (2) the attainment of fiscal responsibility by First World governments to bring down interest rates; (3) a significant raising of wages in Third World countries to enable their workers to finally become consumers in the global marketplace and so avoid global depression; (4) the relaxation of conditions on development loans to permit local government spending on housing, schools, medical care and social-welfare programmes, and so avert social chaos.[63] The philosophical changes in Western political and economic thinking needed to bring about such changes seem remote, at best.

What about the alternative, development that relies solely on self-generated capital, leading to a global economy of basically self-sufficient countries voluntarily trading their surpluses with each other, as shown on the right-hand side of Figure 13.1? Such a path would demand severance, of former external capitalist ties, which often is achieved only after revolution. Independent development almost always entails some form of communal cooperation, which may or may not be accompanied by coercion.

In many emerging nations, socialism is 'natural'. As with the natives of Java, community and sharing are everything and the individualism promised by the capitalist West is threatening, a thing to be resisted, not embraced. As Terry Ratigan, a Peace Corps volunteer, observed in a West African village:

> Western visitors, from colonial administrators to development workers, were said to 'like money too much'. The implied selfishness was hard to deny when living among poor farmers who would gladly share their last grain of rice with a stranger.[64]

In countries with traditional ways, like Sierra Leone and many other African countries, and the Muslim nations of the Middle East, there is little popular inclination toward Western industrial 'progress', and when political leaders push for it they may be strongly resisted. Egypt's president Anwar Sadat was assassinated for just that reason. In such cases, where cultural and religious traditions are strong, 'development' is likely to take quite different forms from those in either the capitalist First World or the communist Second World. It is likely to be self-sufficient, egalitarian and local, rather than dependent, hierarchical and centralised.

A remarkable case in point is Sri Lanka. After 450 years of Portuguese, Dutch and British colonialism, Ceylon (as it then was) was left heavily dependent on a plantation economy. As the terms of trade worsened following its independence in 1948, the country sank into deep poverty, with the usual dual economy of a minority urban elite in charge and a majority of peasant poor; there was even an unsuccessful Marxist-led uprising in 1970. Meantime a Buddhist school teacher and his students began establishing the Sarvodaya movement, which between 1968 and 1981 spread from 100 to 4000 villages. Based on the Buddhist notion of *udaya*, awakening, the movement organised villagers at all levels – youth, mothers, children, workers, the elderly – to create their own health centres, pre-schools, cottage industries, cooperatives, and village technologies. Ultimately the villagers themselves took full political charge of their community.

Even though Sri Lanka remains 'poor', it ranks higher than countries with ten times its *per capita* income in the Physical Quality of Life Index of the Overseas Development Council; the general health and well-being of the people have improved enormously, and the birthrate has dropped. Remarkably, all this social progress has occurred under a pro-capitalist government that fosters exports and tourism to finance giant 'development' projects.[65] Even the modest government support for Sarvodaya has been criticised as 'extravagant' by international economic 'experts'. The Sri Lankan demographer, Tarzie Vittachi, writes:

> It is true ... that the free education and subsidized food policies of Sri Lanka were revolutionary ideas promulgated by resolute government leaders, but they became an integral part of national life and were sustained against the abjurations of the World Bank, the International Monetary Fund and many other 'friends' who frequently pressed succeeding governments to abandon what they saw as extravagances of social policy.[66]

But most of the monetary aid for Sarvodaya has come as small private gifts from overseas, and its main strength comes from the thousands of volunteer Sri Lankan youths whose rewards are spiritual rather than monetary.

When not overwhelmingly interfered with by external or internal forces, small-scale, morally inspired, democratic socialism seems to work well in 'developing' improvement in the lives of people. It is being experimented with in numerous ways, in Tanzania, China, the base communities of Latin America and elsewhere. Autonomous development is discussed further in the following chapters.

POPULATION GROWTH AND CARRYING CAPACITY

In 1960, three scientists predicted the world population would 'go to infinity' on Friday 13 November 2026. Curiously, that population 'scare' has largely faded from public concern, as though the problem were somehow taking care of itself. But it isn't. Despite massive efforts at family planning, including China's controversial 'one-child' family programme, global population is two to three years *ahead of schedule*![67] More than 8000 extra persons are being added every hour to the world's population, most of them in those countries already in serious economic and environmental trouble. Neither political coercion nor 'miracle' contraceptives have reversed the propensity of the hopelessly poor to have more children than are needed to maintain a constant population. Preferred family size is a *social* phenomenon, determined by what people believe is best for them. It is this that must be addressed if the global population explosion is to be brought under control.

Before we turn to these social questions, however, a look at the physical carrying capacity of the Earth is necessary to discover what regions will face major problems in the coming decades.

The Carrying Capacity of Various Regions

Depending on their preferred points of view, different people will tell you that the world has no food shortage into the foreseeable future, or that the exploding global population is about to overrun its food supplies. A recent, regional study of the potential food-growing resources of the world, done for the Food and Agricultural Organisation of the United Nations, throws light on this dispute.[68]

The survey began with a simple premise. If *all* of the potentially arable lands in developing countries were cleared and used to grow *only* the most favourable calorie-producing crops for each local soil and climate, how many people could each region support? The yields per acre, and hence the maximum carrying capacities, were calculated for three levels of energy-subsidy inputs: low inputs, characteristic of Third World subsistence farming; intermediate inputs, which include modest additions of water, fertiliser and mechanical energy, as in China and Eastern Europe; and high energy inputs, which typify industrialised Western agriculture. No arable land is set aside for the following: growing vegetables or fruits for dietary diversity and vitamins; growing fibres such as cotton, linen, silk or wool; growing forests for timber, pulp, or firewood; growing feed for poultry and livestock (which today accounts for one-quarter of arable land in the Third World); soil conservation by fallowing; and, finally, the production of cash crops – sugar, coffee, dates, rubber, bananas, pineapples – to earn foreign currency for capital investments such as roads, mines, dams and factories. Finally, no land is set aside for wilderness, game parks, or other 'uses' not directly related to human subsistence.

Based on these stringent assumptions, the survey calculates the absolute maximum carrying capacity that each region throughout the Third World can sustain *at the barest subsistence levels*, and *without any margin of safety* for climatic extremes, natural calamities, or human acts of land abuse, such as overgrazing, erosion, desertification, salination, and destruction through war. The results are then compared with the projected populations in the five major regions of developing countries in the year 2000, as shown in Table 13.2. (Of course, population growth will not suddenly stop in that year; in fact, the increase in the following quarter century may equal that between 1975 and 2000, leading to further food shortages.)

Most developing countries now use minimal energy subsidies in their agriculture. At low input levels, only South America will be able to feed all its people by the year 2000. This will only be possible *if* large areas of giant haciendas, now lying fallow, are brought into production, and *if* half the remaining rain-forests are cleared for farming, and *if* food is distributed equally throughout the population of each country. Enormous ecological and politicoeconomic changes will be essential in the next decade or so, even in this most favoured region.

The next 'best' region in terms of food self-sufficiency is Southeast Asia (extending from Pakistan to the Philippines, but excluding China, where both population control and ongoing mechanisation of agriculture make predictions difficult). This region has favourable climate and soils, and with an increasing use of irrigation, even with low levels of other inputs, it will be able to keep up with population growth except in a few areas; six of its sixteen countries with 18 per cent of the region's people, will be 'critical' in the year 2000. If medium energy inputs are employed throughout the region, as is likely, then only two countries will be unable to feed their own people. One is Singapore, which could not support itself even at high energy inputs, and *must* live by trading manufactures for food. The other, much more serious case, is Bangladesh, where even at high energy inputs, it could scarcely support its population in the year 2000. Without food aid or mass emigration, rising death rates in Bangladesh seem inevitable. Fortunately, population growth rates in most countries of Southeast Asia are among the lowest in developing nations (except Malaysia which plans to quintuple its population by 2100!).

Table 13.2 Third World countries unable to feed themselves in the year 2000

	Region of the world					
	Africa	Southwest Asia	Southeast Asia	South America	Central America	Total
No. of countries surveyed	51	16	16	13	21	117
1975 – Total population (millions)	380	136	1118	216	106	1956
2000 – Projections Total population (millions)	780	265	1937	393	215	3590
a No. of critical countries b Population affected c % of regional population						
... at low inputs						
a	29.0	15	6.0	—	14.0	64
b	466.0	195	341.0	—	52.0	1054
c	60.0	74	18.0	—	24.0	29
... at medium inputs						
a	13.0	15	2.0	—	7.0	36
b	110.0	195	156.0	—	24.0	486
c	14.0	74	8.0	—	11.0	13
... at high inputs						
a	4.0	12	1.0	—	2.0	19
b	11.0	89	3.0	—	0.7	104
c	1.4	34	0.2	—	0.3	3

Adapted from P. Harrison, 'Land and people: A new framework for the food security equation', *Ceres*, No. 98, Vol. 17(2) (Mar–April 1984) Centrepiece.

Central America, including the island nations of the Caribbean, is more critical, though fewer people are involved. It is short of land and uses only low inputs to its agriculture. In 1975, half these countries faced critical food shortages, and this number will increase to two-thirds by the year 2000, given the high birth rates of the region. Even with medium levels of energy inputs, a third of the countries will face food crises.

In Africa, where population will more than double by 2000, there is considerable unused land, but since available land and population centres do not coincide, over half the countries, with 60 per cent of the African population, will be short of food, since they are unlikely to obtain the resources to increase their present low energy inputs. Even if all countries achieved medium inputs, 14 per cent of Africa's people would not receive enough calories from their own land.

In terms of its local carrying capacity, Southwest Asia is in the most critical position. This area, which includes Turkey, the Middle East and Afghanistan, has the highest birthrate of all, and the poorest land. Even with high energy subsidies, three-quarters of the countries, comprising a third of all the region's people, could not feed themselves.

The data in Table 13.3 are far too optimistic, since they exclude all *other* land uses, and assume that food will be equally distributed throughout an entire population. Given various uncertainties on both the high and the low side (better technology, more irrigation, vs inability to obtain the necessary energy inputs), Harrison concludes that around 80 per cent of Third World countries will face food shortages in the year 2000. His proposed strategies for Third World countries are given here, along with major difficulties in achieving them.

1. Concerted family planning policies. (Difficult in societies where women have low status and children are the only 'old-age security'. Especially difficult in Muslim and Catholic countries.)
2. Increase energy subsidies to agriculture in the form of fertilisers, pesticides and improved seeds. (Difficult to obtain, since they must be imported and paid for with exports that often include cash crops, thus lowering food producing potential.)
3. Land reform, so that underutilised estates can be used by peasants, who for a given level of energy subsidy are more productive. (Requires major political reforms in many Third World countries in order to redistribute land.)
4. Increase irrigation. (In many areas, major gains could be made by irrigation projects, but these often require large capital inputs.)
5. Grow crops better suited to the region than those traditionally grown. (Requires change in farm practices and dietary habits that may be difficult to introduce.)
6. Practice soil conservation, by terracing, crop rotation, and other techniques to prevent erosion that will otherwise decimate 20 per cent of the Third World's farmlands by the year 2000.* (Requires cultural change in land use patterns.)

The key factors for most developing Third World countries are some combination of population control, political reform, *and* social and

---

*About 36 per cent of current farmland is expected to become useless in the next century; since potential new farmland is only 50 per cent more than what is now farmed, the *net gain* will be only 14 per cent, while global population doubles. Moreover, much of this new land consists of easily degraded soils. 'The long-term prospect', says Oxford ecologist P.J. Stewart, 'is that we shall end up with less arable land than we have today' (P.J. Stewart, 'An Alternative to Overloaded Resources', *Ceres 17*(2), pp. 37–42, No. 98, March–April 1984).

technological change. Countries rich in oil or minerals will be able, so long as their resources last, to trade them for food. But a good-dependent country is politically and economically exploitable, and its survival hinges on fluctuating world market prices for both the resources it sells and the food it buys. And when the resources are exhausted – in perhaps a generation or less – there will be nothing to fall back on. Countries which have only cash crops or unskilled labour to exchange for food are likewise in a precarious position, one of essentially permanent dependence on food producing areas. Any country or biogeographic region that aspires to political autonomy must ultimately learn to subsist within its own environmental income on a sustainable basis.

What, then, are the social and political factors that affect population growth and demographic change and hence allow a society to achieve a stable and sustainable population size?

Social Factors Affecting Demographic Change

People have children because they perceive a need for them. In societies which long ago grew too big and lost the security of the extended family, the communal village, and the tribe, one's own children became the only form of economic security, an extra pair of hands to help feed the family and to ensure the parents' welfare in their old age. Under such conditions, numerous children became the social ideal, the respected norm, which was naturally incorporated into religious traditions.

As long as infant mortality rates were high, the encouragement of high birth rates was probably necessary to ensure the survival of society, as well as of the ageing parents. Wherever improved diets and public health measures were introduced, death rates fell. But without concomitant social changes, the traditional needs for children, from the parents' perspective, still remained, leading to rapid population growth. Only when social and cultural patterns changed did birth rates begin to fall.

The nature of the demographic transition

This switch from conditions of stable populations with both high birth rates and high death rates, to stable populations where both birth and death rates are low, is depicted in Figure 13.2. In today's First World countries, for which the Swedish data serve as a model, the entire demographic transition spanned nearly one and a half centuries. During the period when births exceeded deaths, Sweden's population grew rapidly, but not as much as the data suggest because enormous numbers emigrated to the New World to escape increasing misery and hunger at home.

In Third World countries such as Mexico, which often started at higher birth and death rates than did Sweden, the rate of fall in the death rate has been exceedingly abrupt, particularly with the advent of vaccinations and

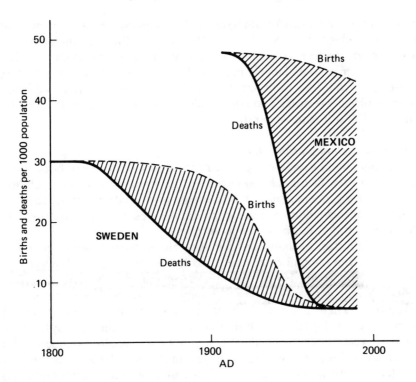

Source: Modified from J.R. Weeks, *Population: An Introduction to Concepts and Issues*, 2nd edn (Belmont, CA: Wadsworth, 1981) Figure 3.5, p. 63.

Figure 13.2 Contrasted demographic transitions: Sweden and Mexico.

Sweden's demographic transition typifies what occurred in most modern industrial countries. Death rates and birth rates, both high around 1800, separated in the first half of the nineteenth century, as death rates began to fall. At this time, when births exceeded deaths, population increased rapidly, until finally, about 100 years later, births also declined, so that these countries today are approaching zero population growth, with low birth and death rates. The whole process thus spanned well over a century.

In a Third World country like Mexico, which started with much higher birth and death rates, between 40 and 50 per 1000 population, death rates began to fall only in recent decades, with the spread of modern medical and health technologies. In many cases, birth rates have fallen only slightly or not at all, leading to exploding populations. Note that what required over a century to occur in the First World, Third World countries are being forced to accomplish in a few decades.

antibiotics. Until recently, traditionally high birth rates have continued almost unabated, leading to population doubling with every generation. In Africa, the fall in death rates, which only began around 1950, was even more rapid and the population growth rate even more explosive.

Meantime, the safety valve of empty lands to which to migrate has disappeared. First World countries have, temporarily at least, closed their doors to the free influx of immigrants. The seemingly massive exodus of perhaps a million 'boat people' from Vietnam and their resettlement else-where made scarcely a dent in the 30 plus million left behind. 'Guest workers' from poor and highly populated southern Europe are less and less welcome in countries to the north. The United States struggles with mil-lions of illegal Mexicans annually, while Mexico in turn tries to stem the influx of Central American refugees.

Causes of declining birth rates

Demographers have long searched for 'causes' of declining birth rates, suggesting such factors as: opportunities for upward economic mobility, especially for women; a substantial decrease in infant death rates; increased literacy; and availability of contraceptives. All these no doubt contribute to declining fertility, but none seems sufficient.

A recent historical analysis of demographic change in fourteen countries – ranging from the United States across Europe to Russia – shows that in every case the decline in birth rate began in conjunction with *freedom from economic and political repression*. It began in America even before the revolution, when a sense of 'freedom' was already felt. In France, it occurred following the revolution of 1789. In Britain, it only came a hundred years later, when the right to vote was finally extended to all commoners, not just those with property. In Russia, it began after the 1917 revolution with its promise of political and economic emancipation. As soon as people believed themselves to be in charge of their own lives, fertility began to decline.[69]

If this thesis is correct, then a rapid fall in birth rates should occur wherever people are *consciously* refashioning their own lives. It further predicts that a government-sponsored push to lower birth rates will be most successful when government policies are perceived to be designed by the people themselves. Strong economic and propaganda incentives have thus been effective in popularly socialist China[70] and in independent Singapore,[71] while they have failed miserably in India, where the elite-run Congress party takes a highly paternalistic attitude towards the populace. Here are the patronising words of Dr D.N. Pai, Director of Family Planning in Bombay, in regard to the coercive male sterilisation programme of the mid-1970s under Indira Gandhi:

> Ninety per cent of the people have no stake in life. How do you motivate them? They have nothing to lose. The only way out of the situation is compulsion.[72]

By contrast, a 1979 world fertility survey found remarkable fertility declines in Costa Rica, Thailand, Sri Lanka, and Colombia, none of which was related to significant economic 'development'. They seemed rather to reflect strides toward economic and political self-reliance.[73] Such findings are encouraging, since they suggest that a high level of material affluence, far beyond the reach of most Third World countries, is not a prerequisite for stabilising populations.

A final but important factor in demographic change is the *size* of the politico-economic unit involved. As pointed out by Sri Lankan demographer Tarzie Vittachi, the most successful Third World members of the British Commonwealth at reducing fertility rates – regardless of economic level – have been relatively small: island states such as Singapore, Trinidad and Tobago, and Sri Lanka; and the state of Kerala in India. In all cases, *local governments* have been closely involved with the people, providing easy access to health care and to schools, helping to create local jobs, and ensuring that wealth is fairly distributed. There is active, democratic participation in the local social enterprise. The people themselves are taking charge.[74]

SOME CONCLUDING THOUGHTS

The world seems to be sitting at a crossroads. Either it will develop into a single, competitive and exploitative, profit-oriented market economy, a sort of industrialised version of Thomas Hobbes's primitive state of man, with 'all against all', a soaring population and plummeting resources. Or it will swing towards local self-sufficiency, based on cooperative community efforts and indigenous resources, with a rapidly stabilising population and reduced political tensions. The latter route, of course, in no way precludes the eventual attainment of higher levels of material throughput through improved technologies nor the trading of genuine surpluses around the globe. It simply constrains the whole process within humanly and environmentally acceptable limits.

The choice between these two outcomes rests largely with the industrialised countries of the North. Even with the most enlightened regimes, many emerging Third World countries will require some degree of assistance if they are to escape chaos, misery and violence. What will the response be? Canada's former Prime Minister, Lester Pearson, headed a report for the United Nations entitled *Partners in Development*, which concluded that: 'A keynote of aid policy should be the achievement of long-term and self-sustaining development'.[75]

If the Pearson Report is right, then the West must consider 'investments' not in terms of short-term profits and interest payments – that is, in the conventional terms of capitalism – but in terms of a future stable global population living within its environmental means. Aid must be defined as the transfer of wealth in forms that assist the self-reliant fulfilment of locally determined goals – 'wealth' particularly in the form of appropriate technical knowledge. Aid should be in the form of gifts, carefully made, for the right reasons. As Joanna Macy puts it:

> [A] new developmental philosophy is essential in our time, for the industrialized countries can no longer serve as models for development. Their growth-oriented policies and high consumption levels deplete the planet's resources, while polarizing the rich and the poor. . . Meanwhile, the industrialized societies themselves exhibit signs of acute social disorder and psychic alienation. In the face of the rigid, unresponsive bureaucracies of big business and big government, people feel anonymous and impotent to alter the drift of events. Indeed, a sense of personal powerlessness appears to be the feature most common to people in both 'have' and 'have-not' nations.[76]

The next chapter addresses the nature of this thing called 'government', that is, the arrangements by which people – or at least people with power – decide the rules by which a society is organised and under which its members live.

14 Politics: Worldviews in Action

Civilisations are in crisis... Modern civilisations are in a transition stage from order based on coercion to an order that is self-sustaining, a transition from positive law and central control to some form of social organisation, the nature of which eludes us, that enables individual development and participatory control. The transition is inevitably a critical stage in evolution and from which civilisations will emerge only if goals are clearly perceived and deliberately pursued.

John W. Burton, 1984[†]

INTRODUCTION

We now come to the crux of our problem: how do people successfully organise themselves into *stable* societies? Today, most societies are far from stable. In many Third World countries, growing economic inequality, exacerbated by exploding populations, means rising instability. The Second World – the USSR and communist East Europe – has remained 'stable' over four decades only through imposing oppressive coercion. Even the industrial democracies, which still seem stable on the surface, face powerful internal tensions in the face of material shortages and an end to economic expansion. Understanding today's world means understanding the factors that determine social stability and instability, including aspects of human nature (gleaned in Part II), and of the Western worldview (from Part III), which now dominates so much of the global economic and political power structure.

Politics, like religion, tends to inflame people's emotions. We naturally expect some fireworks when regional, class or ethnic interests are at stake. Modern politics is the means for arbitrating among special interest groups. But politics goes even further. In its broadest sense, it is the process by which a common understanding about the nature and purpose of an entire society is modified over time. This aspect of politics explains the differences between the Tory and Labour parties in Britain (although scarcely between the almost indistinguishable Republicans and Democrats in the United States). On the whole, political parties stand for more than just warring constituencies; they represent, however diffusely, different social values and social goals. They have different ideals, different Utopias in mind. And when in power, they move society as far as possible in their preferred direction, attempting to make their imagined worldview a reality.

Politics is thus more than just 'changing the laws'; it is the process of changing the sacred institutions we believe in and live by.

Ideological change never occurs in a vacuum; it is always overlaid on an earlier ideology. This explains why Marxism, for example, is quite different in China, the Soviet Union, Mozambique and Nicaragua, and can even be fitted into the Liberation Theology evolving in Catholic Latin America. Images and ideas can take on infinite shapes and meanings, defying the neat compartmentalisation of social analysts.

At this historical moment of unprecedented technological and social change, the beliefs of the parents may no longer seem appropriate to the children of the next generation. We live in an era when old worldviews are suddenly out-of-date, yet, as Margaret Mead observed: 'Children in our own and many other cultures are being reared to an expectation of *change within changelessness*'.[1] No one is bothering to *critique* the old ideologies or seriously trying – in schools, in political dialogue or in the media – to *create* a new one.

Rapid change is not restricted to the industrialised nations. The end of colonialism left most of the world's peoples not only with an expectation of political independence from external controls, but also with a promise of creating their *own* society according to their *own* values. This process is being hampered, however, by the combined effects of neocolonialism, declining resources and burgeoning populations, which can create crisis conditions leading to violence.

Both rapid technical change (particularly regarding our ability for mutual annihilation) and growing demands from the Third World, force humankind generally to face up to the need to address our varied assumptions about who we are and where we are going. As a species, we can no longer afford the luxury of a slow, almost imperceptible evolution of worldviews. Forces which we ourselves have set in motion have caught up with us.

A particular need of our time is to rid ourselves of what Gerald and Patricia Mische call 'national security states'[2] – those impersonal, centrally run polities whose powerless citizens cohere together largely from historical habit. Basing their identities more on military and economic power than on shared cultural values, these competing political entities rigidly block solutions to both global and domestic problems, while alienating their own citizens. To achieve true social stability, people need to *participate* in the social decisions that affect their lives; they need to be active agents in the creation and evolution of their shared values. Only by recognising what kinds of social arrangements human nature *requires* in order to function at its best, do we humans have a chance to escape the crises that loom before us. As Jonathan Schell said in the conclusion to his book, *The Fate of the Earth*:

In sum, the task is nothing less than to reinvent politics: to reinvent the world.[3]

The future seems to demand a global politics that today seems impossible: a multitude of small, more intimate societies, offering individuals far greater scope for personal participation in social decisions, all interacting together within a non-hierarchical global framework. We need to foster everywhere the kinds of grassroots, people-based societies that are tentatively emerging around the globe. Before exploring them, however, it is necessary to ask how political systems function today.

THE BODY POLITIC

A society occupying a circumscribed territory and independent of external political control is today recognised as a 'state' or 'nation'. Yet long before such geopolitical units appeared on maps, cohesive social units existed around the globe. Here, the critical question is: What factors hold such societies together?

The Concept of Government

It is remarkable how, despite abundant evidence to the contrary, we tenaciously cling to the Hobbesian notion that humans are antisocial animals engaged in an instinct-driven competition of 'all against all', that we are saved from catastrophe only by virtue of that higher faculty, 'Reason', which enables us to enter into calculated contracts which whip us into shape and create 'civilised' society. Such a self-image justifies a hierarchical government by authority-bearing rulers who restrain our natural tendencies to antisocial behaviour. We accept the necessity of widespread obedience to laws, backed up by physical force.

Those of us living in large societies automatically assume that without 'government' human societies would degenerate into uncivilised chaos. In their lack of 'law-and-order', precivilised societies seem clearly deficient, lagging behind in the linear march of social 'progress'.[4] What we forget is that our prehuman ancestors were *social* and formed stable social systems long before they were *rational*. As shown in Part II, strong social bonds, far from being something we create in a calculating, 'civilised' manner, are something we deeply need and cannot help forming. We are not social because of rationally perceived individual benefits, such as economic gain or national security, but because our deepest emotional being does not allow us to be otherwise. In his vision of prehistoric man, Hobbes was about as wrong as it was possible to be.

For the first million years of human existence, societies were stabilised by social bonds and shared customs. Changes in the latter occurred slowly, unconsciously, without disrupting social cohesiveness. Whatever was decided, was decided by all. Except that elders were held to be wiser, the concept of governance did not apply. The required level of social conformity was a matter of course. Social disruptiveness was more likely to be treated by care and concern than by physical punishment. Ostracism was unknown, although individuals might change groups, and groups themselves might fission and reform. Although submerged, *all these aspects of our human social natures are still part of us today*.

As societies grew and became less homogeneous, rules multiplied; government emerged as a 'social management' system. Authority, once vested purely in shared tradition – 'That's–the–way–we–do' – became identified with rulers who interpreted and even elaborated on the old traditions and myths. Nevertheless, rulers were constrained by historical precedent and their own religious beliefs from straying too far from their allotted powers. As noted earlier, the worst abuses of these massive totalitarian societies were resisted by the major religions that apparently arose to keep them in check and retain a modicum of respect for the individual. With the brief exception of ancient Greece and Rome, there were no built-in processes for consciously changing either the worldviews or the social institutions that flowed from them. Government and worldview evolved together, imperceptibly slowly. The hierarchical class structure was legitimised through a commonly accepted belief in the 'rightness' of the social order.

By the sixteenth century, however, the myriad feudal princedoms in Western Europe were merging into new nation-states. Authority became unified under a central monarch. As political scientist Robert MacIver notes, this meant redefining the legitimising myths to give the new king the authority to maintain public order:

> So the myth-makers, inspired and sustained by the movements of their time, restated the myth of authority. They were for the centralization of authority. They were for public order against the private rights of feudalism, for public peace against the anarchy of private wars.[5]

So was born the modern notion of authoritarian law-making and the right of a ruling class to use centralised force 'for the public good'. Instead of a worldview based on religious tradition, 'the common good' could now be redefined and modified to suit those in power. Tradition as the measure of legitimacy gave way to the 'right' of rulers to rule – to make and enforce laws.

Yet to maintain order, rulers must have sufficient support from their subjects. If support is not volunteered for the sake of a perceived common good, it must be obtained by threat of force. Clearly, the more convinced citizens are that social institutions are to their benefit, the less force is necessary. This tension between support and coercion is the foundation of modern politics (see Figure 14.1).

Indoctrination vs Coercion

Social stability thus rests on some combination of two factors: (1) voluntary loyalty to the rules and rituals of a shared worldview that gives meaning to one's life – the unselfconscious *belonging* to one's community – and (2) the fear of social retribution that disobedience to the rules entails. Not surprisingly, rulers attempt to convince subjects that their laws and institutions are worthy of support. Whenever possible their values are imparted to the young via the schools and to society through the mass media. The ruling class defines the shared worldview, as Marx well understood:

> The ideas of the ruling class are in every epoch the ruling ideas, i.e. the class which is the ruling *material* force of society, is at the same time its ruling *intellectual* force. The class which has the means of material production at its disposal, has control at the same time over the means of mental production.[6]

In reflecting on how we acquire our worldview with its embodied perceptions of a proper social order, we see why a new ruling class that breaks with old traditions has to create a new, widely accepted ideology before it attains voluntary acceptance. This, too, Marx saw clearly:

> For each new class which puts itself in the place of one ruling before it, is compelled, merely in order to carry through its aim, to represent its interest as the common interest of all the members of society, that is, expressed in ideal form: it has to give its ideas the form of universality, and represent them as the only rational, universally valid ones.[7]

Whoever would rule must first convince society that *their* social vision is more desirable than that of their opponents. This explains the extraordinary *political* roles of education and the media in society; after parents and peer groups, these are the major sources of the worldview held by the young. Successful rulers have influence over both.

Figure 14.1 illustrates the balance of power between rulers and ruled in the modern state. In stable periods of history when worldviews are changing little, the main coercive function of the state is to defend society from the cheat, the common thief, the corrupt public official and the impassioned murderer. But in times like ours, when worldviews are in flux, a new threat to social stability exists – namely, political dissidence. How does 'the state' – that small handful of rulers – cope with those who openly disagree with

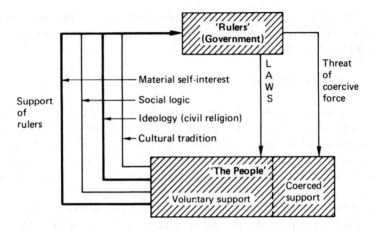

Source: Drawn up from material in a lecture given by Professor H. Janssen (Department of Political Science) in 'Our Global Future', a course at San Diego State University.

Figure 14.1 The basis of political power.

In large hierarchically ordered political systems, those who govern must possess sufficient power to enforce social rules; this comes from a sufficiently large base of popular support to maintain order. Support can be in four (not mutually exclusive) forms:

1. Traditional societies give customary support to religious leaders or kings.
2. Ideological support supplants tradition in modern nationalistic societies.
3. Social logic argues for support on the basis of a reasoned contract.
4. Material self-interest ensures support from those who personally gain under the rulers in power.

When voluntary support fails, rulers threaten coercive force to maintain social order.

In modern capitalist societies, as the text discusses, a small economically powerful elite controls the political process and the majority are persuaded to follow from a combination of mainly ideological and material reasons.

its policies and dispute their moral legitimacy? How does it treat 'the opposition', the 'unpatriotic', the 'dissidents', the 'revisionists'? – the workers who strike for their perceived rights, the youths who refuse to fight in a war they believe unjust, the unemployed who live by barter and so escape payment of taxes? How effective are rhetoric and propaganda? When is it legitimate to control the media and the schools? At what point do the rulers apply force, and what forms of force do they use? Is it psychological, the use of peer pressure to obtain conformity, as employed so successfully in China? Is it economic pressure, whereby taxes, fines and duties are imposed to maintain obedience? Or is it physical force, including exile, imprisonment, torture and outright murder, that ensures a compliant population? The range of force available to rulers for maintaining their privileged position when ideological persuasion fails is broad indeed.

Today the position of rulers is growing less secure as more and more ordinary citizens around the world are questioning the meaning and purpose of the laws and institutions that control their lives. To what extent does 'the state' reflect *my* sense of justice, of purpose in life, of a harmonious society? To what extent, on the other hand, does it reflect the needs of special interests, of a select elite? As political scientist and international relations scholar John Burton puts it:

> We can no longer assume that there is something natural and sanc-
> tified about 'legal' authority and its coercive institutions. We are being
> forced by events to accept a notion of legitimacy based, not on legality,
> but upon .. the degree to which authorities and institutions ... promote
> the identity, development and a sense of fulfilment of the people. Failure
> in this, leading to disaffection, is likely to lead to coercion by authorities,
> making the position worse.[8]

The social institutions and their legitimating myths in our large, central-
ised social systems are no longer satisfying the basic needs for meaning and purpose of the great mass of their citizens. They are becoming unstable.

In an ideal society, conformity is spontaneous; people *want* to partici-
pate because the social system provides meaning and purpose to their lives. Work, ritual and other social activities are performed not for the bribe of money, nor the fear of coercion or of eternal damnation, but for the *intrinsic rewards* of belonging to and participating in a shared human enterprise. In such a society, conformity is subconscious and people perceive themselves as free entities acting out of their own initiative. We suffer only when conformity is forced upon us and whatever we are con-
forming to seems meaningless. This failing tends to grow in proportion to the size of the political system and the degree of centralisation of political power. The larger the system, the greater our alienation. As Simone Weil said: 'Men feel that there is something hideous about a human existence devoid of loyalty', ... but in the modern world, 'there exists nothing, apart from the State to which loyalty can cling.'[9]

The Disaster of Size

Despite their ingenious technological advances, giant social systems – Lewis Mumford's 'megamachines' – have never managed to recapture the human scale, the human values, and the human bondedness of smaller societies. In the West, especially, cooperation has been replaced by competition as the idealised form of human behaviour. Accumulation of wealth has become not a mark of greed, but of virtue, something to be achieved

through ever-increasing 'productivity', regardless of social or environmental consequences. Measurable quantities are the yardsticks of a society's success. Material consumption, money, and all other 'economic' matters are the central focus of society and the sole measure of its progress.

The price of our economic gains, however, has been a social cost whose magnitude is only now being recognised. Competitiveness has spread from business to schools and even into homes, weakening personal bonds. The old extended family has shrunk to the nuclear family, and social workers complain that parents spend too little 'quality' time with their children. Mobility of both workers and factories has destroyed personal security and community solidarity.

This weakening of the one-to-one bonds which once provided both self-identity and physical security forced people to seek substitutes. These they found in social status, in the attainment of wealth and power, in the chance for 'upward mobility', if not for themselves, then for their children. 'Economic growth' and 'progress' became watchwords of society. No wonder economist Lester Thurow could write about human nature: 'Man is an acquisitive animal whose wants cannot be satiated'.[10] He was writing about the only 'man' he knows, Western man, whose worldview teaches that there is never enough.

For those on the lower rungs of society, vicarious identification with the successful is better than nothing. Belonging to an important nation makes up, at least a little, for the meaninglessness of one's personal existence. The same need explains the worship of distant rock stars and sports heroes, as well as the frequent arrogance of petty government officials towards the public they are hired to serve.

The final consequence of impersonal giantism is widespread alienation. As already noted, when people have nothing meaningful to identify with or conform to, they become lost, empty, lonely, a bundle of mysterious, unfulfilled 'needs'. Among the well-to-do, psychiatry is a flourishing profession.

These social symptoms characterise all large, highly centralised societies. The 'communist' Soviet Union and the 'capitalist' United States share almost equally the multiple problems of social giantism and of highly centralised decision-making systems. Since these superpower 'enemies' currently dominate global politics, a closer examination of their ideological bases and the structural problems they face is called for.

Western capitalism

Since it is widely supposed that Western capitalism is synonymous with democracy, it is necessary to examine this assumption, using the United States as a model.

The word 'democracy' comes from the Greek δῆμος (people) and κρατεῖν (to rule): hence, a government where supreme power is retained by the people. That power can be exercised either *directly*, as it was in ancient Athens where every citizen had the right to participate in government, or *indirectly* through elected delegates, as occurs in modern democracies. Consensual, participatory democracy characterised most tribal societies, and the old New England town meetings were a form of it. Direct democracy is best suited to small groups and tends to be dismissed in modern textbooks as 'unresponsive to the demands of civilisation'. Representative democracy is posed as the only option.

Modern representative democracy was born at the time when private property became the central concern of Western social thought, and the function of government was turning more and more toward economic matters. As political scientist Creel Froman points out, the United States Constitution was specifically designed to protect the private property of the few from the poor majority. In *The Federalist*, No. 10, James Madison flatly rejects political (let alone economic) equality, even as an ideal. In analysing Madison's argument, Froman writes:

> The advantage of a republic, then, 'consists in the substitution of representatives whose enlightened views and virtuous sentiments render them superior to local prejudices and to schemes of injustice'. The bias here, as elsewhere throughout the text, is clearly evident in Madison's choice of words for those he favors, the property owners ('enlightened', 'virtuous'), and those he doesn't favor, those who would be likely to compose a majority faction ('prejudice', 'injustice'). [11]

As Froman goes on to show in his book, *The Two American Political Systems*, the state of the national economy has become equated with the common good, and political freedom has been equated with free enterprise and the capitalist's right to the unfettered pursuit of profit. The upshot is a government that supports those who manage the privately owned economic engines for the presumed benefit of all.

This is not the way democracy is described in American textbooks, however, where much attention is paid to the holding of free and open elections, the freedom of the press, the existence of political parties, the Constitutional checks and balances among the departments of government, and the rights of citizens as laid forth in the Bill of Rights. This is the *symbolic* political system, in which the American public places great faith. Nowhere, however, are the socioeconomic consequences of such a government ever made clear; the *substantive* political system remains hidden. Political economist Michael Parenti writes:

> The symbolic system is highly visible, taught in the schools, dissected by academicians, gossiped about by news commentators. The substantive system is seldom heard of or accounted for. [12]

But while schoolchildren are taught to revere the symbolism of American politics, adults are told that society no longer needs an ethical base, since the free market works better than any ethical system ever devised. As Charles Schultz said, when he was chairman of the President's Council of Economic Advisers:

> Market-like arrangements . . . reduce the need for compassion, patriotism, brotherly love, and cultural solidarity as motivating forces behind social improvement . . . Harnessing the 'base' motive of material self-interest to promote the common good is perhaps *the* most important social invention mankind has achieved.[13]

When modern economics becomes the basis of politics and human nature is taken as self-rewarding, calculating and competitive, is it any wonder that so many of America's young exhibit a level of cynicism by which they exploitatively comply with rules they regard as meaningless – a description of modern 'Yuppies'. Self-centred, uncaring Hobbesian man is becoming more and more an unhappy, disaffected reality in modern Western society.

The political consequences of making economics the central focus of government and then dubbing it an irrefutable 'science', effectively prevents serious politial debate about entrenched economic institutions. As British economist Joan Robinson has pointed out, modern economic theory as developed in academia has been designed to support the ruling ideology.

> It is the business of the economists, not to tell us what to do, but to show why what we are doing anyway is in accord with proper principles. Certainly, there seems currently to be a strong tendency for professional advisers to governments in the western economies to justify, on economic grounds, what the political parties in power would wish to do anyway.[14]

This is by no means sheer intellectual corruption on the part of economists. They, too, are trapped within the general Western worldview. The thinking of Soviet economists is equally trapped within the Marxist worldview, so their advice is likewise politically 'acceptable'. In each case, untested assumptions are the basis of the 'value-free, scientific laws' that are claimed to flow from them. It is the *assumptions* that require political debate, which, of course, they almost never receive.

In capitalist nations, 'the economy' is 'scientifically' stated to work best under conditions of a free marketplace and free enterprise, thus carefully placing *outside* the political arena this most central of our political concerns. The major function of government is to keep something which it claims to have no control over, growing and competing. Parenti sums it up thus:

The health of the capitalist economy is treated by policymakers as a necessary condition for the health of the nation, and since it happens that the economy is in the hands of big companies, then presumably government's service to the public is best accomplished by service to these companies. The goals of business ... become the goals of government, and the 'national interest' becomes identified with the dominant capitalist interests.[15]

Political scientist Creel Froman makes the ideological connection:

All societies have their economic values, and private property and free enterprise are the American ones. In terms of the distribution of property and the control of wealth in this country, those who benefit most from our ideology are those who either possess great wealth or control it. Our major ideological pronouncements thus come from the business and corporate sector and from most public educators.[16]

In a free enterprise 'democracy', those holding power are not only beyond popular control; they are in a position to demand that the government supply the infrastructure that maintains a strong private economy: roads, public education, research and development, and a national defence. To be sure, under strong political pressure, capitalist democracies are forced to provide welfare services, to redistribute wealth and to establish minimum working conditions. But the main role is to strengthen 'the private sector'. Parenti concludes from all of this that:

Representative government is a very serviceable form of governance for capitalism...

The only way the state could redirect the wealth of society toward egalitarian goals would be to exercise democratic control over capital investments and capital return, but that would mean, in effect, public ownership of the means of production – a giant step toward *socialism*.[17]

Can such a powerfully centralised political and economic system change democratically? Or, as Reinhold Niebuhr suggested, are the economic forces so powerful that no government can control them without threatening social chaos?[18] If democratic change is to succeed, it will happen only if ordinary people are willing to participate *actively* and *thoughtfully* in their own government. We return to this thread later. First it is necessary to examine that other giant, centralised political system, the Soviet Union.

Eastern Marxism

Marx lived in the nineteenth century, at the peak of the brutalities created by the Industrial Revolution. Although he came from a moderately well-to-do family, he and his devoted wife, Jenny, lived their lives, often in poverty and frequently scorned, on behalf of the dispossessed. These were the urban proletariat, the wage-earners who toiled for bare subsistence wages in the sweat-houses of Europe, yet whose labour, Marx argued, provided the wealth that went to others.

Two basic premises run through Marx's thinking. One is Hegel's notion of the progressive unfolding of history and the other is that economics is all that really matters in human affairs. Hegel, an early nineteenth century philosopher, saw humankind embroiled in an ongoing struggle with Nature on the one hand, and with its own internal contest for power among social classes on the other. For Hegel, the course of history was predetermined, and could be predicted by examining the forces in play at a given time. Through Hegelian thought, Marx became convinced that a *science* of human behaviour exists, and once its laws are perceived, the future can be known, at least in its broad outlines. Furthermore, he believed that human behaviour could be explained in purely economic terms. He wholeheartedly accepted the Enlightenment's ideas that reason and rationality are paramount in determining our actions, and that quantifiable economic goals are the driving force within human societies. For Marx, economic relationships between classes were *the* determinant of history.

> The mode of production of the material subsistence, conditions the social, political and spiritual life-process in general. It is not the consciousness of men which determines their existence, but on the contrary it is their social existence which determines their consciousness.[19]

Marx's thinking has extraordinary parallels with capitalist ideology, for it places materialism and economic production at the apex of political concern and claims a 'scientific certainty' derived from equally untestable assumptions about human nature and the course of human history. Both Marxism and capitalism employ a fallacious form of 'scientism' to validate their arguments and hence preclude the possibility of *political* discussion regarding them.

Marx correctly predicted that capitalism would concentrate wealth in fewer and fewer hands. He also argued that working men of all nations, denied political status by their common capitalist oppressors, would rise up as an international brotherhood of the dispossessed and throw off their chains. Thus was born the concept of international communism, still a great bogeyman of capitalist politicians.

Marx also realised that in the first stages of their new-found power, the proletariat would be faced with many unresolved problems as to how to develop the economy and how equitably to share what it produced. To settle these initial shortcomings and prevent destructive bickering, Marx saw a temporary need for a Dictatorship of the Proletariat. Thus was born the second bogeyman of Marxism, the legitimisation at the outset of an undemocratic, authoritarian government.

The workers' revolts that Marx foretold failed to materialise, owing to the extraordinary technological success of the fossil fuel-driven Industrial Revolution, which gradually raised working-class living standards. Labour laws and welfare institutions further offset the worst abuses of capitalism. As American economist John Kenneth Galbraith pointed out, in the more socialist democracies of twentieth century Europe, most of the major demands of Marx and Engels's *Communist Manifesto* were met through liberal democratic reforms. These included a progressive income tax; public ownership of railways, communications and a number of other industries; free education; and the abolition of child labour. Only public ownership of land, deurbanisation, and a state banking monopoly have so far failed to materialise.[20] These changes, democratically fought for and won by the working classes and their liberal supporters, staved off Marx's predicted revolution. It remains to be seen what will happen in highly centralised, class-structured societies like Great Britain and the United States when economic growth tapers off and reverses – what Burton terms '*un*development'.

As Niebuhr predicted over 50 years ago,[21] the countries where communist revolutions have occurred have all been agrarian societies where impoverished and landless peasants overthrew elite landowners and tax collectors who were living in luxury at their expense. This occurred first in Russia, but communism, as an alternative to the exploitative capitalism of neo-colonialism, has had a natural appeal among peasants in many Third World countries.

Marxist revolutions have thus occurred mainly among politically naive and initially illiterate peoples, largely devoid of modern scientific knowledge or technological skills. In 70 years, the Soviet Union has achieved an amount of technological change that required 200 years in the West. Furthermore, access to education, health care and other forms of social security is now available in the Soviet Union on a far more equal basis than in the United States.

Nevertheless, the Soviet Union shares social symptoms similar to those in the United States. Its achievements have been carried out through a highly autocratic, centralised bureaucracy that has left the great mass of people outside the decision-making process. There has grown up a widespread 'lack of order', signalled by an increase in crime, alcoholism, divorce, abortion and class conflict, together with declining worker productivity which leads to chronic shortages.[22] An illegal underground economy, like bartering in the United States, accounts for 10 per cent of Soviet GNP.[23]

It would be foolish to presume, however, that the rank and file of Soviet citizens would welcome 'liberation' by the West from an 'oppressive'

Table 14.1 Characteristics shared by United States and Soviet societies

	United States	USSR
Economic materialism as central focus of society	'Free enterprise' is the basis of 'democracy'	'Dialectical materialism' is basis of Marxist 'communism'
Ideology based on 'scientific' arguments	Material self-interest 'promotes' the common good	Working class dominance is 'scientifically' inevitable
Centralised, secretive decision-making by an elite class, that excludes masses; elites not elected	Elites in private corporations, supported by government	Elites in Politburo and central planning bureaucracy
Media support ideology	Major media, owned by elites, seldom question concept of 'free enterprise'	Government controls media outright
Underground economy	8–10 per cent of GNP is not reported to IRS* and lies outside legally approved transactions	About 10 per cent of GNP is outside legally allowed transactions
Growing alienation and disillusionment	Unemployed, especially minorities; retired, draft resisters and pacifists. Increased crime, alcoholism, etc.	'Silent bungling'; strong informal peace movement, army careers unattractive. Increased crime, alcoholism, etc.
Significant government coercion	Gaoling of draft resisters, and of church workers offering sanctuary to refugees from Central America, etc. 'Law-and-order' emphasised by politicians of all political stripes	'Law-and-order' emphasised by recent political leaders until Gorbachev

*Internal Revenue Service, the federal taxing authority.

regime,[24] any more than disaffected Americans would welcome the destabilisation of their government by foreign agents, no matter how loudly they complained of it themselves. Rather, we are observing the internal disintegration of two giant social systems, both focused on materialism, and both based on rigid ideologies that have tended toward centralised, secretive and authoritarian decision-making carried out by a small group of elites. The great mass of people have been effectively excluded from meaningful political participation. The mass media within both countries, with but minor exceptions, has uncritically interpreted events according to the dogma of the national worldview.[25] Some of these similarities of the two systems are summarised in Table 14.1.

Some consequences for superpower relations

The years since the Second World War, and particularly those from 1975 to 1985, have witnessed a global politics dominated by incredible tensions between these two superpowers who have few directly conflicting interests, yet have managed to define one another as deadly enemies. Why has such a confrontation been so artificially nourished?

Political scientist John Burton has provided a plausible explanation. He argues that politicians on each side, fearful of internal collapse if any faults were admitted, took to magnifying the threats of the other – the 'evil enemy' – while bolstering their own domestic policies. During this period, each side made more and more of its decisions in secret, telling the public secrecy was necessary on grounds of national security, an argument Burton dismisses as largely baseless.[26] A public accustomed to listening passively and not actively participating easily falls for simplistic rhetoric and readily believes complex problems have easy solutions. Once again, the words of Richard Lamm, former Governor of Colorado, make the point:

> I know some politicians who as captains of the *Titanic* would persuade the passengers they were only stopping for ice.[27]

A major consequence of the internal failings of the two great giants and their ideological inability to address those failings has been a growing global instability. Says Burton:

> It is in this perspective that we need to examine US–USSR relations, and world society in general. Our problems are not mainly international in the sense that there are some unique features of relations between states that lead to tension and to conflict. Our international problems are a spillover of domestic problems . . . Capitalism and communism are a threat to each other, not because the authorities in each threaten each other, but because each feels threatened by its own failings.[28]

In the 1980s, and particularly since 1985, the internal politics of the two superpowers have diverged noticeably. The United States, closely followed by Great Britain, has retreated even further into government secrecy, while simultaneously placing even more power in private corporate hands, thus widening the gap between the symbolic and substantive political systems and removing the people still further from control over their own lives. Meantime, the Soviet Union, hardly a place where people in the past have had much control over their lives, has initiated a revolutionary movement called *perestroika* – 'restructuring' – and begun opening up, democratising and decentralising its society. Details of this are pursued later in this chapter; here we simply remark the concurrent closing up of one side and opening up of the other.

THIRD WORLD CLIENTS: INTERNATIONAL POLITICAL HEGEMONY

On 6 November 1984, the government of a Third World nation suddenly declared a 'state of siege'. Overnight it arrested opposition leaders, raided the offices of dissident organisations, stopped publication of six of the nation's seven newspapers, prohibited all political news on the media, and arrested 6000 impoverished demonstrators in the capital. In the days following, masses of people flocked to the Catholic churches for sanctuary and the government refused to allow the church leader, known as the 'Vicar of Solidarity', to return home from Rome.[29]

American students are astonished to learn that the country in question is not Poland, but Chile.[30] Run by a military regime that, with US help, overthrew the last democratically elected government and assassinated its leader, President Salvadore Allende, in 1973, Chile has alternated between a 'state of siege' and a slightly less severe 'state of emergency' ever since.

The nations of the Third World, most of them emerging from colonialism with the legacy of a dual economy, have provided a convenient stage onto which East and West have projected their ideological struggle. Although there are many historical, cultural and social differences, the general relationship between Third World countries and the two superpowers is that depicted in Figure 14.2. On the whole, 'independence' left in control rulers, supported by urban elites and rural landowners, who live according to Western standards while the great mass of people suffer deprivation, poverty and unemployment. These are the exploding populations of landless peasants who are crowding into the slums of Bombay (11 million), São Paulo (17 million) and Mexico City (21 million).[31] Rulers in such countries, often dictators insecure in their tenure of power, require a strong military to maintain order.

417

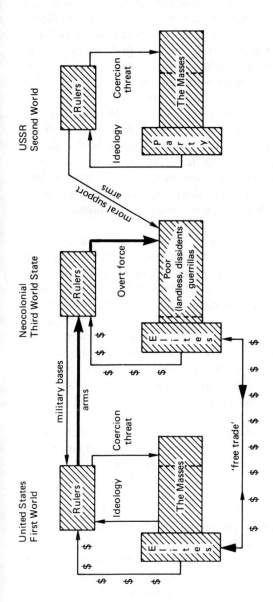

Figure 14.2 Relations between neocolonial states in the Third World and the two superpowers.

The dual economies of many developing nations place economic power in the hands of a wealthy elite who control political power as well. The overt force needed to maintain order among the poor majority is obtained from the First World nations, in return for military bases and access to local markets and a cheap labour force by multinational corporations. Dissident guerrillas are suppressed by massive force, using arms and often training obtained from First World governments. Moral support and sometimes arms for dissidents are supplied by Second World governments.

In cases where socialist governments have come to power in the Third World, the roles of the United States and USSR are largely reversed, except that trade is far less exploitative and there is usually less need for repressive arms to suppress dissidents, who are fewer in number. (These dispossessed elites, however, if supplied arms by the United States or others, can destabilise socialist governments as has been attempted recently in Nicaragua.)

The well-to-do capitalist elites naturally seek support from the United States and Western Europe, offering them access to military bases and other political advantages. In return, they receive not only financial aid and investments, but weapons and training of troops and police to control their own people.[32] The Soviet Union, meantime, lends moral support to the oppressed majority, seeing them as fellow sufferers of the capitalist yoke. They are more circumspect than is usually thought, however, about giving direct military aid until a sizeable guerrilla movement is under way.

In countries where revolutionary socialist governments have come to power, almost without exception they have remained authoritarian dictatorships. Yet on the whole such regimes are more concerned with the health, education and general welfare of their peoples than were their capitalist-oriented predecessors. The basic human needs of poverty-stricken peoples take immediate precedence over direct political participation, as expressed in the following words of a young student from Havana, Cuba:

> So you have two systems – one where everybody eats, has a job, no fear of unemployment, a shabby but free place to live, free medical treatment for all, free schools for all, safe streets, no racism, reasonable equality. And another where some eat well, where there is unemployment and chronic fear of unemployment, a shabby expensive place to live, expensive medical treatment, very poor education for millions, crime in the streets, racism, inequality. Yet the first is worse than the second. Why? Because there are no elections. And what is the purpose of elections? To make sure everybody eats, has a job, no fear of unemployment, a free place to live, free medical treatment and schools, safe streets, no racism, reasonable equality.[33]

But although physical security and social justice may take precedence over direct participation in the decision-making process, such participation cannot be indefinitely postponed without employing coercion.

US Patterns of Interaction

Official US policy toward Third World countries is defined in terms of 'strategic interests'. Pro-American governments are expected to provide military bases and access to such minerals as oil, tin, manganese, platinum and chromium. Yet a change of government, even to one blatantly anti-American, need not mean the loss of strategic bases. Fidel Castro, for example, has been unable to dislodge the Americans from the huge US Naval Reserve on Cuba's Guantanamo Bay. Furthermore, as the Center for Defense Information argues, farflung military bases would be less necessary if the United States were to stockpile vital minerals, as it now does oil. Tiny reductions in defence spending would easily pay for the

necessary stockpiles.[34] The official argument is that access to both bases
and minerals would be lost if client countries were to 'go communist'.
Moreover, whoever opposes a pro-American regime must automatically be
pro-Soviet. All opponents are given the blanket label 'communists', and
identified as part of an international conspiracy. Former US Ambassador
to the United Nations, Jeane Kirkpatrick, states the official American view
of Third World politics quite clearly:

> [N]on-communist autocracies are under pressure from revolutionary
> guerillas. Since Moscow is the aggressive, expansionist power today, it is
> more often than not insurgents, encouraged and armed by the Soviet
> Union, who challenge the status quo.[35]

To yield to the inevitable forces of political reform afoot in the world
would, in Kirkpatrick's view, mean a loss of face 'by aligning us tacitly with
Soviet clients'.

The United States, in fact, seems to be responding to the 'menace' of
social change in developing countries much as the British did to the
American colonists' demands in the 1770s. In each case, the dominant
power deliberately ignores political realities and attempts to impose force
to maintain an untenable *status quo*; both are examples of what historian
Barbara Tuchman labels political 'folly' – the blind pursuit of policy
contrary to self-interest.[36]

Curiously, despite the existence of a 'free press', mainstream American
journalists and media have unquestioningly accepted official government
justifications for foreign interventions. Ignoring alternative interpretations
widely printed in the foreign press, they automatically report world events
in ways that are consistently anti-Soviet and pro-American. As journalism
professor William Dorman observes, this black–white view of the world is
so pervasive throughout American society that even journalists are quite
unaware of their biased vision. Society and media mutually reinforce one
another:

> For the pictures about foreign affairs in the heads of most Americans
> are largely put there by the news media, and these pictures throughout
> the Cold War have been neither reassuring nor substantively at odds
> with official Washington's dark vision of the world . . .
>
> [I]t is precisely because [this] worldview is so widely shared inside and
> outside the journalistic system that it appears nonprejudicial.[37]

The 'teflon' age has ushered in an opportunity less for informing than for
*mis*informing, as it centralises the sources and outlets of information.
Most Americans are quite unaware of the frequency and extent of their

nation's attempts to shape the developing world to suit itself. Over the years, efforts have grown from old-fashioned gunboat diplomacy – 'send in the marines' – to include the less obvious arming and training of 'counter-revolutionary' forces, the covert overthrow of 'unfriendly' regimes and assassination of their leaders, and the spread of disinformation abroad for eventual consumption by the American media at home.

Since the Second World War, the United States has been involved in some 250 military interventions, of which only a few have been much noticed by the public: the 1961 Bay of Pigs disaster, when US-assisted Cuban exiles failed miserably in their attempt to overthrow Castro; the 1965 US invasion of the Dominican Republic that prevented reformist leader Juan Bosch from returning to the presidency; the tragic Vietnam episode; the 1983 lightning invasion of tiny Grenada; the 1986 bombing of Libya; the recent military presence in the Persian Gulf.

The United States is also a major supplier of arms. Of more than $100 billion in military aid to foreign governments in the three decades after 1950, two-thirds went to Third World countries. There were also private commercial sales of small arms, tear-gas grenades and 'crime detection equipment' to police forces of right-wing dictatorships. Tens of thousands of foreign military and police have been trained by US government agents, not only abroad but in American-based police academies, in the techniques of crowd control, torture and terrorism. Among the United States' major clients have been the ten Third World countries most cited by Amnesty International for abusing the human rights of their own citizens: Argentina, Brazil, Guatemala, Indonesia, Iran, Morocco, the Philippines, South Korea, Taiwan, and Thailand. In their book, *Supplying Repression*, Michael Klare and Cynthia Arnson state:

> [T]he evidence suggests that the United States is, and has been, deeply complicit in the proliferation of repression abroad, through authoritarian regimes and through *sales of repressive technology and techniques to those foreign government agencies directly responsible for political terrorism and the suppression of dissent.*[38]

The vast majority of Americans remain unaware that arms and skills supplied by their taxes to repressive regimes throughout Latin America are being used against peasants, priests, nuns, schoolteachers, social workers – anyone who is suspected of trying to organise, or even give assistance and succour to, the poor. They may be arrested and held without charge; often, indeed, they simply 'disappear' forever. Churches are machine-gunned; villages are bombed and napalmed; student groups, unions, doctors' and lawyers' organisations are banned; newspapers are closed; the civil courts are replaced by military tribunals:

On this wreckage of civil association has arisen a vast network of government spies, secret police, and para-police operations with their attendant torture chambers and death squads.[39]

Thus writes prize-winning journalist Penny Lernoux in her heart-rending book, *Cry of the People*. Attempts at self-help are labelled 'communist', even when sponsored by the Catholic Church, and are targets for violent repression.

The United States has also intervened to destabilise 'unfriendly' regimes through the covert operations of the Central Intelligence Agency (CIA): bribes; spreading of propaganda; infiltration of local unions; and outright assassinations by its 'Health Alteration Committee' of such foreign leaders as Diem in Vietnam, Trujillo in the Dominican Republic, and Lumumba in the Congo (now Zaïre). Here are CIA Director Allen Dulles' telegraphed instructions regarding Lumumba:

IN HIGH QUARTERS HERE IT IS THE CLEARCUT CONCLUSION THAT IF LUMUMBA CONTINUES TO HOLD HIGH OFFICE, THE INEVITABLE RESULT WILL AT LEAST BE CHAOS AND AT WORST PAVE THE WAY TO COMMUNIST TAKE-OVER OF THE CONGO . . . HIS REMOVAL MUST BE AN URGENT AND PRIME OBJECTIVE . . . THIS SHOULD BE A HIGH PRIORITY OF OUR COVERT ACTION.[40]

The CIA was also involved in the overthrow of Nkrumah in Ghana, of Mossadeq in Iran, Arbenz in Guatemala, Allende in Chile, Sukarno in Indonesia, and Goulart in Brazil,[41] and probably of Manley in Jamaica.[42] In Chile, the CIA struggled mightily, first to prevent the election of socialist leader Salvadore Allende, then to depose him. During his brief regime, US aid to the military who opposed him rose sharply, while food aid and development loans dropped precipitously. 'The much publicized "Empty Pots March",' Lernoux records, 'a supposedly middle-class women's demonstration against the Allende government was composed principally of the wives of high-salaried employees, managers, senior executives, and industrialists'.[43] The CIA brought 10,000 right-wing Chileans to Virginia for subversion training, and spent $8 million in Chile on bribes alone.[44] After Allende's death, thousands of his supporters were incarcerated in Santiago's National Arena, and hundreds were tortured and killed.

American political scientist Hans Morgenthau sums up US behaviour:

With unfailing consistency, we have since the end of the Second World War intervened on behalf of conservative and fascist repression against revolution and radical reform. In an age when societies are in a revolutionary or prerevolutionary stage, we have become the foremost counterrevolutionary status quo power on earth. Such a policy can only lead to moral and political disaster.[45]

The Iran–Contra scandal, which rocked the American public in 1986–7,[46] is only the latest in a long series of secret, often illegal, and certainly 'unAmerican' covert operations by which the United States government has attempted to mould the world to suit its own interests rather than those of the Third World peoples it manipulates. By seeing a 'communist plot' behind every attempt at land reform, nationalisation of industries and the redistribution of wealth to alleviate poverty and injustice, the United States has committed itself to supporting regimes whose 'enemies' are their own people. As foreign policy analyst Richard Feinberg argues, American behaviour has left many Third World leaders cautious, even pro-Soviet:

> The ideological and cultural biases of U.S. diplomats and policy-makers have contributed to the reflexive identification of U.S. interests with local business elites and – to a less degree – with military officers . . .

> But once U.S. interests are properly narrowed, the fostering of non-alignment, defending of national sovereignty, and tolerating of ideological diversity become the most efficient means of protecting those interests.[47]

The United States, in other words, has yet to discover that its own long-term interests lie in free and autonomous development of nations around the globe, the very thing which in principle it claims to stand for.

Soviet Patterns of Interaction

Despite Marx's vision of a historically inevitable global communist brotherhood, the Soviet Union's recent political exploits have been somewhat less grandiose than those of the United States. It assuredly retains hegemony over Eastern Europe. But outside Cuba, nowhere is 'communism' synonymous with unalloyed allegiance to the Soviet Union. China will never again be subservient to Soviet aims, nor will Yugoslavia, Albania, nor probably most of the Third World communist states. Marxist nationalism is not equivalent to a global communism dominated by the Kremlin. The Soviet Union, like the United States, is thus starkly faced with the real limits of its own potential influence in the world.

Soviet adventures abroad only began seriously after the Second World War. The question of Eastern Europe is postponed to Chapter 15. Elsewhere, despite modest assistance to the Chinese communists in 1946 and more substantial aid to North Korea in the early 1950s, major exports of arms by the USSR really began only in 1955, after the Soviets had acquired a nuclear capability that enabled them to stand up to the United States. Arms went mainly to governments of emerging Third World states hovering between capitalism and socialism: Afghanistan, Iraq, Indonesia. Except for aid to Castro's overthrow of the US-supported Batista dictatorship in Cuba in 1959, early Soviet efforts to back client states in their struggles with the West were largely unsuccessful. By the 1970s, however, the USSR had developed the capability for massive airlifts of arms and troops, which were then used in Angola and Ethiopia. The USSR now provides military training for Third World forces, although many countries are loathe to send trainees there owing to the heavy ideological indoctrination they receive. It supplies military advisers; it even supplies troops, most often Cuban. Since the Second World War, the Soviet Union, like the United States, has been involved, directly or indirectly, in over 250 incidents of conflict around the globe.[48]

The motivations for these interventions are almost always political. The USSR has few economic investments to protect and is not dependent on external supplies of 'strategic' resources; in fact, the support of Cuba is a heavy economic cost to the Soviet people. Although the Soviet Union does supply economic aid, its major thrust has been supplying arms. It is the Third World's foremost weapons supplier, although the combined arms exports of NATO countries slightly exceed it overall. Many countries buy from the USSR because of its easy credit terms, although in recent years the Soviets, like the West, have begun demanding cash. Nevertheless, client states are carefully screened for their political appropriateness prior to sales, and widespread Western belief notwithstanding, support for revolutionary movements is a very minor part of Soviet arms exports. According to the German sovietologist Joachim Krause:

> The fact is that Moscow as a rule seeks to maintain correct relations with governments in power within the context of its military aid policy. Support for guerrilla movements and national liberation movements makes up only a very small part of Soviet arms supplies.[49]

Given its presumed ultimate goal of fostering Soviet-style socialism in its client states, how successful has Soviet foreign policy been? On the whole, less than the expenditures would suggest. As Krause notes, the Soviets have achieved a few foreign bases in exchange for weapons; they have stockpiles of weapons in Libya and Syria that would be useful in a crisis; and they have but three client states so dependent on them that they are more or less controlled by the Kremlin. Thus only a few nations among the Soviet clients are loyal to the USSR. Global 'communism' is scarcely monolithic. Indeed, nationalist socialism, independent of *either* super-power, seems to be on the increase, and Soviet indoctrination is less and less welcome. Third World countries want, besides arms, economic and technical aid which has not been forthcoming. And their loyalty to the USSR as purchasers of arms is fickle; the global arms market is a buyers' market. To hold onto their clients, the Soviets are discovering that they must infuse massive quantities of economic and financial aid as well.

Like the United States, then, the USSR has discovered that global hegemony is receding out of reach. If, as Burton argues, successes abroad are needed to offset domestic failures, the price is escalating and the results are rather poor.

Terrorism

No discussion of the uses of force in international politics would be complete without a reference to terrorism. International terrorist acts are carried out by men and women who feel politically disenfranchised. Their aim is not to eliminate specific persons; their targets are generally unknown to them and without large political influence. Their goal, rather, is to obtain a particular political response: the return of political prisoners, the right to a homeland, or perhaps only the right of political expression within some larger framework. At the very least, they want to draw global attention to themselves.

In the latter, they succeed enormously. The incredible emotional response of a nation of 240 million Americans to the taking hostage of 50 fellow citizens at the American embassy in Teheran, for example, was out of all proportion to the number of people involved. A far greater number of equally identifiable children are annually stolen from their parents and never returned, but this does not cause nightly attention on the news. Terrorism is indeed a potent political force, especially when it is not just a matter of random bombings but the taking of hostages. Although sympathy for the terrorists' cause is scarcely enhanced, practical results are often forthcoming, even when not admitted by the governments affected.

At first, the West saw terrorism as one more Soviet strategy for creating global tension. But it has become clear, as Soviet citizens have also become targets, that it is a global illness that needs attention. Spokesmen for the major powers decry terrorists as 'international outlaws' or 'crazy extremists', that is, people who fall outside all accepted norms of human behaviour. But as John Burton states, political deviance, whether domestic or international, is nearly always a manifestation of the frustration of truly legitimate human needs. Terrorism is one such form of deviance:

It is a symptom of a problem. Behind the terrorist is a community of feeling and a support group. Reprisals and punishments merely create an increased sense of frustration and injustice. Today's terrorists are tomorrow's national leaders. Labels such as deviant, dissident, terrorist, revolutionary and rebel may reasonably apply to emotionally disturbed individuals who need sympathetic treatment and even control. No less they may be labels that apply to progressive and concerned leadership of peoples who have legitimate goals to pursue. It is important to differentiate between the two. Labelling is usually dysfunctional and a defensive device used for political purposes.[50]

The ongoing failure of world opinion to denounce acts of terrorism carried out by oppressive governments (of both left and right) against their own people, while vociferously condemning acts of terrorism by the dispossessed, only postpones even further into the future the eventual decline and elimination of political violence from the Earth.

THE GREENING OF THE WORLD

Green is the colour of hope – the ecologists' colour for the lifegiving green of the Earth and the hue appropriate to the grassroots movements shooting up in the cracks of the arid political arena of human affairs. Giant, top-down political systems, whether capitalist or communist, are simply unworkable. They become centralised, bureaucratic and, in their different ways, coerce and alienate their people, reducing individuals to mere ciphers and destroying the sense of real community human nature so deeply needs.

The remainder of this chapter explores the alternatives – the wide range of human-scale social systems that are already coming into existence around the globe.

Self-reliance: The Ideal Goal *Law of Consecration*

If top-down does not work, then we must look to bottom-up, to the evolution of more self-reliant communities to end both the psychological alienation of modern industrialism and the economic exploitation left over from the age of imperialism. The goal is to bring about the cohesiveness of a culturally distinct group of people who develop in their own way, independent of – although by no means ignorant of – the political, economic and psychological influence of other cultures. Roy Preiswerk, Director of the Institute of Development Studies in Geneva, describes self-reliance thus:

Self-reliance can only function where there is group solidarity. The pursuit of personal advantage, detrimental to others, and the private accumulation of goods are not compatible with collective self-reliance. This means that collective self-reliance can only operate where individual efforts are in favor of it. A self-reliant social system is somewhere in between an individualistic type of society and a centralized and authoritarian one.[51]

In the words of Norwegian sociologist Johan Galtung, self-reliance implies 'local decision-making on the basis of direct democracy';[52] everyone participates, not just politicians or 'experts' or corporate executives. It does not fit any modern ideological scheme, although it contains components of many. Self-reliance combines the fulfilment of basic material needs with a sense of meaningful community. It specifies no 'best' way, leaving unlimited opportunities for cultural diversity and ideological experimentation. As Galtung argues:

> [T]his movement toward self-reliance around the globe will continue, and not only because self-reliance seems to speak so much better to the human condition in all parts of the world than any other short formula in today's arsenal, or supermarket, of political slogans. It is because *self-reliance as a method is entirely compatible with self-reliance as a goal*. It is so different from competitive, social Darwinist capitalism or totalitarian, repression socialism with their visions of liberating human creativity and bringing about the brotherhood/sisterhood of all, while in reality forging new chains of enslavement.[53]

Self-reliance is often taken to mean economic autarky and cultural isolationism. This is incorrect. It is, rather, a way to *preserve* cultural integrity, to *prevent* economic exploitation, and to *protect and maintain* local environments. An excerpt from the 1977 Marrakesh Declaration of the World Future Studies Federation explains the relations in a world made up of self-reliant social units:

> Self-reliance will not reduce and does not aim to reduce exchanges among the various units making up our world. On the contrary, it gives them added scope and vigour, but at the same time it also transforms them, making them more balanced and more just. Conceived thus, self-reliance represents an open, pluralist system that takes account of the diversity of the world, for it is the tolerance and respect of this diversity that can best ensure the unity of all men.[54]

Self-reliance means decentralisation of political and economic power. It means, wherever possible, meeting basic needs from local resources, for example, by using local sources of renewable energy rather than becoming dependent on imported fuels. It implies planning for the future, and hence husbanding resources. It means preserving historical continuity and developing technologies consistent with cultural traditions.

By reinforcing social commitment and cooperation, self-reliance recovers a sense of social purpose. By insisting on widespread participation in decision-making, it eliminates extreme individualism and competition, thus easing social tensions. By meeting local basic needs in an equitable fashion, it dispenses with that curious notion that societies must compete in the global marketplace in order to survive. Self-reliance would thus bring an end to the sort of competition that encourages environmental destruction and to the misuse of science and technology for economic and political power. It would mean an end to secrecy and encourage the open sharing of knowledge. The incredible destructive potential unleashed by modern science makes an end to both competition and secrecy essential for human survival.

Since self-reliance means decentralising economic and political power and redefining political boundaries, its establishment globally will not occur without challenge. It will be resisted by those who most benefit from exploitation of the environment and of other humans. For developing nations, breaking current dependency connections with the two superpowers will not be easy. Nor will it always be easy to encourage the passive, the helpless and the exploited to participate actively. It is wrong to argue, however, as some do, that without the rewards of personal prestige and private wealth, 'human nature' will remain apathetic. This is true only in huge, impersonal societies. Within small communities, a sense of social participation and an opportunity to share are among the most important rewards of life. A former Peace Corps volunteer records the attitudes toward work and community in the villages of Sierra Leone in West Africa:

> The high value placed on work may be seen in the common greeting, 'Thank you for the work'. In many different languages the greeting is used in passing whenever a person is seen working, going to work or coming from work – in short, *constantly*. The important point is this: *there is no implication of ownership or personal benefit in the greeting*. Whoever expressed thanks for the work did not expect any direct benefit from the work being done. It seemed more a cultural recognition that work – any kind of work – contributed to the common wealth of the community. A farmer plowing a field, a blacksmith making a hoe, a woman carrying water, a weaver spinning thread; all deserved and received recognition for their contributions. . .

> The end result of invested labor was certainly important. But more than money, respect was the prized possession. And respect was earned through honest effort, and fulfillment of social responsibilities. A farmer whose rice harvest exceeded all expectations could not expect his bounty to go unnoticed. More visitors than usual could be expected around meal time. Distant relatives may leave their children at the house for months at a time. In short, responsibilities considered impositions in the West were accepted without question as the norm. A rich man still had to earn his respect. He was not envied merely for his wealth.[55]

In summary, self-reliance, far from implying isolated, independent individuals all competing with each other in a dehumanised world, means just the opposite: interpersonal bondedness and reciprocal sharing. In the developed nations, opportunities for sharing have been gradually eliminated. Nuclear families, boxed into four square walls of supposedly self-sufficient privacy, have lost all sense of sharing scarce resources or of feeling concern for the common air, soil and water on which everyone depends. Self-reliance offers the opportunity to reverse the twofold evils that plague the modern world: environmental destruction and the growth of dissatisfied, competitive and often unjust societies.

The Beginnings of Change

If humankind survives, it will probably be because more and more people come to recognise the necessity of actively participating in shaping the future. The greening of the world will be an evolutionary process, occurring in different ways and at different rates around the globe. Self-reliance does not imply any particular size of social or political unit. It implies, rather, self-reliance at many levels: of villages and neighbourhoods; of cities and their surrounding regions; of small developing countries; of larger regions that share a single environment. Self-reliance at one level may mean interdependence at another, but an interdependence between equals.

If self-reliance and reciprocal interdependence are essential to the human future, we can take heart from the multiple beginnings already occurring quite unconnectedly around the world. Some of these experiments may falter or even fail, but surely this is to be expected. How curious are technocrats who persuade us to have faith in our technological cleverness, but deny the possibility for inventing more viable social institutions; supposedly human nature can adapt to new technologies but not to new sets of values. But are human cleverness and adaptability really limited to technological fixes and nothing else? The examples of change already underway suggest otherwise.

The greening of Western Europe

Among industrialised nations, Sweden is the pioneer in self-reliance, developing not only regional democratic control of industry, but a plan for national energy self-sufficiency based on renewable forest biomass and hydropower. And even as interdependency in the European Economic Community grows, local regions are seeking more economic and political autonomy. In Great Britain, for instance, there is persistent pressure from the Scots and Welsh for increased self-determination. But most important are the Green parties springing up across the continent.

Although the first political party based on strong ecological principles began two decades ago in New Zealand, the idea has taken its strongest hold in Belgium and West Germany, with growing parties in Great Britain,

Sweden, France and several other countries. The parties are stronger in some cities and provinces than others, and have diverse interests owing to different local needs, yet they have sent a coalition of Green representatives to the European Parliament, and in Belgium and West Germany, Greens serve in the national parliaments. (This is possible in countries having proportional representation; any party receiving five per cent or more of the vote sends its share of representatives to the central government.)

Europe's Green parties share four central principles: (1) an ecologically sound economic base, and protection of natural resources; (2) a foreign policy based on disarmament and global justice, including cooperation with developing countries; (3) equal economic opportunity for all, including women and the disadvantaged, with work-sharing to ensure full employment; (4) grassroots, participatory democracy. The Greens represent an amalgamation of non-violent peace activists, feminists, environmentalists, and – at least in Germany – Marxists.

Green politics begins locally, with small groups meeting regularly to discuss current issues and agree on appropriate Green policies. One or two members of each local group participate in regional meetings, which may encompass a city and its surroundings. Growth is slow for a number of reasons. People simply are not used to 'doing' politics, to being actively involved. As grassroots democrats, the Greens are dubious of strong, charismatic leaders, and hence are often ignored or even denounced by the media. Finally, disagreements among 'pragmatic' compromisers and 'purist' factions within the parties themselves, especially in West Germany, make the development of a united political strategy a slow process.[56]

Nevertheless, with its motto 'Act locally, think globally', the German Green movement has had remarkable success since its founding in 1980, becoming by 1985 the third most popular party in West Germany, and the only major party not scandalised by illegal corruption.[57] The Greens in Great Britain now attract three million voters.[58] Although some see the Green parties as doomed due to internal differences and their reluctance to form coalitions with other parties, others see their open arguments as a sign of healthy democracy and their educational role as of more long-term political importance than gaining immediate power; it is whether Green ideas gradually become the accepted worldview that really matters.

Another kind of grassroots politics has arisen in Italy, where a chronically unstable central government has literally forced self-reliance on cities. An independent Communist party has linked up with liberal Catholicism to weave together urban politics, economics and culture, creating lively and flourishing cities. As one Italian worker says:

> Italy is a laboratory where Marxism is intertwined with the most creative part of Catholic thought. For both of us politics is part of our concrete lives and involves our social totality.[59]

Politics is a way of working out worldviews, not just a once-a-year trip to the polls. Everyone is expected to participate, not just career politicians. Cities such as Reggio Emilia, Bologna, Perugia, Genoa and Brescia are preserving their ancient centres from technological 'progress', creating strong neighbourhood communities, adapting their economies to regional needs and shifting emphasis from blue-collar to service jobs.

As three German news correspondents who accidentally 'discovered' Bologna in 1977 wrote:

> Direct democracy in Bologna's neighborhoods means the right of all interested citizens to take part in the most important administrative decisions.[60]

The result has been urban renewal and restoration of the medieval city; public housing for the needy; free public transport in rush hours, which eliminated traffic jams; laws to protect consumers and employees; a liberal school system designed to help the disadvantaged; preventive medical care; an end to unemployment, charity and ghettoes; and independence for the aged. The socialist government encourages a mixed, public–private economy and supports cooperatives. In Bologna, the citizens participate every day; governance is everyone's job.

According to the mayor of Brescia, political initiative throughout Europe is shifting away from national governments to the local level,[61] at the same time that the growing supranational government, the European Parliament, makes Europe as a whole more self-reliant. The old rigid boundaries between nation-states are becoming more blurred, less definitive.

The greening of Eastern Europe

'How can "red" be "green"? Isn't such a thing impossible?' Not altogether, perhaps. During the past decade, major changes have been occurring in the Second World which have been widely publicised only since Mikhail Gorbachev came to power in the Soviet Union. Some years ago, leaders in Hungary and Romania began quietly moving toward political independence from Moscow and adopting 'market socialism', based on Lenin's New Economic Policy. In a 1983 policy speech, then Soviet leader Yuri Andropov observed that the USSR could indeed learn something from its 'fraternal nations'. He urged that material rewards should be more closely linked to each citizen's work contribution, and he hinted at greater decentralisation of planning and decision-making. The Dictatorship of the Proletariat should give way to more local control. Finally, he formally opened the door to creative change by stating: 'Marxism is not a dogma but an effective guide to action'.[62]

In the three years since his protégé, Gorbachev, came to power, all the major changes hinted at by Andropov have been initiated. For the first time policy arguments that used to be held in secret in the Politburo are being aired publicly in the press and electronic media, and the public is invited to participate. There is election of managers in enterprises, multiple candidates for public office, encouragement of small-scale private businesses, closure of inefficient operations and, above all, an invitation to the average citizen openly to participate in social change. *Glasnost* (openness), democratisation and decentralisation are the watchwords. As Gorbachev writes, widespread participation is the key:

The original meaning of the concept of 'socialism' . . . as an ideological and political movement of the masses, a grass-roots movement whose strength lies primarily in man's consciousness and activity, has again come to the fore.[63]

The people will become self-reliant by being made *locally* responsible:

We are contemplating democratizing planning. This means that plan-making – not formal but actual – will begin within enterprises and work collectives . . .

We envisage broadening openness at all stages of planning, and introducing wide discussion of state and regional social, economic, scientific, technological and ecological problems.[64]

For the first time in their lives, today's Soviet citizens are learning what is going on in their country and the world without relying on an underground press. Although Westerners may belittle the lack of political parties and 'choice', they might well ponder whether Soviet citizens may not be gaining more democratic control over their economic lives than now exists under capitalism. It is an experiment worth watching.

The greening of the Third World

If greening comprises self-reliance, democratic grassroots decision-making, social justice, *and* a concern for the environment, then perhaps no Third World country fits the description. There are, as already discussed, several reasons why they should so far have failed. There is the widespread apathy left over from the paternalism of the colonial era that can be overcome only by the slow rebuilding of self-respect through grass-roots efforts. Such change is often impeded by the existence of dual economies and an elitist, often oppressive government. With few exceptions, development has been a top-down, autocratic process.

When symptoms of greening have emerged in Third World nations they have often failed to develop because of the kinds of covert and overt interventions, mainly by Western capitalist nations, that have already been described. Any country that nationalises major industries, seeks land reform, or otherwise behaves in a 'non-capitalist' fashion becomes a target for destabilisation. Nevertheless, a few countries have managed to achieve some of the characteristics of greening.

Within three decades after its 1949 revolution, China managed on its own to meet its basic needs. Today, it provides adequate nutrition and health care for a billion people where, a generation before, disease and hunger were prevalent. Despite its monolithic political system, China under Mao provided considerable economic autonomy to the communes, an autonomy that has recently been extended to individual families. Whether more political autonomy will follow remains to be seen. (China and the USSR appear to be on parallel tracks here.) Although China is not ignorant of its environmental problems, it has yet to address them seriously – and the ultimate need to live within its 'environmental income'. Morale in China remains high, however, and an atmosphere of self-criticism combined with flexible experimentation permeates society. Several other socialist countries – Zimbabwe, Tanzania, North Korea, Mozambique – are struggling toward similar goals with partial success. Mozambique is hindered by constant military harrassment from South Africa, and Zimbabwe has made enormous concessions to its ex-colonial white citizens in the hope of maintaining social stability.[65]

In Costa Rica, the most democratic country in Central America, a new Ecology Party, the PEC, has formed that has similar goals to the European Greens. As its founder, geologist Alexander Bonilla, says:

> The PEC will be concerned with the problems of housing, poor distribution of wealth and land, but will orient everything towards development without destruction.[66]

Environmental concern comes none too soon. In the past 25 years, over half the country's forests have been felled and before the year 2000, Costa Rica, like Nigeria, will have to import timber to meet its own needs.

Among the most hopeful, if still embryonic, signs of greening in the Third World are the Christian 'base communities' of Latin America. There is scarcely a country from Mexico to Argentina and Chile which has not recently experienced violent repression of the attempts by the poor to achieve self-reliance and by students, priests and social workers to assist them. Death squads, kidnappings, rape, torture and outright murder have slowly but inexorably turned a large number of Catholic priests, bishops and cardinals into non-violent political activists, helping the people to resist the terror. As Penny Lernoux observes:

> [T]he most dynamic sector of the Church is coming to believe that the only answer is to make over society from the bottom up.[67]

The message of the New Testament is in fact revolutionary, for it is about political and economic liberation. 'Liberation theology' means living out Christ's message on behalf of today's poor: to empower them; to give them self-esteem by teaching them their *own* history; to make them literate and aware of their legal rights; to help them organise into cooperatives and other self-help groups. Instead of believing their destiny is to be poor, the poor are being taught ways to seek, self-reliantly and non-violently, a better life. The *comunidades de base* (Christian grassroots communities) are more than just 'self-help cooperatives', however; they are creating a new social integration among the poor, a sense of identity, of belonging, and of participation.

So great is this movement that it seems to have a momentum independent of Rome. As Blase Bonpane, a former priest who helped introduce liberation theology into Guatemala in the 1960s, writes:

An irreversible dynamic has been released. It is the very catalyst of the hemispheric revolution.[68]

The greening of America

In his 1970 book, *The Greening of America*, Charles Reich predicted a new age.[69] The rebellious youth of the 1960s would bring an end to complacent materialism and revitalise America. But the 'flower children's' dream never arrived. A decade later, other writers claimed a new, transcendent consciousness was afoot that marked a turning point.[70] The key was meditation and group interaction. For several years, stickers with 'Smile' or 'Have a nice day!' were seen everywhere, and the US Post Office still sells 'LOVE' postage stamps. Nevertheless, a lack of community continues to permeate American life, and government is a distant institution that one is powerless to alter. Political scientist Robert MacIver observes that such disconnectedness threatens the very existence of democracy:

Democracy has proved to be unworkable where the majority of the people are politically inert, uneducated, unconscious of their unity or of any binding common interest.[71]

Despite the recent flag-waving and fireworks on the 200th anniversary of the United States Constitution, American political apathy is deep and pervasive. It is well-known that Reagan was elected President by less than a quarter of the eligible voters. Reasons for disenchantment are not hard to find. In a national government of over 3.5 million employees, only 537 are elected officials; the rest are appointed. The system (*not* specified in the Constitution) of electing representatives by districts rather than voting for parties, ensures that each candidate draws support from a multitude of diverse, often conflicting interests. As a result, Republicans and Democrats are scarcely distinguishable, a fact foreigners frequently comment on. Moreover, the plurality system of counting votes precludes any new political party having an electoral voice; they are seen as 'wasted votes'. Most Americans today are satisfied to identify passively with one or another charismatic personality projected on the television screen, for they see no way to participate more meaningfully. The centralised American political system is thus effectively stagnant, unable to offer any significant alternatives to the *status quo*. The solution is not to struggle against the centralised system, but to go around it.

Local, grassroots political change has indeed begun to sprout in various places in the United States. The voters of a number of cities have declared them as 'nuclear-free zones' – off limits to nuclear weapons.[72] Missouri voters have imposed a state sales tax specifically earmarked for conservation of soil and other natural resources,[73] and Oregon now has laws to protect its privately-owned forest lands and waterways from the excesses of development.[74] But perhaps most important is the sprouting of a Green 'bioregional' movement across the United States and Canada. The movement's local groups keep in touch through 'Committees of Correspondence' (named after the activist groups that played an important role in American revolutionary days), and they have begun to hold annual North American Bioregional Conferences at which vibrant discussions take place. The principles so far drafted are similar to those of the European Greens, but political emphasis remains local and regional. The nascent movement is focusing first on local issues that create community, but it certainly is not inclined to isolationism.[75]

By addressing *regional* concerns, where people are tied together by a common environment, a common local economy, and often a similar historic or cultural background, certain otherwise awkward problems could more easily be tackled. The smooth functioning of the St Lawrence Seaway, for example, is more important to Canadian and American communities that border it than it is to either Ottawa or Washington. Likewise, the numerous economic ties and environmental problems that beset the Mexican–American border states might again better be solved locally. Communities on both sides of the border suffer unduly when Mexico City and Washington bicker over such international incidents as the murder of a US narcotics agent in Guadalajara that resulted in closing of the border and much harm to innocent businessmen on either side. As former Congressional aide, Alan Sweedler, has put it:

> We thus have a situation in which the U.S. and Mexican border communities are close to being a country within a country, more similar to each other than their respective countries.[76]

If the cultivation of the public-minded person is the first step in insuring a participatory democracy, where else can it begin but at the local grassroots level? As David Mathews, former US Secretary of Health, Education and Welfare, says:

Being a good citizen is probably something that we cannot think about until we do it.[77]

SOME CONCLUDING THOUGHTS

New and diverse kinds of political organisation and political participation are beginning to occur around the globe. As people reorganise themselves, the old nation-state political boundaries blur, and both smaller, local units and larger, supranational groupings arise. If people are indeed to take charge of their lives, then direct participation in decision processes must become a universal social responsibility. The words of Pericles, spoken in ancient Athens in 430 BC sum up this notion very well:

Our constitution . . . favors the many instead of the few; this is why it is called a democracy . . . Our public men have, besides politics, their private affairs to attend to, and our ordinary citizens, though occupied with pursuits of industry, are still fair judges of public matters . . . We regard him who takes no part in these duties not as unambitious but as useless . . . [I]nstead of evoking discussion as a stumbling block in the way of action, we think it an indispensable preliminary to any action at all.[78]

An ongoing public discussion of the rules by which society is governed, and of the shared values from which those rules arise still seems the surest means for the peaceful, non-violent and just governance of any society. This is true whether the values of a given society are derived from the teachings of great religious leaders – Christ, Muhammad or Gandhi – or are blended and modified by society as a whole from the understandings of its individual constituents. For it is only through the active and universal discussion and sharing of values that any society can long cohere. Traditional societies use 'MOULD-TO-FIT' rituals to ensure the continuity of time-honoured values; societies undergoing change have the more difficult job of 'CRITIQUE/CREATE', which requires universal participation if it is to be universally accepted. In both cases, however, values have to be 're-created' in every new generation, for without them, the human spirit has no purpose, no meaning.

15 Nuclear 'Defence' – or Conflict Resolution

> In a culture where science is practically identified with virtue, and security with power, there is unrelenting pressure to translate vaguely stated political problems into clearly stated military ones.
>
> Anatol Rapoport, 1980s[+]

INTRODUCTION

At 11:00 AM on June 3, 1980, the Air Force officers monitoring the early warning system deep underground inside Cheyenne Mountain, Colorado, were struck with terror... [S]creens connected to the Nova Data General computer were flashing a warning. The Soviet Union had launched a large attack from its land-based missiles and strategic submarines... [T]he missiles would reach their targets in less than ten minutes.[1]

This was not an episode from 'Star Trek' or an update of H. G. Wells's *War of the Worlds*. It was a real alert. Bombers were about to take off; men in silos were ready to turn the keys that would fire off nuclear-loaded rockets. All because of a malfunctioning 'electronic chip, about the size of a dime, which costs forty-six cents'. On such potential failures hangs the fate of the Earth.

Arthur Macy Cox, who spent 20 years with the State Department and CIA, argues in his book *Russian Roulette: The Superpower Game* that not only are nuclear weapons growing more lethal, they are also escaping from our control. And there are more and more of them. The global security system now comprises 18,000 megatons* of TNT equivalents. One could hold the equivalent of a Second World War every day, without stopping, for 16 years, 5 months and 8 days (Figure 15.1). Most of these weapons belong to the United States and the Soviet Union. It takes only 30 minutes for weapons fired off in one country to reach the other, far less if they are fired from offshore submarines. Both sides live on hair-trigger alert. Soviet intelligence was put on full alert for two years during the early Reagan administration, anticipating a surprise attack.[2]

Of course, each side tries to prevent accidentally starting a conflict by having redundant control systems. Even so, both human error and the

---

*A megaton is one million tons; a kiloton is one thousand tons.

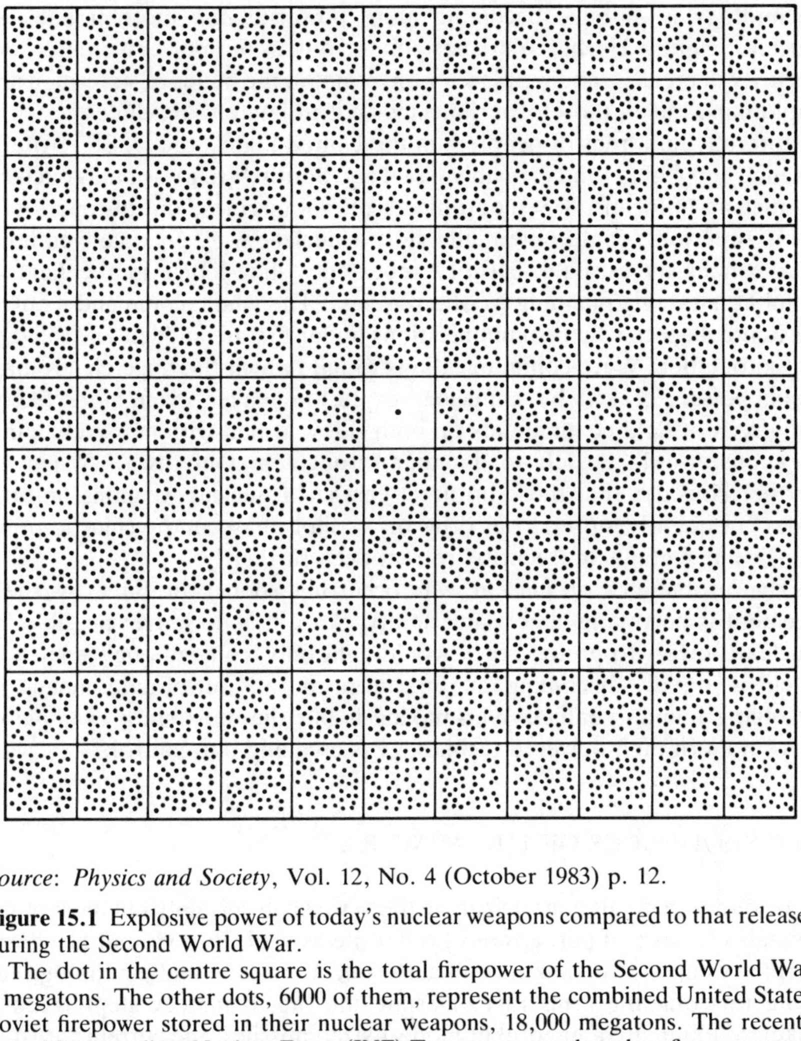

Source: Physics and Society, Vol. 12, No. 4 (October 1983) p. 12.

Figure 15.1 Explosive power of today's nuclear weapons compared to that released during the Second World War.

The dot in the centre square is the total firepower of the Second World War: 3 megatons. The other dots, 6000 of them, represent the combined United States/Soviet firepower stored in their nuclear weapons, 18,000 megatons. The recently signed Intermediate Nuclear Force (INF) Treaty removes less than five per cent of these nuclear weapons (less than six squares).

Even the proposed 50 per cent reduction leaves an equivalent of 3000 Second World Wars' weapons still in existence. Just two squares on the chart (300 megatons) could destroy all the world's large and medium-sized cities.

breakdown of automated controls are sufficiently common events to give rise to deep concern. As recent reminders we have the US Challenger spacecraft explosion in 1986 and the massive accident at the Soviet's Chernobyl nuclear power plant that same year which contaminated large areas of Europe with radioactive fallout. These were both caused by technologically undetected human failures. We live in an era when technology is at once becoming more powerful *and* more fragile. As nuclear scientist Herbert York has pointed out, as our power increases, our security decreases, and we place our lives not in the hands of statesmen and diplomats, but in those of technicians and, ultimately, machines.[3]

Literally hundreds of people – in the United States, the USSR, Britain, France and China (and probably Israel and South Africa) – could touch off a nuclear holocaust.[4] Senior officers on submarines, out-of-touch with their superiors, could potentially start a nuclear war. Or a conventional war can easily escalate. History repeatedly shows that in the heat of combat 'strategic planning' is often useless. The whole command–control system collapses when communications under fire are suddenly garbled or cut altogether. Were communications to fail (which the huge electromagnetic pulse released by a nuclear explosion makes even more likely) hundreds of field commanders on both sides, armed with massively lethal weapons, would be left to flounder through on their own.[5] When we further consider that those in direct charge of nuclear weapons on both sides 'frequently suffer from drug or alcohol abuse or psychiatric problems', our feelings of 'security' retreat even further.[6]

The probability of nuclear war occurring sooner or later by sheer accident is becoming almost a certainty. Since it is no longer possible for nuclear powers to pretend they would fight only 'conventional' wars, war itself is rapidly becoming an anachronism, an out-of-date means of resolving human conflict. War-making is but the most stark example of how modern high-powered technology is escaping our control.

CONSEQUENCES OF THE ARMS RACE

If we detonate all our nuclear weapons we shall certainly kill most of the human species and perhaps most other life as well. But without firing one further weapon, humankind still suffers economically and psychologically from this ongoing arms race. Despite the first laudatory steps recently taken to eliminate a handful of modest-sized missiles, the current thrust of both superpowers is still to upgrade its weapons and out-manoeuvre the other technologically. For each ante of the United States, the Soviet response has been 'See you and raise you'. We are still a long way from having put the threat of violence – which today means *nuclear* violence – behind us. To do so must be one of humankind's most urgent tasks.

The Meaning of a Holocaust

The intentional mass extinction of some eight million innocent Jews is known today as 'the holocaust'. Every thinking person condemns that episode as perhaps the most barbaric in all of human history. But what word would we use to describe the deaths of 80 million, or 800 million, or even billions of our kind in a nuclear war? Are they all belligerents? Should they all have died? What are the 'rules' of life and death in a nuclear age? What conceivable *purpose* would a nuclear war serve? Military men with actual field experience have generally agreed that nuclear weapons may deter war but are of little use in waging it.

Chapter 1 of this book discussed the possibility of a 'nuclear winter' following even a modest exchange of nuclear weapons. Amazingly, disagreements among scientists over exactly how bad the effects on climate might be are being used to discount the dangers of a nuclear war altogether; if nuclear winter is less likely, then a nuclear war again becomes 'thinkable'. Despite years of public lecturing by dedicated members of Physicians for Social Responsibility and others, the average person, unaware of her or his complete dependence on a gigantic yet fragile manmade network to supply water, food, electricity, gasoline, information and emergency medical care, still cannot imagine a world in which that network is largely destroyed. Even the relatively 'mild' events experienced by Hiroshima survivors begin to picture what a world after a nuclear holocaust would be like:

> There were cries for help from women, children, and old people pinned under houses or crushed between pillars. The fire spread so rapidly that two hundred eighty thousand people died from burning, asphyxiation, drowning, and being crushed.
>
> Ken Nakagawa

> On each raft were one noncomissioned officer, three soldiers, fifty injured people, and about twenty of their relatives taking care of them. We could do nothing for the injured people but give them water. A girl spurted blood from her artery when the pressure bandage was taken off. There was a man whose face looked like a broken watermelon whenever he moved his mouth. There was just one medical orderly on five of such rafts!
>
> Yoshimi Hara

> There were few people to be seen in the scorched field. I saw for the first time a pile of burned bodies in a water tank by the entrance to the broadcasting station. Then I was suddenly frightened by a terrible sight ... There was a charred body of a woman standing frozen in a running posture with one leg lifted and her baby clutched in her arms. Who on earth could she be? [see Figure 15.2].
>
> Yasuko Yamagata[7]

Source: Japan Broadcasting Corpn (NHK) (ed.), *Unforgettable Fire: Pictures Drawn by Atomic Bomb Survivors* (New York: Pantheon Books (Random House), 1977) p. 53.

Figure 15.2 Hiroshima after the Bomb.
Reproduction of part of Yasuko Yamagata's painting of her recollections of Hiroshima on 7 August 1945, when she was 17 years old.

As Helen Caldicott, the founder of Physicians for Social Responsibility, reminds us, not only were the bombs dropped on Hiroshima and Nagasaki small by present-day standards. They were dropped in isolation.

> After those bombings and the initial horror, what became crucial to the psychological, as well as the physical, survival of the citizens of those cities was the knowledge that a world existed outside the disaster area: doctors, medical supplies, food, clothing. Today, given the sophisticated weaponry with which we have equipped ourselves, and the inevitability that a small-scale conflagration would escalate into a massive holocaust, nothing but an unanswering wasteland would remain.[8]

We have created a weapon that can *never, ever safely be used* – and yet we go on building them, upgrading them, and replacing them. Incredible! A. Barrie Pittock, an Australian atmospheric scientist, puts it like this:

If all these possibilities do not leave any normal human being feeling that we and our leaders have gone mad, then something is really wrong with human nature.[9]

Economic Effects of the Arms Race

Even without the detonation of a single weapon in anger, the arms race is taking its toll. There is its growing impact on the environment. Both the Soviets and Americans have escalated the accumulation of dangerous garbage orbiting in space by intentionally destroying satellites in antisatellite weapons tests.[10] On Earth itself, in addition to the destruction caused by underground weapons tests, there is the accumulating problem of dealing with the radioactive waste generated by the military. In the United States, this high-level waste, some of it stored since the 1950s in 'temporary' tanks now grown leaky, is posing a major hazard, but the Department of Energy is reluctant to spend part of its budget to solve the $1 billion problem.[11] Yet this is a tiny cost compared to the overall economic cost of the arms race, and it is to this cost that we now turn.

Every year, the world spends $900 billion on military-related activities: soldiers, guns, bombers and the like.[12] This amount, more than the annual income of the two billion people in the world's poorest countries, is essentially 'wasted' because it contributes nothing useful to the lives of people. In fact, many causes of social unrest, conflict and violence could be removed if the resources spent on the military were used for social benefits instead. Ruth Leger Sivard, in her biennial booklets, *World Military and Social Expenditures*,[13] shows that just *one-tenth* of the world's military budget could pay for *all* the following: convert military R & D to civilian needs; give vocational training to disabled persons around the world; provide safe drinking water globally within a decade; provide food for hundreds of millions of hungry people; give assistance to small farm holdings to increase their yields; provide primary schools and train teachers for hundreds of millions of illiterate people; support research on nuclear waste disposal around the world; replenish the world's forests by a fivefold increase in tree planting; provide a worldwide clean-air programme; immunise the remaining 90 per cent of the world's children against killer diseases; provide family planning in Third World countries where demand now exceeds supply; give vocational training to 50 million young people each year.

Sivard also shows that weapons are growing far faster than the quality of life. Since the Second World War, literacy, income, education and suffrage have increased two to five times globally, while the firepower, lethality and destructiveness of weapons have increased between 25 and 250 times. Every year, a greater fraction of the global economy is being deflected for 'defence'.[14]

Nor are the wealthy nations immune from the devastating effects of the arms race. Among the leading ten developed nations of the world, those having the *slowest* annual growth rates in manufacturing productivity are the USSR and the United States, which deflect the largest proportion of their GNP to the military. By competing for scarce resources – raw materials, energy, investment dollars, and scientists and engineers – the military sector deprives the rest of the economy. As shown by economist Lloyd Dumas, military spending in the United States not only deflects resources needed to upgrade productive potential; it also consumes capital that should go toward maintaining infrastructure, such as roads, sewers and water supplies. Furthermore, deficit financing of the military drives up interest rates, exacerbating both national and international debt problems. This leads to inflation, yet unemployment remains high since military production requires far less labour and more capital than civilian production. The arms race thus contributes heavily to 'stagflation', while the non-competitive quality of American industry leads to a growing trade deficit. As Dumas concludes:

> Sustained distractive [military] spending at high levels has caused enormous damage to the economies of *more* developed countries. It has caused substantial economic deterioration in a nation as rich and economically advanced as the United States, playing a major role in ruining the extraordinary economic advantage this nation enjoyed only a few decades ago. It has undermined the capacity of as large and relatively well endowed a socialist nation as the Soviet Union to raise the standard of living of its population even to the level common in Eastern Europe. If distractive spending can do all this, if it can begin to 'undevelop' more developed countries, there is little question that its impact on the less developed nations of the world must be even more devastating.[15]

Or, in economist Kenneth Boulding's terse phrase:

> National defense is now the greatest enemy of national security.[16]

In more personal terms, the American military budget represents a daily transfer of $3.50 from every man, woman and child to the 'military–industrial complex' – the government bureaux, the military and the defence contractors (see Figure 12.1). Together with the research scientists and engineers who come up with ideas for new weapon systems in the first place,[17] this group represents a formidable lobby for the arms race in every weapons-producing industrialised country, even in the USSR. Furthermore, because defence activities are cloaked in a combination of patriotism and secrecy, public discussion which could lay bare the irrationality of the whole process is deeply hindered. Thus does the arms race maintain its own impetus from *within* each country, quite apart from any external threat.

To halt this upward spiralling race, a massive grassroots movement will be needed, with a workable plan for converting from military to civilian production. By recognising that for the same capital investment, non-military enterprises create far more jobs – teaching and health care create about twice as many – than does defence, there should be no fear of increased unemployment once the transition is completed. It is the transition process which requires planning, and this should include all sectors of the community: management, labour, politicians, financiers, educators, urban planners, and ordinary citizens. Together they determine how best to convert military bases, defence plants and the labour force into needed new uses, products and services. A plan is generated to raise the capital needed for converting land and buildings and retraining workers at all levels.[18] If this sounds similar to what was earlier discussed regarding the decentralisation of economies and the regaining of local community control of economic decision-making, it is because demilitarisation is but one part of the entire process.[19]

Nations that wean themselves from a 'permanent war economy' will obviously set free funds for domestic social programmes, such as education and job retraining, health and nutrition programmes, low-income housing and other often desperately needed projects. They will also release scientific and technological resources for improving health, energy, agriculture and our basic understanding of the world. (In the United States, for example, 72 per cent of Federal research support goes toward defence.[20]) Finally, they will reduce the high level of internal secrecy – both in the research sphere where it runs counter to the principle of free scientific inquiry, and in the public sphere where it interferes with the whole notion of an open, democratic society.[21] The present administrations in both the United States and Great Britain have greatly expanded the use of their respective official secrets acts to block public access to information, and openly claim that critics 'endanger' national security.[22] The Soviet government is not the only one that disapproves of 'dissidents'. The arms race, in fact, is causing Western democracies to lose the very things they claim they stand for: freedom of inquiry, freedom of speech, and access to the governmental decision-making process. This is taken up further below.

The Psychic Burden of the Arms Race

Even if we are willing to accept the economic burden of the arms race, there is another burden that is less obvious, but potentially as severe – the psychic burden people bear who live continuously under the threat of a nuclear holocaust. This burden is two-edged – it is a cause as well as a consequence of the military build-up. Gradually and subtly, the very existence of anything as terrible as nuclear weapons dehumanises us, making us more, not less, accepting of the upwardly spiralling arms race. People and nations alike become psychologically addicted to more and more weapons to maintain an acceptable level of 'security'. This addiction, of course, infects virtually all governments, the fears of one reinforcing those of the others.

A widespread consequence of living under the constant threat of a holocaust is psychic repression. When one feels helpless to act, it is impossible to maintain constant feelings of intense fear and retain one's sanity. Suppression is learned in early childhood. Many pre-schoolers by age four or five have vivid images of nuclear extinction – even though their teachers and parents have not discussed nuclear weapons and may not even know the children are aware. Some parents are uncomfortable on discovering their children have heard of such things. Since discussion is not invited, children quickly learn to suppress their fears: nuclear bombs are taboo subjects.[23]

Yet psychologists have shown that re-emergence of fear and the sense of a hopeless future frequently occurs in adolescence. The boxed text on p. 445 gives the responses of twelve-year-old school children to questions about world security and the future. They feel concerned not only about nuclear war itself, but also about adults' refusal to deal with it. Ultimately they suppress their concerns and join the conspiracy of silence, but the psychic impact is enormous. John E, Mack, professor of psychiatry at Harvard, explains:

> It seems that these young people are growing up without the ability to form stable ideals, or the sense of continuity upon which the development of stable personality structure and the formation of serviceable ideals depend. We may find we are raising generations of young people without a basis for making long-term commitments, who are given over, of necessity, to doctrines of impulsiveness and immediacy in their personal relationships or choice of behaviors and activity. At the very least these young people need an opportunity to learn about and participate in decisions on matters which affect their lives so critically.[24]

Similar findings exist for children in other countries.[25] Indeed, for human beings everywhere, psychological continuity between the past and the future is becoming ever harder to maintain. The young who seem best able to cope are those whose parents and teachers openly discuss their own fears and are actively seeking solutions.[26] Openness leads to empowerment.

Besides repression there is an even more subtle and dangerous effect permeating human society: psychic numbing – the turning off of *all* feeling.

RESPONSES OF FOUR SEVENTH GRADERS IN HAMILTON, NEW YORK, DURING A DISCUSSION ABOUT THE PROSPECTS OF THEIR LONG-TERM FUTURE*

How old do you expect to live to be?

- I can hardly say. There is the nuclear war, or something. There's lots of diseases. Maybe I can get frozen or something so I don't have to be in on it, and then I can just come back later.
- I want to live until I'm 95 and be the most famous lawyer in the world, if the world stays and becomes a little better.
- I hope to live to be about 95 or 100 but I know I won't. If the world keeps up as it is and if there is a nuclear war, I want to die before it.

What do you understand about the world that you feel adults do not understand? What do you know that adults don't know?

- I'm sure they realize what's going on, but they don't ever really think about it. Soon we are going to be in their position, and we are going to be solving all the things they are trying to solve now. I don't think they are really preparing us for it, because they don't realize that.
- That's it, I don't think they are preparing a world for us, I think they are just doing what they think they can now and just what comes into their minds. I don't think they are looking ahead or seeing how it is going to be later. I think they just think that one day we'll take over, but they are not helping us.

Are you happy?

- You mean generally or... no, I'm not happy, I don't want the world to be blown up. My God, I think that would be crazy, I mean...
- I'm happy right now, I'm not sure I'll be happy when it gets blown up.
- Yeah, but if the world's going to be blown up, I mean, God.
- Knowing that there's going to be, there might be a nuclear war, I'm not very happy, but if there isn't...
- Well, just thinking of that makes me kind of unhappy about it: thinking about how the world might be. But then again, if we think about how you wish the world would be, how you could help it, makes me feel better about what it's going to be like, or could be like.

---

*Selected quotations from E. Boulding, 'World Security and the Future from the Junior High Perspective', *Peace and Change*, 7 (4), pp. 65–76 (1981). These children are twelve years old.

People often experience it in war. Some of America's Vietnam veterans
still suffer from it years later – inability to experience love or compassion,
or even sorrow. Psychologist Robert Lifton reports one of the survivors of
Hiroshima recalling years later his experience:

> 'You know, I could see what was going on around me, I could see
> people dead or dying, but suddenly I felt nothing'.[27]

In a less severe way, society as a whole has become less compassionate,
more selfish, more angry. The alienation that twentieth century industrial
'efficiency' imposes reaches its peak in that 'highest' technology of all,
nuclear warheads and the missiles that deliver them. By deflecting precious
resources for 'defence' we are knowingly inflicting suffering on our own
poor, as well as aiming an unspeakable threat at others. By choosing to
acquiesce, we *must* dehumanise ourselves in order to suppress an unsup-
portable burden of guilt.

Psychic numbing all too easily leads to 'justified' brutality. The wielding
of massive physical power demands that those against whom it is used be
identified as less than human. This is the basis of wanton cruelty in warfare
– the My Lai village massacre in Vietnam, or President Truman's
justification for dropping atomic bombs on Japanese cities in August, 1945:
'When you have to deal with a beast, you have to treat him as a beast'.[28] In
times of peace, by treating those with whom we merely disagree as less
than human we 'justify' to ourselves the threat of massive force against
them. The possessor of such force imagines himself not only right, but
invincible. 'Who would dare to oppose the virtue that my power gives me?'
As historian Barton Bernstein shows, this was the delusion America
suffered shortly after the Second World War:

> The [atomic] bomb conferred upon policy-makers the mistaken belief
> in the omnipotence of the United States and encouraged them to believe
> that it could achieve their goals: an open door world of peace and
> prosperity, with the Soviet Union yielding to U.S. demands. [Secretary
> of State James F.] Byrnes' great expectations soon withered. Atomic
> diplomacy, though ultimately failing, contributed to the Cold War and to
> the arms race.[29]

Despite its obvious failure, this same bullying 'diplomacy' has been
pursued for over 40 years by both superpowers, and both have found it
necessary to dehumanise the other in order to justify the arms race and the
security it supposedly offers. The tiger we have by the tail is of our own
making, but we do not know how safely to let go. The rest of this chapter
explores this problem.

THE 'LOGIC' OF THE ARMS RACE

Few of us realise that, for all their elegance, the mathematically-based arguments about the 'balance of power', generated by theoreticians sequestered in 'think tanks', are largely beside the point. They are based on contrived abstractions about the military potencies of this or that weapon system – and hence of its ability effectively to threaten an enemy. But these arguments must perforce ignore all those undiscoverable quantities on both sides that determine what *actually* happens during a cold war or a hot one: the response of a propagandised or frightened public; the face-saving behaviour of egotistical leaders under pressure; the sudden breakdown of communications between politicians and military; the failure of trust among supposed allies, and so on. Because none of these can be quantitatively incorporated into the strategic equations, they are disregarded in the making of policy, and the resulting 'logical' arguments for the arms race have nothing to do with the real world.

Curiously enough, the people who belittle ecologically pessimistic futures models, such as 'The Limits to Growth', because they do not incorporate in their calculations the inevitable but unpredictable responses people may make to new conditions, are the very people who are most prone to accept the 'logic' of the arms race. How easily we have fooled ourselves into thinking that quantifiable weapons could substitute for unquantifiable, more diplomatic, more *human* approaches to peacekeeping.

This irrational pursuit of what superficially seems a logical policy no doubt has numerous causes, some of which have already been met in earlier chapters in other contexts. One, of course, is our 'scientism', our misplaced attempts to argue everything in quantifiable terms and to mistrust non-quantitative thinking. Another, perhaps more important, is the *internal* instability that characterises giant modern societies, particularly the United States and the Soviet Union. As John Burton observed, they threaten each other largely 'because each feels threatened by its own failings'.[30] Each government thus feels compelled to emphasise and exaggerate the threats of the other. The situation is exacerbated because each side, having different histories and national temperaments, repeatedly misunderstands the responses of the other. What follows is a brief historical summary of the evolution of East–West antagonism.

A History of the 'Communist Threat'

Denunciation of communism has become a test of patriotism in the West. The antipathy originated among the capitalist ruling classes in the nineteenth century with the writings of Karl Marx, but only rose to fever pitch after the 1917 Bolshevik revolution, which caught the West almost totally by surprise. It had largely ignored events inside Russia, including the failed mutiny of 1905 against the despotic Tsar. Until 1917, Russia and the West had been allies fighting Germany, but Lenin soon displeased the Western governments. To preserve the revolution, he made a separate peace with the Germans and refused to honour the Tsar's war debts, largely owed to the United States. Finally, believing in Marx's prophecy that the proletariat everywhere – the workers, peasants and common soldiers – were ready to overthrow their capitalist 'masters', Lenin vigorously encouraged formation of the Communist International, the Comintern. Yet while governments recoiled, many among the oppressed working classes of the industrialised democracies found new hope. In the winter of 1919–20, a million and a half recent immigrants to the United States from Eastern Europe, hopeful that the revolution meant a new life, applied for return passports.[31]

Not surprisingly, the conservative governments of the Allies, angered by Russia's sudden defection from the war and frightened by the possibility of workers' uprisings at home, reacted precipitately. America denied passports to the would-be returnees and clamped down on the resulting social unrest. Members of the International Workers of the World were tried by the hundreds for sedition. Jane Addams, a leading social worker and pacifist, saw a parallel in the panicky suppression of the discontented working classes in the United States with what had occurred over a hundred years earlier in England at the time of the French revolution.[32]

Lenin's formal offers of friendship with Western governments were ignored. Even before the Armistice, the Allies undertook military action inside Russia to overthrow the revolution. The 100,000 Allied forces already stationed in Russia's northern ports and Siberia joined with 30,000 French troops sent to Odessa on the Black Sea to give military aid to the Don Cossacks, the White Army faction still loyal to the Tsar. Simultaneously the British Navy blockaded the Baltic and Black Seas, Russia's only warm water ports. This Allied Intervention, strongly opposed by many Western liberals, lasted until 1920 when, under the relentless advance of the Red Army, the last foreign forces were ejected. During this time, the USSR was omitted from the Versailles Treaty negotiations and barred from the League of Nations. Few Western schoolchildren are taught about this episode in the history of East–West relations.

Russia has remained an official pariah ever since. Although Americans privately sent $70 million to relieve the Soviet famine of 1921–3, official US foreign policy had already been laid down by Woodrow Wilson. In August 1920, his Secretary of State, Bainbridge Colby, sent a statement outlining the US policy toward the Soviet Union to various foreign governments. Its last sentence read as follows:

> We cannot recognize, hold official relations with, or give friendly reception to the agents of a government which is determined and bound to conspire against our institutions; whose diplomats will be the agitators of dangerous revolt; whose spokesmen say that they sign agreements with no intention of keeping them.[33]

The tone was clearly that of a diplomatic deep freeze.

Many travellers to Russia during the 1920s, however, 'were favorably impressed by the toleration of limited dissent and by the social and educational reforms'.[34] Had Lenin lived another decade or so, it is possible that detente might have come about; US–Soviet trade was rapidly growing, despite the lack of diplomatic relations. But Lenin died prematurely and by 1928 Stalin had become undisputed dictator. His political concerns were totally internal. The Russian economy remained sluggish, so he instituted a 'crash programme' to modernise Soviet agriculture and industry, turning a backward, rural nation into a modern world power at whatever sacrifice.

Although Stalin had neither the intent nor the military strength to intervene abroad, Western fear of the Soviet Union grew with news of the bloody purges he conducted in the 1930s. Western socialists, still sympathetic to Marxist ideas, found it ever more difficult to defend Stalin's actions. The bloody rise and difficult defeat of another brutal dictator, Adolf Hitler, who had overrun most of Europe, reinforced the Western belief that *all* non-democratic regimes were not only brutal, but expansionist.

The end of the Second World War was thus a decisive moment in East–West relations. Soviet troops occupied all of Eastern Europe, including part of the German capital of Berlin. They were there as *allies*. But from 1917 until that moment in 1945 when a cessation of hostilities was imminent, Western leaders had refused to consider the USSR as a legitimate world power. As Sumner Welles, Roosevelt's Undersecretary of State, observed:

> [At the end of the First World War] territories [in Eastern Europe] which had been for centuries under Russian sovereignty and were considered by all nations as integral parts of Greater Russia were disposed of without regard for Russian rights, and without thought of what Russia might eventually do in order to secure their return.[35]

The Treaty of Versailles had treated both Germany *and* Russia as non-nations.

Among the Allied leaders in the Second World War, it was Franklin Roosevelt who accepted the reality of the Soviet nation as a future power. He saw the new Soviet presence as not only inevitable, but also as a balance of power preventing a resurgence of British and French colonialism.

At his meeting with Churchill and Stalin at Yalta in February 1945, just two months before his death, Roosevelt agreed with Stalin that the Soviets should have hegemony in Eastern Europe, which their troops already occupied. Later writers have often described Yalta as Roosevelt's 'sell-out' to the Soviets, ignoring both the historic reasons for Soviet interests in the region and the fact of their massive presence there.[36]

At that critical moment, Westerners were profoundly ignorant of the historical perspectives that made hegemony over Eastern Europe so essential to the Soviets. Exposed across open plains to Asia in the east and Europe in the west, Russia for a thousand years suffered repeated invasions from Mongolian hordes, followed more recently by Napoleon and then Hitler from the opposite direction. To protect themselves in the East, the Tsars had gradually extended Russian domination across Siberia, but no buffer had been erected against the West. In the Second World War, the Soviets suffered 20 million casualties, almost half the global total. By 1945, Russia was staggering from massive destruction and surrounded by antagonistic capitalist regimes. And she was 'backward'. Less than 100 years before, at the time the Americans freed the slaves, some 10 per cent of their population, the Tsar decreed an end to serfdom, freeing 60 per cent of Russia's population from bondage. The problems facing the United States in integrating a formerly suppressed minority were thus multiplied many times in Russia. Nor had the pre-revolutionary Russians any historical experience with democratic institutions on which the new Communist party could build. Regrettable though Stalin's terrible purges of his own people in the 1930s were, and inflammatory though official Soviet rhetoric seemed, both become more understandable given this background, one which few in the West appreciated – including those on whom it fell to make political policy.

In 1945, the USSR faced an outspokenly belligerent world. Roosevelt's death brought an inexperienced Harry Truman into the global spotlight, a man deeply influenced by his advisors. George F. Kennan, his new Ambassador to Moscow, sent a long telegram to the State Department in February, 1946, with this conclusion about 'the Soviet mind':

> Finally, it is seemingly inaccessible to considerations of reality in its basic reactions. For it, the vast fund of objective fact about human society is not, as with us, the measure against which outlook is constantly being tested and re-formed, but a grab bag from which individual items are selected arbitrarily and tendenciously to bolster an outlook already preconceived.[37]

In other words, Soviet thinking is inferior to ours, their worldview is less valid. Kennan also warned of Soviet 'penetration and command of key positions in administration' of foreign countries, and urged the government to educate the public 'to realities of the Russian situation'.

In September, Truman's Special Counsel, the young Clark Clifford, who knew little about the Soviet Union, prepared a top secret report to the President which in effect said: All the Russians will respect is power. There is no point in negotiating. Their drive toward world domination is like dealing with an onrushing river – it must be contained with dams. He invented the Domino Theory, which was to lead to US involvement in Korea and Vietnam. The 'dams' he urged included increased military strength, including 'atomic and biological warfare'; aid to all non-aligned developing countries; and an all-out government effort to 'inform' the American public of 'the record of Soviet evasion, misrepresentation, aggression and militarism'.[38]

In March 1946, Winston Churchill sent a chill through the hearts of the American people when he spoke the following words at Fulton, Missouri.

> From Stettin, in the Baltic, to Trieste, in the Adriatic, an iron curtain has descended across the Continent. Behind that line lie all the capitals of the ancient States of Central and Eastern Europe – Warsaw, Berlin, Prague, Vienna, Budapest, Belgrade, Bucharest, and Sofia. All these famous cities and the populations around them lie in the Soviet sphere, and all are subject in one form or another not only to Soviet influence, but to a very high and increasing measure of control from Moscow. Athens alone, with its immortal glories, is free to decide its future at an election under British, American, and French observation.[39]

It was the first time the world heard the words 'iron curtain', and the effect was instantaneous. Churchill urged an Anglo-American alliance to contain further Soviet expansion, a call he repeated in a speech in New York City three years later:

> I tell you – it's no use arguing with a Communist. It's no good trying to convert a Communist, or persuade him. You can only deal with them on the following basis... you can only do it by having superior force on your side on the matter in question – and they must also be convinced that you will use – you will not hesitate to use – these forces, if necessary, in the most ruthless manner... And that is the greatest chance of peace, the surest road to peace.[40]

Churchill's words encapsulate the political strategy that has been pursued by the United States and its NATO allies ever since. 'Peace' lies in the *believable* threat of unacceptable violence.

Churchill died an unregenerate believer in the inevitable necessity of military force and in the evilness of the Soviet Union. But both Clifford and Kennan came to regret their respective roles in initiating the cold war and the arms race. Clifford called it 'one of the most damn-fool things I ever did', and Kennan, in 1982, chastised the American people for their persistently distorted and jaundiced view of the Soviet Union:

I find the view of the Soviet Union that prevails today in large portions of our governmental and journalistic establishments so extreme, so subjective, so far removed from what any sober scrutiny of external reality would reveal, that it is not only ineffective but dangerous as a guide to political action.

This endless series of distortions and oversimplifications; this systematic dehumanization of the leadership of another great country; this routine exaggeration of Moscow's military capability and of the supposed iniquity of Soviet intentions; this monotonous misrepresentation of the nature and the attitudes of another great people – and a long-suffering people at that, sorely tried by the vicissitudes of this past century; this ignoring of their pride, their hopes – yes, even of their illusions (for they have their illusions, just as we have ours; and illusions, too, deserve respect); this reckless application of the double standard to the judgment of Soviet conduct and our own; this failure to recognize, finally, the communality of many of their problems and ours as we both move inexorably into the modern technological age; and this corresponding tendency to view all aspects of the relationship in terms of a supposed total and irreconcilable conflict of concerns and aims: these, believe me, are not the marks of the maturity and discrimination one expects of the diplomacy of a great power; they are the marks of an intellectual primitivism and naivete unpardonable in a great government...

And we shall not be able to turn these things around as they should be turned, on the plane of military and nuclear rivalry, until we learn to correct these childish distortions – until we correct our tendency to see in the Soviet Union only a mirror in which we look for the reflection of our own virtue – until we consent to see there another great people, one of the world's greatest, in all its complexity and variety, embracing the good with the bad – a people whose life, whose views, whose habits, whose fears and aspirations, whose successes and failures, are the products, just as ours are the products, not of any inherent iniquity but of the relentless discipline of history, tradition, and national experience.[41]

Yet into the late 1980s, Western leaders, particularly those in the United States, have maintained strong anti-Soviet feelings among their followers through a combination of outrageous rhetoric, intentional deceit and illegal acts. The world is familiar with President Ronald Reagan's virulent phrases; he has publicly called the Soviets 'the focus of evil in the modern world', consigned them to 'the ash heap of history', and claimed that communists 'reserve unto themselves the right to commit any crime, to lie, to cheat, in order to obtain' their goals. [42] By dehumanising the enemy, not only does he justify the arms race, but also the telling of lies to his *own* people, lest they lapse into detente or even friendship with the other side.

Any number of exaggerations, half-truths and lies have been used to convince the public not to trust the USSR and to support the arms race. One was the famous 'window of vulnerability' that Reagan told the American people existed between US and Soviet forces, which was based on false or distorted information, yet won him an enormous increase in the defence budget. [43] Another was the US State Department's claim that the Soviets had unleashed outlawed biological weapons against helpless villagers in Kampuchea, but the famous 'yellow rain' turned out to contain toxins of natural origin – namely the faeces of honey bees. [44] When the Soviets shot down a Korean airliner, in August 1983, that they mistook for a spy plane that had clearly violated Soviet airspace, American officials loudly denounced them for wantonly killing civilians. Yet the US intelligence community agrees that the Soviets believed they were 'dealing with an American reconnaissance plane', and others have argued that US listening posts should have detected that the plane was far off course and warned it. [45] Finally, the Reagan Administration has repeatedly stated that the Soviet Union has systematically violated its treaties with the United States, when a careful analysis shows that they have complied with all the principal requirements of SALT I and SALT II, and in order to do so they have dismantled or withdrawn far more strategic weapons than the United States. [46] Yet in each case, in spite of all the evidence against its allegations, the US government has never retracted one of them, preferring to leave the great mass of people misinformed about 'the enemy'.

Besides putting out misinformation, the US government or its officials have repeatedly committed illegal or near-illegal acts in recent years, among them the invasion of Grenada, the bombing of Libya, the mining of Nicaragua's harbours, and the sale of arms to Iran in order to obtain funds for supplying the Nicaraguan 'contras'. [47] Other funds, as noted earlier, were obtained from smuggling drugs, particularly cocaine, from Colombia via a CIA controlled ranch in Costa Rica into American cities. [48]

Curiously little notice has been paid by the so-called 'free press' – especially the mass media – to this gradual erosion of democratic principles. Although a public witchhunt equivalent to the era of Senator Joseph McCarthy has not emerged, labelling critics as 'unpatriotic' or 'communist inspired' seems sufficient to suppress public attention to what they have to say. Meanwhile, the Central Intelligence Agency was 'for the first time authorised to conduct certain kinds of intelligence-gathering operations inside the U.S.', and the Federal Bureau of Investigation now conducts undercover political activity by infiltrating the peace movement and the church groups offering sanctuary to political refugees from Central

America.[49] Inflated fears of a repressive enemy abroad have encouraged acceptance of repression at home. America is embracing a moral double standard: political repression is justified for 'us', but is not all right for 'them'.

With the December 1987 signing by Reagan and Gorbachev of the Intermediate-range Nuclear Force (INF) Agreement, which for the first time substantially decreases the world's nuclear arsenal, a possible new beginning has been made. At their widely publicised summit meetings, both leaders have been seen behaving toward one another as human beings. The image of a dehumanised enemy, essential for justifying nuclear weapons, seems to be giving way. A humanised, rational enemy makes weapons of mass destruction less acceptable.

'Containment' and 'MAD' as Security Strategies

The US global security strategy envisioned by Clark Clifford has undergone remarkably little change over four decades. On the assumption that America and other capitalist nations possess both the moral right and the physical strength to contain communism and to shape the developing world in their own image, the United States and its allies have intervened overtly and covertly around the globe, both militarily and economically. The period since the Second World War has indeed been one of 'neo-colonialism', as discussed in earlier chapters. Any emerging nation with reformist or socialist tendencies that might thereby become resistant to free access by Western corporations, has been an immediate target for 'communist containment'.

In several cases, of course, developing nations actually *were* communist, but as already noted, among them only Cuba has remained closely allied with the USSR. While the rhetoric of Marxism has been one of universal brotherhood of the down-trodden, the reality has been one of independent nationalism. Clifford's Domino Theory, in terms of Soviet political hegemony, has failed to materialise. Although Soviet arms have gone to support socialist movements, except for Afghanistan Soviet troops have not been involved. Efforts to 'contain' the global spread of Marxism or of Soviet 'expansionism' have largely been tilting at windmills – unnecessary exercises that have wasted untold resources.

The American military adventure in Vietnam, one would think, should have proved that such intervention was both futile and unnecessary. US forces could not 'win' in Vietnam because most Vietnamese people did not wish them to win, and they need not have been there in the first place because a communist Vietnam was not a threat to the United States. As historian Leften Stavrianos pointed out:

The basic purpose of United States policy in Vietnam was to demonstrate that social revolution in the Third World is not feasible because it can and will be crushed by American military power.[50]

But it did not work. While in Vietnam, Americans discovered the ultimate paradox. 'We had to destroy it in order to save it', said a US military adviser surveying the ruins of the village of Ben Tre in the Mekong Delta.[51] The Americans defoliated forests with toxic defoliants and sprayed napalm on villagers, and then (wrongly, as it turned out) accused the Soviets of biological warfare in Kampuchea!

Afghanistan appears to be the Soviets' Vietnam. The Afghan people have their own notions of what kind of society they want. The guerrilla patriots are fighting a 'jihad', an Islamic holy war, against their own communist government which is assisted by the Soviet troops sent by Moscow. (Curiously enough, the United States, which finds the waging of a 'holy war' by Iranian leaders despicable, has gladly armed the Islamic fundamentalists in Afghanistan because they were resisting 'Soviet aggression'.) The USSR is now in the process of extracting itself from its untenable position.

What appears to be slowly dawning on the leaders of both the First and Second Worlds is that neither superpower has the physical strength nor the moral backing globally to coerce unwilling peoples to accede to their wishes. Containment as a policy has moved beyond the reach of both sides, and hence become unnecessary for both as well. There is no monolithic enemy to contain. But what of MAD? How do we deal with the nuclear arms race between the superpowers?

The arms race may have begun as a contest to see which 'system' – capitalist or communist – would control the world, but since it is now clear that neither can, the whole purpose of its continuance seems in doubt. After dropping its atomic bombs on Japan, the United States believed for a short time that it could control political events globally. But only four years later, the Soviets exploded 'Joe 1', and the race was on. From then on, every American advance in weaponry was matched by the USSR, with an ever shortening time-lag. In 1956, the Soviets launched Sputnik, the world's first manmade satellite, followed two years later by Yuri Gagarin, the world's first astronaut. From then on, neither side has held a decisive technological lead over the other for more than a few months.

At the time of the Cuban missile crisis of 1962, when the United States confronted Soviet ships carrying missiles with nuclear warheads for deployment in Cuba and, through threat of nuclear retaliation, forced them to turn back, America had sufficient nuclear weapons stockpiled to intimidate the Soviet Premier, Nikita Khrushchev. Shortly after, the Soviets increased their weapons stockpile. Instead of one-sided deterrence, there was now MAD, *M*utual *A*ssured *D*estruction. Each side could destroy the other, no matter who struck first.

This nuclear stalemate has now existed for a quarter century, yet in the West – and perhaps in the Soviet Union – it has not been accepted as such. Instead, there has developed a curious psychological ambivalence toward the arms race, even by those in charge. Outwardly there is the belief that 'we can win – we will "decapitate"* the Soviets and destroy their ability to coordinate their forces'. Inwardly, there is the realisation that the very existence of nuclear weapons makes no logical sense. As Robert Jervis, a professor of international relations at Columbia University, concludes in his book, *The Illogic of American Nuclear Strategy*:

> [M]utual assured destruction exists as a fact irrespective of policy . . . Once each side can destroy the other, any crisis brings up the possibility of this disastrous outcome.[52]

If nuclear weapons have *no* strategic usefulness, then why do both sides build more and keep upgrading them? Psychologist Stephen Kull may have discovered the answer. Through in-depth interviews with former nuclear policy makers in the United States and the USSR, he uncovered three layers of response to the reasoning behind the arms race. The initial response was the conventional one: the strongest bully wins and therefore weapons *must* be matched. But when asked if a nuclear war was winnable, the response was 'What do you think I am, crazy?' Finally, when pushed to defend the need for a balance, the response was 'We've got to maintain an *illusion* of superiority – or at least equality'. The critical element in nuclear policy-making has been the decision to go along with this illusion as though it were reality.[53]

The Failure of the 'Technological Fix'

Both 'containment' and the notion of nuclear superiority have turned out to be myths, bankrupting illusions pursued by both sides for over four decades. The Soviet nuclear weapons did not help retain control over Yugoslavia or China, or hinder the United States from interfering in the internal politics of various countries. Those of the Americans did not stop their defeat in Vietnam, or regain their hostages from Iran, or keep Soviet troops out of Afghanistan. Militants on each side have grown more and

---

*'Decapitate' means blowing up the central government, the command and control centres, of the other side in order to put its war machine out of action. The wisdom of this is doubtful, since a 'headless' nuclear attack force instantly becomes 'irrational', firing off its weapons in quite unpredictable ways. Any major nuclear engagement almost instantly loses the ability to pursue the purposes for which it supposedly is being fought.

more frustrated at their apparent powerlessness to control global events, while those who want an end to the arms race fear its burgeoning spiral. In seeking a solution to satisfy both constituencies, President Reagan hit upon the ultimate technological fix: 'Star Wars'.

The Strategic Defense Initiative

On 23 March 1983, Reagan offered the American public the glowing vision of a world forever free from the terror of nuclear war, not by abolishing nuclear weapons, but by defending against them.

> What if free people could live secure in the knowledge that their security did not rest upon the threat of instant US retaliation to deter a Soviet attack; that we could intercept and destroy strategic ballistic missiles before they reached our own soil or that of our allies? . . .
>
> I call upon the scientific community who gave us nuclear weapons to turn their great talents to the cause of mankind and world peace: to give us the means of rendering these nuclear weapons impotent and obsolete.[54]

To ordinary Americans, ready believers in the unlimited prowess of their scientists and technologists and unaware or uncaring of the $1 trillion plus price tag,[55] the Strategic Defense Initiative (SDI) sounded grand. They could finally overcome their fears without having to trust the 'untrustworthy' Russians. And for the many scientists, universities, corporations and communities that would benefit from an almost unlimited flow of R & D dollars, it was hard to admit the flaws in the scheme. Yet the overwhelming majority of scientists, even many whose research would directly benefit from proceeding with SDI, are agreed that it poses an insurmountable combination of technical difficulties. These have been outlined at length in numerous non-technical publications.[56] Even its proponents agree that SDI will work only if the USSR 'cooperates', either by limiting its new weapons or by building its own Star Wars defence.[57]

Beyond the improbability of its functioning effectively and its astronomic price tag, the SDI proposal threatens to destabilise the current balance between the superpowers. The testing of Star Wars components would clearly breach the 1967 Outer Space Treaty and the more important Anti-Ballistic Missile (ABM) Treaty signed in 1972.[58] The latter is the necessary precondition to all limitations and reductions on strategic offensive weapons: SALT I, SALT II, the INF Treaty, and any future treaties that may be negotiated.

If, as the scientific experts have concluded,

> the prospect that emerging 'Star Wars' technologies, when further developed, will provide a perfect or near-perfect defense system . . . is so

remote that it should not serve as the basis of public expectation or national policy,[59]

then why do high level officials in the Reagan administration pursue it? Psychologist Kull cites three reasons: it creates the perception America is marching forward, it makes the Russians nervous, and it makes Americans *feel* more secure. Star Wars is but one more illusion to which people cling for comfort.

The unending technological alternative

For some thinkers, the answer to nuclear madness is the abolition of nuclear weapons and a return to 'conventional warfare'.[60] Yet this, too, is an illusion. Big explosive weapons are by no means the only way to threaten others. Almost every piece of scientific knowledge can be turned toward the destruction of fellow human beings. A few examples will suffice.

First of all, large bombs are not necessary to cause havoc to an enemy, as today's numerous guerrilla and terrorist actions attest. Most of these are carried out with minimum financing or technological skills, but a mighty nation in a nuclear-free world could easily cause havoc by smuggling in small yet powerful conventional weapons and simultaneously detonating them at multiple strategic sites – communications centres, power plants, major dams, critical bridges, and so on. In the absence of the mutual deterrent of MAD, an on-going cold war would create enormous pressures to develop a guerrilla-type attack potential against an enemy. Protection against that would essentially require the institution of a police state within each threatened nation. No one, effectively, could really be trusted. The equivalent of the anti-terrorist inspections now carried out at airports would expand throughout society.

Besides compact explosive weapons, there are all the chemical and biological weapons now being actively researched. The Reagan administration has asked for $8 billion to develop a new generation of nerve gases.[61] The first nerve gases, used in the First World War, included cyanide, chlorine, phosgene and mustard gas, which either asphyxiated outright or stripped the mucus membranes from the throat and lungs, preventing breathing. Between the wars, new nerve poisons were developed that are among the most toxic substances known. An almost invisible amount falling on the skin causes death by asphyxiation, as one's breathing muscles go into spasm. In 1983, Congress voted to develop new 'binary nerve gas' weapons, where two non-lethal substances, stored separately, are placed side-by-side just before a weapon is fired; on firing, they mix chemically to form the lethal substance. Although the use of such weapons is outlawed by the 1925 Geneva Protocol, their production and stockpiling are not. Both sides presumably are now developing new forms of nerve-gas bombs.[62]

In addition to nerve gases (and all the 'lesser' substances such as napalm, tear gas, and so on) there are *biological agents* that are even more lethal. Some are various toxins, produced by bacteria or other organisms. Others are living germs that cause potentially lethal diseases: anthrax, plague, cholera, Q fever, typhus, dengue fever and encephalitis. Throughout history, diseases have caused as many deaths during wars as weapons, and military strategy has often encouraged their spread among the enemy. Toxins and infectious agents causing diseases in crops and domesticated animals have also been used, from the defoliants in Vietnam, to viruses causing fever in pigs, which the Cubans accused the CIA of introducing onto their island.[63]

Despite the Biological and Toxin Weapons Convention of 1972, which bans both the development and stockpiling of biological weapons, the United States, and presumably the more secretive United Kingdom and USSR, are busily researching potential agents of biological warfare. Under the guise of preparing for defence against biological attack, the US Department of Defense is pushing to develop new, genetically-engineered bacteria and toxins at its Utah CBW test site. In 1984, the US Army advertised for scientists to submit research proposals on toxins, and on bacterial and parasitic diseases of military importance.[64] The goal is to produce killer agents against which an enemy would have no vaccines or antibiotics. As Jonathan King, a molecular biologist at the Massachusetts Institute of Technology, has said:

> The notion of [a recombinant-DNA] R & D program that is strictly defensive is a fundamental misrepresentation.[65]

As long as physical violence is the 'last resort' for governments who disagree, there will always be new technologies for developing new kinds of weapons and new, small delivery systems to escape whatever defences have been devised, and there will always be some scientists willing to do the job. Technology can never be its own 'solution'. The illusion of technical-based security can never become a reality. Real security demands a quite different approach, what Kull calls 'the next evolutionary step as the species tries to adapt to a changed reality'.

NEW MODES OF THINKING

Ours is an age marked by the expectation of unstoppable *technological progress*. Computers, robots and Star Wars are seen as the next crowning achievements in the intellectual evolution of humankind. Any other kind of progress – social, political or moral – when not utterly ignored, is given a distant second place to the supposedly inevitable march of technology.

Yet technological progress, untempered by wisdom, is threatening the survival of our species. In the words of historian Walter McDougall:

> The fallacy of the early Space Age was that the pursuit of power, especially through science and technology, could absolve modern man from his duty to examine, affirm, or alter his own values and behavior.[66]

It was E. F. Schumacher who frequently pointed out how violent most modern technologies are: they distort human lives to accommodate them; they place unusual power in the hands of a few individuals or, in the case of weapons, a few nations. And power, the ability to exert force, whether over Nature or each other, when untempered has the ability to corrupt the human spirit. The ardent French social philosopher, Simone Weil, vividly expressed the effect that force has on the human soul:

> Force is as pitiless to the man who possesses it, or thinks he does, as it is to its victims; the second it crushes, the first it intoxicates... [T]hose who have force on loan from fate count on it too much and are destroyed.[67]

Force, Weil argues, dehumanises – 'turns into things' – both victor and vanquished. Our current failure to understand this and to temper our conduct accordingly, is the central deficiency of our age. To regain our humanity, we need to 'learn not to admire force, not to hate the enemy, nor to scorn the unfortunate'.[68] We urgently need to learn the art of tempering force, and so side-step the violence that dehumanises us all and today threatens our very survival. Our problem is not technological, but moral and psychological. How do two adversaries, pointing loaded guns at each other, discover a safe means to lay down their weapons and resolve their differences?

The Psychology of Negotiation

In the absence of direct prior violence, labelling another as an 'enemy' demands an act of the imagination. In fact, virtually none of the mistrust between today's superpowers can be ascribed to the use of force by one against the other. The entire arms race is based on innuendo, supposition and rhetoric. One way to tackle this psychological barrier is to ask under what conditions has earlier progress been made? When have negotiations worked?

As was noted earlier, those who govern large societies rely on some mixture of coercion and voluntary approval to maintain social order. In negotiations between the United States and the USSR, Soviet leaders ultimately have the option of greater flexibility, since for them order is relatively more coerced. American governments, being more responsive to public opinion, must first persuade the masses of the correctness of their actions. The on-again, off-again relations between the two superpowers have largely been due to vacillations in US public policy.[69] From the beginning, the professed goal of the Kremlin has been to reach and maintain military parity with the West.

Periods of detente, when the 1963 Partial Test Ban Treaty and the 1972 ABM Treaty were signed and ratified, came about only after massive pressure from the American people.[70] Most of the time, however, public fear of the 'enemy' has been sufficiently high to prevent significant further progress. This fear has been maintained for a number of reasons. Neither political party has dared to seem 'soft on communism'. Both Republican and Democratic presidents have continually pictured the Soviet Union as dangerous, a view with which the Pentagon heartily concurs. The powerful military–industrial complex, with a vested interest in the arms race, persistently lobbies against negotiations, always claiming US deficiencies in one or another area of defence. Even the ploy of designing new weapons merely as 'bargaining chips' has always failed; the United States never cashes them in, but proceeds to build them, and the Soviets, far from capitulating, obstinately struggle to catch up.[71]

With a new attitude toward the arms race developing in the Kremlin, it remains for public opinion in the West to review its own assumptions about East–West relations if negotiations are ever to succeed.

Learning to negotiate

The Western view of life as a competitive game in which there are inevitable winners and losers offers a poor climate for negotiations. Leo Durocher, one-time coach of the Brooklyn Dodgers, once quipped: 'Nice guys finish last'. The same attitude is echoed by the Pentagon: 'Negotiate from strength'. Settling differences becomes a matter of force, strength, intimidation and willpower. Negotiation – like life in general – is viewed as a game of power.

Bargaining under such conditions obviously invites strategy, suspicion, disappointment and the inevitable hard feelings that make subsequent disputes even more difficult to resolve. This confrontational approach is so deeply ingrained in Western culture that we scarcely notice how it stultifies our thinking. Even our idea of 'fairness' – of 'presenting both sides' – immediately establishes a confrontational situation that precludes asking whether the problem itself has even been correctly defined. This is a suicidal approach in a world where technology has made sheer force a disutility.

A way out of this bad habit of positional bargaining has been explored by the Harvard Negotiation Project.[72] By means of 'principled negotiation' or 'negotiation on the merits of the case', participants focus not on 'winning' but on achieving a wise outcome for all concerned. People are *not* the problem. There are no positions, only interests which need exploring. Once all interests are identified, as many options as possible for resolving the differences among them are put forth. Nothing is omitted, no matter how odd it seems. During this period of 'brainstorming', the problem is sometimes broken down into smaller, more manageable parts, or restated in a more negotiable format. Finally, the decision as to which option is best is based on objective principles to which all parties have agreed. Fairness and reasonableness – not 'honour', 'positions', or even 'trust' – become the fundamental criteria. A comparison of the tactics of

Table 15.1 Positional bargaining vs principled negotiation

PROBLEM		**SOLUTION**
Positional bargaining: which game should you play?		*Change the game – negotiate on the merits*
SOFT	**HARD**	**PRINCIPLED**
Participants are friends	Participants are adversaries	Participants are problem-solvers
The goal is agreement	The goal is victory	The goal is a wise outcome reached efficiently and amicably
Make concessions to cultivate the relationship	Demand concessions as a condition of the relationship	**Separate the people from the problem**
Be soft on the people and the problem	Be hard on the problem and the people	Be soft on the people, hard on the problem
Trust others	Distrust others	Proceed independent of trust
Change your position easily	Dig in to your position	**Focus on interests, not positions**
Make offers	Make threats	Explore interests
Disclose your bottom line	Mislead as to your bottom line	Avoid having a bottom line
Accept one-sided losses to reach agreement	Demand one-sided gains as the price of agreement	**Invent options for mutual gain**
Search for the single answer: the one *they* will accept	Search for the single answer: the one *you* will accept	Develop multiple options to choose from; decide later
Insist on agreement	Insist on your position	**Insist on objective criteria**
Try to avoid a contest of will	Try to win a contest of will	Try to reach a result based on standards independent of will
Yield to presure	Apply pressure	Reason and be open to reason; yield to principle, not pressure

Source: R. Fisher and W. Ury, *Getting to Yes: Negotiating Agreement without Giving In* (Boston: Houghton, Mifflin, 1981); Table on p. 13.

this approach with that of the more usual positional bargaining is shown in Table 15.1. In principled negotiation, the concerned parties no longer confront each other but become partners in solving a common problem.

It is, of course, difficult for governments, especially in a democracy, to undertake principled negotiation as long as the public still demands positional bargaining. Here two possibilities exist. The first is to educate people in the theory and practice of principled negotiation, a long and difficult process. The second is to reduce confrontation by whittling away at the image of a 'dehumanised enemy'.

Some hopeful beginnings

Establishing good working relations among nations is a delicate and often illogical process. Since the start of the Second World War, 180 degree swings in popular Anglo-American attitudes have occurred toward Germany, Japan, Russia (twice) and China (twice). China, in particular, went from pariah to friend almost overnight, several years before any internal political change took place.

Many non-government groups in the West have privately established contacts with counterparts in the USSR. Most publicised has been the International Physicians for the Prevention of Nuclear War, a global federation of doctors and laypersons from 41 countries, including both East and West. In December 1985, it was awarded the Nobel Peace Prize. But there are also groups of scientists, businessmen, artists, civic leaders, church members and so on.

One group of American seismologists has made excellent progress toward making it possible to verify a future ban on underground nuclear weapons tests. New weapons can be developed only if they can be tested, but without 'on-site' inspection, which the Soviets had long refused, the United States has been unwilling to agree to stop testing. Sponsored by the Natural Resources Defense Council, a private, non-profit organisation, the seismologists worked with colleagues in the USSR Academy of Sciences to set up monitoring stations in both countries that are capable of distinguishing between weapons tests and natural earthquakes.[73] Says Soviet seismologist Oleg Stolyarov:

> Given the monitoring technology used today in the USSR, the United States, and other countries, it is impossible to hide even low-yield nuclear tests.[74]

Even though governments fail, opportunities clearly exist for real progress to be made by non-governmental organisations and even private citizens. Indeed, such people are fulfilling President Eisenhower's prophecy:

> I like to believe that people in the long run are going to do more to promote peace than our governments. Indeed, I think that people want

peace so much that one of these days governments had better get out of the way and let them have it.[75]

International Forums

On 22 May 1984, a remarkable statement appeared simultaneously in Athens, Buenos Aires, Dar es Salaam, Mexico City, New Delhi and Stockholm. It said:

> We come from different parts of the globe, with differences in religion, culture and political systems. But we are united in the conviction that there must not be another world war... We urge, as a necessary first step, the United States and the Soviet Union, as well as the United Kingdom, France and China, to halt all testing, production and deployment of nuclear weapons... We will do everything in our power to facilitate agreement among the nuclear weapons states.[76]

It was signed by Raul Alfonsín, President of Argentina, Indira Gandhi, Prime Minister of India, Miguel de la Madrid, President of Mexico, Julius Nyerere, President of Tanzania, Olaf Palme, Prime Minister of Sweden, and Andreas Papandreou, Prime Minister of Greece. This 'Five Continent Peace Initiative', the first cooperative statement by heads of state around the world, was followed by a second one in 1985, the Delhi Declaration, that opened the door for the American seismologists to enter the Soviet Union.

Meantime, 'nuclear allergy' is spreading among the nations of the world. As reported in the *Bulletin of the Atomic Scientists*, more and more countries are banning nuclear weapons from their territories:

> The scope of antinuclear activity is vast. Greece has pledged to remove U.S. nuclear weapons and bases from its soil by 1989. New Zealand declared itself a nuclear-free zone and banned U.S. nuclear weapons and nuclear-powered ships from its water. Spain and Canada succeeded in having U.S. nuclear warheads removed from their soil in 1979 and 1984, respectively. The Netherlands is debating a process to reduce its nuclear 'tasks' within NATO. Norway and Denmark halted their national funding for U.S. nuclear missile bases in Europe. Third World countries such as Egypt, Sri Lanka, and India prohibit nuclear ships in their territorial waters. Even Soviet allies have expressed displeasure about the superpower nuclear prerogatives. Romania stated in 1981 that it would not accept Soviet missiles. And Bulgaria denies that it bases Soviet nuclear weapons on its soil in peacetime.[77]

The Labour Party in Britain has promised to dismantle Britain's nuclear weapons and close US bases if it is returned to power,[78] and larger international nuclear-free zones in the South Pacific, in the Nordic countries, and in the Balkans are beginning to coalesce. The nations of the world are learning to say 'No' to nuclear weapons and to the military infrastructure which supports them.

The Ultimate Alternative: Non-violence

Sooner or later, if lasting peace is to become a reality, governments will need to put away their weapons of destruction and find permanent alternatives to violence. There are, essentially, two complementary options: diplomacy – based on the sort of discussion and negotiation described by the Harvard Negotiation Project – and non-violence.

Achieving effective diplomacy is a matter of training people in the skills of conducting negotiations between nations. This has been a largely haphazard affair in the United States, as some of its recent efforts in diplomacy attest. Although in 1984 a new National Peace Institute was founded to offset America's dozen war colleges and train negotiators in the arts of diplomacy, it has so far been funded at trivial levels and is likely to have little impact.[79] More hopeful are the burgeoning interdisciplinary programmes in Peace Studies and Conflict Resolution now springing up in universities across the nation. Once again it seems that grassroots efforts are what make the difference.

If diplomacy and negotiation fail – as sometimes they must – how then shall nations defend themselves against a violent adversary unless they threaten violence in return? How can the world move beyond MAD? A common suggestion is a permanent international peace-keeping force, but the question of who, exactly, would control that power has never been satisfactorily resolved. Today's United Nations peace-keeping forces (which have seldom brought true peace) comprise contingents of the *national* forces of various member states, armed with their own weapons. Ultimately, however, the nations of the world need an alternative to violence as an effective deterrent to violence. The answer lies in the techniques of non-violence.

Non-violence does *not* mean giving in, nor is it a form of weakness or cowardice. On the contrary, it takes far more courage to practice non-violence than violence. And, when properly used, non-violence is the more powerful, for it does not dehumanise. Although non-violence has been practised frequently by individuals down the ages – Jesus was perhaps its greatest practitioner – in this century it has been developed into a political force to bring great social change to oppressed people.

The strategy of non-violent action is to oppose injustice – whether from one's own government or from a conquering enemy – by mounting massive *civil disobedience* against those in power. Through non-cooperation, strikes, demonstrations, sit-ins, non-violent actions of all sorts, oppression is opposed at every turn. The theory and history of non-violent political change are well documented in the writings of Gene Sharp at Harvard University.[80] The requirements for success against a violent opponent are massive involvement and thorough training of the population.

The man who first thought out and taught the concepts of political non-violent resistance was Mohandas Gandhi. As a young man, Gandhi travelled to London to study law and then went to live for a time in South Africa. It was there that he developed his non-violent strategy, empowering his fellow non-whites to resist, if not overcome, the oppressive racist government. He then took his new-found power back home to India and, some 30 years later, in 1947, the British troops voluntarily left – on a friendly basis.

Gandhi's methods involved massive, coordinated and disciplined non-violent civil disobedience – a *satyagraha*. During one famous *satyagraha*, row upon row of unarmed volunteers slowly walked toward the police who barred their way. 'The British beat the Indians with batons and rifle butts. The Indians neither cringed nor complained nor retreated. That made England powerless and India invincible'.[81]

It always seems odd how ready Gandhi's detractors are to denounce the effectiveness of non-violence because Gandhi was unable to prevent the terrible bloodshed between Muslims and Hindus after the British finally left India. In their eyes, non-violence 'failed' because this first great experiment was not able to sustain itself beyond the peaceful liberation of India from 200 years of degrading British rule. But has the father whose strength permits him to save only one of his children from their burning home 'failed' because the other perished? The cultural chaos left in India as a result of British dominance overwhelmed even Gandhi's strength.

> Gandhi was probably the first person in history to lift the love ethic of Jesus above mere interaction between individuals to a powerful and effective social force on a large scale... It was in this Gandhian emphasis on love and nonviolence that I discovered the method for social reform that I had been seeking for so many months.[82]

Thus wrote Martin Luther King, who so effectively applied Gandhi's methods in the American South to win great social advances for black people during the 1960s. In his powerfully moving letter to his fellow clergymen, written while he was gaoled in Birmingham, Alabama, King eloquently explained the moral rightness *and necessity* of 'illegal' non-violent action to attain justice and human dignity:

> [L]aw and order exist for the purpose of establishing justice, and... when they fail to do this they become the dangerously structured dams that block the flow of social progress...

I have tried to make it clear that it is wrong to use immoral means [e.g., violence] to attain moral ends. But now I must affirm that it is just as wrong, or even more so, to use moral means to preserve immoral ends.[83]

In King's view, a peace-loving, law-abiding citizen was not exempt from guilt if the laws themselves were unjust.

Another great exponent of non-violence is Danilo Dolci, who left a promising career as an architect to spend his life among the impoverished peasants of Sicily, oppressed on one side by the Mafiosi and on the other by corrupt and uncaring petty officials. For Dolci, it was not enough to give charity to the down-trodden; it was necessary to help them discover for themselves how to create a new culture and a new existence. His technique he calls *autoanalisi popolare* or 'popular self-analysis'. He gets people to start telling each other what kind of changes in their lives they most want and then to discover strengths within themselves to bring these about in a non-violent manner. When the need was for employment, Dolci encouraged the peasants 'to strike in reverse' by building roads and dams in defiance of local authorities; these improvements greatly helped the villages' economies. He constantly taught both adults and children how to organise their own political and economic lives.[84] There is a deep connection between Dolci's, King's and Gandhi's teachings, Liberation Theology and the base communities in Latin America, and the kind of grassroots democracy described earlier.

Despite these remarkable examples, non-violence as a political strategy has hardly been explored. It played a brief role in the 1986 overthrow of the dictatorial president of the Philippines, Ferdinand Marcos, by which Corazon Aquino assumed her rightful place as the duly elected head of state. Following the disputed election, the faction of the army still loyal to Marcos could not bring itself to fire on the unarmed masses non-violently protesting in the streets of Manila. (Unfortunately, the complex situation in the Philippines remains tense owing to many factors, not the least of which has been the failure of the wealthy nations to give Aquino the financial support needed to begin correcting grave social injustices.) Non-violence is not for the cowardly or the bully, for whom guns seem to offer far greater 'security'. Yet given the magnitude of carnage that modern weapons can inflict, non-violence is by far the sounder path.

SOME CONCLUDING THOUGHTS

Over the past few centuries, the Western mind has become extraordinarily clever, but it has not become comparably wiser. It has learned to build the most disastrous weapons, and now that it realises that they really are far too dangerous and therefore quite useless, it finds itself floundering about looking for a way to rid itself of them. What seems to be missing is an awareness that it is not just the weapons that must be got rid of, but the 'logic', the mindset that brought them into existence in the first place. It is in the search for absolute security, either through the threat of an overwhelming violence or through a perfect technological defence against such violence, that we keep going wrong. It is no good wanting to go back to 'just conventional weapons', which is like wanting to go back to a hunting–gathering economy in order to escape our ecological disasters. Neither is possible. We humans must be more inventive than that. We have to come up with quite new arguments, at last paying serious attention to the moral underpinnings of our existence and their implications for our vision of appropriate human behaviour.

Peace – real peace – cannot simply be wished for. We must teach ourselves how to make it happen. It is far more than the mere absence of war. Peace is the striving for an end to *all* violence among human beings. This requires profound education, globally. It requires consciously eliminating our present worship of strength and force, and of competition. It requires learning to avoid a hierarchical ordering of people, and teaching our children to discover their own worth without denying that of others. It requires consciously teaching them to disdain and deny many of those things we were taught to admire and pursue: wealth, power, status, stoicism, independence to the point of alienation. We have built into ourselves a militarised intellect that we must now, slowly, painfully, but unrelentingly expunge and replace with a new vision.

Conclusion:
Emerging from the Labyrinth

Finally, we begin to see how we might emerge from the labyrinth. Chapter 16, 'Humankind at the Crossroads', which in fact is a summary of the whole book, lays out three essential attributes towards which *all* societies must strive: ecological harmony, cultural meaning, and acceptance of cultural diversity. The Western world, in particular, must redefine its economic purpose and rediscover the value of cooperative social action and of a sense of relatedness to the biosphere. It needs to understand the proper uses of science and, most important of all, truly to educate its young in the skills of participatory democracy through community dialogue.

16 Humankind at the Crossroads

Today we live in a globally interconnected world in which biological, psychological, social and environmental phenomena are all interdependent. To describe this world appropriately we need a new paradigm, a new vision of reality – a fundamental change in our thoughts, perceptions and values. The beginnings of this change, or the shift from the mechanistic to the holistic conception of reality, are already visible ... The gravity and global extent of our crisis indicates that the current changes are likely to result in a transformation of unprecedented dimension, a turning point for the planet as a whole.

Fritjof Capra, 1982[†]

INTRODUCTION

And so we arrive at the turning point, the crossroads, the decisive epoch in human history that will determine whether or not our species survives. Will the present generation be able to extract itself from the dilemmas that human history has created for it, dilemmas that only humans, unlike all other species, can consciously perceive and respond to? Is humankind up to forestalling its own self-inflicted extinction or not? For the disembodied scholar, there can be no more fascinating intellectual drama. For five billion flesh-and-blood human beings, the question and its answer are far more immediate.

Our forebears knew intuitively that, despite their own mortality and the follies of their age, some of their children and grandchildren would survive to carry on the values not only of their own era but of the deeper past. There was continuity throughout the ages: When I am gone, what I and my forefathers knew and believed in will still exist. The life I lived, the values I knew, will survive. It is easy to die if all you knew, lived for and believed in will continue when you are gone. This is why young zealots gladly give their lives for their people.

But as today's adolescents move into adulthood, they well know they are displacing their parents' world with their own – which is soon to be displaced by that of their offspring. As generational immortality declines, individual mortality becomes more terrifying. One's connectedness with both past and future grows tenuous; life loses its deeper meaning and becomes constructed of instantaneous events and highly personal goals. Life in this temporary world requires constant distractions to deflect

471

attention from its emptiness: the passive entertainments of television and spectator sports; a series of sexual encounters; the pursuit of wealth; an escape into drugs.

Our young, however, are not deceived; they know something is wrong, they sense their rootlessness, but they are not being given the tools of analysis and action that would empower them. There is a conspiracy of silence regarding the values of modern society and the assumptions on which they are based. 'Growth', 'progress', 'scientific objectivity', 'competitive individualism', 'proprietorship of Earth' – all remain unexamined and hence stand as barriers to action.

Part of the failure arises from the way we over-compartmentalise knowledge about the world into 'disciplines'. Educators and students alike end up resembling the blind men who described the same elephant variously as a tree trunk, a rope, a snake, a wall, a large leaf flapping in the wind, and a highly polished rock. Formal learning has become a meaningless vaccination process, and the information transmitted is next to useless for properly understanding the world.

Furthermore, formal education eschews any mention of values; they are a 'private' matter for the individual's conscience. Morality becomes a matter of personal opinion and, when it is formally tackled, almost never addresses major global issues. Since the values of one's own society are never critically examined, it becomes impossible for the young to perceive which values may need adjustment in light of current global conditions. Even worse, they cannot truly appreciate how historically other societies have arrived at quite different sets of values. 'Tolerance' is thus reduced to an empty buzzword and the giant problem of the West's ignorance of its own ethnocentrism remains unaddressed.

This state of affairs is no longer tenable. In ages past, worldviews evolved gradually, often imperceptibly. But the enormous powers now unleashed by the science and technology that emerged out of the Western worldview are creating such a rate of environmental change that we can no longer rely on the old, imperceptible kind of evolution in our thinking. 'Progress' will not be made by pursuing technology, but by reordering our goals. All worldviews are likely to require some degree of adjustment if our species is to survive, but the one most in need of overhaul is that of the dominant West,* whose enormous military, technological and economic power impinges upon the entire globe. It is the West that is most in need of the 'new modes of thinking' that Einstein demanded.

The modern belief that the direction of change is 'inevitable', that 'technological progress' cannot be stopped, is not based on any laws of Nature but is purely the result of the dominant values of our Western

---

*'West' here refers to both centralised capitalism and communism, since both emerged from and contain similar values to the Enlightenment and the Industrial Revolution that followed.

worldview. Yet this worldview did not emerge out of thin air. It is based on values that have historical continuity. Their evolution, like any form of evolution, was a contingent process, dependent upon the inter-action of two events: what was inherited from the past to work with, and the circumstances to which adaptation was being made at each moment in history. In this it precisely resembles biological evolution. If the old forms have no useable traits that will 'make do' in new circumstances, they simply go extinct. Had there been no lobe-finned fishes among all the swamp-dwelling fishes caught in the drying inland seas hundreds of millions of years ago, with appendages that could gradually be adapted for walking on dry land, human beings would not have evolved. The current Western worldview, as this book has shown, has grown maladaptive, and it remains to be seen whether it contains sufficient valuable traits that can be adapted to meet new circumstances.

Cultural evolution, of course, has far greater latitude than biological evolution. It cannot change the past, but it can reflect on it. It can change what it *emphasises* from the past, thus making the present appear in a different light. And it can take place far, far more rapidly. Yet that required rate of change in our thinking today is so enormous that it is by no means certain that we shall succeed. A too rapid rate of change can alienate people and destroy social cohesion. This is why violent ideological revolutions so often fall back into the old social patterns they supposedly displaced. Cultures and worldviews are not old clothes to be casually discarded for new. They must be nurtured, treated with concern, and lovingly moulded by society as a whole if they are to be changed without destroying society itself. This means seeking within each tradition those values that are still valid within the changed circumstances, the new environment. It is sorting out what characters to save and develop, like the lobe-fins and lungs of ancient fishes, and which to discard, such as the gills and the scales. Are we capable of the rapid cultural evolution our own technological advances now demand of us? Who can say? Surely it is worth giving it our best try.

These final pages, as they seek an exit from the labyrinth, are not intended to provide specific 'answers', but rather to offer both some new ways of thinking and some new methods of facilitating cultural change. They begin by identifying three attributes – three qualities – that *all* future cultures must somehow incorporate into their worldviews, no matter how diverse they otherwise may be, if humankind is to survive.

ESSENTIAL CULTURAL ATTRIBUTES

The three qualities summarised here were all developed earlier and are restated now along with the reasons why each is universally required.

(1) Each society must live within its ecological 'income'. (2) Each cultural vision must engender a passionate engagement in life. (3) Each worldview must see itself as but one of a diverse group of human experiments, and come to welcome and support this diversity.

Ecological Harmony

The peoples of the planet are borrowing from the future in different ways. In the North, a wealthy minority constantly expects to consume more each year than last, while in the impoverished South, people for whom bare subsistence is a struggle, continue in their hopelessness to bear more children than their lands can support. Both in their different ways are overdrawing on their Environmental 'Credit Card' Accounts. But the two phenomena are intimately related. The total sustainable material income of the world, if properly distributed, could very probably maintain today's global population in modest comfort indefinitely. The fact that this wealth is not more equitably shared is largely a consequence of the colonial era and its neocolonial sequel. There is built into the worldviews of the North, particularly of the capitalist First World, the belief that indefinite growth in material consumption is possible, and that today's grossly unequal distribution of global wealth is somehow 'justified'; the wealthy, it is said, have 'earned it', although no one ever makes clear just how that earning took place.

What is missing from too many worldviews today, whether in the First, Second or Third Worlds, is the connectedness of human beings with the land from which all of us – today and tomorrow – *must* find our sustenance. Each culture must retrieve within its vision what many primitive peoples understood about their relations with the soil, air and water of Earth. In the words of N. Scott Momaday, a Kiowa Indian:

> You say that I *use* the land and I reply, yes, it is true; but it is not the first truth. The first truth is that I *love* the land; I see it is beautiful; I delight in it; I am alive in it.[1]

Without this deep appreciation of the beauty of the land that nourishes us, we cannot long survive. Yet such appreciation is not limited to the visions of primitive societies; it exists in the sacred texts of every major religion. We need but search our own heritage to recapture a sense of ecological harmony.

Cultural Meaning

Each person, in order to retain those attributes we recognise as human, must live in relation to others, within a social context, a culture which gives meaning to individual existence. This need for cultural meaning is at once the *sine qua non* of human existence and the source of our greatest danger. As pointed out by geographer Bernard Nietschmann, almost all today's global conflicts are between ethnic groups, or 'nations', not between the 'states' identifiable on maps. They are struggles for 'territory, resources and identity' – for freedom and independence of cultural groups:

> Unlike a state, a nation does not require a central military–political bureaucracy to create nationality, nationalism or national territory.[2]

Imposed on some 3000 or more culturally distinct nations are the arbitrary boundaries of 168 political units called states; of these, more than 95 per cent are multinational, many imposing the will – and culture – of one particular group on several others.

The present trend toward political giantism has other destabilising consequences besides the enforced dominance of one ethnic people by another. It leads to what anthropologist John Bodley calls the 'disease of civilisation',[3] the extreme social stratification and specialisation of labour that has marked all huge societies over the past 5000 years. They inevitably develop internal symptoms of alienation and social decay. As John Burton has noted, leaders then try to head off political breakdown by inventing an external enemy and whipping up an artificial 'nationalism' to reconstitute a sense of social meaning and cohesiveness.

It is apparent that people everywhere, as they struggle to adjust their traditional worldviews to meet changing circumstances, must take care that they do not throw out the 'baby' of cultural meaning and bondedness with the 'bath water' of maladaptive institutions, lest they end up with new institutions that are destructive of the human psyche itself.

Global Acceptance of Cultural Diversity

A great many dedicated people are busy trying to bring about 'world order'. There are any number of transnational organisations whose central purpose is to discover a workable political policing system that would *enforce* order. Although concerned with global justice and understanding, their thrust is generally toward a uniform set of laws, and institutions to impose them. This seems however to compound rather than resolve our dilemmas. Neither the League of Nations nor its successor, the United Nations, found a means to enforce global security (although each provided a needed international forum for dialogue and established useful non-political agencies for the nations of the world to draw on). The solution may not lie in centralised 'government' at all, but in quite the opposite direction, in fostering *diversity*.

This is likely for several reasons. First, since giant societies are inherently alienating and unstable, requiring the threat of violence in order to cohere, then the goal of a single global society cannot make sense. How shall one individual find meaning in a social system organised to encompass some 5 – soon to be 8 or 10 or 12 – billion people? If in today's giant societies, individuals feel lost and powerless, why will they feel better in an even *larger* system?

Second, if there is to be a single 'world community', how shall it come about? Whose rules shall we adopt? Shall we search about among the world's present offerings of worldviews – Islam, Confucianism, Marxism, Liberation Theology, Swedish Socialism, *Sarvodaya* in Sri Lanka – for a single solution? By what criteria shall we know which is best? No single society seems to accommodate all our current global problems, let alone to offer promise for the indefinite future.

Finally, even if one worldview seemed 'ideal', how could we get every-one to accept it? If cultural imperialism is our main source of conflict today, how much worse it would be if we tried to impose a single global culture. It seems far more sensible to conclude that not only is 'one world' impossible – it is extremely unwise. The other option, the decentralisation of political power and the positive fostering of multiple autonomous cultures, has the immediate benefit of reducing rather than increasing the potential for global violence. But it has another, very important long-term advantage as well.

As already noted, human evolution proceeds not only by the impercept-ibly slow process of genetic selection that it shares with all other life forms, but by the far more rapid, flexible and even reversible process of cultural change. As evolutionist Stephen J. Gould reminds us:

> Cultural evolution is not only rapid; it is also readily reversible because its products are not coded in our genes.[4]

Our big brains, which have led us into so much trouble, are also our way out.

Yet as discussed earlier in the book, although our minds are capable of constructing imagined worlds that have never existed – indeed, this is the basis of our creativity – it is not possible to invent an entire new worldview from scratch. Each individual can create only in the context of a culturally transmitted, functioning image of the world. In a culturally diverse world, then, there exist side by side hundreds of working images of the world.

We have here, I believe, a strong parallel with the rich genetic diversity that plays such a critical role in biological evolution. Sexual reproduction, common to all complex organisms, ensures a constant remixing of genetic traits, so that as conditions change, there are always likely to be some individuals pre-adapted to meet the new situation. In a similar way, the presence of cultural diversity, with the constant exchanging, through travel, trade, intermarriage and so on, of cultural traits, offers humankind a rich source of new adaptive possibilities for cultural evolution to meet changing conditions in the future. Our adaptability depends not on genetic diversity, but on *cultural* diversity; relative to the rates of genetic change, we can rapidly change, even reverse, the direction of what we define as 'progress'.

How foolish humankind would be to try to erase the very diversity upon which its future survival depends. It is for this reason that we should view with alarm the technological and cultural imperialism that draws all societies into a single, competitive global market place and imposes upon them a single, highly centralised technology with a single set of accompanying goals and values. Such wreaking of cultural destruction, such breaking of 'the cups of clay from which each people drink their life', should be resisted just as strongly as if it were outright political imperialism. Preservation of cultural diversity, far from being perceived as a threat to human survival owing to rivalries and differences, needs rather to be respected and fostered by all cultures. *Our global goal must not be simply tolerance of diversity, but its positive nurturing.* Each culture represents an important, perhaps crucial, experiment in the unfolding drama of human life on Earth.

To sum up, in a future world where all societies respected and lived within their environmental incomes, where each had evolved a social meaning that passionately engaged its members, and where every society understood the absolute values for all humankind of a grand diversity of cultural worldviews, we could at last have believable hopes for an indefinite, albeit ever-changing and problematic, human existence. Insofar as I can see, these three cultural universals threaten none of the major ethical or religious visions that exist; they threaten only those who, at the moment, profit either psychologically or economically from one or another form of injustice and exploitation. Since *any* outright violence potentially threatens *all* humankind, all societies, acting in concert, have a duty to resist, morally and economically, all potential forms of aggression. A primary role of international organisations must be to arouse, mediate and focus global moral action – to become humankind's vocal conscience and cultural healer.

Finally, if these three attributes are to be absorbed into and given a prominent place in cultural visions around the globe, teachers, preachers and others responsible for preserving, interpreting and developing worldviews will need to promote an understanding and acceptance of them.

'Environmental concern', 'sacred meaning', and 'cultural tolerance' need to be on an equal footing in the American rhetorical lexicon with 'freedom' and 'democracy', and in the Soviet with 'socialism' and 'brotherhood'.

REDEFINING THE SECULAR

In discussing the readjustment of worldviews, it is the worldview of the West to which attention must mainly be directed, since for the past half century or more it has been that particular vision of reality and virtue that has deeply affected not only the readers of this book, but a considerable portion of humankind. This discussion is divided into two parts: redefining the secular, and redefining the sacred. By 'secular' is meant those aspects of social vision that have to do with the mechanics of survival. In a technological society, this goes far beyond mere subsistence to include the entire panoply of institutions having to do with the production and consumption of wealth. The 'sacred' is taken in its broadest sense of something revered and honoured – that which matters deeply and gives life its meaning. It may be, but is not necessarily, divine in a religious sense.

Now in Western society, and particularly in the United States, the shared *social* notion of what is sacred – what matters most and has top priority in social discourse – largely coincides with what is defined above as secular. As discussed in Part III, this reduction of ultimate meaning to ordinary ends has contributed in large part to a sense of alienation and emptiness in modern industrial societies. In pursuit of freedom of conscience in religious matters, virtually all sacred meaning has been banished from public dialogue – with the odd exceptions of exactly how life should begin and end.

Of course, there are no sharp boundaries between secular and sacred. In many societies food and other necessities can also symbolise the sacred. It is a matter of how things are valued, of how people stand *in relation to things*, that determines to which realm they belong.

Modifying our worldviews is something we all must do together. No 'expert' has anything more than a piece of the answer. It is a skill of which we have little experience, since our political life has largely been so distant and undemanding; no doubt we shall be clumsy at first. Our first task is learning to ask: 'Are we on the right path? By what maps and signposts are we to find our way?' We begin with what Westerners call secular but treat in a peculiar manner as sacred: 'the economy'.

Economic Purpose

In every society, economic production is a social enterprise in which all participate in some manner or other, and one's economic contribution is thus an important part of one's sense of belonging to the community. It is no accident that the people of Sierra Leone greet each other with the phrase: 'Thank you for the work'. Using our hands and minds creatively to contribute to the well-being of those with whom we share our lives is, as E.F. Schumacher so well knew, a necessary – indeed, a sacred – element in experiencing a deeply rewarding life. The products of such work bear this same mark of the creative human spirit, and are intended to be both beautiful and long-lasting.

This sacred nature of work has all but disappeared in modern industrial society. As earlier chapters have shown, the logic of efficiency, special-isation, and mass production have destroyed the connection between the worker and his work. Labour has lost its social meaning and instead become a burden, to be compensated by wages. The products of work are disembodied 'things', sold in an impersonal marketplace and consumed by persons unknown. This 'efficiency' in production spills over from the factory into all economic activity: hospitals, universities, fast food palaces. All are driven toward depersonalisation and alienation. Instead of the bonds of a community of fellow workers, today's wage-earner must console himself or herself through consuming goods and services, for this is the promised path to well-being and satisfaction.

Consumer *dissatisfaction* however is necessary to the modern Western economy – and is assured by ubiquitous advertising – in order to generate the levels of profit the system demands. Both profit and dissatisfaction necessitate a constantly increasing throughput, which puts increasing stress on the underlying ecosystem that supports the economy. Meantime, efficiency demands concentrating economic power and decision-making in fewer and fewer hands. The entire society is dependent on a gigantic economic machine, which despite the application of Keynesian controls remains largely beyond the reach of popularly elected governments. Little wonder that ordinary citizens feel quite powerless to control their own destiny. Furthermore, the impetus for economic growth overflows beyond national boundaries, encompassing the globe, exploiting both resources and labour, without producing significant economic benefits for the great majority of poor, and without concern for the welfare of future genera-tions. The modern 'economy', on which we completely depend yet over which we have virtually no control, has in a very odd way become 'sacred'. It is the centre of our attention. It takes precedence in the headlines; the military defend it; governments are elected to keep it 'healthy'. In short, we worship it.

Yet around the world there are signs, small but hopeful, that women and men are abandoning the materialist, throughput logic of growth eco-nomics. The creative response to a massively centralised, totalitarian economy is local self-reliance, where people meet their collectively perceived needs though their own local environmental incomes. They

Table 16.1 A comparison of Buddhist and Western attitudes toward work

	Buddhist	Western
Labour	Seek Right Livelihood in order to develop one's faculties and to contribute to society in a way that reflects well on oneself	A disutility, to be eliminated (the employer wishes output without workers; the employee, income without work). If work cannot be eliminated, then specialise and reduce 'inefficiency'
Leisure	Is complementary to work; both are necessary	Leisure is preferable to work
Technology	Tools are to help humans do creative work	Technology is a means for abolishing human work
Trade	A sign of local economic imbalance and failure	A sign of economic progress
Goal of life	To perfect one's character through good work which nourishes the spirit	To accumulate wealth to satisfy unlimited wants
Unemployment	Is unacceptable; all who want jobs should have them; mothering is a socially esteemed profession	Is tolerable; one who is not employed is probably 'lazy'; mothering is not socially useful work, since it is not paid
Quality of work	Should be simple, non-violent, sparing of resources, use local materials, and provide satisfaction	Is energy-consuming, high-pressure, competitive, anxiety-creating; often employs imported materials
Quality of life	Consumption is incidental to living; attachment to wealth interferes with satisfaction; one's role is to blend with the environment, to protect it and to reverence life	Consumption levels measure 'Standard-of-living'; nature is to be conquered and controlled; one should consume whatever comes to hand – one is a fool not to
Material goods	Should be simple, long-lasting, beautiful, unique, and as few as possible to live well	Should be complex, mass-produced, cheaply made, short-lived, and as numerous as possible

Source: Adapted from E.F. Schumacher, Small is Beautiful (London: Abacus, Sphere Books, 1974) pp. 45ff.

recapture a sense of being in charge of their own lives and rediscover meaningful community. Such an economy, the purposes of which are compared in Table 16.1 with Western economic attitudes, was well-understood by E.F. Schumacher. He called it 'Buddhist economics'.[5] It does not deny technology, but insists that it harmonise with more humanistic and ecologically sound economic purposes.

The Purpose of Technology and Science

Because it has made 'the economy' the central force controlling human relationships, the Western worldview gives an important second place to technology and science, the engines believed to drive the economic system. As science writer David Dickson recently observed:

> The last few years have seen the reemergence in the U.S. of an almost religious belief – dormant for much of the 1970s – in the power of science-based technology. Industrial leaders argue that only scientific and technological supremacy over the rest of the world will allow the country to prosper economically.[6]

Technology, however, is not a sudden addition to human culture. The earliest hominids used objects as tools, gradually learning to shape them or otherwise modify them. Even long periods of apparent 'sameness', such as the so-called Dark Ages, were punctuated with innovations. What is different today is not the use of technology *per se*, but the rate of technological change that imposes itself on human social institutions, and the vastness of the military and economic power that modern technology places in relatively few hands. Although human history has often been determined in the past by the increasing technological dominance of one society over another, evolutionarily we have virtually reached the end of this path. Its further pursuit will be omnicidal, or at the very least, highly maladaptive. When every society has the power to kill off every other (as well as itself), what will military 'dominance' *mean*? And economic competition, pursued through ever-greater exploitation of our environmental support system, is equally maladaptive.

To say all this is by no means to imply an end to technological innovation or to an increasing understanding of ourselves and our world through ongoing scientific inquiry. It is all too easy for those who support untrammeled technological 'progress' to shout 'Luddites!' at those who urge more caution in technological development. Rather, the concerns are directed at the social purposes of science and technology. What, in fact, is technology *for*? Is it, indeed, to compete and dominate? Or is it to foster the general well-being of humankind and an assured future for the species? What is science *for*? Is it mainly to provide more powerful technologies, or rather to offer wisdom about how human beings can live in harmony with each other and their surroundings? If humankind is to survive, we need to shift from a competitive to a cooperative mode. The massive power of modern technology leaves us no option.

Consequences of the competitive mode

As already shown, the Western belief that competition is 'natural' and 'healthy' stems from a prior belief in the individualistic nature of human beings. Social institutions beyond the bare minimum needed to keep order only hinder the free pursuit of self-interest, life's ultimate goal. Furthermore, the 'invisible hand' of self-interest, functioning in a 'free market', will produce the economically most 'efficient' and *ipso facto* the 'best' possible society. Competitive individualism is therefore in everyone's interest.

Like all beliefs, those of the West rest on unproved assumptions, yet they are treated as immutable 'laws' of human behaviour. In this respect, the creed of competitive capitalism is just as dogmatic as any religion, or as Marxism, which claims to be based on 'scientific laws of economic history', but in fact derives from Hegel's quite non-scientific dialectical metaphysic. If there are any general 'laws' of human society, they will be found in the repeated reinvestigation of history – of *all* cultures – and in continuous observation of living societies – *all* societies. And a comprehensive observation of the present strongly suggests that in a world of powerful technologies competition, far from being a benefit, is becoming a dire threat to human existence. Competition implies dominance. Dominance implies power and efficiency. And these, in turn, are anti-democratic, creating secrecy, psychic damage and economic instability, not to mention the military threat of annihilation.

The problem of secrecy 'Winning' in a competition means knowing better how to do something. If others copy you, you lose your advantage; if they improve on you, they gain an advantage. In such a situation, keeping your advantage secret is essential to staying ahead. Or, if the knowledge cannot be kept secret, claiming it as 'private property', something owned for which potential competitors must pay rents and royalties, assures your competitive edge. Most industrial research and almost all military research is highly secretive. Moreover, 'proprietary rights' and 'national security' classifications are now being extended to ever wider areas of scientific and technological knowledge, which means that the giant powers derived from new knowledge are controlled by an ever smaller group of elites, whose identity is often unknown.[7] Where secrecy fails, knowledge is converted to a commodity. There is a great rush to patent processes, to patent computer software, to patent hybrid seeds, even to patent human genes. 'Ownership' of knowledge precludes access by the poor to useful information, and secrecy precludes public discussion of the possible social and environmental consequences of new technologies. The private possession of knowledge demanded by competition simultaneously blocks the open exchange of information essential to a sound scientific enterprise, prolongs social inequality and denies democracy, which can exist only if knowledge is accessible to all.

Headlong competition also forces the pace of technological change, leading to exponentially growing overheads of innovation. 'Keeping up with the Joneses' in R & D places increasingly costly demands on the resources of any technologically competing society. The soaring cost of military R & D is well known, but the same process is at work in the production of ever 'higher-tech' non-military products. The price of an 'average' family car has risen about twice as fast as the average wage. The costs of 'advanced' health care technologies are threatening to bankrupt medical insurance programmes. These growing overheads of our technological race into the future, however, are conveniently overlooked in the general hubris.

The psychic damage   Our competitive high-tech system also has detrimental effects on the human psyche, one of which is the extraordinary rate and degree of social change. 'Science finds – Industry applies – Man conforms'. This slogan of the 1933 Chicago World's fair is powerfully alive in the West today. Technocrats are blithely predicting that a child born in the 1980s will be completely retrained three times during its working life, so rapidly will technological change render a person's skills obsolete. No mention is made of the costs of retraining, of the social disruptions implied, or of the psychic impacts on people. There will be no sense of connectedness between past and future, on which life's meaning ultimately rests. Instead of being in control of life, an active participant *in* and shaper *of*, social change, the average citizen in the technocrats' vision of tomorrow, regardless of any right to vote, to read uncensored newspapers, and to speak openly to others, will be a slave to the 'technological imperative'.

In this future technologically determined world, culture comes to lack all human meaning. As the English writer Arnold Pacey notes: '[We] go on inventing, developing and producing regardless of society's needs'.[8] The psychic emptiness of this all pervasive consumerism is hard to escape. The highest purpose of society is to produce and consume and the highest purpose of the individual is to facilitate these goals. Manfred Stanley describes well the curse of a technologically-driven scientism on the human psyche:

It is by now a Sunday-supplement commonplace that the social, economic and technological modernization of the world is accompanied by a spiritual malaise that has come to be called alienation. At its most fundamental level, the diagnosis of alienation is based on the view that modernization forces upon us a world that, although baptized as real by science, is denuded of all humanly recognizable qualities: beauty and ugliness, love and hate, passion and fulfillment, salvation and damnation. It is not, of course, being claimed that such matters are not part of the existential realities of human life. It is rather that the scientific worldview makes it illegitimate to speak of them as being 'objectively' part of the world, forcing us instead to define such evaluation and such emotional experience as 'merely subjective' projections of people's inner lives.

The world, once an 'enchanted garden,' to use Max Weber's memorable phrase, has now become disenchanted, deprived of purpose and direction, bereft – in these senses – of life itself. All that which is allegedly basic to the specifically human status in nature, comes to be forced back upon the precincts of the 'subjective' which, in turn, is pushed by the modern scientific view ever more into the province of dreams and illusions.[9]

We have turned science into a religion; we have made technology sacred. But for life to have meaning, it must be something other than mere consumption. How will all this escalating technological change make life more meaningful? That is one of the best-kept secrets of modern existence.

The vulnerability of 'efficiency'　　Another fact that is seldom appreciated is that high-tech economies are remarkably fragile. Their components are too complex, they interact over too long distances, and they lack redundancy–back-up systems. Complex technologies composed of highly specialised parts are prone to breakdown, and the cost of keeping them running increases with their complexity. This is as true of keeping up the automated family car as it is of successfully launching a spacecraft. The dangers of being dependent on distant resources were clearly brought home to Americans by the long gas-lines of the 1970s' oil crises and by the chaos caused by New York City's two major blackouts. Finally, when a system fails, due to complexity or distance, there is seldom a readily available back-up. Redundancy, in today's short-term thinking, is 'uneconomic'.

While for the sake of industrial efficiency we have centralised our supplies of water and energy, and located production in gigantic factories, relying in each case on the most highly complex technology available, we have attained agricultural 'efficiency' through monoculture. Hundreds of square miles are planted to a single crop, an easy target for pests. Furthermore, as shown earlier, the most 'efficient' farming techniques, measured in yields per acre, are destructive of the land. Again, efficiency and overspecialisation lead to vulnerability, this time of an ecological sort.

Modern societies are also vulnerable in another way. A hundred years ago both men and women possessed a wide variety of 'survival' skills: between them, a husband and wife could raise crops, tend animals, preserve foods, build houses, make candles, soap and cloth, repair tools, cast metal objects, treat common illnesses, fashion furniture and shoes. Knowledge was redundantly spread throughout the population. Today, most people – especially our 'best' educated college graduates – can do none of these things. This may be the 'information age', but modern information is useful only in the context of a highly complex, highly centralised, super-managed system. Anything that disrupts the orderly functioning of that system will leave the millions who depend on it in a totally helpless state. In addition to a nuclear holocaust, modern high-tech economies could be severely dislocated by a sudden collapse of faith in the financial system, by a sudden shortage of oil supplies, or by acts of terrorists. All of these become more probable the longer we persist in ignoring the underlying factors that could bring them about.

Over and above these undesirable costs – secrecy, alienation, and vulnerability – of a society designed around the concept of 'competitive efficiency', there is the ironic fact that what we presume is efficient is in fact often quite inefficient. Many of our 'advances' have depended on substituting undervalued fossil fuel energy for human time and labour. Our agricultural yields per acre may be high, but per calorie of energy input they yield considerably *less* than several forms of subsistence agriculture. In industry, although it is true that the energy inputs into productive processes *per se* have frequently decreased through the application of new knowledge, this decrease has often been more than offset by the energy and the social costs of creating and maintaining high-tech production. It costs more and more to train the technologists; it costs more and more to coordinate the complex economic system, with its energy-consuming transportation networks, intricate banking and investment systems, inevitable government agencies to regulate all this activity, and private consultants and lawyers to advise corporations and carry out litigation. In the United States, less than six per cent of the labour force actually produces things. Most of the rest perform the overhead jobs of a complex system. To the costs of these latter activities (which are labelled 'benefits' and added to the GNP), should be added the costs in social welfare for those dislocated by efficiency, as well as all the unpaid costs we are charging up to the Environmental 'Credit Card' Account to be paid back sometime in the future. When all the *actual* costs of our modern 'efficient' system are added up in the way they should be, the whole notion of 'efficiency' becomes extremely doubtful. What we clearly need is a totally new kind of social accounting system, based on a host of values presently excluded from our calculus. We need to create a new set of signposts by which to find our way.

Towards a cooperative mode

Despite their supposed 'backwardness', contemporary pre-scientific cultures often possess a remarkable wisdom about both human nature and about the workings of the environment. Of course, societies in the past have made mistakes, particularly *vis-à-vis* the environment, but that is no reason for dismissing all that might be learned from them. In fact, some of the ancient truths that have stood the test of countless generations are now being 'discovered' by modern science – from physics to psychology.

Both our sense of belonging to 'a people' and our deep aesthetic pleasure in our surroundings are not just chance values popping up in this or that culture. They are an intrinsic part of our humanness, built into the human psyche through millions of years of evolution. Cultures can, and of course do, modify, manipulate and even attempt to suppress these attributes of bondedness and relatedness. Yet cultures that survive neither totally obliterate social cooperation nor extinguish sensitivity to their surroundings. If these are, indeed, essential traits for human survival, where do science and technology fit in? What are their proper purposes?

Science is for wisdom In Darwin's day, science was an avocation, something rich men did for amusement. Today, it is a vocation, a profession where people are paid to pursue particular kinds of knowledge that some segment of society is interested in. As a consequence, almost all of the new knowledge being learned about the world is *directed* toward some preconceived practical purpose, most often for improving military or industrial technological proficiency. Only a tiny fraction of resources is spent on researching how Nature works in a general way, let alone on understanding contemporary human behaviour. This huge disparity between our efforts aimed at competition and conquest and those directed toward wisdom and understanding raised a question in the mind of William Carey, the former Executive Officer of the American Association for the Advancement of Science:

> Will we assume simplistically that our political technology is advancing in wisdom and competence sufficiently to pace a national economy that will be scaled at $3 trillion to $4 trillion? . . . Will we take a different view of justice and generosity relative to the distance we are putting between ourselves and the Third World? These things matter just as much as the strength of our forces or the edge we regain in international markets.[10]

The fundamental function of science must be to contribute to our wisdom. Yet so poorly is this recognised today that important scientific knowledge has yet to 'filter down' to economists and other 'experts' who advise decision-makers. General understanding about the laws of energy, the behaviour of ecosystems, the meaning of carrying capacity is essential for continued human survival in this age of extreme technology. We must ask scientists for more information about how our support system works, not to gain power over it but to understand how safely to live within its boundaries. Science is for discovering the limits of the natural world and the laws by which it proceeds and within which we are free to act. This aspect of science can add greatly to the maps and signposts we need to guide us into the future.

On the other hand, the belief that the reductionist tools of the physicists and chemists can ultimately explain human societies is becoming recognised throughout the human sciences as a false start. Although some aspects of our behaviour are no doubt explainable in reductionist terms, the most important aspects of ourselves are to be understood from pursuing history, anthropology, philosophy and the arts – those approaches to self-understanding that 'objective' science cannot tackle.

Meantime we are accumulating such vast quantities of detailed 'knowledge' that we are in danger of drowning in a tidal wave of 'information'. In his inimitable way, Erwin Chargaff, molecular biologist and probing critic of modern science, states our predicament:

> The institutionalization of all intellectual activities; a misunderstood and misapplied scientism; a crude reductionism exerted on what cannot be reduced; a galloping expertitis, degree- and prestige-drunk; the general persuasion that anything new automatically deposes anything old – all those agents have caused scholarship nearly to vanish after having been in a slowly accelerating decline for the past 100 years.[11]

And with regard to the mountains of accumulating facts that no one seems capable of absorbing, he writes:

> The principal general recipient of, and beneficiary from, new knowledge appears, therefore, to be the computer.[12]

It is apparent that research, knowledge, science and wisdom all float rather loosely about in our thinking. We tend automatically to equate them when, in practice, they are not at all the same thing. Sooner or later we shall have to fit the scientific enterprise into a new kind of integrative, holistic way of thinking that is more appropriate to an interactive, cooperative mode of being – what Gregory Bateson called 'an ecology of mind'.

Technology is for reducing throughput It was argued earlier that the purpose of an economy is to provide (1) those goods that all people need a continuous supply of to survive: food, drink, fuel for heating and cooking; and (2) the capital goods such as shoes, houses, tools, implements, books, equipment, whose purpose is to be used over and over again and whose destruction in use, far from being a benefit, is a deplorable waste. Yet a large proportion of the capital goods turned out by modern industries is designed not for permanence through maintenance and repair, but rather to be replaced at frequent intervals by 'improved' models. Furthermore, much of what is consumed is of questionable value for either sustaining life or lending it significant meaning. The consequence is an unnecessarily high economic throughput.

If humankind is to survive, such excessive economic throughput can no longer be considered morally acceptable, since it consumes resources other people (including future generations) require simply to stay alive. Its purpose, quite simply, is to provide unending profits to a small segment of society by encouraging people to consume not only what they need but what they merely 'want', because they have been carefully and expensively taught to want it. E. F. Schumacher clearly understood not only the environmental but also the social destructiveness of this contrived head-long rush to consume:

> The cultivation and expansion of needs is the antithesis of wisdom. It is also the antithesis of freedom and peace. Every increase of needs tends to increase one's dependence on outside forces over which one cannot have control, and therefore increases existential fear. Only by a reduction of needs can one promote a genuine reduction in those tensions which are the ultimate causes of strife and war. [13]

Unnecessary economic throughput destroys not only our support system, but also our sense of self-worth and self-sufficiency. Our technologies are too often violent, destructive, inhuman.

If we progress from today's massive throughputs of intentionally short-lived products to the production of a modest quantity of high-quality, long-lived capital goods to grace our lives, then our technologies will shift radically. If our goal also includes humanistic means in the production of those goods, then the appropriate technologies will shift even further. Most of today's centralised, 'efficient' productive systems will be seen as destructive, and many currently 'inefficient' arrangements will seem right. Everything revolves around how products are valued. When non-dollar factors begin to count – Will the product last a long, long time? Can it be easily maintained? Was it made at the expense of another human being, or of our shared environment? Will my life really be enhanced by it or is it a trivial 'need'? – then the buyer begins to control technology.

'The culture of technology'   Societies around the world, as they learn to live within their environmental 'incomes', will not all find the same

technologies 'appropriate'. As Table 16.1 suggests, local resources are preferable to distant ones, and demand different technological approaches. Building materials and building styles would thus vary according to the locale, as would the design of cities. Different climates and availabilities of energy and water would all affect what technologies were adopted. Opportunities for innovation are endless. The point is that 'fitting-in' with what Earth itself provides, the goal of such outstanding architects as Ian McHarg and Frank Lloyd Wright, would guide the choices of technologies in each region. The result would be a rich diversity of communities with the pride of uniqueness to each area, rather than the deadly sameness that is rapidly taking over the world's cities and is beginning to spread even into rural landscapes.

But fitting-in with the local environment, although necessary, is not a sufficient criterion for choosing technologies. These must also reflect the local cultural values of each people. Almost all Western aid to developing countries has been defective in this respect. Too often attempts have been made to graft imported, high-tech 'solutions' onto societies where they were utterly inappropriate. Hundreds of examples of such technological failures exist, from rusting tractors in West Africa, to babies dying of diarrhoea in many countries after being fed on manufactured infant formula.

It was E. F. Schumacher who introduced the term 'appropriate' – or 'intermediate' – technology and made great strides in helping to design manufacturing processes that local people already saw a need for and could immediately implement on their own. In his thoughtful book, *The Culture of Technology*, Arnold Pacey describes an incident that took place in the sub-Saharan state of Mali. Visiting 'experts', working with local craftsmen, were able to modify the construction of native mud-walled grain bins by lining them with ferrocement so they could be used for storing precious rainwater runoff from the housetops in that drought-ridden country:

> The men who introduced the ferrocement concept worked out the details with local craftsmen in a collaborative effort. That led them to rethink some of their western values, so that they came to see their work not as modernization but as part of 'the organic development of a traditional society.'[14]

Technological innovation through community discussion and decision-making thus ensures that technology is 'appropriate' for the institutions and values of a particular people; it 'fits in'. When people also possess or are given ecological information, then new technologies can enhance their lives with minimal dislocation of social arrangements and without additional stress upon the local ecosystem.

This ideal form of development is by no means restricted to peasant communities in developing countries. It equally meets the need for more environmentally sensitive and democratically controlled technologies in *all* societies. The complex, highly centralised institutions of modern industrial societies, however, are not amenable to this kind of public, grassroots, democratic decision-making, and form a formidable barrier to the development of a sustainable, humanistic economy. It is not so much the 'backward' countries that need to make major cultural changes, but the 'forward' countries, the overdeveloped industrialised nations. If humankind succeeds, it will be because today's dominant Western worldview has been displaced by – or has evolved into – biologically more reasonable world visions, each adapted to the needs of decentralised regional polities.

REDEFINING THE SACRED

The struggle to provide meaning through identity with a sacred set of shared values, so easy in small, traditional societies, has always been the nemesis of large civilisations. The hierarchical organisations necessary for the management of such 'megamachines', as Lewis Mumford dubbed them, has been maintained through loyalty to a mystical image that serves as the binding force. Where once religious dogma promised spiritual rewards in Heaven, today the 'rational' dogma of economic growth and national power promise material rewards in this life. But no megamachine can continue to function if the promised reward fails to compensate for the absence of a daily sense of fulfilment and belonging. No amount of coercion, of imposition of 'law-and-order', can hold together discontented people, made antisocial by their very alienation.

As described in Part III, alienation in the West has been growing, not declining, in the presence of increasing material wealth. In line with capitalism's need for 'freely competing individuals', the original notion of political freedom has grown into a pathological desire for freedom from all social commitments. With the loss of meaning in one's personal life has come a search for vicarious meaning by identification with the symbols of heroic nationalism and through passive engagement in spectator sports and the mesmerising entertainments of television. Yet these same individuals are told that this 'freedom' is what is needed for their lives miraculously to blossom, each to its fullest potential, when in fact, without connectedness, the flower fails to open, wilting in the bud from boredom and emptiness. The hoped for beautiful self becomes instead a vessel of unfulfilled 'needs', which no amount of material consumption can meet.

Ironically, this vaunted freedom, gained at the cost of isolation and alienation, is fictitious, a sham. For our lonely existence depends on fitting ourselves to serve a giant, socioeconomic machine that we neither comprehend nor control. Hence we anxiously train ourselves – and our children –

to acquire some specialised skill that is the necessary passport to a particular niche in some corner of its massive, impersonal soma.

That which we claim is secular, we have made sacred. As a society, we are focused on the wrong things: individualism, competition, materialism, consumerism, technological power and efficiency. As Arnold Pacey has put it, it is 'time to change the topic of conversation'. We have long since used up the benefits to be achieved from the old perspective. It is time critically to assess our values, saving what still seems appropriate, while casting about – in history, among other contemporary cultures, among our present-day 'dissidents', and within our own imaginations – for a new ordering of social values.

Regaining Relatedness

Intense individualism is not something we are born with, which in other societies has been ruthlessly suppressed. Rather, we are taught it. For a person raised in the West, it is sometimes hard to perceive how this conditioning takes place. Indeed, the foremost developmental psychologists – Freud, Piaget and Kohlberg – have all told us that children 'naturally' begin as self-centred beings who must painfully progress through a series of developmental stages to arrive at a principled and compassionate adulthood. What these experts forgot to notice was that they were looking at only half the human race – for they all concentrated on males. Females, as Carol Gilligan reveals in her book, *In a Different Voice*, usually grow up from the start as nurturing human beings, deeply concerned for others. Moreover, unlike men who tend to perceive social responsibility in a negative way, as something that requires them to restrain their individual impulses, women tend to perceive it positively, as a means of expanding even further their active concern for others. On the one hand, life is seen as an egocentric free-for-all, played by agreed-upon groundrules – the Hobbesian vision; on the other, life is a communal sharing based on feelings of connectedness – the ancient hunting–gathering vision. Although exaggerated here, these stereotypes really do describe gender-based modal differences in outlook.

There are, of course, biological bases for some of the differences in male–female psychologies, some of which are only now being revealed. Some, indeed, may be present from birth. However, the emphasis between them upon which a given society builds its values, its worldview, its picture of 'human nature' can vary (and has in the past) from one pole to the other. But once the choice is made, children of both sexes are reared to accept the dominant attitude as the yardstick of the mature, successful person. Thus in the West, women's tendencies for emotional attachments are seen as a deficiency, some sort of gender inadequacy to be overcome if they are to function in positions of importance. 'Leadership' demands of both men and women a kind of 'John Wayne' emotional detachment. As nuclear activist Dr Helen Caldicott observed, the men who hold the pivotal positions of power in the major nations of the world tend to share a particular set of qualities:

> [T]ypically these men never show emotion, never admit mistakes, and are very dependent upon others of the same sex for peer-group approval. They are always sure of themselves, are always right, and above all they are always tough and strong.[15]

And any woman who aspires to higher office must prove herself equally tough. Margaret Thatcher dubbed herself the 'Iron Maiden' when she became Britain's prime minister, and Geraldine Ferraro, when running for vice-president of the United States, swore she would 'push the nuclear button' if necessary.

Yet there is something terrifying about this overemphasis on the autonomous individual as the exemplar of a social being. It denies too much that is human in us. As one of Gilligan's female interviewees, a medical student, said:

> By yourself, there is little sense to things. It is like the sound of one hand clapping, the sound of one man or one woman, there is something lacking. It is the collective that is important to me, and that collective is based on certain guiding principles, one of which is that everybody belongs to it and that you all come from it. You have to love someone else, because while you may not like them, you are inseparable from them. In a way, it is like loving your right hand. *They are part of you*; that other person is part of that giant collection of people that you are connected to.[16]

Whether the differences between male and female attitudes are biologically or culturally driven is beside the point. Societies can, and do, vary widely on which they choose to emphasise in generating their worldviews. And surely men in our highly individualistic Western society suffer greatly from the exaggerated expectations of aloofness and emotional independence placed upon them. We urgently need to reinstate feelings of relatedness and community into our social vision.

Relatedness must extend even further, however, from relatedness to each other to a renewed relatedness with the world that supports us. After Bacon, Newton and other Enlightenment thinkers had turned God's mysterious creation into a mechanism understandable to human minds and God into the master clockmaker, it was but a short step to imagine humans as God's agents, proprietors over Nature. It needed only Darwin's elucidation of evolution as a process of 'selection of the fittest' for another small leap of the imagination to see Nature as hostile, a fit object for control, manipulation and exploitation.

This hubristic image of the ultimate conquest of Nature overlooks the total dependence of all humankind – of whatever degree of 'development' – on vast natural forces over which it can have no significant control. Our failure in the West to retain our ancestors' sense of connectedness and relatedness with the world we inhabit, our penchant for viewing existence as a competitive game, leads us simultaneously to destroy Nature and to threaten one another with annihilation. As British philosopher Mary Midgley says:

> On every side now we can see people busily engaged in sawing off the branches on which they (along with many others) are sitting, intent only on getting those branches to market before the price of timber falls.[17]

Whatever social purposes the notions of competition and ownership of the Earth may once have served, their promotion to the pinnacle of our values hierarchy has brought us to the brink of evolutionary disaster. We need to recognise that people depend absolutely on each other and on their environment. It is as simple as that. We did not invent ourselves nor the universe. What we understand of its details is minuscule compared to its vastness and complexity. Our proper relationship is one of awe, not of dominance and plunder; it is one of cooperation and connectedness, not of competition and alienation; it is one of temperance and wisdom rather than aggressiveness and power. As Midgley says: 'Exclusive self-interest tends by its very nature not to be enlightened'.[18]

We must also change our attitudes about what is 'emotional' and what is 'rational'. The Enlightenment made much of discarding one and lauding the other, inventing for us the high-sounding distinction between 'subjective' and 'objective'. But we have fallen into a logical fallacy. We have labelled things that can be quantified – material consumption, military strength – as scientifically sound and therefore unimpeachably reasonable, while stamping those things that cannot – our attachments to others, the beauties of Nature – as suspect, merely matters of 'feelings'. What is overlooked is that it was a choice among 'feelings' that led to this dichotomy in the first place. Inner motives, not immutable laws of Nature, determine our hierarchy of values. By arbitrarily calling our *feelings* of self-interest 'rational' and our *feelings* of relatedness 'irrational', our scientism has led us into dangerous waters. We have let one set of feelings completely override another, yet both, presumably, are permanently anchored in our hypothalamuses as general behaviour guides selected for over millions of years.[19]

Survival will require that we in the West begin to take seriously our human feelings of community: community with the land and community with one another. Neither is alien to our natures. We have a basic attachment to 'home', that small corner of the world where we grew up. We have all experienced an upwelling of emotion on hearing songs or poems about our native land. In its distorted way, the arms race is supposed to protect this sacred place. Yet we ignore our own treatment of this very land we profess to love. We exploit it, bulldoze it, pave it over, flood it, consume it, and dominate it. It is time to do an about face in our attitudes, to heed the words of ecologist Aldo Leopold when he said:

> That land is a community is the basic concept of ecology, but that land is to be loved and respected is an extension of ethics. That land yields a cultural harvest is a fact long known, but latterly often forgotten.[20]

Agricultural economist Richard Norgaard stresses that societies everywhere, rich and poor, will need to 'coevolve' with their local environments, choosing technologies and creating social institutions that permit the environment to sustain society indefinitely.[21] This implies decentralised communities, each adapted to its own environmental income. There is no one 'best' path to human progress.

It is also part of our natures to be bonded to one another. As the advertising world has discovered, the opinions of others matter more than almost anything else. Without social attention, life is meaningless. And if that attention is not to be had through acceptance and belonging, it is acquired through domination and violence. As with our love of the land, however, the Western worldview pays lip-service to bondedness – America's professed devotion to 'applepie and motherhood', for example – while in reality emphasising isolation and competition. We physically isolate ourselves in separate houses and private automobiles, and sit in psychic isolation for long hours in front of the television screen. From preschool on, we are told that social acceptance – 'success' – derives from position and power. The resulting alienation spills over into growing psychological despair and social violence, ranging from child abuse at home, to the structural violence of an increasingly polarised economy, to the continuation of the arms race and a growing willingness to bomb or invade smaller countries overseas.

Our difficulty lies in part in our distortion of the meaning of 'freedom'. The West has lost sight of the fact that culture and tradition, far from constraining us, provide a meaningful guide for our actions. Laws merely put specific constraints on us: 'pay your taxes'; 'don't shoot your neighbour'. Culture, by contrast, tells us what it is *meaningful* to do. It provides a framework, a map, within which the individual is free to act purposefully. Anthropologist Dorothy Lee clearly explains the role of culture as expressed in native American societies:

The intricate set of [cultural] regulations is like a map which affords freedom to proceed to a man lost in the jungle . . .

Every individual in the group, male and female, young or old, has his proper place and role in the organisation of the community . . . with duties and rights commensurate with their age and status.[22]

Yet no constraints are placed on the individual:

Thus an individual can decide to what extent he will fill the responsibility which is his privilege . . .

It is evident, then, that a tremendous respect and trust is accorded to the individual since no provision is made for a man's failure, neglect, error.[23]

By confusing the important but limited freedom from the often oppressive rules of large hierarchical societies – the sort of thing that democracy is meant to cure – with freedom from *all* shared social values that might give meaning to our lives, we deny ourselves the deep social involvement and bondedness that our natures demand. We were socially bonded beings before we were human,[24] and socialisation has been a far more powerful force than tool-making in our evolutionary success. Indeed, the invention of tools could scarcely have occurred in the absence of culture.

To create in today's 'megasocieties' the social institutions that would permit the depth of social bondedness that characterised so much of our past is probably impossible. Once again, it is the decentralisation of our over-centralised political and social institutions that offers a solution. Self-sufficient regions, even self-reliant cities, can become foci for meaningful participation and personal identity. Through commitment to an identifiable *local* community, people could again experience the sharing of work *with* and *for* known others. With the possibility of direct, grassroots democracy, people could personally participate in political and economic decision-making. But self-directed communities offer far more, namely the opportunity to regenerate a shared cultural meaning, to recreate a sense of the sacred in life.

In the past, sacred cultural meaning has been bound up with traditions that were passed intact and uncritically from one generation to the next. Fragments of these traditions lie scattered about in religious and ethnic enclaves within the pluralistic hotchpotch of our modern megamachines. Yet even these are only compartments of people's lives, diluted by having to accommodate to the dominant secular values of the greater society in which they are embedded. The struggles of native Americans to retain their cultural integrity are but the best-known example of the tenuous nature of many sacred traditions in today's world. In some of the decentralised communities of the future, it may be possible to retain much from these ancient traditions, which are by no means as static as is often assumed. But in the majority of communities arising from the ruins of the modern industrial nations it will be necessary to rediscover the sacred – in a way, to re-invent it. For the first time in human history, people around the globe will need to learn and practise the art of 'critique/create'.

'Critiquing' Today and Creating Tomorrow

At relatively few times in history have a people consciously set about creating a new social vision. In the West, the Golden Age of Pericles in Athens, which spawned Socrates, Plato and Aristotle, was such a time. Dialogue, debate and argument about what a society *should* be were part of daily life, discussed openly in the public square. Tradition no longer was the arbitrary yardstick for determining values. These debates continued, with waning strength, through the days of republican Rome, after which public discourse on matters of social ordering virtually ceased.

A second period of creative thought in the West about the nature and purpose of society began some sixteen centuries later with the Protestant reformation, followed by the Enlightenment and the rise of scientific enquiry. A whole new crop of philosophers was needed to explain what it meant for humans to 'discover' and 'know' so much about the workings of the world. The human position in the cosmos subtly changed under the force of their arguments, culminating in the worldview still widely held today: 'competitive individualism' defines our human relations and 'private property rights' our relations to the land. Overlaid on these images is a sense of linearly directed change, of a kind of evolutionary 'progress', by which the future is not only to be different from the past, but inevitably better.

This vision, crowned within its own definitions by ever-increasing success, has led us to the brink of catastrophe, and demands that we undertake the new modes of thinking that Einstein urged upon us years ago. Rethinking is a process that will surely occur globally, although by no means in the concerted monolithic fashion that some futurists would like to see. Each people's history, local environment, and current institutions will lead it to a singular approach to this common task. It behoves us in the industrialised West to concentrate on our own efforts rather than presuming, as is now our tendency, to impose our particular vision on others. This does not mean that these multiple reconstructive experiments will occur in isolation from one another. Not only do they need to harmonise in matters of global concern, such as protecting the oceans and the atmosphere. Each society will also need to borrow ideas and techniques from others in an ongoing series of cultural cross-fertilisations. There are thus two requirements for the process of 'critiquing' and creating a worldview: a tolerance of diversity among all cultures, and a tolerance of criticism and change within one's own.

The process of 'Critique/Create'

The great challenge before us – perhaps the greatest challenge humankind has ever faced – is learning how to accommodate dispute and diverse opinion without losing our basic social cohesiveness; how to eliminate the rigidity of dogma without losing the sense of shared meaning. It is a challenge many societies face, and we can all learn from one another the skills needed for the task.

New worldviews, new social institutions, do not just drop from the sky. They evolve from previous visions. In fact, as William Ophuls reminds us, our first tendency when things go wrong 'is to redouble effort and belief in support of the current paradigm, which is after all a kind of civil religion'.[25] Although such rigid clinging to the past is counterproductive, it is out of the past that the future of society must be constructed. Indeed, every society, if it hopes to retain its cohesiveness, will have to extract the best from its historic past while simultaneously criticising the present. In this way, criticism intended for the good of all becomes not a sign of cultural weakness, but of strength.

A good way to begin is to start teaching history as it actually happened, as a series of ideological struggles that zig-zagged their way to the present. The past of every modern nation offers its children insight into the processes their forebears used to shape worldviews, provided one stops glamourising that past and seriously analyses it. Take America as example. Her children need to know that their government did not emerge full-blown after just one hot summer of debate in 1787. Says Ophuls:

> We sometimes forget, for example, that our Constitution was the culmination of several decades of intense and sustained political discussion and action by our founding fathers.[26]

There were two factions (both composed of well-to-do men) who had distinctly different visions of democracy. The Federalist vision championed by Madison and Hamilton won out, creating not the popular grassroots democracy envisioned by Jefferson, but the highly centralised, representative government that exists today. Resurrecting this dialogue in the schools would provide American children not only with an historical example of the *process* of public dialogue, but an alternative social myth from their past on which to base their visions of the future. Parallel examples can be found in the histories of all nations. In a non-trivial sense, the study of history is to our civic self-image what theology is to religious faith – a means for redefining our worldview; it provides the opportunity to discover new insights.

Although historical re-examination can initiate the process, 'critique/ create' further requires a general skill and willingness among the members of a society to undertake the necessary public dialogue. Periods of ideological flux in the past have too often been marked by bitterness and violence, traits our military cleverness has now put off-limits. No doubt it will be difficult at first to openly argue about our deeply held values without rancour, since we have not been taught how to do this. The European Greens, dedicated as they are to grassroots, consensual politics, are experiencing severe differences among themselves. As individuals, they find it hard to erase the cultural burdens from their past: male chauvinism and jealousy of emerging leaders. But more than that, this party, struggling to establish trust and political consensus among its members, is being born into a hostile world of national power politics and acute international mistrust. It has, as Rudolf Bahro, a member of West Germany's Green Party, says, a double task:

> The Green party has developed with the pretense that it is the political arm of a movement and a new culture. But that culture hardly exists; it is in the embryonic state. There is a stream of humanistic psychology running through society now, but this, too, is merely part of the emerging culture. The Greens in West Germany are more closely related to the new culture than are political forces in any other country, but it is to the party's disadvantage that its own development is more advanced than the countercultural network [that elects it]. For example, there is only a very small communitarian movement. We now must also do the sort of more fundamental work that should have preceded the party.[27]

Nevertheless, in the small, grassroots neighourhood meetings of Greens, both in Europe and America, people are gradually training themselves in the first and most important skill of cultural change, learning to give and accept criticism in non-destructive ways.

Most difficult of all is learning the technique of healthy self-criticism. In Western societies it is almost unheard of. People seem unable to separate honest mistakes and misapprehensions from a guilt-ridden sense of either incompetence, ignorance or deliberate wrong-doing. There is massive social insecurity, brought on by excessive individualism and constant competition in all aspects of life. Absence of self-criticism is particularly marked among Western politicians. Those who admit to any problems are immediately labelled as 'weak' or 'indecisive' by their opponents. As another German Green, Gabi Potthast, says:

> In power politics there is a lust to destroy; many people get energy by beating others down.[28]

Oddly enough, it is not the political leaders of the Western democracies who are beginning to show signs of self-criticism and are freely criticising the policies of their countries' recent past. It is leaders in China and more recently the Soviet Union who have publicly announced major changes in direction. Although these changes were not initiated from the grassroots, they are widely popular, and are certainly in marked contrast to the so far inflexible ideologies of the West. These two important experiments bear watching.

Besides learning to cope with criticism, societies who are attempting to create new world visions need new and effective ways to cope with conflict. Other than violence, there seem to be only two solutions. Either the majority rules or consensus is achieved. All too often, the rule of the majority is tyrannical over the legitimate interests of a minority. But achieving consensus, at least to the uninformed, seems utterly impractical. Since disputes are widely perceived as 'zero-sum' situations – if one side wins, the other must lose by an equal amount – one is thrown back on majority rule. This quickly invites the use of intimidation and even violence, as 'elections' in all too many Third World nations show. The alternative is a seemingly (to the Western mind) improbable solution, namely cooperative problem solving. Writes political scientist Robert Axelrod:

> We are used to thinking about competitions in which there is only one winner, competitions such as football or chess. But the world is rarely like that. In a vast range of situations mutual cooperation can be better for *both* sides than mutual defection. The key to doing well lies not in overcoming others, but in eliciting their cooperation.[29]

Cooperative problem solving is quite distinct from the usual notion of 'compromise', in which each side grudgingly gives in a little. Compromise invites horse-trading, wheeling and dealing and pork-barrelling – the least savoury aspects of modern power politics. The secret of cooperative problem solving, as discussed in the previous chapter, is not to cast the problem in terms of status and position, of 'winning' and 'losing', as is commonly done when the popular media air debates on public issues. Instead, the parties involved approach it as a *joint* problem, something they share between them. This kind of approach to settling differences requires the following: (1) a means for mutually agreeing on 'the merits of a case'; (2) learning how to set aside confrontational attitudes among factions; and (3) a method for establishing security when negotiations appear to be failing and violence is threatened. These three ingredients are essential for avoiding destructive behaviours whether the differences are within or between those political entities known as states.

Seeking the 'merits of the case' Whenever there is disagreement, the parties involved must first agree on what is in dispute. Otherwise their

chances of reaching a mutually agreeable solution are virtually zero.
Unlike textbook problems, where the issue is clearly defined and all the
data for solving it are given, real-life problems are often vaguely stated.
They are usually oversimplified, treated like a mechanical breakdown, a
reparable *glitsch* in the social machine that merely needs to be identified –
like a flat tyre or a leaky gasket – in order to solve the problem. But as
social psychologist Ian Mitroff points out, a complex issue is viewed
differently by parties with different vested interests. Each holds different
assumptions about what is true or untrue, important or unimportant. The
essential first step is to sort out all the various assumptions.[30]

Now assumptions are just that; they are our beliefs about reality. They
are our personal experiences as interpreted and explained by our cultural
worldview. Indeed, it is only through such a synthesis that we can make
sense of the world we live in. By identifying and evaluating the assumptions of all participants, the issue-resolving process moves from the
superficial to the basic, becoming a mutual exploration of diverse
worldviews. People begin to explore the depths of each other's minds,
learning to share each other's visions, and ultimately forming the shared
vision that paves the way for mutual resolution of the problem at hand.
Although there are certain parallels with the Hegelian dialectic, there is no
assertion that some permanently 'higher' state has been reached. Rather,
the new vision is, like all evolutionary 'solutions', adapted to the moment,
susceptible to reevaluation as new data or changing circumstances demand.

In summary, establishing the 'merits of the case' requires that all
stakeholders become active participants in its resolution. By analysing
their different assumptions and working to establish mutually agreed
principles, the parties can hope for a consensual solution without a residue
of bitterness and resentment. To participate successfully in this kind of
conflict resolution requires some training, however.

Education: learning to participate   Education for 'critique/create' must
perform several functions, all of which are deficient to some degree in most
Western countries. Equipping a people to participate fully in the shaping of
their own destiny obviously requires learning the skills of *community
dialogue*. But before such a dialogue can be effective, there must also be
shared *information* about the broad spectrum of cultural values that exist in
the world, and students must acquire the capacity for *critical thinking*.

Cultural perspective is doubly necessary. It not only provides an understanding of the values held by other cultures with which one's own must
co-exist; it also provides what has been called 'the anthropological
imagination' – the ability to analyse one's own culture better from having a
background of alternatives to reflect it against. In Europe, at least a
beginning is being made in developing cultural perspective, and some
knowledge of global history and geography can be assumed of the average

European. Not so in America. In discussing how his fellow Americans might discover a new mission in the world, one where they would fit in *with* rather than dominate *over* other peoples, social philosopher William Irwin Thompson responded:

But how can we do that when [we] North Americans are so incredibly chauvinistic and simply not interested in the world? There is never any world news except disasters on American television. And if you look at *USA Today* or your local newspapers, there is no world information, unless you go to *The New York Times*, and then the news is focused through a distorting lens. Americans don't speak any other languages, and they just don't seem to care. They see the world as a collection of old cultures that need to be Americanized after the fashion of Disneyworld's Epcot.[31]

America clearly lags far behind in its global cultural perspective. Yet it is doubtful if many European students are taught to examine the origins of the value systems held by others in order that they might think more clearly about their own. Approaching history as the comparative study of the evolution of worldviews could perhaps do more than anything else to develop the understanding needed to identify 'the merits of the case'. Tracing the origins of beliefs and assumptions, both others' and our own, is a first step in making the 'critique/create' mode of social change a real possibility. It provides the framework for understanding and therefore accepting differences among conflicting groups of any size, be they members of a single family or of diverse cultures within the family of nations. Differences come to be seen as natural, necessary historical outcomes; they may also begin to be seen as desirable, as providing a richness to our global society and a constant source of new perspectives.

This educational approach also allows us to perceive that cultural change is in fact *evolutionary*, not revolutionary. Bloody – or even bloodless – political upheavals seldom change the way people think. They simply seek a new set of institutions that 'fits in' with the same internalised worldview, and often the new institutions bear a close resemblance to the old. Imposing a new worldview on others is thus seldom feasible, and is certain to fail if it comes entirely from outside the old cultural context. The changing of a worldview occurs *only through active engagement of the people in general*. Neither coercive threats, legal or otherwise, nor technological fixes, however clever, can permanently correct maladaptive social patterns. The best such methods can achieve is a temporary postponement of crisis.

Teaching history as something more than a long sequence of political facts with a few comments on 'daily life' thrown in to keep students' attention is a major step in developing critical thinking generally. 'Critical thinking' is not, as some might suppose, learning to apply formal logic. It goes far deeper, teaching the critical examination of the underlying assumptions on which cultural values are based. In that sense, it becomes a threat to those who benefit from the currently held values. Such persons, most often those with political and economic power, prefer that education should simply socialise children to the *status quo*, on the assumption that 'our social vision is complete'. Flaws in the system are seen as failures of people, not of the shared worldview and its institutions. Criticism of shared beliefs and assumptions is regarded as unpatriotic, especially in North America, and as downright seditious throughout much of Latin America.

But change, self-actuated, democratic change, cannot occur without liberatory learning; it demands the ability to think critically. Stanley Aronowitz, an American educator, describes the consequences of the failure of analytical thought in mass culture.

> The issue is the capacity for theoretical or conceptual thought itself. When people lack such competence, social action that transcends the struggle for justice within the empirically given rules of social organisation and discourse is impossible ... Since critical thinking is the fundamental precondition for an autonomous and self-motivated public or citizenry, its decline would threaten the future of democratic social, cultural and political forms.[32]

A society whose educational system omits the critical analysis of values cannot be said to be educating its young at all; it is merely *training* them. Meantime, major social decisions are left in the hands of professional politicians. But politics, the reformulating of our shared beliefs and assumptions, is *not* a 'specialty'. If a nation's worldview is to evolve and adapt – as so many must if their societies are to survive – everyone must participate in the political process of change.

Throughout his life, the American philosopher and educator John Dewey argued that political participation must be learned. It is not something we are born knowing how to do. To him, the whole purpose of modern education was the 'reconstruction of experience', developing in the individual the ability to consciously reconstruct her or his inner worldview. Wrote Dewey:

> The keynote of democracy as a way of life may be expressed, it seems to me, as the necessity for the participation of every mature human being in formation of the values that regulate the living of men together ...

> The school cannot be a preparation for social life excepting as it reproduces, within itself, the typical conditions of social life.[33]

This kind of guided practice in being an active participant in one's society scarcely occurs in schools today. In America, for a brief time, a number of Dewey's disciples developed cooperative programs that taught group problem solving and participatory citizenship, and to a lesser extent critical thinking about values. Now, after decades of creating alienated adults in a fiercely competitive school system, educators are beginning to discover that cooperative learning is far more effective, even when the educational goal is merely to train a more effective workforce.[34] The next steps are to introduce cultural perspective and critical thinking, in order to make the 'critique/create' mode of social change a reality. In the spectrum of means for achieving social stability in a changing world, universal active participation seems the one sure answer to coercion and violence.

Security through non-violence

> Insistence that the use of violent force is *inevitable* limits the use of available intelligence ... Commitment to inevitability is always the fruit of dogma; intelligence does not pretend to *know* save as a result of experimentation ... Moreover, acceptance in advance of the inevitability of violence tends to produce the use of violence in cases where peaceful methods might otherwise avail.[35]

Once again, Dewey speaks to us from half a century ago with words that are even more cogent today. Our continuing belief in the inevitability of violence, 'as a last resort', has brought us to the irrational state of having enough weapons to destroy all life many times over. The effort and energy involved in this insane preparation for terminal violence imperils any hope for future security by inducing a global sense of fear and futility. 'If there is no future, why bother trying to make things better?' The arms race also diverts massive resources from correcting those social ills around the globe that lead to violence: hunger, poverty, injustice, ignorance, disease. Worst of all, it paralyses our imaginations. Our persistent belief that there is no effective alternative to violence blocks the search for creative ways of settling disputes and defending ourselves.

Scepticism about relying on non-violent means of security seems premature. Since the emergence of massively organised violence around 5000 to 8000 years ago, the efforts exerted on behalf of violent one-upmanship have so greatly outstripped the search for workable alternatives that any intelligent person, as Dewey points out, can only conclude that the experiment has never been properly tried. In the United States alone, not only is the art of war making taught in most major universities, but there are also a dozen service-run war colleges. An official United States Institute of Peace, established in 1984 after prolonged public pressure, is still a tiny, impotent organisation funded at less than $5 million a year. More hopeful are the Peace Studies programmes blossoming on more and more university compuses. Yet the United States, and most other countries as well, are a long way from pursuing the study and practice of peace throughout society as a whole.

Such study would require a comprehensive look not just at international diplomacy and disarmament (which is what most 'peace studies' programmes concentrate on), but at all the causes of violence between and within societies: ethnic disputes and cultural imperialism; racism; economic exploitation and class inequality; alienating and dehumanising factories and other social institutions; the persistent valuing of 'competition' over 'cooperation', of 'efficiency' over 'meaning'. These are the factors underlying everything from guerrilla wars and terrorism to child abuse, rape, drug addiction and suicide. Such overt violence is thus a symptom of persistent social violence: physical, economic or psychological.

There are two ways for a society to handle its symptomatic overt violence. One is to employ counter-violence: bomb the Libyans; gaol the cocaine addicts. But this all too often only makes the problem worse. Besides, it is extraordinarily costly, as already seen. It siphons off human and material wealth that could in fact be used to redress many of the underlying causes of violence. The world is currently on an upwardly spiralling armaments binge that threatens to bankrupt it – not just financially, but morally, socially and environmentally. The alternative is non-violence.

Like any other form of human behaviour, including organised warfare, non-violence has to be *learned*. Educating people in the methods of peaceful negotiation and conflict resolution based on 'the merits of the case'; educating people about the underlying social causes of violence; and educating people to participate actively in creative social change – all these are critical steps in putting an end to the escalating threats of violence that now engulf the globe. But they are long, slow processes. Something is needed in the meantime to break the upward spiral of 'violence begetting violence'. Those who suffer from injustice and institutional violence – which is what ultimately must be eliminated – can begin to break the chain by the practice of *active* non-violent resistance.

Already touched on in the last chapter, the methods of Gandhi have been used effectively on numerous occasions by carefully trained activists. Political scientist-historian Gene Sharp has made a lifetime study of non-violent action, documenting his ideas with massive historical evidence that it can and does work.[36] Sharp's dictum: Governments do not have power *over* people; all they have is violence. Rather, it is the people who have power over governments. Violence, of any sort, is the consequence of a *loss* of power, and the greater the threat of violence, the less powerful, the less to be admired, is the person who wields it. The bully has neither courage nor morality to recommend him.

Yet despite Sharp's overwhelming evidence, the theory of non-violence is regularly ignored in academic circles. It is not taught in schools and colleges. Governments fail to fund research in non-violence. There is a tacit global conspiracy to deny its very possibility. There is indeed some truth in the saying that people who behave like victims often become victims. A world that continues to be intimidated by violence and refuses to consider non-violence as even a possible political and social alternative will naturally remain unskilled in the tactics of non-violence, a ready victim to the ever higher-tech forms of violence. A world that believes in the necessity and legitimacy of war and violence consigns itself to become the victim of its own beliefs. No one has stated our predicament better than General Omar N. Bradley, one of America's heroes from the Second World War, in his 11 November 1948, Armistice Day address:

> With the monstrous weapons man already has, humanity is in danger of being trapped in this world by its moral adolescence. Our knowledge of science has clearly outstripped our capacity to control it. We have too many men of science, too few men of god. We have grasped the mystery of the atom and rejected the Sermon on the Mount. Man is stumbling blindly through spiritual darkness while toying with the precarious secrets of life and death ... The world has achieved brilliance without wisdom, power without conscience. Ours is a world of nuclear giants and ethical infants. We know more about war than we know about peace; more about killing, than we know about living.[37]

A BRIEF SUMMING UP

It is clear that unless we begin the widespread teaching of the techniques and skills of peace we shall never succeed in outgrowing our moral adolescence. This must include not only the techniques of conflict resolution, of tolerance and participation, and of non-violent action, but perhaps more important, the teaching of the whole nature of human societies, of their intrinsic rootedness in our universal need to be bonded, of the historical evolution of their varied worldviews, of the means by which we acquire our values and beliefs and the limits they place on our ability to empathise with those holding different visions. As we all know, we now physically threaten ourselves with instantaneous annihilation in a nuclear holocaust; fewer recognise that we also threaten ourselves with no less certain annihilation by destruction of our habitat. Until we teach each other and our children that these things are so, and begin to examine together *why* they are so, they will continue to be accepted as inevitable; humankind will remain paralysed, unable to create a viable future world of diverse societies, enhancing rather than threatening each other.

'Peace' for the moment may have to remain defined as 'the absence of war', but ultimately we will need to think in terms of Michael Nagler's definition: 'Peace is that state in which parties spontaneously desire one another's welfare'.[38] And it will come about not when every college and university around the world has a major in 'peace studies', but when every discipline is expounded in relation to the insights this book – and others like it – have tried to provide. If we are to succeed and not succumb, then slowly but surely these visions will grow throughout our global society, and become in unique ways part of every worldview, replacing the now universally held belief that war and violence are inevitable. When non-violence, toward one another and toward the Earth as well, finally becomes 'natural' human behaviour, as much a part of our automatic thinking as is our acceptance of violence today, then humankind will finally have made it safely out of the labyrinth, past the turning point, having chosen the viable pathway to the future and established the new modes of thinking Einstein urged us to seek.

Notes and References

1 The Future: a Search for Values

†   Loren Eiseley, *The Firmament of Time* (New York: Atheneum Publishers, 1962) p. 117.
1.   Giscard d'Estaing, quoted by Richard Lamm, Governor of Colorado in 'Promoting Finitude', *The Amicus Journal* (Summer 1980) p. 7.
2.   R.P. Turco, O.B. Toon, T.P. Ackerman, J.B. Pollack and C. Sagan, 'Global Atmospheric Consequences of Nuclear War', *Science*, 222 (1983) pp. 1283–92; P.R. Ehrlich, M.A. Harwell, P.H. Raven, C. Sagan, G.M. Woodwell *et al.*, 'The Long-Term Biological Consequences of Nuclear War', *Science*, 222 (1983) pp. 1293–1300; C. Sagan, 'Nuclear War and Climatic Catastrophe: Some Policy Implications', *Foreign Affairs* (Winter 1984) pp. 257–92. In late 1983, the United States Federal Emergency Management Agency (FEMA) released a report arguing that agriculture would 'emerge in relatively good shape' after a major nuclear war; the report received scathing ridicule (see *Science*, 222 (1983) p. 1308).
3.   J. Schell, *The Fate of the Earth* (New York: Alfred A. Knopf, 1982).
4.   W. Alvarez, E.G. Kauffman, R. Surlyk, L.W. Alvarez, F. Asaro and H.V. Michel, 'Impact Theory of Mass Extinctions and the Invertebrate Fossil Record', *Science*, 223 (1984) pp. 1135–41; D.A. Russell, 'The Mass Extinctions of the Late Mesozoic', *Scientific American* (January 1982) pp. 58–65.
5.   Turco *et al.*, 'Global Atmospheric Consequences'.
6.   Ehrlich *et al.*, 'The Long-Term Biological Consequences', p. 1299.
7.   Sagan, 'Nuclear War', p. 264 f.
8.   For a recent update on the scientific consensus on nuclear winter see M.A. Harwell and C.C. Harwell, 'Updating the "nuclear winter" debate', *Bulletin of the Atomic Scientists* (October 1987) pp. 42–4. Forty scientists from around the world conclude that four-fifths of the world's population might well starve after a major nuclear war.
9.   C. Zraket, quoted by R.J. Smith, 'Nuclear Winter Attracts Additional Scrutiny', *Science*, 225 (1984) p. 32.
10. D.H. Meadows, D.L. Meadows, J. Randers and W.W. Behrens, III, *The Limits to Growth* (New York: Universe Books, 1972).
11. Meadows *et al.*, *The Limits to Growth*, p. 169.
12. H.S. Cole, C. Freeman, M. Jahoda and K.L.R. Pavitt (eds), *Models of Doom* (New York: Universe Books, 1973).
13. *The Global 2000 Report to the President, Vol. 1, Entering the Twenty-First Century* (Washington, DC: US Government Printing Office, 1980) p. iii. See also, *Vol. 2, The Technical Report; Vol. 3, Documentation on the Government's Global Sectoral Models: The Government's 'Global Model'* and *Global Future: Time to Act* (1981).
14. *The Global 2000 Report, Vol. 1*, p. iii.
15. *The Global 2000 Report, Vol. 1*, pp. 3–5.
16. J. Simon, 'Resources, Population, Environment: An Oversupply of False Bad News', *Science*, 208 (1980) pp. 1431–37; quote is from p. 1434.
17. J. Simon, *The Ultimate Resource* (Princeton, NJ: Princeton University Press, 1981).
18. J. Simon, *The Ultimate Resource*, p. 27.

19. J. Simon, *The Ultimate Resource*, p. 90.
20. For a good example of this kind of argument, see H.J. Barnett, G.M. van Muiswinkel, M. Shechter and J.G. Myers, 'Global Trends in Non-Fuel Minerals', Chapter 11 in J. Simon and H. Kahn (eds), *The Resourceful Earth: A Response to Global 2000* (New York: Basil Blackwell, 1984) pp. 316–38.
21. H.J. Barnett and C. Morse, *Scarcity and Growth: The Economics of Natural Resource Availability* (Baltimore: Johns Hopkins Press, 1963) pp. 11–12; sentences quoted are slightly out of sequence.
22. L.R. Brown, *State of the World: 1984; A Worldwatch Institute Report on Progress Toward a Sustainable Society* (New York: W.W. Norton, 1984); each year new editions appear, researching different global topics.
23. E. Hyams, *Soil and Civilization* (New York: Harper and Row, 1976 (© 1952)) p. 16.
24. P. Siekevitz, 'Not Every Problem Can Be Solved with a Technological Fix', *The Sciences*, 18 (February 1978) pp. 29–30.
25. M.E. Clark and L. Holler, 'Teaching Science within the Limits of Science', *Journal of College Science Teaching* (September–October 1986) pp. 8–9, 56–7.
26. M. Polanyi, *Personal Knowledge: Towards a Post-Critical Philosophy* (Chicago: University of Chicago Press, 1962) pp. 134 ff.
27. P. Alpert, 'The Boulder and the Sphere', unpublished essay (1981); Dr Alpert is Assistant Professor of Biology in the Department of Botany at the University of Massachusetts at Amherst.
28. T. Kuhn, *The Nature of Scientific Revolutions*, 2nd edn (Chicago: University of Chicago Press, 1970).

2 Energy and Exponentials

†   Henry Adams, *A Letter to American Teachers of History* (privately published by the author, 1910) pp. 131, 155.
1.   For a discussion of the multiple environmental consequences that may result from the building of very large dams, see C. Sterling 'The Aswan Disaster', *Environmental Journal, National Parks and Conservation Magazine*, 45 (1971) pp. 10–13; numerous articles in M.T. Farvar and J. Milton (eds), *The Careless Technology* (Garden City, NY: Natural History Press, 1972); and D. Deudney, 'Hydropower: An Old Technology for a New Era', *Environment* (September 1981) pp. 17–20, 37–45.
2.   R. Stobaugh and D. Yergin (eds), *Energy Future: Report of the Energy Project at the Harvard Business School* (New York: Random House, 1979) p. 232.
3.   E. Teller, *Energy from Heaven and Earth* (San Francisco: W.H. Freeman, 1979) pp. 292–3.
4.   B. Russell, quoted in S.M. Blinder, *Advanced Physical Chemistry* (London: Macmillan, 1969) p. 300.
5.   W.L. Thomas, Jr (ed.), *Man's Role in Changing the Face of the Earth*, vol. 1, (Chicago: University of Chicago Press, 1956). See p. 188 f for Europe, p. 416 f for China. This book and its companion vol. 2 are among the most comprehensive and fascinating accounts of the role humans have played from prehistory to the present in altering ecosystems, communities of organisms and even climate, by assorted conscious and unconscious manipulations. It is drawn on further in Chapters 3 and 4.
6.   A.A. Bartlett, 'Forgotten Fundamentals of the Energy Crisis', *American Journal of Physics*, 46 (1978) pp. 876–88. A central article for this chapter.

7.   Bartlett, 'Forgotten Fundamentals', p. 876.
8.   Thomas, *Man's Role*, p. 200.
9.   See S. Goodwin, 'Hubbert's curve', *Country Journal* (November 1980) pp. 56–60, for an account of the history of Hubbert's predictions. Hubbert's data are available as follows: M. King Hubbert, Chapter 8 in *Resources and Man*, issued by the National Academy of Science and the National Research Council (San Francisco: W.H. Freeman, 1969); *Energy Resources of the Earth* (San Francisco: W.H. Freeman, 1971); A National Fuels and Energy Policy Study, Serial 93–40 (92–75) Part I (Washington, DC: US Government Printing Office, 1973).
10.   W. Ophuls, *Ecology and the Politics of Scarcity* (San Francisco: W.H. Freeman, 1977) pp. 88 ff. Ophuls outlines the environmental perils and costs of going after low-grade fossil fuels. For recent estimates of US oil reserves, see *Science*, 226 (1984) p. 426, and *Science*, 228 (1985) p. 974. Amounts remaining to be found will last 10–15 years at present rates of consumption.
11.   Bartlett, 'Forgotten Fundamentals', quotes on p. 881 from ERDA (Energy Research and Development Agency) report of 1975 and 1976: 'Coal reserves far exceed supplies of oil and gas, and yet coal supplies only 18% of our total energy. To maintain even this contribution we will need to increase coal production by 70% by 1985, but the real goal, to increase coal's share of the energy market will require a staggering growth rate'.
12.   L.R. Brown, *State of the World, 1984* (New York: W.W. Norton, 1984) p. 36.
13.   D.L. Meadows, W.W. Behrens, III, D.H. Meadows, R.F. Naill, J. Randers and E.K.O. Zahn, *The Dynamics of Growth in a Finite World* (Cambridge, MA: Wright–Allen, 1974). See also Figure 5.1.
14.   H. Kahn, W. Brown and L. Martel, *The Next 200 Years* (New York: Morrow, 1976) p. 101 f.
15.   From the entropy of mixing, defined by the Second Law of Thermodynamics, we know that the minimal energy required to separate a mole of pure substance is inversely proportional to the natural logarithm of its mole fraction in the mixture.
16.   C.A.S. Hall and C.J. Cleveland, 'Petroleum Drilling and Production in the United States: Yield Per Effort and Net Energy Analysis', *Science*, 211 (1981) pp. 576–9. The same may be true for natural gas – see C.B. Hatfield, *Science*, 219 (1983) p. 10, and J. Cason, *Science*, 220 (1983) p. 359.
17.   For use of steam to express oil see P.R. Riva, Jr, *World Petroleum Resources and Reserves* (Boulder, CO: Westview Press, 1983) pp. 89–109. Up to 35 per cent of the energy recovered may have been expended in generating the steam to force it out of the ground. For use of compressed (supercritical) CO_2 to express oil, see P.H. Abelson, 'Oil Recovery with Supercritical CO_2', *Science*, 221 (1983) p. 815.
18.   R.H. Romer, *Energy – An Introduction to Physics* (San Francisco: W.H. Freeman, 1976) p. 5. A readable text for anyone who wishes to grasp the meaning of energy and work.
19.   R.L. Sivard, *World Energy Survey* (Leesburg, VA: World Priorities, 1981) p. 5. A massive amount of information from diverse sources has been put into easily understandable graphs, charts and tables.
20.   Compare Table 14.1, p. 292, in Teller, *Energy*, with Table 8.1, p. 232, in Stobaugh and Yergin, *Energy Future*. Although Teller favours nuclear energy while the Harvard group pushes eventually for more conservation and more solar and other renewable resources, their short-term predictions about the mix of energy resources and of energy consumption in the United States are almost identical.

21. T.B. Johansson, P. Steen, E. Borgren and R. Fredriksson, 'Sweden Beyond Oil: The Efficient Use of Energy', *Science*, 219 (1983) pp. 355–61. See also the revised 1980 edition of Stobaugh and Yergin, *Energy Future*, published by Ballantine, New York, which emphasises conservation.

22. Stobaugh and Yergin, *Energy Future*, p. 157 f.

23. Stobaugh and Yergin, *Energy Future*, p. 159.

24. N. Seldman and J. Heels, 'Waste Management: Beyond the Throwaway Ethic', *Environment* (November 1981) p. 32 f.

25. P.H. Abelson, 'World Energy in Transition', *Science*, 210 (1980) p. 1311.

26. C. Norman, 'Boom and Bust in Energy', *Science*, 221 (1983) p. 443.

27. Costs of reclaiming land in terms of a fraction of the energy contained in the extracted coal vary widely according to the topography and richness of the seams being mined and their depth. 'Guesstimates' range from 1 per cent to perhaps 30. For dollar costs see *Reclamation Cost Inputs for the Resource Allocation and Mine Costing Model*, 30 November 1984 (DOE Paper/OR/20837–2, Washington DC: US Department of Energy). As pointed out by Denis Hayes (*Rays of Hope: The Transition to a Post-Petroleum World*, New York: W.W. Norton, 1977, p. 40), '[E]ven reclaimed land, while ransomed from aesthetic oblivion, is often worth less... than in its virgin state'. It is useful for pastures, but seldom crops.

28. J. Hansen, D. Johnson, A. Lacis, S. Lebedeff, P. Lee, D. Rind and G. Russell, 'Climate Impact of Increasing Carbon Dioxide', *Science*, 213 (1981) p. 966.

29. R.D. Lipschutz, *Radioactive Waste: Politics, Technology and Risk* (Cambridge, MA: Ballinger, 1980) pp. 34–6, 135–8.

30. D. Ford, *The Cult of the Atom* (New York: Simon & Schuster, 1982) p. 238.

31. D. Egger, 'Chernobyl's Cup Runneth Over: West Germany Pours Hot Milk', *The Nation* (28 March 1987) pp. 392–3. This describes the attempts in Western Europe to sell radioactive foods to unsuspecting Africans and others. An early Soviet account of the Chernobyl disaster appeared in *Soviet Life* (September 1986) pp. 34–41. Nearby wells have been permanently closed.

32. A.M. Weinberg and I. Spiewak, 'Inherently Safe Reactors and a Second Nuclear Era', *Science*, 224 (1984) pp. 1398–1402.

33. Lipschutz, *Radioactive Waste*. This book amplifies clearly and unemotionally this entire problem.

34. M.L. Wald, 'Steel Turned Brittle by Radiation Called a Peril at 13 Nuclear Plants', *New York Times* (27 September 1981) p. 1; E. Marshall, 'Reactor Safety and the Research Budget', *Science*, 214 (1981) pp. 766–8.

35. C. Norman, 'A Long-Term Problem for the Nuclear Industry', *Science*, 215 (1982) pp. 376–9; C. Norman, 'Isotopes the Nuclear Industry Overlooked', *Science*, 215 (1982) p. 377.

36. Marshall, 'Reactor Safety'.

37. International Energy Agency, *World Energy Outlook* (Paris: OECD/IEA, 1982) p. 338 f.

38. See Ford, *The Cult of the Atom*, p. 54 f. However, as Edward Teller points out in his book (*Energy*, pp. 190 ff), another nuclear alternative is possible. Canada has developed a fission process that uses thorium as fuel and deuterium (in heavy water) to absorb neutrons and control the rate of the reaction. Once under way, the thorium cycle is self-regenerating, but is less dangerous than the plutonium breeder cycle. Since thorium is more abundant than ^{235}U, the fuel problem could be postponed, perhaps for several centuries. All the other drawbacks of nuclear power would of course remain.

39. I. Barbour, H. Brooks, S. Lakoff and J. Opie, *Energy and American Values* (New York: Praeger, 1982) p. 112. See also Hayes, *Rays of Hope*, pp. 61–74.
40. Hayes, *Rays of Hope*, p. 207.
41. Hayes, *Rays of Hope*, p. 180. See also Deudney, 'Hydropower', for citation of problems.
42. Hayes, *Rays of Hope*, p. 207.
43. Hayes, *Rays of Hope*, p. 158.
44. Y. Hamakawa, 'Photovoltaic Power', *Scientific American* (April 1987) pp. 87–92.
45. J.B. Tucker, 'Biogas Systems in India: Is the Technology Appropriate?', *Environment*, 24 (October 1982) pp. 12–20, 39.
46. P.H. Abelson, 'Energy and Chemicals from Biomass', *Science*, 213 (1981) p. 605.
47. S. Ferrey and M.-C. Baker, 'United States May Block Solutions to World Energy Problem', *Albuquerque Journal* (27 October 1981) p. A–5.
48. Quanta, 'Magma Power', *The Sciences* (May–June 1982) p. 4.
49. Haynes, *Rays of Hope*, p. 159.

3 The Economics of Spaceship Earth

† Masanobu Fukuoka, *The One-Straw Revolution: An Introduction to Natural Farming* (Emmaus, PA: Rodale Press, 1978) p. 172.
1. G.V. Jacks and R.O. Whyte, *Vanishing Lands: A World Survey of Soil Erosion* (New York: Doubleday, 1939) p. 4.
2. For theories of conditions on prebiotic Earth, see L.E. Orgel, *The Origins of Life* (New York: Wiley, 1973); C. Ponnamperuma, *The Origins of Life* (London: Thames & Hudson, 1977); and L. Margulis, *Symbiosis in Cell Evolution: Life and Its Environment on Early Earth* (San Francisco: W.H. Freeman, 1981).
3. D.R. Rust, 'Solar Flares, Proton Showers, and the Space Shuttle', *Science*, 216 (1982) pp. 939–46.
4. For a discussion of various theories of species diversity and ecosystem stability, see Chapter 11, 'The Integration of Ecosystem Structure and Function', especially pp. 502 ff, in B. Collier, G. Cox, A. Johnson and P. Miller, *Dynamic Ecology* (Englewood Cliffs, NJ: Prentice-Hall, 1973).
5. F.H. Borman and G.E. Likens, 'The Nutrient Cycles of an Ecosystem', *Scientific American*, Offprint No. 1202 (San Francisco: W.H. Freeman, 1970). The authors describe a major experiment showing the important role of living plants in retaining nutrients in an ecosystem.
6. F.T. Turner, 'Soil Nitrification Retardation by Rice Pesticides', *Soil Science Society of America, Journal* 43 (1979) pp. 955–7; J.W. Doran, 'Microbiological Changes Associated with Residue Management with Reduced Tillage', *Soil Science Society of America, Journal* 44 (1980) pp. 518–24.
7. Collier, *et al. Dynamic Ecology*, Table 9.1, p. 378. While agricultural lands have a moderately high primary productivity, this depends on addition of extra energy by the farmers, who also attempt to exclude all predators on the plants except the humans themselves and their domesticated animals. The total biomass that agriculture supports is therefore less than that supported by almost all other types of ecosystems. Even with all the energy added by humans, agricultural lands seldom ever exceed the total productivity achieved in the most productive natural ecosystems under the same general

climatic conditions. Probably the only significant exception is the artificial productivity achieved in irrigated deserts, but even there, if the energy required to import the water is subtracted, the added productivity is not nearly so remarkable.

8. See 'American Scientist Interviews: Peter Ellison', *American Scientist*, 75 (1987) pp. 622–7. Demographer/endocrinologist Ellison's recent work on exercise, nutrition and lactation as they relate to fertility among African women confirms that frequent bouts of lactation suppress ovulation through production of prolactin and that both strenuous exercise and periods of weight loss can interfere with luteal progesterone production, preventing successful implantation in the uterine wall.

9. R.V. Short, 'The Evolution of Human Reproduction', *Proceedings of the Royal Society, London, B*, 195 (1976) p. 17.

10. N. Maxwell, 'Medical Secrets of the Amazon', *Américas*, 29 (6–7) (1977) pp. 2–8.

11. D. Dumond, 'The Limitation of Human Population: A Natural History', *Science*, 187 (1975) pp. 713–20.

12. M. Harris, *Cannibals and Kings: The Origins of Cultures* (New York: Random House, 1977) pp. 57 ff.

13. Harris, *Cannibals and Kings*, p. 51.

14. Harris, *Cannibals and Kings*, p. 13.

15. Harris, *Cannibals and Kings*, p. 12.

16. C. Sauer, in W.L. Thomas, Jr (ed.), *Man's Role in Changing the Face of the Earth*, vol. 1 (Chicago: University of Chicago Press, 1956) p. 54.

17. C.F. Cooper, 'The Ecology of Fire', *Scientific American* (April 1961) p. 150.

18. P.S. Martin, 'Pleistocene Overkill', *Natural History*, December 1967, pp. 32–8; P.S. Martin and H.E. Wright, Jr, *Pleistocene Extinctions: The Search for a Cause* (New Haven, CT: Yale University Press, 1967).

19. Harris, *Cannibals and Kings*, p. 17.

20. For Middle East, see Harris, *Cannibals and Kings*, pp. 36–7; for Nile, see S. Iker, 'The Seeds of Agriculture', *Mosaic* (July–August, 1982), pp. 3–9.

21. R. Lewin, 'Disease Clue to Dawn of Agriculture', *Science*, 211 (1981) p. 41. For life-expectancy of hunter–gatherers, see N. Howell, 'The Population of the Dobe Area !Kung', in R. Lee and I. DeVore (eds), *Kalahari Hunter–Gatherers: Studies of the !Kung San and Their Neighbors* (Cambridge, MA: Harvard University Press, 1976) pp. 137–51.

22. M.N. Cohen, *The Food Crisis in Prehistory* (New Haven, CT: Yale University Press, 1977).

23. Cohen, *The Food Crisis*, p. 53.

24. F. Wendorf, A.E. Close and R. Schild, 'Prehistoric Settlements in the Nubian Desert', *American Scientist*, 73 (1985) pp. 132–41.

25. E. Hyams, *Soil and Civilization* (New York: Harper & Row, 1952, 1976).

26. W.C. Lowdermilk, *Conquest of the Land through Seven Thousand Years* (Washington, DC: US Department of Agriculture, Soil Conservation Service, Agriculture Information Bulletin No. 99, 1975 (revised). (GPO #001–000–03446)) p. 5. This pamphlet gives a succinct overview of the whole history of human-caused soil erosion as deduced by one soil scientist.

27. Hyams, *Soil and Civilization*, p. 65 f.

28. Hyams, *Soil and Civilization*, p. 67.

29. L.M. Bahr, Jr, R. Costanza, J.W. Day, Jr, S.E. Bayley, C. Neill, S.G. Leibowitz and J. Fruci, *Ecological Characterization of the Mississippi Deltaic Plain Region: A Narrative with Management Recommendations* (Washington,

DC: Fish and Wildlife Service, Division of Biological Services, FWS/OBS–82/69, 1983).

30. Hyams, *Soil and Civilization*, p. 168.

31. E.P. Eckholm, *Losing Ground: Environmental Stress and World Food Prospects* (New York: W.W. Norton, 1976). Chapter 4, 'Encroaching Deserts', gives a description of present-day desertification throughout the world. The effects of overgrazing on the American Southwest are described by Carl Sauer, p. 60, in Thomas (ed.), *Man's Role*, and also by Aldo Leopold in his *Sand Country Almanac, with Essays on Conservation from Round River* (New York: Ballantine Books, 1982) (© 1949, 1953, 1966) p. 242 and p. 276. For recent changes in the American Southwest, see J.R. Hastings and R.M. Turner, *The Changing Mile* (Tucson: University of Arizona Press, 1965).

32. Eckholm, *Losing Ground*, pp. 26–7.

33. Plato, *Critias* 111B.

34. Hyams, *Soil and Civilization*, p. 230 f.

35. J. Palmer, 'Farmland at Risk of Erosion', *Manchester Guardian Weekly* (19 July 1987) p. 5. Sixty million acres of European soils, including over one third of Britain's farmlands, are endangered from intensive farming methods.

36. H.H. Bartlett, 'Fire, Primitive Agriculture, and Grazing in the Tropics', in Thomas (ed.), *Man's Role*, pp. 699–700.

37. See discussion in Harris, *Cannibals and Kings*, p. 130 f. Also, B.L. Turner II and P.D. Harrison, 'Prehistoric Raised-Field Agriculture in the Maya Lowlands', *Science*, 213 (1981) pp. 399–405. Recently scientists have found that when properly carried out, raised-field agriculture is indefinitely sustainable in tropical and semi-tropical forests.

4   Our Environmental 'Credit Card' Account Becomes Due

†   This letter, purported to have been sent in 1854 by Chief Seattle to President Franklin Pierce, who had offered to 'buy' a large area of Indian lands and resettle the tribe on a 'reservation', was reprinted in *Outdoor California*, November–December 1976, pp. 12–13, a publication of the State of California Department of Fish and Game.

1.   A.C. Fisher and F.M. Peterson, 'The Environment in Economics: A Survey', *Journal of Economic Literature* (March 1976) pp. 1–33, quoted in H. Daly, *Steady-State Economics* (San Francisco: W.H. Freeman, 1977) p. 125.

2.   Daly, *Steady-State Economics*, p. 125.

3.   N. Guppy, 'Tropical Deforestation: A Global View', *Foreign Affairs* (Spring 1984) pp. 928–65; quotes are from p. 932 and p. 943.

4.   H.E. Goeller and A. Zucker, 'Infinite Resources: The Ultimate Strategy', *Science*, 223 (1984) pp. 456–62.

5.   R.A. Rappaport, *Pigs for the Ancestors* (New Haven, CT: Yale University Press, 1968) pp. 42 ff.

6.   J. Wessel and M. Hantman, *Trading the Future: Farm Exports and the Concentration of Economic Power in Our Food Economy* (San Francisco: Institute for Food and Development Policy, 1983) pp. 84, 85.

7.   J.S. Steinhart and C.E. Steinhart, 'Energy Use in the U.S. Food System', *Science*, 184 (1974) pp. 307–16. Note that this number of calories does not include indirect costs, such as agricultural research and development, government regulation of the use of pesticides, food additives and other chemicals, all of which represent food subsidies paid via taxes.

8. Much of this material comes from five sources: H.E. Thomas, 'Changes in Quantities and Qualities of Surface Waters', in W.L. Thomas, Jr (ed.), *Man's Role in Changing the Face of the Earth* (Chicago: University of Chicago Press, 1956) pp. 542–63; D. Sheridan, 'The Desert Blooms – At a Price', *Environment*, 23 (3) (1981) pp. 6–20, 38–41; J. Adler, 'The Browning of America', *Newsweek* (23 February 1981) pp. 26–37; J. Boslough, 'Rationing a River', *Science 81* (June 1981) pp. 26–43; M. Reisner, *Cadillac Desert* (New York: Viking Penguin, 1986).

9. Sheridan, 'The Desert Blooms', p. 12.

10. Adler, 'The Browning of America', p. 30.

11. Adler, 'The Browning of America', p. 30.

12. C. Bowden, *Killing the Hidden Waters* (Austin, TX: University of Texas Press, 1977) p. 7.

13. D.E. Burmaster, 'The New Pollution: Groundwater Contamination', *Environment*, 24 (March 1982) pp. 6–13, 33–36. For detailed studies on this subject, see V.I. Pye, R. Patrick and J. Quarles, *Groundwater Contamination in the United States* (Philadelphia: University of Pennsylvania Press, 1983); and Geophysics Research Forum, *Studies in Geophysics: Groundwater Contamination* (Washington, DC: National Academy Press, 1984).

14. Sheridan, 'The Desert Blooms', pp. 8–9.

15. Sheridan, 'The Desert Blooms', p. 12.

16. Sheridan, 'The Desert Blooms', p. 13.

17. Quoted in P. Gwynne and M.J. Kukic, 'Israel: Good to the Last Drop', *Newsweek* (23 February 1981) p. 35.

18. R.P. Ambroggi, 'Underground Reservoirs to Control the Water Cycle', *Scientific American* (May 1977) pp. 21–8.

19. R.D. Zweig, 'Ecological Perspectives: Use of Freshwater Resources', *New Alchemy Quarterly* (Fall 1981) pp. 4–8.

20. W.C. Lowdermilk, *Conquest of the Land through Seven Thousand Years* (Washington, DC: US Department of Agriculture, Soil Conservation Service, Agriculture Information Bulletin No. 99, 1975 (revised) (GPO #001–000–03446–4)) p. 26.

21. J. Aucoin, 'The Irrigation Revolution and Its Environmental Consequences', *Environment* (October 1979) pp. 17–20, 38–40.

22. L.R. Brown and E.C. Wolf, *Soil Erosion: Quiet Crisis in the World Economy*, Worldwatch Paper 60 (Washington, DC: Worldwatch Institute, 1984) p. 15.

23. E.P. Eckholm, *Losing Ground: Environmental Stress and World Food Prospects* (New York: W.W. Norton, 1976) p. 17.

24. Eckholm, *Losing Ground*, p. 60f.

25. Council on Environmental Quality, 'Environment in the '80s – The Carter CEQ Says Farewell', *Sierra*, (September–October 1981) p. 35. This article, prepared by the outgoing CEQ staff whom the Reagan Administration cut from 60 to 6, summarises major global environmental concerns.

26. I. McMillan, 'Farming the Farm Program', *Hi Sierran* (San Diego: Sierra Club, February 1984) pp. 1, 14.

27. W. Bertsch, quoted by Sheridan, 'The Desert Blooms', p. 11.

28. Council on Environmental Quality, 'Environment in the '80s', p. 35. See also J. Riser, 'A Renewed Threat of Soil Erosion: It's Worse Than the Dust Bowl', *Smithsonian* (March 1981) pp. 120–31.

29. W.E. Larson, F.J. Pierce and R.H. Dowdy, 'The Threat of Soil Erosion to Long-Term Crop Production', *Science*, 219 (1983) pp. 458–65; Soil Con-

servation Service, *Potential Croplands Study* (Washington, DC: US Department of Agriculture, 1975).

30. American Farmland Trust communication, 1717 Massachusetts Ave. NW, Washington, DC, 20036.

31. Brown and Wolf, *Soil Erosion*, pp. 7–8. For the original data on which these figures are based, see Table 10.9 in 'World Soil Resources and Excessive Soil Loss, 1980, with Projections to 2000', in L.R. Brown *et al.*, *State of the World, 1984* (New York: W.W. Norton, 1984) p. 190.

32. R.L. Sivard, *World Military and Social Expenditures, 1983* (Leesburg, Va: World Priorities, 1983).

33. Recent general articles include: E. Salati and P.B. Vose, 'Amazon Basin: A System in Equilibrium', *Science*, 225 (1984) pp. 129–38; N. Guppy, 'Tropical Deforestation'; C. Caufield, 'The Rain Forests', *New Yorker* (14 January 1985) pp. 44–101.

34. Guppy, 'Tropical Deforestation', p. 934.

35. Guppy, 'Tropical Deforestation', p. 941.

36. Guppy, 'Tropical Deforestation', p. 949.

37. Guppy, 'Tropical Deforestation', p. 941.

38. H.E. Daly, 'Brazil's Leading Environmentalist: Interview with Jose Lutzenberger', *Not Man Apart* (March 1985) pp. 12–13, 23.

39. Salati and Vose, 'Amazon Basin', p. 137.

40. See L.A. Hammergren, 'Peruvian Political and Administrative Response to El Nino: Organizational, Ideological, and Political Constraints', in M.H. Glantz and J.D. Thompson (eds), *Resource Management and Environmental Uncertainty: Lessons from Coastal Upwelling Fisheries* (New York: Wiley–Interscience, 1981) pp. 317–50.

41. J.H. Ryther, 'Photosynthesis and Fish Production in the Sea', *Science*, 166 (1969) pp. 72–6. The 1982–83 *El Niño* appears to have decimated the shrimp population on which fish populations depend. See G. Alexander, 'Krill, a Staple of Marine Animals, Disappear from Arctic Water', *Los Angeles Times* (23 May 1984) Part I, p. 3.

42. J.W. Grier, 'Ban of DDT and Subsequent Recovery of Reproduction in Bald Eagles', *Science*, 218 (1982) pp. 1232–5.

43. J. Newell, 'New Research Links Agent Orange with Cancers', *New Scientist* (7 February 1985) p. 6.

44. M. Sun, 'Missouri's Costly Dioxin Lesson', *Science*, 219 (1983) pp. 357–9; J. Aucoin, 'Dioxin in Missouri: The Search Continues for a Cleanup Strategy', *Sierra* (January–February 1984) pp. 22–4, 26.

45. J. Crossland, 'Fallout from the Disaster', *Environment*, 21 (7) (1979) pp. 6–14; E. Chen, 'Most Residents of Michigan Still Contaminated by PBB', *Los Angeles Times* (16 April 1982) Part I, p. 6.

46. The 29 May 1987 issue of the *Manchester Guardian Weekly* reported 50,000 to 180,000 excess deaths among elderly living downwind from the Three Mile Island nuclear reactor during the three years following the partial meltdown accident there.

47. R.J. Smith, 'Hawaiian Milk Contamination Creates Alarm', *Science*, 217 (1982) pp. 137–40. For an update, see W.S. Merwin, 'Hawaii Wakes Up to Pesticides', *The Nation* (2 March 1985) pp. 235–7.

48. M. Sun, 'In Search of Salmonella's Smoking Gun', *Science*, 226 (1984) pp. 30–2; M. Sun, 'Use of Antibiotics in Animal Feed Challenged', *Science*, 226 (1984) pp. 144–6.

49. *Los Angeles Times* (18 November 1987) Part I, p. 7.

50. Studies by the Environmental Protection Agency on effects of lead on

circulatory disease were reported in the *Washington Post*, 3 March 1985. The 29 May 1987 *Manchester Guardian Weekly* reported that most urban children in Britain have lead levels sufficient to hinder mental development.

51.   D.E. Burmaster, 'The New Pollution: Groundwater Contamination', *Environment*, 24 (2) (1982) pp. 6–13, 33–6; J.C. Fine, 'A Crisis of Contamination', *The Sciences*, 23 (3) (1984) pp. 20–4.

52.   A.G. Levine, *Love Canal: Science, Politics and People* (Lexington, MA.: Heath Press, 1982). See also Fine, 'A Crisis of Contamination'.

53.   P.H. Abelson, 'Waste Management', *Science*, 228 (1985) p. 1145. See also M. Sun, 'Superfund Needs More Funding', *Science*, 227 (1985) p. 1447; Office of Technology Assessment, *Superfund Strategy* (Washington, DC: US Government Printing Office, OTA-ITE-252, April 1985); M. Dowling, 'Defining and Classifying Hazardous Wastes', *Environment* (April 1985) pp. 18–19, 36–41 (this last article deals with the problem on an international level).

54.   C.B. Officer, R.B. Biggs, J.L. Taft, L.E. Cronin, M.A. Tyler and W.R. Boynton, 'Chesapeake Bay Anoxia: Origin, Development, and Significance', *Science*, 223 (1984) pp. 22–7.

55.   P. Borrelli, 'To Dredge or Not to Dredge: A Hudson River Saga', *The Amicus Journal* (Spring 1985) pp. 14–26.

56.   Personal observation of the author. See P.R. Miller and A.A. Millecan, 'Extent of Oxidant Air Pollution Damage to Some Pines and Other Conifers in California', *Plant Disease Reporter*, 55 (1971) pp. 555–9.

57.   Environment Canada, *Downwind: The Acid Rain Story* (Ottawa: Minister of Supply and Services Canada, 1981); L.R. Ember, 'Acid Pollutants: Hitchhikers Ride the Wind', *Chemical and Engineering News* (14 September 1981) pp. 20–31; C.K. Graves, 'Rain of Troubles', *Science 80* (July–August 1980) pp. 79–80; W.Y. Brown, *Acid Rain and Wildlife Conservation* (Washington, DC: Environmental Defense Fund Paper, August 1981); M. Johnson, G.E. Likens, M.C. Feller and C.T. Driscoll, 'Acid Rain and Soil Chemistry', *Science*, 225 (1984) pp. 1424–5; M. Sun, 'Possible Acid Rain Woes in the West', *Science*, 228 (1985) pp. 34–5; K.A. Rahn and D.H. Lowenthal, 'Pollution Aerosol in the Northeast: Northeastern-Midwestern Contributions', *Science* 228 (1985) pp. 275–84; K. Bird and M. Holland, 'Europe: A Hard Rain Falls', *The Nation* (8 December 1984) p. 609.

58.   F. Colon, *The Life of Christopher Columbus by His Son Ferdinand*, translated and annotated by B. Keen (New Brunswick, NJ: Rutgers University Press, 1959) pp. 142–3. For an analysis of the multiple ways humans could alter climate see *Inadvertent Climate Modification* (A Report of the Study of Man's Impact on Climate, Stockholm, 1970) (Cambridge, MA: MIT Press, 1971).

59.   J. Shukla and Y. Mintz, 'Influence of Land-Surface Evapotranspiration on the Earth's Climate', *Science*, 213 (1982) pp. 1498–501.

60.   R.A. Kerr, 'Has Stratospheric Ozone Started to Disappear?' *Science*, 277 (1987) pp. 131–2.

61.   Panel on Atmospheric Chemistry of the Committee on Impacts of Stratospheric Change, *Halocarbons: Effects on Stratospheric Ozone* (Washington, DC: National Academy of Sciences, 1976) p. 172. See also p. 173 for effect of decreased ozone on increased ultraviolet light and its consequences for life on Earth. Whether the 15 September 1987 internation pact to regulate production of ozone-destroying chemicals will be effective in time is discussed by J. Gliedman, 'The Ozone Follies: Is the Pact Too Little, Too Late?', *The Nation* (10 October 1987) pp. 376–8.

62. S. Seidel and D. Keyes, 'Can We Delay a Greenhouse Warming?'
(Washington, DC: US Environmental Protection Agency Report, November 1983) p. 7–7. Evidence that warming is already occurring is found in:
World Climate Program, Proceedings from an Internation Conference,
Villach, Austria, 1985 (Geneva: WMO no. 661, World Meteorological
Organization, 1986); P.D. Jones, T.M.L. Wigley and P.B. Wright, 'Global
Temperature Variations between 1861 and 1984', *Nature*, 322 (1986)
pp. 430–4. Increases in CO_2 could also alter global patterns of rainfall: R.S.
Bradley, H.F. Diaz, J.K. Eischeid, P.D. Jones, P.M. Kelly and C.M.
Goodess, 'Precipitation Fluctuations over Northern Hemisphere Land Areas
Since the Mid-19th Century', *Science*, 237 (1987) pp. 171–5.
63. J.S. Boyer, 'Plant Productivity and Environment', *Science*, 218 (1982)
pp. 443–8; J. Walsh, 'Germplasm Resources Are Losing Ground', *Science*,
214 (1981) pp. 421–3; M.S. Swaminthan, 'Rice', *Scientific American* (January 1984) pp. 81–93; D.L. Plucknett, N.J.H. Smith, J.T. Williams and N.M.
Anishetty, 'Crop Germplasm Conservation and Developing Countries',
Science, 220 (1983) pp. 163–9.
64. A. Leopold, 'The Land Ethic', in *A Sand County Almanac and Sketches
Here and There* (London: Oxford University Press, 1975) pp. 201–26; quote
is on pp. 203–4.
65. E.F. Connor and D. Simberloff, 'Neutral Models of Species' Co-occurrence
Patterns', Chapter 18, pp. 316–31 in D.R. Strong, Jr. D. Simberloff, L.G.
Abele and A.B. Thistle (eds), *Ecological Communities: Conceptual Issues
and the Evidence* (Princeton, NJ: Princeton University Press, 1984). See also
E.G. Lawrence, 'Postscript to General Introduction', pp. 247–8 in E.G.
Lawrence and P.H. Raven (eds), *Coevolution of Plants and Animals*, 2nd
edn (Austin, TX: University of Texas Press, 1980). Lawrence argues that
coevolution of *guilds* of pollinators and flowering plants – he calls it 'diffuse
coevolution' – could explain the sudden explosive appearance of flowering
plants in the Cretaceous.
66. R.C. Lewontin, 'Gene, Organism and Environment', Chapter 14, pp. 273–
85 in D.S. Bendall (ed.), *Evolution from Molecules to Men* (Cambridge:
Cambridge University Press, 1983) 594pp; quotes are from pp. 281, 282, 284.
67. R. Costanza, 'Embodied Energy and Economic Valuation', *Science*, 210
(1980) pp. 1219–24.
68. C. Cleveland, R. Costanza, C.A.S. Hall and R. Kaufman, 'Energy and the
U.S. Economy: A Biophysical Perspective', *Science*, 225 (1984) pp. 890–7.
69. F. Soddy, *Wealth, Virtual Wealth and Debt: The Solution of the Economic
Paradox* (New York: E.P. Dutton, 1926); quote on p. 56.
70. Costanza, 'Embodied Energy', p. 1224.
71. Leopold, 'The Land Ethic', pp. 225–6.

5 The Emergence of Human Nature

† Theodosius Dobzhansky, *The Biology of Ultimate Concern* (New York: New
American Library, 1967) p. 10.
1. T. Hobbes, *Leviathan* (Oxford: Basil Blackwell, 1946) p. 82. See especially
'The First Part: Of Man'.
2. R. Ardrey, *African Genesis* (New York: Atheneum, 1961); R. Ardrey, *The
Territorial Imperative* (New York: Atheneum, 1966); D. Morris, *The Naked
Ape* (New York: McGraw-Hill, 1967); K. Lorenz, *On Aggression* (New
York: Harcourt, Brace & World, 1966). More recently, Alex Haley, in his
reconstruction of black Americans' African past, has helped to dispel from

the popular mind such misconceptions of life among so-called primitive people. See A. Haley, *Roots* (Garden City, NY: Doubleday, 1976).

3. The worst case, of course, was Adolph Hitler's claim of Aryan racial supremacy and its terrible consequences.

4. E.O. Wilson, *Sociobiology: The New Synthesis* (Cambridge, MA: Belknap Press of Harvard University Press, 1975). For further extensions of his ideas, see also his *On Human Nature* (Cambridge, MA: Harvard University Press, 1978) and, with C.J. Lumsden, *Genes, Mind and Culture* (Cambridge, MA: Harvard University Press, 1981).

5. For a comparison of these views, see Mary Midgley, 'Rival Fatalisms: The Hollowness of the Sociobiology Debate', pp. 15–38 in A. Montagu (ed.), *Sociobiology Examined* (New York: Oxford University Press, 1980).

6. B.F. Skinner, *Beyond Freedom and Dignity* (New York: Knopf, 1971).

7. L.L. Gatlin, *Information Theory and the Living System* (New York: Columbia University Press, 1977).

8. For a discussion of the conditions under which DNA would spontaneously form, see C. Ponnamperuma, *The Origins of Life* (New York: E.P. Dutton, 1972) pp. 78–95.

9. J. Maynard Smith, *Theory of Evolution*, 3rd edn (Baltimore: Penguin Books, 1975) and T. Dobzhansky, F.J. Ayala, G.L. Stebbins and J.W. Valentine, *Evolution* (San Francisco: W.H. Freeman, 1975).

10. S.M. Stanley, *Macroevolution* (San Francisco: W.H. Freeman, 1979); N. Eldredge and S.J. Gould, 'Punctuated Equilibria: An Alternative to Phyletic Gradualism', in T.J.M. Schopf (ed.), *Models in Paleobiology* (San Francisco: Freeman, Cooper, 1974) pp. 82–115.

11. A. Montagu, *Darwin: Competition and Cooperation* (New York: Henry Schuman, 1952) p. 48.

12. See R. Dawkins, *The Selfish Gene* (New York: Oxford University Press, 1976) and E.O. Wilson, 'The Morality of the Gene', Chapter 1 in *Sociobiology*. Quote is from Dawkins, p. ix. A particularly good rebuttal to these ideas is Mary Midgley's 'Gene-Juggling', pp. 108–34 in Montagu, *Sociobiology Examined*.

13. G.G. Simpson, *The Meaning of Evolution* (New Haven, CT: Yale University Press, 1949) pp. 221–2. See also J. MacMahon and C. Fowler, 'Selective Extinction and Speciation: Their Influence on the Structure and Functioning of Communities and Ecosystems', *American Naturalist*, 119 (1982) pp. 480–98. They discuss the importance of community structure in an ecosystem and the network of interactions among species as a major determinant in evolution.

14. M. Midgley, *Beast and Man* (Ithaca, NY: Cornell University Press, 1978) p. 164.

15. Midgley, *Beast and Man*, p. 158.

16. M.-C. King and A.C. Wilson, 'Evolution at Two Levels in Humans and Chimpanzees', *Science*, 188 (1975) pp. 107–16.

17. A. Jolly, 'The Evolution of Primate Behavior', *American Scientist*, 73 (1985) p. 230.

18. Jolly, 'The Evolution of Primate Behavior', p. 232.

19. S.J. Gould, 'The Child as Man's Real Father', in S.J. Gould, *Ever Since Darwin* (New York: W.W. Norton, 1977) pp. 63–9.

20. Jolly, 'The Evolution of Primate Behavior', p. 236. For data on Japanese macaques, see M. Kawai, 'Newly Acquired Pre-Cultural Behavior of the Natural Troop of Japanese Monkeys on Koshima Islet', *Primates*, 6 (1965) pp. 1–30.

21. M. Ghiglieri, 'The Social Ecology of Chimpanzees', *Scientific American* (June 1985) pp. 102–13; quote is on p. 110.

22. W.D. Hamilton, 'The Genetical Evolution of Social Behavior', *Journal of Theoretical Biology*, 70 (1964) pp. 1–52; and D.S. Wilson, 'A Theory of Group Selection', *Proceedings of the National Academy of Sciences*, 72 (1964) pp. 143–6.

23. One of the greatest exponents of contingency in the evolutionary process is Stephen Jay Gould. Only what history has already provided can be further shaped by natural selection. He also champions the need to understand evolution in its appropriate context, as integrated organism in its total surroundings. In the case of hominids, the social group *is* the major environmental factor. For respective arguments, see 'The Panda's Thumb', pp. 19–26 and 'Caring Groups and Selfish Genes', pp. 85–92, in S.J. Gould, *The Panda's Thumb* (New York: W.W. Norton, 1980).

24. Much of the information and argument described in this section is to be found in: R.E. Leakey and R. Lewin, *Origins* (New York: E.P. Dutton, 1977) and R.E. Leakey, *The Making of Mankind* (New York: E.P. Dutton, 1981). See also S.J. Gould, 'Our Greatest Evolutionary Step', pp. 125–33 in *The Panda's Thumb*.

25. J.B. Lancaster, 'Carrying and Sharing in Human Evolution', *Human Nature* (February 1978) pp. 82–9.

26. A.J. Jelinek, 'The Tabun Cave and Paleolithic Man in the Levant', *Science*, 216 (1982) pp. 1369–75.

27. S.P. Springer and G. Deutsch, *Left Brain, Right Brain* (San Francisco: W.H. Freeman, 1981).

28. P.D. MacLean, *A Triune Concept of Brain and Behavior* (Toronto: University of Toronto Press, 1973) pp. 7, 21. Unfortunately, this reductionist interpretation of human behaviour has been simplified even further for popular consumption. See C. Sagan, *The Dragons of Eden* (New York: Ballantine Books, 1977) Chapter 3.

29. J. Olds, 'Hypothalamic Substrates of Reward', *Physiological Reviews*, 42 (1962) pp. 554–604.

30. J.Z. Young, *Programs of the Brain* (Oxford: Oxford University Press, 1978) p. 76.

31. R.G. Northcutt, 'Evolution of the Telencephalon in Nonmammals', *Annual Review of Neuroscience*, 4 (1981) pp. 301–50.

32. M. Midgley, *Heart and Mind: The Varieties of Moral Experience* (New York: St Martin's Press, 1981) pp. 8–9.

33. Midgley, *Heart and Mind*, p. 9. Much of the following discussion is drawn from Midgley, *Beast and Man*.

34. Plato, *The Republic*, 9 571C, translated by Allan Bloom (New York: Basic Books, 1968) pp. 251–2.

35. N. Tinbergen, *The Study of Instinct* (Oxford: The Clarendon Press, 1969) pp. 50, 113–21, 210.

36. M. Midgley, *Beast and Man*, p. 27 footnote and Chapters 3 and 11.

37. M. Midgley, *Beast and Man*, Chapter 11. Read especially her interpretation of the sermons of Bishop Joseph Butler.

38. M. Midgley, *Beast and Man*, p. 272.

39. H.F. Harlow, 'Love in Infant Monkeys', *Scientific American*, Offprint #429, 8 pp. (First appeared in June 1959 issue of *Scientific American*.)

40. N. Tinbergen, 'Ethology and Stress Disease', *Science*, 185 (1974) pp. 20–7. See also E.A. Tinbergen and N. Tinbergen, 'Early Childhood Autism: An

Ethological Approach' (Berlin: Parey, 1972). (Supplement #10 to *Journal of Comparative Ethology*.)

41. D. Harley, Letter to the Editor, *Science*, 217 (1982) p. 296.
42. I. Eibl-Eibesfeldt, *Love and Hate* (New York: Rinehart & Winston, 1942) pp. 143–4. It has also been argued that the shape of the human breast evolved not, as so many today seem to think, as a sexual object, but to insure eye-to-eye contact between the nursing infant and its mother; see V. Reynolds, *The Biology of Human Action* (San Francisco: W.H. Freeman, 1975) p. 163. Finally, Jane Goodall has described life-long bonds among chimpanzees which, however, do not exhibit pair-bonding. See *The Chimpanzees of Gombe* (Cambridge, MA: Harvard University Press, 1986).
43. R. Benedict, *Patterns of Culture* (Boston: Houghton Mifflin, 1934) pp. 23–4.
44. Compare langurs' behaviour in north and central India (P. Jay, 'The Common Langur of North India', pp. 197–249 in I. DeVore (ed.), *Primate Behavior: Field Studies of Monkeys and Apes* (New York: Holt, Rinehart & Winston, 1965)) with that in south India (Y. Sugiyama, 'Social Organization of Hanuman Langurs', pp. 221–36 in S.A. Altman (ed.), *Social Communication among Primates* (Chicago: University of Chicago Press, 1967)).
45. M. Harris, 'Proteins and the Fierce People', Chapter 5 in *Cannibals and Kings* (New York: Vintage Books, Random House, 1977) pp. 67–78.
46. R.F. Salisbury, *From Stone to Steel: Economic Consequences of a Technological Change in New Guinea* (Melbourne: Melbourne University Press, 1962) pp. 112–22. See also M. Meggitt, *Blood Is Their Argument* (Palo Alto, CA: Mayfield, 1977).
47. J. Goodall, A. Bandora, E. Bergmann, C. Busse, H. Mitama, E. Mpongo, A. Pierce and D. Riss, 'Intercommunity Interactions in the Chimpanzee Population of the Gombe National Park', in D.A. Hamburg and R. McCown (eds), *The Great Apes* (Menlo Park, CA: Benjamin/Cummings, 1979) pp. 13–53.
48. S. Zuckerman, *The Social Life of Monkeys and Apes* (New York: Harcourt, Brace, 1932) pp. 215–31.
49. B.G. Trigger, *The Huron: Farmers of the North* (New York: Holt, Rinehart & Winston, 1969) pp. 42–53.
50. L. Mumford, *The Myth of the Machine, Volume I, Technics and Human Development* (New York: Harcourt, Brace & World, 1967) p. 212 f.

6 The Cultural Spectrum

† Ruth Benedict, *Patterns of Culture* (Boston: Houghton Mifflin, 1934) p. 21. Benedict's insights are still immensely valuable.
1. M. Midgley, *Beast and Man* (Ithaca, NY: Cornell University Press, 1978) p. 332.
2. W. Durant, quoted by J.L. Christian in *Philosophy: An Introduction to the Art of Wondering* (New York: Holt, Rinehart & Winston, 1977) p. 384.
3. R. Graves, *The White Goddess*, 3rd edn, (London: Faber & Faber, 1953) p. 27.
4. E.R. Service, 'The Ghosts of Our Ancestors', in *Primitive Worlds: People Lost in Time* (Washington, DC: National Geographic Society, 1973) pp. 8–16
5. R. Benedict, 'The Diversity of Cultures', Chapter 2 in *Patterns of Culture* (Boston: Houghton Mifflin, 1934, 1959) pp. 21–44. Most of the following discussion is based on this source, but see also, A. Gennep, *The Rites of Passage* (Chicago: University of Chicago Press, 1960) especially p. 84 f.

6. K. Birket-Smith, *Primitive Man and His Ways* (New York: Mentor Books, New American Library, 1963) pp. 45–48.
7. R.E. Leakey and R. Lewin, *Origins of Mankind* (New York: E.P. Dutton, 1977) p. 242.
8. J.C. Goodale, 'An Example of Ritual Change among the Tiwi of Melville Island', in A.R. Pilling and R.A. Waterman (eds), *Diprotodon to Detribalization: Studies of Change among Australian Aborigines* (East Lansing, MI: Michigan State University Press, 1970) pp. 350–66.
9. R. Benedict, *Patterns of Culture*, pp. 188–211; quote is from p. 211.
10. R.B. Lee and I. DeVore (eds), *Kalahari Hunter–Gatherers: Studies of the !Kung San and Their Neighbors*, (Cambridge, MA: Harvard University Press, 1976). This book contains summaries of most of the work that had been written before 1976 on these people.
11. R.B. Lee, 'Eating Christmas in the Kalahari', *Natural History* (December 1969) pp. 14–22, 60–63.
12. A.H. Maslow and J.J. Honigmann, compilers, 'Synergy: Some Notes of Ruth Benedict', *American Anthropologist*, 72 (1970) pp. 320–33.
13. Reported in L.S. Stavrianos, *The Promise of the Coming Dark Age* (San Francisco: W.H. Freeman, 1976) pp. 154–5.
14. Benedict, *Patterns of Culture*, pp. 130–72.
15. N. Chagnon, *Yanomamö: The Fierce People* (New York: Holt, Rinehart & Winston, 1968).
16. Chagnon, *Yanomamö*, p. 48.
17. Chagnon, *Yanomamö*, p. 118.
18. Chagnon, *Yanomamö*, p. 91.
19. Benedict, *Patterns of Culture*, pp. 57–239.
20. Benedict, *Patterns of Culture*, p. 89.
21. Personal communication, Barton A. Wright, ethnologist, 4143 Gelding St, Phoenix, AZ, 85023.
22. Benedict, *Patterns of Culture*, p. 103.
23. L. Sharp, 'Steel Axes for Stone Age Australians', Case 5 in E.H. Spicer (ed.), *Human Problems in Technological Change* (New York: John Wiley & Sons, 1952) pp. 69–90. A similar disruption from the introduction of steel implements into New Guinea has been blamed for increased warfare. See reference 46 in Chapter 5.
24. C.M. Turnbull, *The Mountain People* (New York: Simon & Schuster, 1972).
25. Turnbull, *The Mountain People*, p. 282. For a concise statement of how aid can be utterly destructive of a way of life, read pp. 281–2.
26. F.M. Lappé and J. Collins, *Food First: Beyond the Myth of Scarcity*, revised edn (New York: Ballantine Books, 1978) p. 141. See pp. 134–64 for the total extent of the impact of the Green Revolution.
27. For a comprehensive description, see M. Mead (ed.), *Cultural Patterns and Technical Change: A Manual Prepared by the World Federation for Mental Health* (Paris: UNESCO, 1953).
28. Lappé and Collins, *Food First*, pp. 61–6; 330–5; 336–8.
29. M. Mead, *New Lives for Old: Cultural Transformation – Manus, 1928–1953* (New York: William Morrow, 1966).
30. M.C. Bateson, Chapter X in *With a Daughter's Eye* (New York: Washington Square Press, Simon & Schuster, 1984) pp. 174–96.
31. T. Kochman, '"Rapping" in the Black Ghetto', *Trans*action, 6 (February 1969) pp. 26–34; K.R. Johnson, 'The Vocabulary of Race', in T. Kochman (ed.), *Rappin' and Stylin' Out* (Urbana, IL: University of Illinois Press, 1972) pp. 140–51. See also other articles in this book.

32. Benedict's views in this area are to be found in her book *Patterns of Culture*, and more especially in the notes from her lectures at Bryn Mawr, compiled by Maslow and Honigmann, 'Synergy'.

33. J. Jaynes, *The Origins of Consciousness in the Breakdown of the Bicameral Mind* (Boston: Houghton Mifflin, 1982). I cannot subscribe to Jaynes's thesis that primitive people regularly had auditory hallucinations that they believed were gods speaking to them, however.

34. J.H. Bodley, *Anthropology and Contemporary Human Problems* (Menlo Park, CA: Cummings, 1976); quote is from R.L. Heilbroner, *The Great Ascent: The Struggle for Economic Development in Our Time* (New York: Harper & Row Torchbooks, 1963) p. 53.

35. Bodley, *Anthropology*, pp. 22–3. See also R.B. Lee and I. DeVore (eds), *Man the Hunter* (Chicago: Aldine, 1968).

7 Religions and Worldviews

†   Erich Fromm, *On Disobedience* (New York: Seabury Press, 1981) p. 43.

1.   W. Hazlitt, quoted by W.H. Auden, p. 13 in A. Freemantle (ed.), *The Protestant Mystics* (New York: New American Library, Mentor, 1964).

2.   The Oxford English Dictionary gives both *relegere*, to read over again or recite, and *religāre*, to bind or constrain, as possible etymological origins of 'religion'.

3.   Inscription found at Eleusis; translation by S. Angus, *The Mystery Religions and Christianity* (New York: C. Scribner, 1925) p. 140.

4.   See Yu-lan Fung, *Chuang-Tzŭ*, 2nd edn (New York: Paragon Book Reprint Corporation, 1964) 'Introduction', pp. 3–23, for an interpretation of the essence of Tao.

5.   F. Waters, 'Part I. The Myths: Creation of the Four Worlds', in *Book of the Hopi* (New York: Penguin, 1963, 1970 printing) pp. 3–27.

6.   E. Hyams, *Soil and Civilization* (New York: Harper & Row, 1976) p. 276.

7.   A. Aveni, 'Tropical Archeoastronomy', *Science*, 213 (1981) pp. 161–71.

8.   K. Birket-Smith, *The Paths of Culture* (Madison: University of Wisconsin Press, 1965) p. 403.

9.   R.A. Williamson, H.J. Fisher and D. O'Flynn, 'Anasazi Solar Observatories', Chapter 14 in A.F. Aveni (ed.), *Native American Astronomy* (Austin, TX: University of Texas Press, 1975) p. 212.

10. K. Frazier, 'The Anasazi Sundagger', *Science 80* (November–December 1979) pp. 56–67.

11. B. Cobo, *Historia del Nuevo Mundo* (1653) (Madrid: Bibliotica Autores Españoles, 1956) vols 91 and 92, pp. 173–4. Translated by A. Aveni, *Science*, 213 (1981) p. 166.

12. K. Birket-Smith, *The Paths of Culture*, pp. 403–5.

13. J.T. Fraser (ed.), *The Voices of Time* (New York: G. Braziller, 1966) p. 388.

14. J. Friberg, 'Numbers and Measures in the Earliest Written Records', *Scientific American* (February 1984) pp. 110–18.

15. H. Jensen, *Sign, Symbol and Script* (New York: G.P. Putnam's Sons, 1969) p. 27.

16. Jensen, *Sign, Symbol and Script*, pp. 54–73.

17. H. von Wissmann, 'On the Role of Nature and Man in Changing the Face of the Dry Belt of Asia', in W.L. Thomas, Jr (ed.), *Man's Role in Changing the Face of the Earth* (Chicago: University of Chicago Press, 1956) p. 286 f.

18. K. Birket-Smith, *The Paths of Culture*, p. 200.

19. M.D. Sahlins, 'Poor Man, Rich Man, Big-Man, Chief', *Comparative Studies*

in Society and History, 5 (1963) pp. 285–303. See also: L. Mumford, *The Myth of the Machine*, vol. I (New York: Harcourt, Brace & World, 1967) p. 168 f.

20. Mumford, *The Myth of the Machine*, p. 215 f.
21. Mumford, *The Myth of the Machine*, p. 259 f
22. Attributed to Chief Seattle in an 1854 message said to have been sent to President Franklin Pierce, and reprinted in *Outdoor California* (California State Department of Fish and Game) (November–December 1976) pp. 12–13.
23. E. Langton, *Good and Evil Spirits* (London: Society for Promoting Christian Knowledge, 1942) pp. 90–4. For a general discussion of animism, see Sir James G. Frazer, *The Golden Bough*, abridged edn in two volumes (New York: Macmillan, 1951).
24. S.M. Zwemer, *The Influence of Animism on Islam: An Account of Popular Superstition* (New York: Macmillan, 1920) p. 14. There are similar examples throughout the book.
25. R. Cavendish, *The Great Religions* (New York: Arco, 1980) p. 59.
26. H.G. Creel, *Chinese Thought from Confucius to Mao Tsê-tung* (Chicago, University of Chicago Press, 1953) p. 17.
27. Creel, *Chinese Thought*, p. 98.
28. From *The Writings of Kwang-zze (Chuang Tzŭ)*, translated by James Legge, in *Sacred Books of the East*, XL (Book XXII, Part II, Section XV, 2) (Delhi: Motilal Banarsidass, 1966) pp. 60–1.
29. Cavendish, *The Great Religions*, p. 135.
30. *The Holy Bible*, Psalms viii, 4–9.
31. Cavendish, *The Great Religions*, p. 149.
32. R. Shaull, *Heralds of a New Reformation: The Poor of South and North America* (New York: Orbis Books, Maryknoll, 1984) p. 49.
33. M. Midgley, personal communication. See also R. Graves, *The White Goddess* (New York: Octagon Books, 1972) p. 393 f.
34. Cavendish, *The Great Religions*, p. 239.
35. These words appear in John A.T. Robinson, Bishop of Woolich, *Honest to God* (Philadelphia: Westminster Press, 1966). In this hotly debated treatise, Robinson is attempting to do away with the concept of a supernatural deity. See also James A. Pike, Bishop of California, *If This Be Heresy* (New York: Harper & Row, 1967).
36. L. White, 'The Historical Roots of Our Ecologic Crisis', *Science*, 155 (1967) pp. 1203–7. This article has been rebutted and its argument modified by L.W. Moncrief, 'The Cultural Basis for Our Environmental Crisis', *Science*, 170 (1970) pp. 508–12. For further Biblical references see *The Holy Bible*, Genesis, especially i, 27.
37. R.N. Bellah, R. Madsen, W.M. Sullivan, A. Swidler and S.M. Tipton, *Habits of the Heart* (Berkeley, CA: University of California Press, 1985).
38. P. Tillich, *The Protestant Era* (Chicago: University of Chicago Press, 1957) p. xi.
39. J. Bronowski, *Science and Human Values* (New York: Harper Colophon, 1965) p. 70.
40. D.H. Meadows, 'What Will Your College Bring to the Future?', Keynote Address to the 27th National Institute, Council of Independent Colleges, 14 June 1982. Dr Meadows is at the New Policy Center, Dartmouth College, Hanover, NH.
41. R. Heilbroner, *Marxism: For and Against* (New York: W.W. Norton, 1980) p. 86.

42. Shaull, *Heralds of a New Reformation*, p. 39.
43. Robert McAfee Brown, *Making Peace in the Global Village* (Philadelphia: Westminster Press, 1980).
44. *Economic Justice for All: Catholic Social Teaching and the U.S. Economy* (Washington, DC: US Catholic Conference, 1986).
45. P. Lernoux, *Cry of the People* (Harmondsworth, England: Penguin, 1982) p. 31.
46. Shaull, *Heralds of a New Reformation*, p. 20; quote is of O. Gottwald.
47. Shaull, *Heralds of a New Reformation*, p. 13; quote is of Pablo Richard. I.F. Stone, writing in *The Nation* (26 September 1987) p. 293, observes that Christian metaphysics – the Trinity, the Eucharist, the virgin birth and original sin – are not to be found anywhere in the New Testament gospels which are strictly concerned with justice for the oppressed.
48. Shaull, *Heralds of a New Reformation*, p. 56.
49. Lernoux, *Cry of the People*, p. 137.
50. E. Mortimer, *Faith and Power: The Politics of Islam* (New York: Vintage Books, Random House, 1982) pp. 360 ff.
51. J.A. Bill, 'Resurgent Islam in the Persian Gulf', *Foreign Affairs* (Fall 1984) pp. 108–27; quote is from p. 108.
52. Bill, 'Resurgent Islam', p. 127.

8 On Acquiring a Worldview

† E.F. Schumacher, *Small is Beautiful: A Study of Economics as If People Mattered* (London: Sphere Books, 1974) p. 83.
1. H. Jensen, *Sign, Symbol and Script* (New York: G.P. Putnam's Sons, 1969) pp. 177–8.
2. M. Polanyi, *The Study of Man* (Chicago: Phoenix Books, University of Chicago Press, 1958) p. 60.
3. Saint Thomas Aquinas, *Summa Theologica*, I 23,3; I–II, 109, 6; I–II 112,2,3. Translation from J.L. Christian, *Philosophy: An Introduction to the Art of Wondering*, 2nd edn (New York: Holt, Rinehart & Winston, 1977) p. 279.
4. R. Cavendish, *The Great Religions* (New York: Arco, 1980) p. 234.
5. B. Russell, 'On the Notion of Cause', in *Our Knowledge of the External World*, 2nd edn (New York: W.W. Norton, 1929) p. 240.
6. L. Brillouin, *Science and Information Theory*, 2nd edn (New York: Academic Press, 1962), p. 314. See Chapter 21, especially pp. 302–20.
7. The Laplacean fallacy has been refuted by various philosophers. See Mary Midgley's readable comment in *Beast and Man* (Ithaca, NY: Cornell University Press, 1978) p. 87, where she writes 'The need for many different methods [of inquiry] is not going to go away, dissolved in a quasi-physical heaven where all serious work is quantitative'. See also p. 105 f and p. 141 in *Beast and Man*, as well as p. 17 in her *Heart and Mind* (New York: St Martin's Press, 1981).

Philosopher–scientist Michael Polanyi deals with this problem on p. 139 f of *Personal Knowledge: Towards a Post-Critical Philosophy* (Chicago: University of Chicago Press, 1962). On p. 141 he writes: 'The tremendous intellectual feat conjured up by Laplace's imagination has diverted attention (in a manner commonly practised by conjurers) from the decisive sleight of hand by which he substitutes a knowledge of all experience for a knowledge of all atomic data. Once you refuse this deceptive substitution, you immediately see that the Laplacean mind understands precisely nothing and that whatever it knows means precisely nothing. Yet the spell of the Laplacean delusion remains unbroken to this day ... For the time being, ... the peril to the true values of science does not lie in any overt reaction against science. It lies in the very acceptance of a scientific outlook based on the Laplacean fallacy as a guide to human affairs. Its reductive programme, applied to politics, entails the idea that political action is necessarily shaped by force, motivated by greed and fear, with morality used as a screen to delude the victims ... [T]he scientific method [has become] the supreme interpreter of human affairs'.

8. B.F. Skinner, *Beyond Freedom and Dignity* (New York: Bantam Books, 1971, 1980) p. 12.

9. Skinner, *Beyond Freedom and Dignity*, p. 205.

10. B.F. Skinner in R. Epstein (ed.), *Skinner for the Classroom. Selected Papers* (Champaign, IL: Research Press, 1982) p. 217. Skinner argues here that learning demands reinforcement from others; since the teacher cannot provide this to his/her many charges, machines are the only solution.

11. J. Piaget, *Biology and Knowledge* (Chicago: University of Chicago Press, 1971) pp. 48–9.

12. A. Einstein, quoted by J.L. Christian, *Philosophy*, p. 228.

13. T.S. Kuhn, *The Structure of Scientific Revolutions* (Chicago: University of Chicago Press, 1962).

14. M. Midgley, *Beast and Man* (Ithaca, NY: Cornell University Press, 1978) pp. 64–5.

15. For a readable summary of Piaget's work on cognitive development, see M.A.S. Pulaski, *Understanding Piaget* (New York: Harper & Row, 1971).

16. M. Hunt, *The Universe Within: New Science Explores the Human Mind* (New York: Simon & Schuster, 1982) p. 136f. (For a resounding rejection of the Skinnerian explanation of human learning see Chapter 2, 'The Great Black Box Debate'.)

17. S. Levine, 'Stimulation in Infancy', *Scientific American* (May 1960) pp. 80–6; C. Blakemore and G.F. Cooper, 'Development of the Brain Depends on the Visual Environment', *Nature*, 223 (1970) pp. 477–8; M. Rosensweig, E.L. Bennett and M.C. Diamond, 'Brain Changes in Response to Experience', *Scientific American* (February 1972) pp. 22–9.

18. H. Dreyfus and S. Dreyfus, 'Mindless Machines: Computers Don't Think Like Experts and Never Will', *The Sciences* (November–December 1984) pp. 18–22; M.M. Waldrop, 'Natural Language Understanding', *Science*, 224 (1984) pp. 372–4. See also H. Dreyfus, S. Dreyfus and T. Athanasiou, *Mind Over Machine: The Power of Human Intuition and Expertise in the Era of the Computer* (New York: The Free Press, 1986).

19. G. Brown and C. Desforges, *Piaget's Theory: A Psychological Critique* (London: Routledge & Kegan Paul, 1979) p. 20.

20. J. Mander, *Four Arguments for the Elimination of Television* (New York: William Morrow, 1978) pp. 163–9. See also pp. 205–11 for the effect of watching television on brainwaves.

21. I. Opie and P. Opie, *The Lore and Language of Schoolchildren* (Oxford: Clarendon Press, 1959) Chapter 8.

22. D. Boyd and L. Kohlberg, 'The Is–Ought Problem: A Developmental Perspective', *Zygon*, 8 (1973) pp. 358–73. (The stages are defined on pp. 363–4.) For a critical discussion of the relative contributions of internal development and social learning, see the reprint of a 'Symposium on Moral Development', R.L. Simon (ed.), *Ethics*, 92 (1982) pp. 407–532.

23. Ever since 1969, when Arthur Jensen published his bombshell paper 'How

much can we boost IQ and scholastic achievement?', *Harvard Educational Review*, 39, pp. 1–123, in which he claimed that 'blacks' in the United States had hereditarily lower learning capacities than 'whites', the issue of IQ and race has drawn heated debate. Numerous books reviewing the subject have appeared, including J.C. Loehlin, G. Lindzey and J.N. Spuhler, *Race Differences in Intelligence* (San Francisco: W.H. Freeman, 1975). It concludes (as have most other recent studies) that within-group variations far and away exceed between-group variations. For a carefully controlled study of IQ differences among children in United States cities, see P.L. Nichols and V.E. Anderson, 'Intellectual Performance, Race and Socio-Economic Status', *Social Biology*, 20 (1973) pp. 367–74.

24.   Opie and Opie, *The Lore and Language*, p. 4 f.
25.   C.J. Lumsden and E.O. Wilson, *Genes, Mind and Culture: The Co-evolutionary Process* (Cambridge, MA: Harvard University Press, 1981).
26.   C.R. Cloninger and S. Yokoyama, 'The Channeling of Social Behavior', *Science*, 213 (1981) pp. 749–51.
27.   L.L. Cavalli-Sforza and M.W. Feldman, *Cultural Transmission and Evolution: A Quantitative Approach* (Princeton, NJ: Princeton University Press, 1981).
28.   L.L. Cavalli-Sforza, M.W. Feldman, K.H. Chen and S.M. Dornbusch, 'Theory and Observation in Cultural Transmission', *Science*, 218 (1982), pp. 19–27.
29.   M. Mead, *Culture and Commitment: A Study of the Generation Gap* (Garden City, NY: Natural History Press, Doubleday, 1970) p. 1.
30.   J.A. Hostetler, *Hutterite Society* (Baltimore: Johns Hopkins University Press, 1974).
31.   Mead, *Culture and Commitment*, p. 61.
32.   Mead, *Culture and Commitment*, p. 61.
33.   For an excellent essay on the difference in the stakes involved between playing games and living one's life, see Mary Midgley, 'The Game Game', Chapter 8 in *Heart and Mind* (New York: St Martin's Press, 1981).
34.   Mead, *Culture and Commitment*, p. 87.
35.   T. Marshall, 'In Britain Today, Fair Play is Passé', *Los Angeles Times* (4 August 1987) pp. 1, 15. Social critic George Mikes is quoted as saying: 'Rather than being good but fair losers, I believe Britain today would rather be a land of shame-faced winners'.
36.   R. Lamm, Governor of Colorado, 'Promoting Finitude', *The Amicus Journal* (Summer 1980) p. 7.
37.   N.H. Baynes, *The Speeches of Adolf Hitler*, vol. 1 (New York: Howard Fertig, 1969) p. 333.
38.   G. Orwell, *Nineteen Eighty–Four* (London: Martin Secker & Warburg, 1949). Throughout his life, Orwell was highly critical of the manner in which politicians used language to manipulate the public. In this masterpiece – his last – he foresaw how the coming technology of television could become the supreme agency of political control.
39.   P. Wright (with P. Greengrass), *Spycatcher* (New York: Viking Press, 1987). This book by a former member of the British secret service was banned in Britain by Prime Minister Thatcher's government on the grounds of containing official 'secret' information, despite its widespread publication abroad.
40.   N. Cousins, 'Iran–Contra: The No. 1 Question', *Christian Science Monitor* (19 August 1987) p. 13.
41.   The 25th anniversary issue of the *Columbia Journalism Review* (November–

December 1986) was devoted to the recent history and current status of American journalism. Much of what follows is gleaned from an article by James Boylan, 'Declarations of Independence', pp. 29–45.

42. Boylan, 'Declarations', quoted on pp. 34–5. Lippmann spoke these words at the International Press Institute in London, 27 May 1965.

43. Boylan, 'Declarations', p. 44. O'Neill's address as retiring president of the American Society of Newspaper Editors was given in May 1982.

44. Boylan, 'Declarations', p. 44.

45. M. Parenti, *Inventing Reality* (New York: St Martin's Press, 1986) pp. 232–4.

46. This information came out in the testimony of Secretary of State George Shultz during the televised Congressional Iran–Contra hearings in the summer of 1987.

47. Boylan, 'Declarations', p. 45.

48. For an account of the degeneration in quality of televised news on CBS see Peter McCabe, *Bad News at Black Rock: The Sell-Out of CBS News* (New York: Arbor House, 1987).

49. Quoted in Parenti, *Inventing Reality*, pp. 52–3.

50. H. Mayer, 'The Oppressed Press', *The Threepenny Review* (Winter 1987) p. 4.

51. B. Mehl, 'Ancient Greece: The Search for Community', Chapter 4 in *Classical Educational Ideas* (Columbus, OH: C.E. Merrill, 1972) pp. 29–55; quote is from p. 33.

52. L.S. Stavrianos, *The Promise of the Coming Dark Age* (San Francisco: W.H. Freeman, 1976) pp. 4 ff.

53. H.S. Commager, *The Empire of Reason* (London: Weidenfeld & Nicolson, 1978) p. 226.

54. P.L. Ford (ed.), *The Writings of Thomas Jefferson* (New York: G.P. Putnam's Sons, 1892–1899) vol. 10, p. 161. Letter to William C. Jarvis, 28 September 1820.

55. S.S. Wolin, 'Higher Education and the Politics of Knowledge', *democracy*, 1(2) (1981) pp. 38–52.

56. Many Americans are shocked to discover Jefferson's vision of a grassroots democracy was not quite as they supposed. Wolin, 'Higher Education', discusses the elitist attitudes of the various Founding Fathers (p. 41). For Jefferson's ideas, see Ford (ed.), *The Writings of Thomas Jefferson*, vol. 3, pp. 235–55, 'Notes on the State of Virginia. Query XIV. The administration of justice and the description of the laws?' (especially p. 252).

57. Wolin, 'Higher Education', p. 44.

58. P.C. Violas, *The Training of the Urban Working Class: A History of Twentieth Century American Education* (Chicago: Rand McNally, 1978) p. 230. A revealing account of how Americans have been educated.

59. M. Parenti, *Democracy for the Few* (New York: St Martin's Press, 1983) p. 42.

60. F. FitzGerald, *America Revised* (Boston: Atlantic–Little, Brown, 1979) p. 151.

61. Sloan Commission on Government and Higher Education, *A Program for Renewed Partnership* (Boston: Ballinger, 1980) p. 3. In the 1980s, similar events have begun to take place in local schools, where corporations, banks and other private organisations have been offering financial support and teaching aids, all of them stamped in some fashion with the company's name or logo. See Parenti, *Democracy for the Few*, p. 42, footnote 4, and also *Science*, 225 (1984) p. 1456, for further citations of the role of corporations in the American classroom.

62.   A. Bloom, *The Closing of the American Mind* (Chicago: University of Chicago Press, 1987).
63.   J. Marmor, 'Psychiatry and the Survival of Man', *Saturday Review* (22 May 1971) pp. 18–19, 53, 77.

9   From God to Man: Origins of the Western Worldview

†   Alexander Pope, 'Essay on Man, Epistle III', *From Chaucer to Gray*, The Harvard Classics (Sanbury, CT: Grolier Press, 1980) p. 430.
1.   R.H. Tawney, *Religion and the Rise of Capitalism* (Gloucester, MA: Peter Smith, 1962) p. 8. Many of Tawney's ideas are modifications of the earlier work of Max Weber, *The Protestant Ethic and the Spirit of Capitalism*, translated by T. Parsons (New York: C. Scribner's Sons, 1958) (first published in German in 1905). It was Weber's brilliant spadework that permitted Tawney's (in my view) more correct analysis.
2.   D. Lee, *Freedom and Culture* (Englewood Cliffs, NJ: Prentice-Hall, 1959) p. 11.
3.   Lee, *Freedom and Culture*, p. 7.
4.   L. Mumford, *The Myth of the Machine, Vol. I, Technics and Human Development* (New York: Harcourt, Brace & World, 1967) plates 16 and 17.
5.   Aristotle, *Politics*, 1253, a., translated by Jowett. Quoted by T.D. Weldon, *States and Morals* (London: John Murray, 1946, reissued, 1962) p. 69.
6.   Aristotle, *Politics*, 1284, quoted by Weldon, *States and Morals*, pp. 77–8.
7.   Tawney, *Religion*, p. 20.
8.   From T. Arnold (ed.), *Select English Works of John Wyclif*, vol. iii, 1871, pp. 130, 131, 132, 134, 143. Quoted by Tawney, *Religion*, p. 25.
9.   Tawney, *Religion*, pp. 22–3.
10.   P.H. Reaney, *The Origin of English Surnames* (New York: Barnes & Noble, 1967) p. 296 f. Before the Conquest, Anglo-Saxons of all classes had but one name, and the universal use of heritable family names did not occur until the thirteenth to fifteenth centuries; in Sweden, it occurred only in the early part of this century.
11.   B. Tuchman, *A Distant Mirror: The Calamitous Fourteenth Century* (New York: Alfred A. Knopf, 1978) p. 374.
12.   That Robin Hood was an historical character of the fourteenth century has been documented by J.W. Walker, *Yorkshire Archaeological Journal*, No. 141, 1944. He modelled himself and his wife Matilda, whom he renamed 'Maid Marian', on the mythical figures of Robin Hood and Mary Gipsy (the latter being the centre of a Mary-worship cult brought from the Holy Land by the returned Crusaders) whom the peasants soon identified with the pagan May Bride. (See Robert Graves, *The White Goddess* (London: Faber, 1947) pp. 394–6.) Sir Walter Scott unhistorically transposes Robin Hood and his band to the twelfth century in his famous novel *Ivanhoe*. (For those interested in the absorption of one worldview by another, Graves's whole book underlines the retention of pagan worship and its interweaving with Catholic beliefs among the great masses of people throughout Medieval Europe.)
13.   A parallel phenomenon is occurring today in numerous Third World countries, particularly in Latin America, where large landowners have converted rental lands where peasants once farmed into grazing lands for cattle or into the production of luxury crops, both for export to rich nations. Peasants without land rights are driven into the stony, unfertile highlands, or

into the burgeoning ghettos of the cities. See F.M. Lappé and J. Collins, *Food First: Beyond the Myth of Scarcity* (New York: Ballantine Books, 1978) pp. 15–17, 103–8, 209 ff.

14. Quoted by Tawney, *Religion*, p. 150.
15. Tawney, *Religion*, p. 85. As historian Barbara Tuchman has so clearly shown in Section 3 of *The March of Folly* (New York: Alfred A. Knopf, 1984), it was the extreme moral decay of the Church and particularly the Vatican that gave rise to the sense of social disease and the attempt of the Puritan reformists to correct it.
16. Tawney, *Religion*, p. 92.
17. Tawney, *Religion*, p. 101.
18. Quoted by Tawney, *Religion*, p. 105.
19. Tawney, *Religion*, p. 131.
20. This summary of Robert Browne's 1581 tracts on a democratic constitution for Christianity comes from W. Durant and A. Durant, *The Story of Civilization, Vol. VII, The Age of Reason Begins* (New York: Simon & Schuster, 1961) p. 25.
21. Tawney, *Religion*, p. 219.
22. Tract of 1713, quoted by Tawney, *Religion*, p. 205.
23. Tawney, *Religion*, p. 239.
24. Tawney, *Religion*, two quotes, pp. 240, 241.
25. Tawney, *Religion*, p. 267.
26. W. Durant and A. Durant, *The Story of Civilization, Vol. VIII, The Age of Louis XIV* (New York: Simon & Schuster, 1963) pp. 598–9. See also Chapters XVIII and XIX.
27. C.S. Lewis, *The Discarded Image* (Cambridge: Cambridge University Press, 1964) p. 220. In this book, Lewis, with skilled selections from Medieval and Renaissance literature, recreates the all-permeating worldview that the Age of Reason was to discard, giving us a sense of its wholeness and appropriateness for people living then.
28. E. Zilsel, 'The Genesis of the Concept of Scientific Progress', *Journal of the History of Ideas*, 6 (1945) pp. 325–49; quote is from p. 325. Zilsel argues that modern thinking about scientific and technological progress was born in the minds of Renaissance artisans and from them taken up by the thinkers of the age. Documentation for and modification of this idea are to be found in S. Lilley, 'Robert Recorde and the Idea of Progress', *Renaissance and Modern Studies*, 2 (1958) pp. 3–37.
29. T. Hobbes, *Leviathan* (Oxford: Basil Blackwell, 1946) p. 82.
30. C.B. MacPherson, *The Political Theory of Possessive Individualism* (Oxford: Clarendon Press, 1962) p. 81.
31. MacPherson, *The Political Theory*, p. 265.
32. Hobbes, *Leviathan*, Chapter 10, p. 57; Chapter 24, p. 161.
33. J. Locke, *Second Treatise on Civil Government* (Cambridge: Peter Laslett edn, 1960) section 32.
34. Locke, *Second Treatise*, section 31.
35. Locke, *Second Treatise*, section 33.
36. Locke, *Second Treatise*, section 36.
37. J. Locke, Bodleian Library, M.S. Locke, c. 30, F. 18, quoted by MacPherson, *The Political Theory*, p. 207.
38. Locke, *Second Treatise*, section 41.
39. J. Locke, *Works*, 6th edn, London, 1759, ii, 585–6. Cf. *Human Understanding* bk iv, ch. 20, sections 2–3; quoted by MacPherson, *The Political Theory*, pp. 224–5.

40. MacPherson, *The Political Theory*, pp. 227–8, 229.
41. A. Smith, *An Inquiry into the Nature and Causes of the Wealth of Nations* (Chicago: Encyclopaedia Britannica, 1952) p. 280.
42. Smith, *An Inquiry*, p. 194.
43. Smith, *An Inquiry*.
44. J. Bentham (1789), 'The Principle of Utility', from *An Introduction to the Principles of Morals and Legislation* (Oxford: Clarendon Press, 1879). Reprinted in C. Brinton (ed.), *The Portable Age of Reason Reader* (New York: Viking, 1956) pp. 93–7.
45. B. Russell, *A History of Western Philosophy* (New York: Simon & Schuster, 1945) p. 778.
46. Russell, *A History*, p. 776.
47. *American Declaration of Independence*, para. 2.
48. See E. Boykin (ed.), *The Wisdom of Thomas Jefferson* (Garden City, NY: Garden City Publishers, 1941) 'Law of the Majority', p. 54 and 'Our States Should Oppose Federal Encroachment', p. 35.
49. Letter from Jefferson to Dupont de Nemours, 1816, quoted on p. 147 of Boykin (ed.), *The Wisdom of Thomas Jefferson*.
50. MacPherson, *The Political Theory*, pp. 280–1; data are from Gregory King, first published in 1696.
51. Tawney, *Religion*, p. 266.
52. For a discussion of this point see MacPherson, *The Political Theory*, p. 274 f.
53. For a description of this issue see C. Nader, 'Controlling Environmental Health Hazards: Corporate Power, Individual Freedom and Social Control', *Annals of the New York Academy of Science*, 329 (1979) pp. 213–20.
54. M. Midgley, *Beast and Man* (Ithaca, NY: Cornell University Press, 1978) p. 228.
55. Lee, *Freedom and Culture*. This illuminating series of essays demonstrates by means of numerous examples the social freedom and independence of action that primitive peoples had within the well-defined and uniformly held world-views of their own societies. Revealing in its complete negation of Hobbes's assumptions of human nature and its requirement for coercive government.

10   The Cult of Efficiency

†   Frederick W. Taylor, *Scientific Management* (New York: Harper, 1934). 'Taylor's Testimony Before the Special House Committee', was cited from the public document 'Hearings Before Special Committee of the House of Representatives to Investigate the Taylor and Other Systems of Shop Management Under Authority of House Resolution 90; Vol. III, pp. 1377–1508', pp. 222–3 of this section of the book.
1.   H.A. Lloyd, 'Timekeepers – An Historical Sketch', in J.T. Fraser (ed.), *The Voices of Time* (New York: George Braziller, 1966) p. 389 f.
2.   F.W. Taylor, *Scientific Management* (New York: Harper, 1934) 'The Principles of Scientific Management' (first edn, 1911). (Note: The 1934 Book comprises three separately paginated sections.)
3.   C. Darwin, *The Origin of Species by Means of Natural Selection or the Preservation of Favoured Races in the Struggle for Life* (New York: J.A. Hill, 1904) p. 97 f. 'On the degree to which Organization tends to advance'. Here we see Darwin's ideas of the evolution from the 'lower' to the 'higher', from whence the notion of evolutionary 'progress' arises.
4.   W.S. Jevons, quoted by M. Lutz and K. Lux, *The Challenge of Humanistic Economics* (Menlo Park, CA.: Benjamin/Cummings, 1979) p. 49.

5. The theory of how a traditional or sacred society could have progressed through a period of normative rationality to one of full-blown means rationality was developed by the German sociologist Max Weber in his *General Economic History*, translated by F.H. Knight (Glencoe, IL: The Free Press, 1927) p. 354 f.

6. A. Smith, *An Inquiry into the Nature and Causes of the Wealth of Nations* (Chicago: Encyclopaedia Britannica, 1952) p. 3.

7. J.K. Galbraith, *The New Industrial State*, 3rd edn, revised (Boston: Houghton-Mifflin, 1978) pp. 37–8.

8. This notion was cleverly burlesqued at the turn of the century by Thorsten Veblen. In *The Theory of the Leisure Class* (New York: Kelley, 1965) he piercingly shows that the capitalist *nouveau riche* required Conspicuous Leisure and Conspicuous Consumption to advertise their new status in society. Fashion has always flowed from the top down and so it is natural that conspicuous consumption has become the goal of the masses.

9. S.S. Wolin, 'Higher Education and the Politics of Knowledge', *democracy*, 1(2) (1981) p. 46.

10. 'Teaching Drops Sharply As Career Choice: Poll of College Freshmen Shows Materialistic Aims', *San Diego Union* (23 January 1983) Part I, p. 3.

11. O.W. Holmes, 'The Deacon's Masterpiece', in R.J. Cook, compiler, *One Hundred and One Famous Poems* (Chicago: Contemporary Books, 1958) pp. 67–70.

12. D.A. Hounshell, *The Development of Manufacturing Technology in the United States* (Baltimore: Johns Hopkins University Press, 1984).

13. Galbraith, *The New Industrial State*, p. 101 f.

14. J. Costello and T. Hughes, *The Concorde Conspiracy* (New York: Charles Scribner's Sons, 1976) pp. 65, 189.

15. Data on the lifespan of small companies is not easy to obtain. The estimate of first-year failures is from p. 18 of F. Swain and B. Phillips, 'The Relationship Between Business Failures and Economic Growth, 1978–1984' (Washington, DC: US Small Business Administration, 1985). That for five year failure rate from personal communication with L. Reynolds of Small Business Advisory, Washington, DC.

16. S.B. Williams, Jr, 'Protection of Plant Varieties and Parts as Intellectual Property', *Science*, 225 (1984) pp. 18–23; D. Dickson, 'Chemical Giants Push for Patents on Plants', *Science*, 228 (1985) 1290–1.

17. Office of Technology Assessment, *Superfund Strategy* (DTA–ITE–252, Washington, DC, April 1985), quoted by P.H. Abelson, 'Waste Management', *Science*, 228 (1985) p. 1145; Commission on Economic Development, *Poisoning Prosperity* (Sacramento: Office of the Lt Governor, State Capitol, Rm 1028, 1986).

18. The Environmental Protection Agency alone spends $90 million monitoring air quality (Budget of the US Government: Fiscal 1988). This does not include costs of individual states' air monitoring stations, which in California amount to $7 million, plus local community costs.

19. E. Cagle, Jr. 'The Day Texas City Died', *Southwest Airlines Magazine* (April 1984) pp. 92–100.

20. R. Mokhiber, 'Paying for Bhopal: Union Carbide's Campaign to Limit Its Liability', *Multinational Monitor*, 31 July 1985, pp. 1–5. The more recent figures on the number of dead and injured are from *Los Angeles Times* (18 November 1987) Part I, p. 7.

21. D. Ford, *The Cult of the Atom: The Secret Papers of the Atomic Energy Commission* (New York: Simon & Schuster, 1982) pp. 45–6.

22. J. Ellul, *The Technological Society* (New York: Alfred A. Knopf, 1964) p. 105. This book and its sequel, *The Technological System* (New York: Continuum, 1980) are elaborate and rather pessimistic descriptions of the way in which the technological imperative has taken over Western society.

23. J. Miller and V. Valvano, 'The Profits of Doom: Crisis Forecasts Yield Bad Policies', *Dollars and Sense* (October 1987) pp. 6–9.

24. Galbraith, *The New Industrial State*.

25. Galbraith, *The New Industrial State*, pp. 95–6.

26. D.F. Noble, *America by Design: Science, Technology and the Rise of Corporate Capitalism* (New York: Oxford University Press, 1977) pp. 56 ff.

27. Galbraith, *The New Industrial State*, p. 294.

28. Galbraith, *The New Industrial State*, p. 92.

29. Galbraith, *The New Industrial State*, pp. 71–81. See also the excellent if dated article by Carl Kaysen, 'The Corporation, How Much Power? What Scope?', in E.S. Mason (ed.), *The Corporation in Modern Society* (Cambridge, MA: Harvard University Press, 1959) pp. 85–105. (This entire book gives insights on the subject matter of the chapter. On corporate oligopoly and power, see especially Mason's summary, pp. 5 ff.)

30. F.M. Lappé and J. Collins, *Food First: Beyond the Myth of Scarcity* (New York: Ballantine, 1978). For the control of the food giants over agriculture see pp. 302 ff. Safeway, for example, owns 141,000 acres of cheaply irrigated land in the United States (p. 204) and along with Southland (Seven–Eleven) has huge, AID-supported contracts in South America (p. 425).

31. M. Tanzer and S. Zorn, *Energy Update: Oil in the Late Twentieth Century* (New York: Monthly Review Press, 1985) pp. 125–6.

32. R. McIntyre, 'The Failure of Corporate Tax Incentives', *Multinational Monitor* (October–November 1984) pp. 3–10.

33. Galbraith, *The New Industrial State*, Chapter 17.

34. J. Mander, *Four Arguments for the Elimination of Television* (New York: William Morrow, 1978) p. 126.

35. Mander, *Four Arguments*, p. 131.

36. Galbraith, *The New Industrial State*, p. 219.

37. Mander, *Four Arguments*, p. 129.

38. Galbraith, *The New Industrial State*, p. 217.

39. P.A. Samuelson, *Economics*, 8th edn (New York: McGraw–Hill, 1970) p. 40 f.

40. S.S. Wolin, 'The New Public Philosophy', *democracy*, 1(4) (1981) pp. 27–8.

41. J.M. Keynes, *The General Theory of Employment, Interest and Money* (New York: Harcourt, Brace & World, 1964, © 1936) p. 372 f.

42. *Economic Report of the President*, transmitted to the Congress, February 1985 (Washington, DC: US Government Printing Office, 1985) pp. 66, 320.

43. Galbraith, *The New Industrial State*, Chapters 26 and 27: 'The Planning System and the State'. The government's role in education and in research are described elsewhere in his book.

44. J. Walsh, 'Supercompeting over Supercomputers', *Science*, 220 (1983) pp. 581–4.

45. Galbraith, *The New Industrial State*, pp. 367–8.

46. Galbraith, *The New Industrial State*, p. 334. In his first edition, Galbraith used the word 'industrial' where 'planning' now appears. In the years between 1967 and 1978, organisational planning had extended from industry into all major social institutions, and so Galbraith has substituted the more inclusive term in this later edition.

47. H. Skolimowski, 'Philosophy of Technology as a Philosophy of Man',

pp. 325–36 in G. Buliarello and D.B. Doner (eds), *The History and Philosophy of Technology* (Urbana, IL: University of Illinois Press, 1979) p. 334. This excellent article brings our technological–quantoid worldview into clear focus.

48. 'Risk/Benefit Analysis in the Legislative Process', Joint Hearings before Committees of Congress, 24, 25 July 1979. (No. 71, Committee on Science and Technology) (Washington, DC: US Government Printing Office, Publ. 54–509, 1980). See especially pp. 63–96.

49. A. Pacey, *The Culture of Technology* (Cambridge, MA: MIT Press, 1983) p. 10; S.S. Wolin, 'The New Public Philosophy', p. 28.

50. Skolimowski, 'Philosophy of Technology', p. 332.

51. L. Mumford, *The Myth of the Machine: The Pentagon of Power* (New York: Harcourt, Brace, Jovanovich, 1970).

52. J.R. Chiles, 'Learning from the Big Blackouts', *American Heritage of Invention and Technology*, 1(2) (1985) p. 30. (This three-times-a-year glossy magazine is sponsored solely by advertisements from General Motors and sent free of charge to university scientists and others.)

53. Galbraith, *The New Industrial State*, p. 408.

54. Galbraith, *The New Industrial State*, p. 393.

55. A. Miller, 'All My Sons', in Harlan Hatcher (ed.), *Modern American Dramas* (New York: Harcourt, Brace & World, 1949) pp. 277–323.

56. K. Vandivier, 'Case Study – The Aircraft Brake Scandal', in T. Donaldson and P.H. Werhane (eds), *Ethical Issues in Business: A Philosophical Approach* (Englewood Cliffs, NJ: Prentice–Hall, 1979) pp. 11–24. Vandivier was one of the engineers who lost his job; he is now a newspaper reporter. The textbook in which this case history appears is a valuable sourcebook for further study of these questions.

57. Ford, *The Cult of the Atom*. A complete picture of the sort of incompetent management of technological problems when *secrecy* is imposed on decision-making can be had by a complete reading of this book.

58. Ford, *The Cult of the Atom*, pp. 160–4, 167–9, 172–3.

59. J. Ladd, 'Morality and the Ideal of Rationality in Formal Organization', in Donaldson and Werhane, *Ethical Issues*, pp. 102–13.

60. K.E. Goodpaster, 'Morality and Organizations', in Donaldson and Werhane, *Ethical Issues*, pp. 114–22.

61. B. Siegel, 'Trial Makes History, Stirs Controversy', *Los Angeles Times* (16 September 1985) pp. 1 and 8. As of early 1988, this case was still on appeal in Illinois courts. More recently, the US Environmental Protection Agency has begun criminal prosecution of corporate directors who knowingly submit the public to illegal risks. See D. Wann, 'Environmental Crime: Putting Offenders Behind Bars', *Environment*, 29(8) (1987) pp. 5, 44–5.

62. D. Dickson and D. Noble, 'The New Corporate Technocrats', *The Nation* (12 September 1981) p. 212. This fiery article vividly describes corporate efforts to control the American political system.

63. Federal Election Commission, '1986 Congressional Spending Tops $450 Million', Report from FEC Press Office, 10 May 1987; figures are cited from p. 1.

11 Alienation: The Loss of the Sacred

†   Roger Hausheer, 'Introduction' to Isaiah Berlin, *Against the Current: Essays in the History of Ideas* (New York: Viking, 1980) pp. xxxvi–vii.

1.   P. Watchel, *The Poverty of Affluence* (New York: The Free Press,

Macmillan, 1983) p. 199. Quotes are from P. Rieff, *The Triumph of the Therapeutic* (New York: Harper & Row, 1966) p. 22.

2. A.Wheelis, *On Not Knowing How to Live* (New York: Harper & Row, Colophon Books, 1975) pp. 40–1.

3. *The Complete Plays of Gilbert and Sullivan* (New York: The Modern Library, 1936 edn) p. 352.

4. R.S. Kempe and C.H. Kempe, *The Common Secret: Sexual Abuse of Children and Adolescents* (San Francisco: W.H. Freeman, 1984).

5. M.N. Nagler, *America Without Violence* (Covelo, CA: Island Press, 1982) pp. ix, 3, 13, 20. See also *Sourcebook of Criminal Justice Statistics* (Albany, NY: US Department of Justice, Criminal Justice Research Center, published annually).

6. Cited in *The Washington Spectator* (1 December 1984) p. 4.

7. *Vital Statistics of the United States* (Washington, DC: US Department of Health and Human Services, published annually.) See also ' "Suicide Epidemic" among Young Cited', *San Diego Union* (7 February 1985) p. A–6.

8. D.T. Lunde, *Murder and Madness* (San Francisco: San Francisco Book Co., 1976) pp. 4 ff.

9. A.P. Goldstein, 'Needed: A War on Aggression', *National Forum* (Fall 1983) pp. 14–15.

10. Goldstein, 'Needed: A War on Aggression', p. 14.

11. M.W. Cannon, 'Contentious and Burdensome Litigation: A Need for Alternatives', *National Forum* (Fall 1983) pp. 10–12.

12. D. Yankelovich, *New Rules: Searching for Self-Fulfillment in a World Turned Upside Down* (New York: Random House, 1981) pp. 44, 212–13.

13. Yankelovich, *New Rules*, pp. 183–4.

14. A. Campbell, *The Sense-of-Well-Being in America* (New York: McGraw-Hill, 1981) pp. 28–9.

15. Yankelovich, *New Rules*, pp. 188 ff.

16. E.L. Bassuk, 'The Homelessness Problem', *Scientific American* (July 1984) pp. 40–5.

17. 'Teaching Drops Sharply as Career Choice: Poll of College Freshmen Shows Materialistic Aims', *San Diego Union* (23 January 1983) Part I, p. 3.

18. Yankelovich, *New Rules*, Chapter 5.

19. R.N. Bellah, R. Madsen, W.M. Sullivan, A. Swidler and S.M. Tipton, *Habits of the Heart* (Berkeley, CA: University of California Press, 1985); cited are pp. 118 and 111.

20. Yankelovich, *New Rules*, Chapter 23. While Maslow's 'highest needs' are indeed for aesthetics and altruistic behaviour, his assumption that these can be created from scratch, in the absence of a cultural framework in which these traits are recognised and valued is unsound. The notion that 'self-actualisation' can occur in a cultural vacuum is merely a perpetuation of the problem itself: the notion that a society is nothing but a collection of autonomous individuals. For further discussion, see Wachtel, *The Poverty of Affluence*, Chapters 6–10.

21. Bellah *et al.*, *Habits of the Heart*, pp. 28–31. See also H. Arendt, *On Revolution* (New York: Viking, 1963) pp. 22 ff.

22. Bellah *et al.*, *Habits of the Heart*, p. 135.

23. D. Lee, *Freedom and Culture* (New York: Spectrum, Prentice-Hall, 1959). See also John C. Neihart, *Black Elk Speaks* (New York: Simon & Schuster, 1972, © 1932), for a description of life in such a culture.

24. Wachtel, *The Poverty of Affluence*, pp. 60 ff.

25. Wachtel, *The Poverty of Affluence*, p. 120.

26. See L. Mumford, *The Myth of the Machine, Volume I, Technics and Human Development* (1967) and *Volume II, The Pentagon of Power* (1970) (New York: Harcourt, Brace, Jovanovich) for a discussion of the thrust of Western technological society.

27. S.B. Sarason, *Psychology Misdirected* (New York: The Free Press, Macmillan, 1981); quotes are respectively from p. 26, p. 25, p. 26.

28. Yankelovich, *New Rules*, pp. 226–7. See also Jane Jacobs, *The Death and Life of Great American Cities* (New York: Random House, 1961) for an excellent description of the integrated way that socially stable neighbourhoods function. She uses as her primary model Greenwich Village in Manhattan. The importance of neighbourhood structure was graphically illustrated when a commercial airplane crashed into a residential area of San Diego in 1978; the local Mom-and-Pop grocery store instantly became the checkpoint for discovering who had been killed or needed help.

29. E. Fromm, *The Sane Society* (Greenwich, CT: Fawcett, 1955) p. 27.

30. Fromm, *The Sane Society*, p. 28.

31. S. Burenstam-Linder, *The Harried Leisure Class* (New York: Columbia University Press, 1970).

32. J. Mander, *Four Arguments for the Elimination of Television* (New York: William Morrow, 1978) p. 265.

33. T. Gladwin, *Slaves of the White Myth: The Psychology of Neocolonialism* (Atlantic Highlands, NJ: Humanities Press, 1980) pp. 27–8.

34. Sarason, *Psychology Misdirected*, p. 173.

35. This tendency toward excessive reductionism is true not only of psychology but of all modern scientific disciplines, and it leads to fallacious, sometimes dangerous conclusions when applied to practical problems, since it presumes to understand the *relationships* of parts by looking at them in isolation. Both social and natural scientists are only now awakening to the pitfalls of such a reductionist approach.

36. A.S. Neill, *Summerhill* (New York: Hart, 1960) p. 297.

37. C. Bereiter, 'Moral Alternatives to Education', *Interchange*, 3 (1972) pp. 25–42; quotes are from p. 25 and pp. 26–7.

38. L. Kohlberg, *Essays on Moral Development, Vol. I. The Philosophy of Moral Development: Moral States and the Idea of Justice* (San Francisco: Harper & Row, 1981) p. 71.

39. Kohlberg, *Essays*, p. 75.

40. L. Kohlberg, 'The Moral Atmosphere of the School', in N. Overley (ed.), *The Unstudied Curriculum: Its Impact on Children* (Washington, DC: Association for Supervision and Curriculum Development, ASCD Elementary Council, National Education Association, 1970); quote is on p. 104.

41. Kohlberg, *Essays*, p. 294.

42. Kohlberg, *Essays*, p. 295.

43. Y. Sapozhkov, 'Teaching Morality in Byelorussia's Classroom', *Soviet Life* (September 1982) p. 16.

44. H.I. Schiller, *The Mind Managers* (Boston: Beacon Press, 1973) pp. 20–1.

45. Mander, *Four Arguments*, pp. 34–5.

46. The more recent United States invasion of the tiny island of Grenada and the bombing of Libya were accompanied by similar media 'facts' of dubious reality.

47. Mander, *Four Arguments*, p. 308. Television manufacturers, having discovered the public's widespread aversion to commercials, now sell remote control devices with all TVs that permit one to shut off the sound (or the

picture as well) at the touch of a finger. It tends to give one back a tiny sense of power.

48. Nagler, *America Without Violence*, pp. 18 ff.
49. H. London, quoted in *The Washington Spectator* (15 October 1987) p. 3.
50. Mander, *Four Arguments*, p. 167–8.
51. Yankelovich, *New Rules*, p. 121.
52. This is a major point that Lewis Mumford makes throughout his two volume treatise, cited above. See also 'Nationalism: Past Neglect and Present Power', in Isaiah Berlin, *Against the Current: Essays in the History of Ideas* (New York: Viking, 1980) pp. 333–55; see especially p. 351.
53. L.D. Holler, 'In Search of a Whole-System Ethic', *Journal of Religious Ethics*, 12 (1984) pp. 236–7.
54. Holler, 'In Search', p. 229.
55. Mander, *Four Arguments*, p. 123. Bellah *et al.*, *Habits of the Heart*, is essentially an analysis of this entire problem.

12 Rethinking Economics

†   Kenneth Boulding, 'Income or Welfare', *Review of Economic Studies*, 17 (1949) pp. 77–86.
1.   S.S. Wolin, 'The New Public Philosophy', *democracy*, 1(4) (1981) pp. 27–8.
2.   M.A. Lutz and K. Lux, *The Challenge of Humanistic Economics* (Menlo Park, CA: Benjamin/Cummings, 1979) p. 77.
3.   L. Thurow, *The Zero-Sum Society* (Harmondsworth, England: Penguin Books, 1981) p. 118. Mr Thurow made a similar statement in testimony to Congress: 'The Implications of Zero Economic Growth', in *U.S. Prospects for Growth, Vol. 5, The Steady-State Economy*, Joint Economic Committee of Congress, (Washington, DC: US Government Printing Office, 2 December 1976) p. 46.
4.   H.E. Daly, *Steady-State Economics* (San Francisco: W.H. Freeman, 1977) pp. 118–19. For detailed economic analysis of many of the major points in this chapter this text is an excellent resource.
5.   D. Harrop, 'Who Makes What $', *People* (25 March 1985) pp. 92–101.
6.   F. Soddy, *Wealth, Virtual Wealth and Debt* (New York: E.P. Dutton, 1926) pp. 69–70. A valuable resource for understanding the relationships between money and wealth.
7.   Soddy, *Wealth*, p. 106.
8.   Soddy, *Wealth*, p. 157.
9.   Soddy, *Wealth*, p. 123.
10.   Soddy, *Wealth*, p. 152.
11.   A.L. Malabre, Jr, *Beyond Our Means* (New York: Random House, 1987) p. 133.
12.   Malabre, *Beyond our Means*, pp. 155–6; figures cited in preceding paragraphs are from this book.
13.   A.N. Whitehead, *Process and Reality* (New York: Free Press, 1978). See pp. 7, 20, 93, 94. The fallacy of misplaced concreteness, he says, results from 'neglecting the degree of abstraction involved when an actual entity is considered merely so far as it exemplifies certain categories of thought', p. 7. In other words, we mistake our simplified models for the whole of reality.
14.   H.E. Daly, 'The Circular Flow of Exchange Value and the Linear Throughput of Matter/Energy: A Case of Misplaced Concreteness', *Review of Social Economy*, 43 (1985) pp. 279–97. The text quoted is R. Heilbroner

and L. Thurow, *The Economic Problem* (New York: Prentice–Hall, 1981) p. 127 and p. 135, respectively.

15.   S. Burns, quoted by Hazel Henderson, *The Politics of the Solar Age* (Garden City, NY: Anchor Press, Doubleday, 1981) p. 9.

16.   K. Valaskakis, P. Sindell, J.G. Smith and M.I. Fitzpatrick, *The Conserver Society* (New York: Harper & Row, 1978) p. 181.

17.   For an analysis of the ways in which GNP and other money measures of economic value are regularly distorted, see the unpublished article 'Money Value vs. Economic Value: Why Aggregate Economic Indices Mislead', by L. J. Dumas, Dept. of Economics, University of Texas, Dallas.

18.   Henderson, *The Politics of the Solar Age*, pp. 100–101.

19.   Lutz and Lux, *The Challenge*, p. 226.

20.   'High-Tech: The Limits to Growth', *Dollars and Sense* (September 1985) p. 5.

21.   C. Froman, *The Two American Political Systems: Society, Economics, and Politics* (Englewood Cliffs, NJ: Prentice-Hall, 1984) p. 64.

22.   Froman, *The Two American Political Systems*, p. 68 and p. 78. Dye's estimate is taken from his book *Who's Running America? The Carter Years* (Englewood Cliffs, NJ: Prentice-Hall, 1979).

23.   D. Noble, 'Tools of Repression', *Dollars & Sense* (October 1984) p. 16. See also D. Noble, *Forces of Production: A Social History of Industrial Automation* (New York: Alfred A. Knopf, 1985).

24.   Henderson, *The Politics of the Solar Age*, p. 245.

25.   J. Maynard Keynes, *The General Theory of Employment, Interest, and Money* (New York: Harcourt, Brace & World, 1964, first published 1937) p. 221.

26.   Keynes, *The General Theory*, p. 376.

27.   Froman, *The Two American Political Systems*, Chapter 1, 'Politics and Property', pp. 1–17.

28.   *Economic Justice for All: Catholic Social Teaching and the U.S. Economy* (Washington, DC: US Catholic Conference, 1986) 204 pp.

29.   Malabre, *Beyond Our Means*, Chapter 8, 'Beyond Our Means', pp. 144–62.

30.   W. Leontief, 'Letter: Academic Economics', *Science*, 217 (1982) pp. 104–7. See also follow-up correspondence in *Science* (8 October 1982) p. 108, (10 December 1982) p. 1070 and (25 February 1983) p. 904. Martin Baily of the Brookings Institution also severely criticises the short-comings of modern economic thinking (*Science*, 216 (1982) pp. 859–62). Like Leontief, he is highly dubious of the validity of General Equilibrium Theory. For a lighter treatment of economists' theoretical curves, see Martin Gardner, 'Mathematical Games: The Laffer Curve and Other Laughs in Current Economics', *Scientific American* (December 1981) pp. 18–31C.

31.   R.L. Heilbroner, *An Inquiry into the Human Prospect* (New York: W.W. Norton, 1975) p. 170.

32.   See D. Ford, *The Cult of the Atom: The Secret Papers of the Atomic Energy Commission* (New York: Simon & Schuster, 1982) p. 183 f.

33.   D.M. Rust, 'Solar Flares, Proton Showers and the Space Shuttle', *Science* 216 (1982) pp. 939–46; 'Weakness Found in Space Rats', *San Diego Union* (7 September 1985) p. A–11.

34.   Froman, *The Two American Political Systems*, Chapter 8, 'The Individual Political System', pp. 123–44.

35.   See the *Samizdat* book by 'Boris Komarov', *The Destruction of Nature in the Soviet Union* (White Plains, NY: Sharpe, 1980) and P. Pryde, 'The Decade of the Environment in the U.S.S.R.', *Science*, 220 (1983) pp. 274–9.

36. C. Lindblom, *Politics and Markets* (New York: Basic Books, 1977) p. 356.
37. Soddy, *Wealth*, p. 270 f. As already noted (references 25 and 26) John Maynard Keynes looked forward to the day when capital would not be privately 'owned'. More recently, Cornell University economist Jaroslav Vanek has proposed this scheme in detail as a means for financing worker-managed cooperatives in *Self-Management (Economic Liberation of Man)* (Baltimore: Penguin, 1975).
38. G. Alperovitz and J. Faux, *Rebuilding America* (New York: Pantheon Books, 1984) p. 262.
39. Alperovitz and Faux, *Rebuilding America*, pp. 56–7.
40. Lutz and Lux, *The Challenge*, pp. 261–3. In the United States, the cooperative movement has been in the forefront of economic democracy and community self-sufficiency.
41. E.F. Schumacher, *Small is Beautiful* (London: Sphere Books, Abacus, 1974) pp. 76–83.
42. F.M. Lappé, 'Sweden's Third Way to Worker Ownership', *The Nation* (19 February 1983) pp. 203–4; H.M. Christman, 'Swedish Buy-Out', *The Nation* (4 February 1984) p. 117.
43. G. Medard (with others), *Regenerating America: Meeting the Challenge of Building Local Economies* (Emmaus, PA: Rodale Press, 1985). See also G.J. Coates, *Resettling America: The Movement Toward Local Self Reliance* (Andover, MA: Brick House Publishing, 1982).
44. The following articles in the unconventional press indicate this so far little publicised trend: B. Schmiechen, L. Daressa and L. Adelman, 'Waking from the American Dream', *The Nation* (3 March 1984) p. 241 f. (quote on p. 258: 'How can we utilize the existing resources of the area and the skills of the people in order to produce things society needs?'); C. Bass and P. Bass, 'Employee Buy-Out Stops Plant Closing', *In These Times* (30 January–5 February 1985) p. 5 (quote: 'This gives people the security of being in charge of their own destiny'.); C. Cox, 'Living Our Dreams: How Communities Are Creating an Economic Base for Their Values', *Building Economic Alternatives* (A quarterly publication of Co-op America) (Summer 1985) pp. 18–21 (quote p. 21: 'Building a strong community [as in Minneapolis' West Bank] will take a commitment both to values and to change'.); D. Morse, 'The Campaign to Save Dorothy Six', *The Nation* (7 September 1985) pp. 174–6 (quote p. 176: 'We like to refer to this as reindustrialization from below'.); J. Gilbrecht, 'Sewing on Their Own: A Corporate Campaign Nixes Plant Closing', *Dollars & Sense* (September 1985) pp. 12–14 (quote p. 14: 'The concept of a worker co-op captured the imagination of the local community. Suddenly, worker ownership was featured on the six o'clock news and as the subject of in depth articles in major papers'.); L. Compa, 'Fighting Back: Workers Challenge Plant Shut-Downs', *The Progressive* (October 1985) pp. 32–4 (quote p. 34: 'Unfortunately, corporate executives still hold the trump card – the legal right to shut down facilities regardless of objections from workers, unions, or communities. Unions lost the combative influence of communists, socialists, and other radicals in their Cold War rush to conformity, and now the mainstream labor movement fights over workers' share of the economic pie without challenging the system that bakes it'.) See also Daniel Zwerdling's *Democracy at Work*, published by the Association for Self-Management, 1414 Spring Rd NW, Washington, DC, 20010, for numerous similar case histories of community economic self-help.
45. For information on conversion projects see *WIN Magazine*, 1 July 1981.
46. British-born Michael Linton invented the Local Exchange Trading System

(LETS) and first established it in British Columbia. For information write him c/o Landsman Community Services Ltd., 304576 England Ave., Courtenay, BC, V9N 5M7, Canada.

47.   Schumacher, *Small is Beautiful*, p. 184.
48.   H. Arendt, *Between Past and Future: Eight Exercises in Political Thought* (New York: Viking Press, 1954) p. 9.

13   Defusing the Global Powder Keg

†   Gustavo Esteva in S.H. Mendlovitz and R.B.J. Walker (eds), *Towards a Just World Peace* (London: Butterworths, 1987) p. 280.
1.   B. Nietschmann, 'The Third World War', *Cultural Survival Quarterly*, 11(3) (1987) pp. 1–16.
2.   L.S. Stavrianos, *Global Rift: The Third World Comes of Age* (New York: William Morrow, 1981) p. 53.
3.   Stavrianos, *Global Rift*, p. 256. In his Chapter 13, 'Era of Monopoly Capitalism and Global Colonialism', (pp. 256–77), Stavrianos cogently links together the growth of capitalism and the growth of imperialism.
4.   J.S. Saul (ed.), *A Difficult Road: The Transition to Socialism in Mozambique* (New York: Monthly Review Press, 1985) p. 41.
5.   T. Gladwin, *Slaves of the White Myth: The Psychology of Neocolonialism* (Atlantic Highlands, NJ: Human Press, 1980) p. 45 f.
6.   Saul, *A Difficult Road*, p. 45.
7.   Gladwin, *Slaves*, pp. iv–v.
8.   S. Wawrytko, 'Chinese Philosophy on Its Way to the Twenty-First Century: Meeting the Challenge of Cultural Imperialism', *Proceedings of the First Conference in Chinese Philosophy, 1984* (Taichung, Taiwan: Tunghai University Press, 1985) pp. 715–36; quote is from pp. 716–17.
9.   Stavrianos, *Global Rift*, p. 118.
10.   J. Robinson, *Aspects of Development and Underdevelopment* (Cambridge: Cambridge University Press, 1979) p. 103.
11.   E. Easwaran, *Gandhi the Man* (Petaluma, CA: Nilgiri Press, 1978) pp. 76–81.
12.   Stavrianos, *Global Rift*, p. 39.
13.   Stavrianos, *Global Rift*, p. 80 f.
14.   Robinson, *Aspects of Development*, p. 46.
15.   Gladwin, *Slaves*, p. 8 f; Stavrianos, *Global Rift*, p. 181 f.
16.   Stavrianos, *Global Rift*, p. 182.
17.   Stavrianos, *Global Rift*, p. 233.
18.   V. Mehta, 'Personal History', *New Yorker* (22 August 1983) p. 39, pp. 44–50.
19.   J.J. Boeke, *The Structure of Netherlands Indian Economy* (New York: International Secretariat, Institute of Pacific Relations, 1942) pp. 18–19.
20.   Boeke, *The Structure of Netherlands Indian Economy*, p. 25.
21.   Boeke, *The Structure of Netherlands Indian Economy*, p. 74.
22.   Boeke, *The Structure of Netherlands Indian Economy*, p. 68.
23.   Boeke, *The Structure of Netherlands Indian Economy*, p. 72.
24.   J.F. Ade Ajayi and I. Espie, *A Thousand Years of West African History* (New York: Humanities Press, 1972) p. 256 f.
25.   Nietschmann, 'The Third World War', p. 5 and p. 1.
26.   Ajayi and Espie, *A Thousand Years*, end papers.
27.   Stavrianos, *Global Rift*, p. 38.
28.   Stavrianos, *Global Rift*, p. 274.

29. S.A. Hewlett, *The Cruel Dilemma of Development: Twentieth Century Brazil* (New York: Basic Books, 1980) p. 215 f.
30. Robinson, *Aspects of Development*, p. 86 (dams); C.J. Lancaster, 'Africa: Economics and Politics of Development', *Bulletin of the Atomic Scientists* (September 1985) p. 27 (steel mills); D. Avery, 'U.S. Farm Dilemma: The Global Bad News Is Wrong', *Science*, 230 (1985) p. 411 (general preference for industrialisation over agricultural development).
31. Robinson, *Aspects of Development*, p. 40.
32. J. Maynard Keynes, quoted in Robinson, *Aspects of Development*, p. 33.
33. Robinson, *Aspects of Development*, p. 33.
34. UNCTAD (United Nations Conference on Trade and Development) IV TD 183, paragraph 13, quoted in Robinson, *Aspects of Development*, p. 80.
35. Robinson, *Aspects of Development*, p. 52.
36. E. Mortimer, *Faith and Power: The Politics of Islam* (New York: Vintage Books, Random House, 1982); see especially p. 293 f. See also James A. Bill, 'Resurgent Islam in the Persian Gulf', *Foreign Affairs* (Fall 1984) pp. 108–27.
37. Robinson, *Aspects of Development*, p. 68.
38. Robinson, *Aspects of Development*, pp. 73–4.
39. Robinson, *Aspects of Development*, p. 61.
40. F.M. Lappé and J. Collins, *Food First: Beyond the Myth of Scarcity* (New York: Ballantine Books, 1978) p. 280 f; Stavrianos, *Global Rift*, p. 444.
41. A. Dobrin, 'The Vanishing Herds', *Food Monitor* (May–June 1978) p. 23 (quoted by Stavrianos, *Global Rift*, pp. 676–7.)
42. Robinson, *Aspects of Development*, p. 88.
43. Hewlett, *The Cruel Dilemma*, p. 19; T.J. Thompson, II, 'Commentary', *Multinational Monitor* (February 1984) p. 9.
44. Gladwin, *Slaves*, p. 97, 98. See also R.J. Barnet and R.E. Müller, *Global Reach: The Power of the Multinational Corporations* (New York: Simon & Schuster, 1974) p. 178; J. Gay, 'Sweet Darlings in the Media: How Foreign Corporations Sell Western Images of Women in the Third World', *Multinational Monitor* (August 1983) pp. 19–21.
45. M. Clark, *Contemporary Biology*, 2nd edn (Philadelphia: W.B. Saunders, 1979) pp. 200, 398–9.
46. Lappé and Collins, *Food First*, p. 336 f. See also Barnet and Müller, *Global Reach*, p. 183.
47. Stavrianos, *Global Rift*, p. 447.
48. W. Loehr and J.P. Powelson, *Threat to Development: Pitfalls of the NIEO* (Boulder, CO: Westview Press, 1983) p. 157 f.
49. O.G. Wichard, 'U.S. Direct Investment in 1979', *Survey of Current Business* (August 1980) (Washington, DC: US Department of Commerce) pp. 16–37.
50. Loehr and Powelson, *Threat to Development*, p. 82 f.
51. C. Payer, *The Debt Trap: The IMF and the Third World* (New York: Monthly Review Press, 1974) pp. 215–16. See also C. Payer, *The World Bank: A Critical Analysis* (New York: Monthly Review Press, 1982); Loehr and Powelson, *Threat to Development*, p. 96 f.
52. Stavrianos, *Global Rift*, p. 674. He describes the corrupt use of foreign aid by the US-backed dictator of Zaire, Joseph Mobutu.
53. J.W. Mellor, *The New Economics of Growth: A Strategy for India and the Developing World* (Ithaca, NY: Cornell University Press, 1976).
54. Robinson, *Aspects of Development*, p. 93.
55. H.H. Humphrey, quoted by Stavrianos, *Global Rift*, p. 443.

56. Stavrianos, *Global Rift*, p. 442 f.
57. Robinson, *Aspects of Development*, Table, p. 93.
58. Payer, *The Debt Trap*, pp. ix–xi; quote from p. ix.
59. 'The Buck Never Stops: International Lending Out of Control', *Dollars & Sense* (December 1982).
60. Robinson, *Aspects of Development*, p. 97 f.
61. N. Chomsky, *On Power and Ideology* (Boston: South End Press, 1986) p. 7.
62. Robinson, *Aspects of Development*, p. 88 f.
63. These points are elaborated in the following: H.M. Wachtel, *The Money Mandarins: The Making of a New Supranational Economic Order* (New York: Pantheon, 1986); H.M. Watchel, 'The Global Funny Money Game', *The Nation* (26 December 1987) pp. 784–90; W.R. Mead, 'After Hegemony', *New Perspectives Quarterly* (Fall 1987) pp. 42–7. Significantly, in the late 1980s, the World Bank, recognising the general failure of past investments in developing countries, has begun to give loans for environmentally and socially more sensitive projects.
64. T. Ratigan, former Peace Corps Volunteer, personal communication.
65. J. Macy, *Dharma and Development: Religion as Resource in the Sarvodaya Self-Help Movement* (West Hartford, CT: Kumarian Press, 1983) p. 22 f. This offers good insight into how a religious worldview informs all human activities, including economics.
66. T. Vittachi, 'Clues to Development in the Commonwealth Isles', *People* (IPPF Review of Population and Development), 12(4) (1985) pp. 4–6.
67. S.T. Umpleby, 'World Population: Still Ahead of Schedule', *Science*, 237 (1987) pp. 1555–6.
68. P. Harrison, 'Land and People: A New Framework for the Food Security Equation', *Ceres*, No. 98, Vol. 17(2) (March–April 1984), Centrepiece. This report was prepared for the United Nations Food and Agricultural Organization (FAO), based on the most recent LandSat and other global information on soil, climate, and crop yields, and the predicted populations in 2000 by region and by country.
    Contrast this FAO report with a highly optimistic one sponsored by the Bureau of Intelligence and Research, US Department of State, which relies almost exclusively on World Bank assessments of global agricultural outputs and ignores both population growth and virtually all FAO and other United Nations data: K. Avery, 'U.S. Farm Dilemma: The Global Bad News Is Wrong', *Science*, 230 (1985) pp. 408–12.
69. C. Bolton and J.W. Leasure, 'Political Development and Decline of Fertility in the West', Article No. 11, pp. 84–102, in J.W. Leasure *et al.*, *Population and the Social Sciences* (San Diego, CA: San Diego State University Press, 1978).
70. See *Population and Other Problems: China Today (1)*, pamphlet published by *Beijing Review*, China Publications Centre, P.O. Box 339, Beijing, China (1981) pp. 7–33. In addition to economic incentives, powerful peer pressure is brought to bear on Chinese women; those pregnant with a second child are strongly urged by friends and co-workers to seek abortion. Bearing too many children is now considered socially irresponsible.
71. J.R. Weeks, *Population: An Introduction to Concepts and Issues* (Belmont, CA: Wadsworth, 1981) p. 361.
72. Reported by S. Rosenhause, 'India Taking Drastic Birth Control Step', *Los Angeles Times* (25 September 1976) (quoted in Weeks, *Population*, p. 357). Nearly a decade later, both birth and death rates have declined only marginally, leaving the doubling time unaffected. Shortly after the year

2000, India's population will overtake that of China. In 1985 Prime Minister Rajiv Ghandi announced a 'war' on fertility, with major monetary incentives for two-child families and sterilisation (*Popline* (May 1985) pp. 1–2).

73.   M. Kendall, 'The World Fertility Survey: Current Status and Findings', *Population Reports* (Population Information Program of Johns Hopkins University, M–3 (July 1979)).

74.   Vittachi, 'Clues to Development', pp. 4–6.

75.   Robinson, *Aspects of Development*, p. 98.

76.   Macy, *Dharma and Development*, p. 74. For other works on self-reliant development, see J. Galtung, P. O'Brien and R. Preiswerk (eds), *Self-Reliance: A Strategy for Development* (Geneva: Institute of Development Studies, 1980); O. Giarini, *Dialogue on Wealth and Welfare: An Alternative View of World Capital Formation* (Oxford: Pergamon Press, 1980). Both books make the point that 'wealth' comprises not only matter–energy, *per se*, nor even the technical knowledge of how to coopt it for human use, but the wisdom to make such use *serve*, not dominate, the diversity of cultural values that humans have evolved.

14   Politics: Worldviews in Action

†   John W. Burton, *Global Conflict: The Domestic Sources of International Crisis* (Brighton: Wheatsheaf Books, John Spears (distributed in US by The Center for International Development, University of Maryland, College Park, MD), 1984) p. 36.

1.   M. Mead, *Culture and Commitment: A Study of the Generation Gap* (Garden City, NY: Natural History Press, Doubleday, 1970) p. 61.

2.   G. Mische and P. Mische, 'The National Security State', Chapter 3 in *Toward a Human World Order* (New York: Paulist Press, 1977) pp. 44–68.

3.   J. Schell, *The Fate of the Earth* (first published in *The New Yorker*, 1, 8, 15 February 1982); quote is from 15 February 1982, p. 103.

4.   I. Berlin, 'The Counter-Enlightenment', Chapter 1 in *Against the Current: Essays in the History of Ideas* (New York: Viking Press, 1980) pp. 1–24.

5.   R.M. MacIver, *The Web of Government* (New York: The Free Press, Macmillan, 1965) p. 37.

6.   K. Marx and F. Engels, *The German Ideology* (New York: International Publishers, 1947) p. 39.

7.   Marx and Engels, *The German Ideology*, pp. 40–1.

8.   Burton, *Global Conflict*, p. 12.

9.   S. Weil, *The Need for Roots*, translated by A. Wills (New York: Putnam, 1952) pp. 123, 127; quoted by C. Lasch, 'Mass Culture Reconsidered', *democracy* (October 1981) p. 22.

10.   L.C. Thurow, *The Zero-Sum Society: Distribution and the Possibilities for Economic Change* (Harmondsworth, England: Penguin, 1981) p. 120.

11.   C. Froman, *The Two American Political Systems: Society, Economics, and Politics* (Englewood Cliffs, NJ: Prentice-Hall, 1984) p. 14.

12.   M. Parenti, *Democracy for the Few*, 4th edn (New York: St Martin's Press, 1983) p. 329.

13.   C.L. Schultz, *The Public Use of Private Interest* (Washington, DC: The Brookings Institution, 1977) p. 18; quoted in S.S. Wolin, 'The New Public Philosophy', *democracy*, 1(4) (October 1981) p. 33.

14.   J. Robinson, *Economic Heresies: Some Old-Fashioned Questions in Economic Theory* (New York: Basic Books, 1971) p. 25; quoted by Burton, *Global Conflict*, pp. 53–4.

15. Parenti, *Democracy for the Few*, p. 330.
16. Froman, *The Two American Political Systems*, p. 105.
17. Parenti, *Democracy for the Few*, pp. 344, 335.
18. R. Niebuhr, *Moral Man and Immoral Society* (New York: Charles Scribner's Sons, 1932) pp. 15, 89 ff.
19. K. Marx, *Capital, I*, Ben Fowkes edn (New York: Vintage Books, 1977) p. 929.
20. J.K. Galbraith, *The Age of Uncertainty* (Boston: Houghton Mifflin, 1977) p. 93.
21. Niebuhr (*Moral Man*, p. 187 f.) points out that the Russian revolution was due mainly to the stupidity and brutality of the Church and aristocracy rather than to the industrial–economic forces predicted by Marx.
22. P. Reddaway, 'Waiting for Gorbachev', *New York Review of Books* (10 October 1985) pp. 5–10.
23. L.C. Thurow, 'The Dishonest Economy', *New York Review of Books* (21 November 1985) pp. 34–7.
24. Burton, *Global Conflict*, p. 51.
25. The frank governmental control of the media in the USSR is public knowledge. The control of the media in America is far more subtle, but in that sense, far more dangerous for giving the appearance of 'openness'. For a detailed analysis of this point, see Herbert Schiller, *The Mind Managers* (Boston: Beacon Press, 1973), especially Chapter 1, 'Manipulation and the Packaged Consciousness', pp. 8–31.
26. Burton, *Global Conflict*, p. 11.
27. R. Lamm, Governor of Colorado, 'Promoting Finitude', *The Amicus Journal* (Summer 1980) p. 7.
28. Burton, *Global Conflict*, p. 13.
29. C.G. Brown, 'Chile's Road to Crisis', *The Nation* (8 December 1984) pp. 601, 615–18.
30. The similarity of the situations in Poland and Chile can be gleaned from two articles that occurred in the *New York Review of Books* (27 June 1985); T.G. Ash, 'Poland: The Uses of Adversity', pp. 5–9; M.A. Uhlig, 'Pinochet's Tyranny', pp. 35–40.
31. The 1990 United Nations projections for these cities are: Bombay, 11.8 million, Sao Paolo, 17.5 million, and Mexico City, 21.8 million: *U.N. Estimates and Projections of Urban, Rural and City Populations, 1950–2025: The 1980 Assessment* (New York: United Nations, 1982) Table 8, p. 61.
32. M.T. Klare and C. Arnson, *Supplying Repression: U.S. Support for Authoritarian Regimes Abroad* (Washington, DC: Institute for Policy Studies, 1981).
33. T. Gladwin, *Slaves of the White Myth: The Psychology of Neocolonialism* (Atlantic Highlands, NJ: Humanities Press, 1980) pp. 202–3.
34. *The Defense Monitor*, XIV(9) (1985) pp. 1–9. This entire issue of this publication from the Center for Defense Information is devoted to 'The Myth of American Mineral Vulnerability'. See also arguments for stockpiling by MacGeorge Bundy, 'The Inevitability of the Unexpected', in E.N. Castle and K.A. Price (eds), *U.S. Interests and Global Natural Resources* (Baltimore: Resources for the Future, Johns Hopkins University Press, 1983) pp. 132–3.
35. J.J. Kirkpatrick, *Dictatorships and Double Standards: Rationalism and Reason in Politics* (New York: American Enterprise Institute, Simon & Schuster, 1982) p. 41.
36. B. Tuchman, *The March of Folly: From Troy to Vietnam* (New York: Alfred A. Knopf, 1984). Her entire book is devoted to this notion.

37. W.A. Dorman, 'The Media: Playing the Government's Game', *Bulletin of the Atomic Scientists* (August 1985) pp. 118–24; quotes are on pp. 118, 123. See also the preceding article in the same issue by Morton H. Halperin, 'Secrecy and National Security', pp. 114–17. This entire 40th anniversary issue is a valuable resource.

38. Klare and Arnson, *Supplying Repression*; the quote is from pp. 3–4, and the data in the preceding paragraphs are from p. 118 f. Emphasis in original.

39. P. Lernoux, *Cry of the People: The Struggle for Human Rights in Latin America – The Catholic Church in Conflict with U.S. Policy* (New York: Penguin Books, 1982) p. 10. Ms Lernoux is recipient of the Columbia University Maria Moors Cabot award (1980) for her writing on Latin America, and of the Sidney Hillman Foundation book award for 1981 for *Cry of the People*.

40. *Alleged Assassination Plots Involving Foreign Leaders: An Interim Report of the Select Committee to Study Governmental Operations with Respect to Intelligence Activities*, US Senate, 94th Congress, 1st Session (Washington, DC: US Government Printing Office, 1975) p. 15 (quoted by Leften S. Stavrianos, *Global Rift: The Third World Comes of Age* (New York: William Morrow, 1981) pp. 671–2.).

41. Stavrianos, *Global Rift*, pp. 670, 464.

42. See M. Manley, *Jamaica: Struggle in the Periphery* (London: Third World Media (distributed in US by W.W. Norton, New York), 1982).

43. Lernoux, *Cry of the People*, p. 293.

44. M. Gurtov, *The United States against the Third World: Anti-Nationalism and Intervention* (New York: Praeger, 1974) p. 125; Stavrianos, *Global Rift*, pp. 475, 466.

45. H. Morgenthau, cited in the *New York Times* (21 July 1980) (quoted by Stavrianos, *Global Rift*, p. 464).

46. Two official reports on the Iran–Contra affair are available from the US Superintendent of Documents, Washington, DC: The Tower Commission Report and the Report of the Joint Committees of Congress. 'The Affidavit of Daniel Sheehan', available from the Christic Institute (1324 North Capitol St, NW, Washington, DC, 20002), an interfaith public interest law firm, further charges that profits from drug smuggling have been used to finance CIA-backed extralegal activities: heroin in Southeast Asia and cocaine in Central America.

47. R.E. Feinberg, *The Intemperate Zone: The Third World Challenge to U.S. Foreign Policy* (New York: W.W. Norton, 1983) pp. 241, 249.

48. 'A World at War: Small Wars and Superpower Interventions', *The Defense Monitor*, VIII (10) (1979). The entire issue is devoted to this topic. For updates, see *The Defense Monitor*, XII (1) (1983), 'A World at War, 1983'; Stephen D. Goose, 'Armed Conflicts in 1986, and the Iraq–Iran War', pp. 297–317 in *SIPRI Yearbook* (Stockholm, 1987); *The Defense Monitor*, XV (5) (1986), 'Soviet Geopolitical Momentum: Myth or Menace?'.

49. J. Krause, 'Soviet Military Aid to the Third World, *Aussenpolitik* (German Foreign Affairs Review) No. 4 (1983) p. 397. Krause also spoke at the Institute for Global Conflict and Cooperation, University of California, San Diego (November 1985) on this topic.

50. Burton, *Global Conflict*, p. 48.

51. R. Preiswerk, 'Sources of Resistance to Self-Reliance', Chapter 19 in J. Galtung, P. O'Brien and R. Preiswerk (eds), *Self-Reliance: A Strategy for Development* (Geneva: Institute for Development Studies (London: Bogle–L'Ouverture Publications), 1980) p. 341.

52. J. Galtung, 'The Politics of Self-Reliance', Chapter 20 in Galtung *et al.* (eds), *Self Reliance*, p. 363.
53. Galtung *et al.* (eds), *Self Reliance*, p. 378.
54. 'The Marrakesh Declaration (1977)', in Galtung *et al.* (eds), *Self Reliance*, p. 420.
55. T. Ratigan, former Peace Corps Volunteer, unpublished ms.
56. For a discussion of the Green parties' development, see F. Capra and C. Spretnak, *Green Politics* (New York: E.P. Dutton, 1984). See also H. Mewes, 'The Green Party Comes of Age', *Environment* (June 1985) pp. 13–17, 33–8.
57. J. Strawn and C.G. Hogan, 'Democracy on the Take: Flick Scandal Shakes West German Politics', *Multinational Monitor* (December–January 1985) pp. 13–15.
58. J. Lewis, 'Up and Up Go the Dole Queues', *Manchester Guardian Weekly* (24 August 1986) p. 3.
59. Quoted by Harold Baron, 'Innovation in Italian Politics Is in the Cities', *In These Times* (30 January–5 February 1985) p. 11.
60. M. Jaggi, R. Müller and S. Schmid, *Red Bologna* (London: Writers and Readers Publishing Cooperative, 1977) p. 36. This unusual book, written by three German news correspondents who stumbled onto the existence of Bologna's political system while covering a nearby bomb incident, is valuable reading for everyone interested in modern local politics.
61. See Harold Baron, 'A New Approach to Politics Is Forming in the Cities of Italy', *In These Times* (23–29 January 1985) p. 17. See also Jane Jacobs, *Cities and the Wealth of Nations* (New York: Vintage Books, 1985), for an economic argument for more decentralisation of political power. Jacobs unfortunately focuses mainly on economics, *per se*, paying little attention to natural resources on the one hand or to cultural needs on the other.
62. Y. Andropov, 'The Teaching of Karl Marx and Some Questions of Building Socialism in the USSR', *Soviet Life* (May 1983) pp. 6–9, 44; reprinted from *Kommunist*.
63. M. Gorbachev, *Perestroika: New Thinking for Our Country and the World* (New York: Harper & Row, 1987) p. 54.
64. Gorbachev, *Perestroika*, p. 90.
65. R. Harwood, 'Zimbabwe's Gilded Cage', *Manchester Guardian Weekly* (29 November 1987) p. 18.
66. M. Sequeira, 'Costa Rica: Green Alternative', *Multinational Monitor* (July 1984) p. 8.
67. Lernoux, *Cry of the People*, p. 366.
68. B. Bonpane, *Guerrillas of Peace: Liberation Theology and the Central American Revolution* (Boston: South End Press, 1985) 119 pp; quote is on p. 18. See also Lernoux, *Cry of the People*. For descriptions of the Vatican's response to Liberation Theology, see articles in *The Nation* by T.M. Pasca (2 June 1984 and 26 January 1985), and by P. Lernoux (27 August 1988).
69. C. Reich, *The Greening of America* (New York: Random House, 1970).
70. M. Ferguson, *The Aquarian Conspiracy* (Los Angeles: J.P. Tarcher, 1980).
71. MacIver, *The Web of Government*, p. 143.
72. According to Nuclear Free America (325 East 25th Street, Baltimore, MD, 21218) as of the end of 1987, 150 US communities had declared themselves 'nuclear free zones', barring the siting or construction of atomic weapons within their boundaries.
73. J.L. Aucoin, 'Missouri's Resourceful Sales Tax', *Environment* (May 1985) p. 44.

74.   T. McCall, 'Daddy Domino: Oregon's Land Use Program at Its Critical Juncture', *The Amicus Journal* (Spring 1982) pp. 55–60.

75.   For a fuller history see Kirkpatrick Sale, 'Bioregionalism – A Sense of Place', *The Nation* (12 October 1985) and his book, *Dwellers in the Land: The Bioregional Vision* (San Francisco: Sierra Club Books, 1985).

76.   A. Sweedler, 'The U.S.–Mexico Border: Caught between Washington and Mexico City', an unpublished manuscript available from the author (Physics Department, San Diego State University, San Diego, CA, 92182); quote is on p. 4.

77.   D. Mathews, 'We the People. . .', *National Forum* (Phi Kappa Phi Journal) (Fall 1984) pp. 46–8, 63–4; quote is from p. 63.

78.   Pericles, quoted by Mathews, 'We the People. . .', p. 64. For an interesting proposal that in the United States, legislators ought to be chosen by lot, as are jurors, and serve as a matter of responsible citizenship rather than as competitors for political office and public power, see E. Callenbach and M. Phillips, *A Citizen Legislature* (Berkeley/Bodega, CA: Banyan Tree Books/ Clear Glass, 1985). Those who stand for office among Green party candidates in Western Europe are often ordinary members of the party, not professional politicians.

15   Nuclear 'Defence' – or Conflict Resolution?

†   Anatol Rapoport, an American mathematician, is quoted by Roy Morrison in a review of G. Hart and W.S. Lind, *America Can Win: The Case for Military Reform*, in the *Bulletin of the Atomic Scientists* (October 1987) p. 51.

1.   A.M. Cox, *Russian Roulette: The Super Power Game*. (New York: Times Books, 1982) p. 1. See also *The Defense Monitor*, XV(7) (1986) for a discussion of accidents with nuclear weapons systems.

2.   M. Marder, 'Russia Expected U.S. Attacks, Says Gordievsky', *Manchester Guardian Weekly* (17 August 1986) pp. 15–16.

3.   H.F. York, *Race to Oblivion: A Participant's View of the Arms Race* (New York: Simon & Schuster, 1970) p. 228.

4.   *The Defense Monitor*, XIV(3) (1985). The entire issue is concerned with who has immediate control over nuclear weapons.

5.   See Pugwash Workshop statement, inside front cover of the *Bulletin of the Atomic Scientists* (April 1987).

6.   H.L. Abrams, 'Who's Minding the Missiles? Too Often, People with Drug, Drinking, or Psychiatric Problems', *The Sciences*, 26(4) (1986) pp. 22–8.

7.   Quotes are from *Unforgettable Fire: Pictures Drawn by Atomic Bomb Survivors*, edited by Japanese Broadcasting Corporation (NHK) (New York: Pantheon, 1977) pp. 90, 83, 52, respectively.

8.   H.M. Caldicott, 'The Final Epidemic: Physicians Could Never Reverse the Ravages of Nuclear War', *The Sciences* (March 1987) p. 21.

9.   A.B. Pittock, in a review of a National Research Council report, 'The Effects on the Atmosphere of a Major Nuclear Exchange', *Environment* (April 1985) p. 28.

10.   E. Marshall, 'Space Junk Grows with Weapons Tests', *Science*, 230 (1985) pp. 424–5.

11.   E. Marshall, 'The Buried Cost of the Savannah River Plant', *Science*, 233 (1986) pp. 613–14.

12.   'Permanent "Live Aid"', editorial in *Defense and Disarmament News* (August–September 1985) p. 1. (Newsletter of the Institute for Defense and Disarmament Studies, 2001 Beacon St, Brookline, MA 02146.)

13. R.L. Sivard, *World Military and Social Expenditures* (Leesburg, VA: World Priorities, 1981, 1983, 1985, 1986, 1987); information cited is from various editions, data in this paragraph are from 1981, p. 23 and 1985 p. 6.

14. Sivard, *World Military and Social Expenditures*, 1985, p. 23.

15. L.J. Dumas, *The Overburdened Economy* (Berkeley: University of California Press, 1986); quote is from p. 242.

16. Quoted in Greg Mitchell, 'Real Security: What Is It? How Can We Get It?', *Nuclear Times* (May–June 1986) p. 23.

17. See Lord Solly Zuckerman, 'Arming for Armageddon: Defense Scientists – Not Generals – Set the Pace', *The Sciences* (March 1982) pp. 10–15.

18. See, for example, Suzanne Gordon and David McFadden (eds), *Economic Conversion: Revitalizing America's Economy* (Cambridge, MA: Ballinger, 1984).

19. As pointed out by Gordon Adams ('Economic Conversion Misses the Point', *Bulletin of the Atomic Scientists* (February 1986) pp. 24–9), *military* economic conversion is only part of the larger problem of local economic diversity and autonomy that has been discussed in earlier chapters.

20. F.A. Long, 'Federal R & D Budget: Guns versus Butter', *Science*, 223 (1984) p. 1133; C. Norman, 'The Science Budget: A Dose of Austerity', *Science* 227 (1985) pp. 726–8; C. Norman, 'Science Escapes Brunt of Budget Ax', *Science*, 231 (1986) pp. 785–8.

21. For a discussion of secrecy in US science see: G. Kolata 'Chess-Playing Computer Seized by Customs', *Science*, 216 (1982) p. 1392; C. Norman, 'Security Problems Plague Scientific Meetings', *Science*, 228 (1985) pp. 471–2; R.J. Smith, 'X-ray Laser Budget Grows and Public Information Declines', *Science*, 232 (1986) pp. 152–3; R.J. Smith, 'Court Gives CIA Broad Security Rights', *Science*, 228 (1985) p. 566.

22. M.H. Halperin, 'Secrecy and National Security', *Bulletin of the Atomic Scientists* (August 1985) pp. 114–17; W.A. Dorman, 'The Media: Playing the Government's Game', *Bulletin of the Atomic Scientists* (August 1985) pp. 118–24; M.T. Klare, 'Building a Fortress America', *The Nation* (23 March 1985) pp. 321, 337–9.

23. B. Friedman, 'Preschoolers' Awareness of Nuclear Threat', California Association for the Education of Young Children, *Newsletter* (Winter 1984) pp. 4–5.

24. J.E. Mack, 'Psychosocial Trauma', in Ruth Adams and Susan Cullen (eds), *The Final Epidemic: Physicians and Scientists on Nuclear War* (Chicago: Educational Foundation for Nuclear Science, 1981) p. 26. For other studies on children see Robert Jay Lifton, 'In a Dark Time...', in Adams and Cullen (eds), *The Final Epidemic*, pp. 7–20, and Elise Boulding, 'World Security and the Future from the Junior High Perspective', *Peace and Change* (Fall 1981) pp. 65–76.

25. See 'Nuclear Fears in Finland', reported in *Lancet* (7 April 1984) pp. 784–5; 'Children in Soviet Express War Fear', *The New York Times* (13 October 1983) p. A–19.

26. M. Yudkin, 'When Kids Think the Unthinkable', *Psychology Today* (April 1984) pp. 13–25. For a thoughtful article on ways to empower people into resisting the arms race, see Peter M. Sandman and JoAnn M. Valenti, 'Scared Stiff – or Scared into Action', *Bulletin of the Atomic Scientists* (January 1986) pp. 12–16.

27. R.J. Lifton, in Adams and Cullen (eds), *The Final Epidemic*, p. 18. For an authoritative description of the Vietnam experience see William Mahedy, *Out of the Night: The Spiritual Journey of Vietnam Vets* (New York: Ballantine, 1986).

28. 'Harry S. Truman to Samuel McCrea Cavert' (11 August 1945) OF 596A; and Truman Letter OF 692A, Truman Library; quoted by Barton J. Bernstein, 'Shatterer of Worlds: Hiroshima and Nagasaki', *Bulletin of the Atomic Scientists* (December 1975) p. 18.

29. Bernstein, 'Shatterer of Worlds', p. 21.

30. J.W. Burton, *Global Conflict: The Domestic Sources of International Crisis* (Brighton: Wheatsheaf Books, John Spears (distributed in US by The Center for International Development, University of Maryland, College Park, MD) 1984) p. 13.

31. J. Addams, *Peace and Bread in Time of War* (New York: Macmillan, 1922). (Reprinted by National Association of Social Workers, Inc., Silver Spring, MD, 1983) p. 183 f.

32. Addams, *Peace and Bread*, p. 191.

33. B.L. Grayson (ed.), *The American Image of Russia: 1917–1977* (New York: Ungar, 1978) p. 59.

34. Grayson, *The American Image*, p. 10.

35. S. Welles, *The Time for Decision* (New York: Harper & Brothers, 1944) p. 12.

36. D. Clemens, *Yalta* (New York: Oxford University Press, 1970). For tripartite power struggle leading up to Yalta see Chapter 2, 'The Setting', pp. 63–104.

37. G.F. Kennan, 'Moscow Embassy Telegram #511: "The Long Telegram"' (22 February 1946). *Foreign Relations of the United States: 1946*, VI, pp. 696–709. See also George F. Kennan, *Memoirs 1925–1950* (Boston: Little, Brown, 1967) pp. 292–5.

38. C.M. Clifford, 'American Relations with the Soviet Union: A Report to the President by the Special Counsel to the President, September 24, 1946', Harry S. Truman Papers, Harry S. Truman Library, Independence, MO.

39. W. Churchill, *Sir Winston Churchill: A Self-Portrait* (London: Eyre & Spottiswoode, 1954) p. 81.

40. Churchill, *Sir Winston Churchill*, p. 78.

41. Clifford's remark was quoted by Arthur Macy Cox at the *Whittier Institute: Roots of Conflict, US–USSR*, July, 1982, taped by On-Site Taping Services, 6942 Cantaloupe Ave, Van Nuys, CA 91405, Tape #13, Side 1. Kennan's remarks are in *The New York Review of Books* (21 January 1982) pp. 8, 10, 12; quote is on p. 10.

42. These quotes, respectively, are from Reagan's speech to the National Association of Evangelicals, Orlando, Florida (8 March 1983); to the British Parliament, London (8 June 1982) and his first press conference (29 January 1981).

43. See *The Defense Monitor*, IX(5) and (8) (1980) and XIV(7) (1985); Carl Levin, 'U.S. Arms Debate: The Other Side...', *San Diego Union* (14 August 1983) p. C–4.

44. E. Guyot, '"Yellow Rain": The Case Is Not Proved', *The Nation* (10 November 1984) pp. 465, 478–84; T.C. Seeley, J.W. Nowicke, M. Meselson, J. Guillemin and P. Akratanakul, 'Yellow Rain', *Scientific American* (September 1985) pp. 128–37.

45. D. Pearson, 'K.A.L. 007: What the U.S. Knew and When We Knew It', *The Nation* (18–25 August 1984); D. Pearson, J. Keppl, D. Corn *et al.*, 'K.A.L. 007: Unanswered Questions', *The Nation* (17–24 August 1985); Seymour Hersh, *The Target is Destroyed* (New York: Random House, 1986). For more recent evidence about what the US government knew see *Los Angeles Times* (13 January 1988) p. 1.

46. See *The Defense Monitor*, XVI(2) (1987).

47. Discussions of the legality of some of these events are found in: J.J. Stone, 'Libya and the War Powers', *Federation of American Scientists Newsletter* (June–July 1986) – see especially, p. 15; G. Greve and E. Warren, 'Secret Army Unit Allegedly Flies in Latin America', *The Philadelphia Inquirer* (16 December 1984).

    The illegality of private American citizens acting militarily in Central America is discussed in *The Washington Spectator* (1 September 1986) p. 1. Details on this subject are to be found in Jaqueline Sharkey, 'Disturbing the Peace', *Common Cause Magazine* (September–October 1985) pp. 20–32.

48. The Christic Institute brought suit in a US District Court in Miami on behalf of two journalists against 29 defendants, charging that the CIA and others were involved in drug smuggling and other illegal acts in the course of furthering the cause of the Nicaraguan 'contras'. After granting the plaintiffs powers of subpoena, the judge peremptorily dismissed the case in June 1988.

49. *New York Times*, quoted in *The Washington Spectator* (15 August 1982) p. 3. The FBI was also urged to infiltrate 'politically suspicious' organisations, even those acting totally within the law, according to testimony by its director, William Webster, before the Senate Subcommittee on Security and Terrorism (24 June 1982). See also the 'Affidavit of Daniel P. Sheehan', cited in Chapter 14, reference 46.

50. Quoted in Leften S. Stavrianos, *Global Rift: The Third World Comes of Age* (New York: William Morrow, 1981) p. 726.

51. Stavrianos, *Global Rift*, p. 723.

52. R. Jervis, *The Illogic of American Nuclear Strategy* (Ithaca, NY: Cornell University Press, 1984) p. 164. See also Thomas Schelling, *The Strategy of Conflict* (New York: Oxford University Press, 1960).

53. Taped lecture by Stephen Kull (11 November 1987), 'Psychology of the Arms Race: What Keeps the Arms Racers Running'. Available for $5 from Physicians for Social Responsibility, P.O. Box. 8185, LaJolla CA, 92038. Kull's book, *Minds at War: The Inner Conflict of Defense Policy Makers*, is to be published in 1988 by Basic Books.

54. Quoted in E.P. Thompson (ed.), *Star Wars* (New York: Pantheon Books, 1985) p. 26.

55. Most estimates of the Strategic Defense Initiative's R & D costs alone are in the $75–$100 billion range, but Richard D. DeLaue, when Under-secretary of Defense for Research and Engineering, testified before Congress that the R & D costs could exceed $500 billion (cited in *Chicago Sun Times* (17 December 1983)).

56. See for example, H.A. Bethe, R.L. Garwin, K. Gottfried and H.W. Kendall, 'Space-Based Ballistic-Missile Defense', *Scientific American* (October 1984) pp. 47–9; The Union of Concerned Scientists, *The Fallacy of Star Wars: Why Space Weapons Can't Protect Us* (New York: Vintage Books, 1984); Federation of American Scientists, 'The Four Faces of Star Wars: Anatomy of a Debate', *F.A.S. Public Interest Report* (March 1985); George W. Ball, 'The War for Star Wars', *New York Review of Books* (11 April 1985) pp. 38–44; Center for Defense information, 'Star Wars: Vision and Reality', *The Defense Monitor* XV(2) (1986).

57. See the Fletcher Committee Report to Congress, statement of James Fletcher, before the Subcommittee on R & D, Committee on Armed Services, House of Representatives, 98th Congress, 2nd session (1 March 1984) p. 3.

58. R.J. Smith, 'Antisatellite Weapon Sets Dangerous Course', *Science*, 222 (1983) pp. 140–2; R.J. Smith, 'Administration Resists Demand for ASAT Ban', *Science*, 222 (1983) pp. 394–6; John Tirman, 'Star Wars Technology Threatens Satellites', *Bulletin of the Atomic Scientists* (May 1986) pp. 28–32.

59. Bethe *et al.*, 'Space-Based Ballistic-Missile Defense', p. 49.

60. J. Schell, *The Abolition* (New York: Alfred A. Knopf, 1984), and F. Dyson, *Weapons and Hope* (New York: Harper & Row, 1984).

61. S. Murphy, A. Hay and R. Steven, *No Fire, No Thunder: The Threat of Chemical and Biological Weapons* (New York: Monthly Review Press, 1984) p. 6. A useful source of information on alternative weapons. See also Center for Defense Information, *The Defense Monitor*, IX(10) (1980).

62. Progress on Big Eye, the American binary nerve-gas bomb, has been plagued with snags. See *Los Angeles Times* (11 August 1986) Part II, p. 6.

63. Murphy *et al.*, *No Fire*, p. 95.

64. See the advertisement in *Science*, 225 (1984) pp. 879–80, 882.

65. J. King, quoted in R.J. Smith, 'The Dark Side of Biotechnology', *Science*, 224 (1984) pp. 1215–16. For further information see: R. Asinof, 'Averting Genetic Warfare', *Environmental Action* (June 1984) pp. 16–22; S. Wright and R.L. Sinsheimer, 'Recombinant DNA and Biological Warfare', *Bulletin of the Atomic Scientists* (November 1983) pp. 20–6; S. Wright, 'The Military and the New Biology', *Bulletin of the Atomic Scientists* (May 1985) pp. 10–16.

66. W.A. McDougall, 'Technocracy and Statecraft in the Space Age – Toward The History of a Saltation', *American Historical Review*, 82 (1982) pp. 1010–40.

67. Simone Weil, *The Iliad or the Poem of Force*, Pendle Hill Pamphlet No. 91 (Wallingford PA: Pendle Hill Publications, 1957) pp. 11, 14.

68. Weil, *The Iliad*, p. 37.

69. R.R. Combs, Jr, Deputy Secretary of State for Soviet Affairs, *Whittier Institute: Roots of Conflict US–USSR*, July 1982, Tape #8.

70. Jeremy J. Stone, Lecture at the Institute for Global Conflict and Cooperation, University of California, San Diego (1 October 1984).

71. J. Tirman, 'A Way to Break the Arms Deadlock', *The Nation* (16 February 1985) pp. 167–71. President Gerald Ford, at a talk to the Institute for Global Conflict and Cooperation at UCSD (4 February 1986) noted that agreement of SALT II was 95 per cent complete during his administration, but was blocked by the Department of Defense and a few high-ranking Republicans. See also, R. Jeffrey Smith, 'Weapons Labs Influence Test Ban Debate', *Science*, 229 (1985) pp. 1067–9.

72. R. Fisher and W. Ury, *Getting to Yes: Negotiating Agreement Without Giving In* (Boston: Houghton Mifflin, 1981).

73. D. Hafemeister, J.J. Romm and K. Tsipis, 'The Verification of Compliance with Arms-Control Agreements', *Scientific American* (March 1985) pp. 39–45; J.F. Evernden, 'Politics, Technology, and the Test Ban', *Bulletin of the Atomic Scientists*, (March 1985) pp. 9–12; Natural Resource Defense Council, 'NRDC Announces Nuclear Test Monitoring Pact with Soviets', *NRDC Newsline* (July–August 1986) pp. 1, 4; A. Grigoryev, 'U.S. Seismologists in Kazakhstan', *Soviet Life* (October 1986) pp. 2–3.

74. Cited by Grigoryev, 'U.S. Seismologists', p. 9.

75. D.D. Eisenhower, reprinted in Duane Sweeney (ed.), *The Peace Catalog* (Seattle: Press for Peace, 1984) p. 70.

76. Reprinted in *ENDpapers Eight*, *Spokesman 46* (Summer 1984) pp. 93–4 (published by the Bertrand Russell Peace Foundation, Ltd, Bertrand Russell House, Gamble Street, Nottingham NG7 4ET).

77. W.M. Arkin and R.W. Fieldhouse, 'Focus on the Nuclear Infrastructure', *Bulletin of the Atomic Scientists*, (June–July, 1985) pp. 11–15; quote is on p. 11. See also other articles in this issue.
78. J. Langdon, 'Labour Aims for Non-Nuclear Defence Strategy within Nato', *The Guardian* (26 July 1984) p. 3.
79. For a history of the Institute of Peace, see Charles D. Smith (ed.), *The Hundred Percent Challenge* (Cabin John, MD: Seven Locks Press, 1985).
80. G. Sharp, *The Politics of Nonviolent Action* (1973) and *Social Power and Political Freedom* (1980), both published by Porter Sargent, Boston. See also his *Making Europe Unconquerable: The Potential of Civilian-Based Deterrence and Defence* (New York: Ballinger, 1985).
81. L. Fisher, *Life of Mahatma Gandhi* (New York: Macmillan, 1962) pp. 272–9.
82. M.L. King, Jr, *Stride Toward Freedom: The Montgomery Story* (New York: Harper & Row, 1964) pp. 78–9.
83. M.L. King, Jr, 'Letter from Birmingham Jail' (16 April 1963) © 1963, 1964, Martin Luther King, Jr. Excerpted by permission Ms Joan Daves.
84. M. McNeish, *Fire Under the Ashes: The Life of Danilo Dolci* (London: Hodder, 1965) pp. 102–30.

16 Humankind at the Crossroads

†   Fritjof Capra, 'A New Vision of Reality', *New Age* (February 1982) pp. 29–30.
1.   N.S. Momaday, 'A First American Views His Land', *National Geographic* 150(1) (July 1976) pp. 13–19.
2.   B. Nietschmann, 'The Third World War,' *Cultural Survival Quarterly*, 11(3) (1987) pp. 1-16; quote is from p. 1.
3.   J.H. Bodley, *Anthropology and Contemporary Human Problems* (Menlo Park, CA: Cummings, 1976) p. 214 f.
4.   S.J. Gould, *The Mismeasure of Man* (New York: W.W. Norton, 1981) p. 325.
5.   See E.F. Schumacher, *Small Is Beautiful: Economics As If People Mattered* (London: Abacus, Sphere Books, 1974); Hazel Henderson, *The Politics of the Solar Age: Alternatives to Economics* (Garden City, NY: Anchor Press, Doubleday, 1981).
6.   D. Dickson, *The New Politics of Science* (New York: Pantheon, 1984) p. 3.
7.   M.H. Halperin, 'Secrecy and National Security', *Bulletin of the Atomic Scientists*, (August 1985) pp. 114–17.
8.   A. Pacey, *The Culture of Technology* (Cambridge, MA: MIT Press, 1983) p. 171.
9.   M. Stanley, 'Three Post Political Futures', *The Humanist*, 33 (6) (November/December 1973) pp. 28–31. Quote from p. 28.
10. W.D. Carey, 'Charting a Course for Science', *Science*, 212 (1981) p. 1455.
11. E. Chargaff, 'Knowledge without Wisdom', *Harper's* (May 1980) p. 41.
12. Chargaff, 'Knowledge without Wisdom', p. 47.
13. Schumacher, *Small Is Beautiful*, pp. 26–7.
14. Pacey, *The Culture of Technology*, p. 151.
15. H. Caldicott, *Missile Envy* (New York: William Morrow, 1984) p. 315 f.
16. C. Gilligan, *In a Different Voice* (Cambridge, MA: Harvard University Press, 1982) p. 160.
17. M. Midgley, *Evolution as a Religion* (London: Methuen, 1985) p. 135.

18. Midgley, *Evolution*, p. 142.
19. For a complete discussion of these ideas, see Mary Midgley, *Heart and Mind* (New York: St Martin's Press, 1981).
20. A. Leopold, *A Sand County Almanac* (London: Oxford University Press, 1949); quote is from Foreword, pp. viii–ix, 1948.
21. R.B. Norgaard, 'Coevolutionary Agricultural Development', *Economic Development and Cultural Change*, 32 (1984) pp. 525–46; 'Bureaucracy, Systems Management, and the Mythology of Science', Giannini Foundation of Agricultural Economics Working Paper No. 297, University of California, Berkeley, 1984; 'Environmental Economics: An Evolutionary Critique and a Plea for Pluralism', *Journal of Environmental Economics and Management*, 12 (1985) pp. 382–94.
22. D. Lee, *Freedom and Culture* (New York: Spectrum, Prentice-Hall, 1959) pp. 11, 23.
23. Lee, *Freedom and Culture*, pp. 24, 25.
24. See J. Goodall, *In the Shadow of Man* (Boston: Houghton Mifflin, 1971). Dr Goodall's studies of our nearest primate relatives, the chimpanzees, indicate the kind of social bondedness our shared common ancestors must have had.
25. W. Ophuls, *Ecology and the Politics of Scarcity* (San Francisco: W.H. Freeman, 1977) p. 224.
26. Ophuls, *Ecology*, p. 224.
27. Quoted in Fritjof Capra and Charlene Spretnak, *Green Politics* (New York: E.P. Dutton, 1984) p. 156.
28. Capra and Spretnak, *Green Politics*, p. 157.
29. R. Axelrod, *The Evolution of Cooperation* (New York: Basic Books, 1984) p. 190.
30. I.I. Mitroff, 'Beyond Experimentation: New Methods for a New Age', Chapter 8 in E. Seidman (ed.), *Handbook of Social Intervention* (Beverly Hills, CA: Sage Publications, 1983) p. 177.
31. 'A Conversation between David Spangler and William Irwin Thompson', *Annals of Earth*, IV(1) (1986) p. 6. (Published by Ocean Arks, Inc., 10 Shanks Pond Road, Falmouth, MA, 01540.)
32. S. Aronowitz, 'Mass Culture and the Eclipse of Reason: The Implications for Pedagogy', *College English* (April 1977) p. 768.
33. J. Dewey, *The Wit and Wisdom of John Dewey*, A.H. Johnson (ed.) (Boston: Beacon Press, 1949) pp. 90, 103, respectively.
34. R.T. Johnson and D.W. Johnson, 'Cooperative Learning and the Achievement and Socialization Crises in Science and Mathematics Classrooms', in A.B. Champagne and L.E. Hornig (eds), *Students and Science Learning* (Washington, DC: American Association for the Advancement of Science, 1987) pp. 67–93.
35. Dewey, *The Wit and Wisdom*, p. 93.
36. G. Sharp, *The Politics of Nonviolent Action* (1973) and *Social Power and Political Freedom* (1980), both published by Porter Sargent, Boston.
37. O.N. Bradley, quoted in Charles Duryea Smith (ed.), *The Hundred Percent Challenge* (Cabin John, MD: Seven Locks Press, 1985) p. 208.
38. M. Nagler, 'Redefining Peace', *San Francisco Sunday Examiner and Chronicle* (30 October 1983).

Index

553